Westinghouse J46 Axial Turbojet Family

Development History and Technical Profiles

Westinghouse J46 Axial Turbojet Family

Development History and Technical Profiles

Paul J. Christiansen

Bleeg Publishing LLC

Bleeg Publishing LLC
Olney, MD
pj.chris@verizon.net

ISBN-13: 978-0692764886
ISBN-10: 0692764887

First Edition

Front and rear cover illustrations: Insets showing front and rear sections of the J46-WE-8, Westinghouse photo P46977, May 20, 1952, Drawing 58J-790. *Courtesy Hagley Museum and Library*

Dedication

To my daughters, who went to the airshows and held their ears as the jets screamed by while their Dad stood mesmerized by the sights and sounds.
I cannot imagine life without you both in it.

Author's Notes:

I have left the data primarily in the U.S. English measurements of the original sources. House engines were known by a number as "No. X" or "#X", and sometimes by a serial number, but it was not possible to trace them through their builds and re-builds, as application of a new number was inconsistently applied. The terminology and even the spelling of words related to the technology was in a state of flux during the period covered by the project and I have incorporated most the original language with all of its sometimes awkward phrasing to give the reader a sense of the emerging lexicon of the time period. The citations preserve the exact wording of the document's subject lines and are good examples. Where a term or its use is obscure in modern parlance I have provided a translation in the Glossary. The Hagley Museum and Library photo collection of J46 program related images is in the form of B&W cropped prints mounted on photo-library index cards for the Essington photo library. The original negatives no longer exist. The result is that many of the images used in the book are slightly clipped. Where full sized prints were located in reports, these have been used if the quality was not badly degraded and the added detail important. The photo library for the Kansas City plant has not been preserved. Only Kansas City photos attached to reports are known. Hopefully the images selected still enable the reader to gain an understanding of the construction details of the various components and the core engine itself.

The chapters are organized mainly as a chronological sequence for the various engine models but overlap between models, particularly the Navy/Air Force versions and the J46-WE-2/-18 models was unavoidable.

It is fortunate that the project basis for every engine model was the specification, since all changes to it had to be requested and approved along with the various document deliverables required by a specific contract. As a result, sufficient technical data has survived within the BuAer contract records to recover most of the technical development story. Considerable effort was required to extract this data following a long period of organization of the surviving records to arrive at a more or less sequential series of events for each model. Any errors in interpreting the material are mine alone.

Acknowledgements:

My deepest thanks and appreciation go to my wife who edited this work through its numerous iterations, not an easy job given the technical nature of the material. To the many professionals at the National Archives at College Park, MD go my thanks for their assistance in locating the many containers of documents pertaining to the many contracts involved. As well go my special thanks to the professional archivists at the Hagley Museum and Library, stewards of the Westinghouse Electric Gas Turbine Division photographic library, for allowing me to scan and use many J46 and "PD" related photos from their collection. My thanks go to Tommy Thomasson and Bill Spidle for sharing their data on the F7U aircraft from the Vought Heritage Foundation archives. Al Casby of *Project Cutlass* was most helpful with photos of his engines, accessories and allowing me to copy his Chance Vought Service J46 Instruction Manual. Lastly my thanks go to the Aircraft Engine Historical Society for its support and encouragement to produce yet another volume on a Westinghouse engine.

Contents

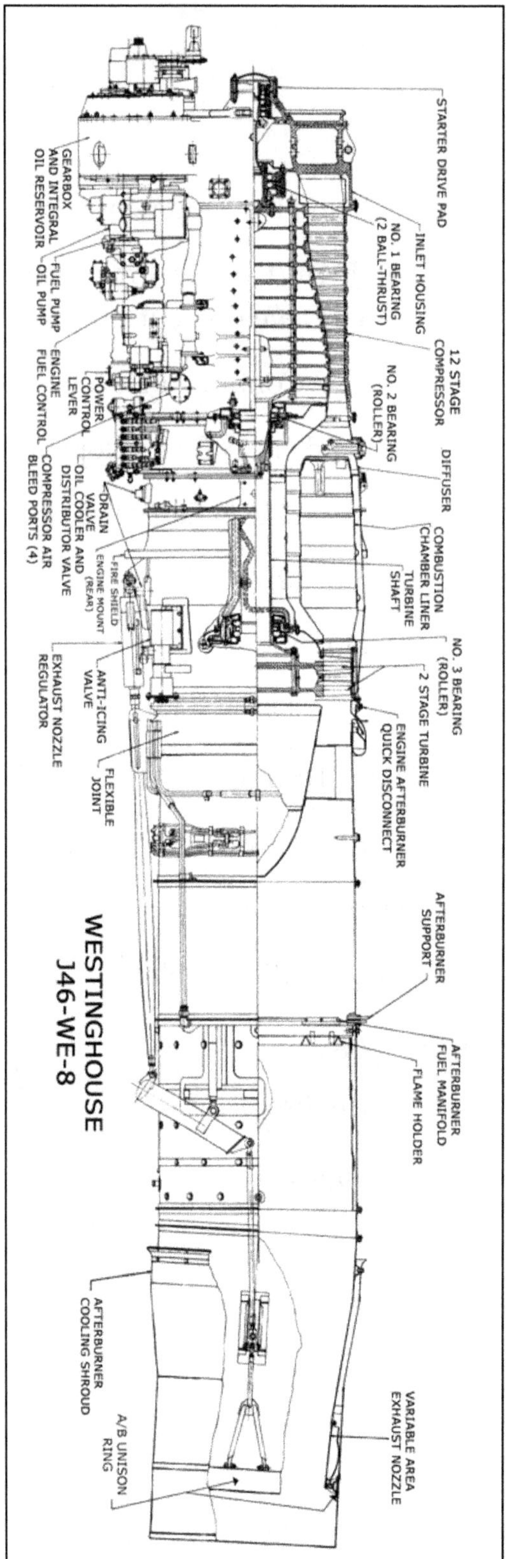

J46-WE-8 construction showing the overall layout of the engine. This general layout was used for all versions of the engine, the A/B versions varying in length, the amount of cant and internal details. P46977, 5/20/1952, Drawing 58J-790, (*Retraced by the author). Courtesy Hagley Museum and Library*

Introduction

The Westinghouse Steam Turbine Division's involvement in jet engine development began in 1941 with a National Advisory Committee for Aeronautics (NACA) briefing on the theory behind such engines. In November of 1942 they received a contract from the U.S. Navy's Bureau of Aeronautics (BuAer) to investigate the design of an axial turbojet. There was no hardware deliverable involved, but two demonstration engines were successfully constructed and operated, the first running on March 19, 1943. Nicknamed the "Yankee" after the successful demonstration, BuAer immediately began asking for higher thrust production versions with the military designation of J30. This engine produced 1,600 pounds of thrust. It was a single shaft three bearing design featuring a six stage axial compressor with a pressure ratio of 3.0:1, a dual annular combustor and a single stage turbine.

Even though assisted by Pratt and Whitney, deliveries were held up by delays in redesign of the engine for production. In the end, 129 J30 engines were produced by P&W and Westinghouse. A cutaway of a P&W built J30 engine is in the New England Air Museum collection in Windsor Locks, Connecticut but is not currently on display.

Beginning what was to become a regular design approach, the new Westinghouse Aviation Gas Turbine sub-division (WAGT), produced a scaled down version (.6 scale compressor diameter) of the J30, the J32. It was built in two versions and very small quantities (44) for several drone programs. At 260/275 pounds of thrust, these failed to find other buyers and were dropped. A cutaway display of a J32 is in the National Air and Space Museum collection and is currently on display at the Steven F. Udvar-Hazy Center in Chantilly, Virginia.

Next, a scaled up version of the original 19.5 inch diameter intake Yankee engine with a 24 inch compressor intake, 11 stage axial compressor, duel annular combustor and a two stage turbine was designed and placed in production with the military designation of J34. This engine, at approximately 3,000 pounds initial thrust in early versions, began to be built initially in the Essington facilities, then in large numbers at the new Westinghouse plant in Kansas City. Many early variants of the J34 came into existence, mostly tailored to minor installation changes to piping and wiring to accommodate individual airframe requirements.

As 1947 began, production of the J30-WE-5 was coming to an end at WAGT, Essington, PA and Pratt and Whitney. WAGT was well into planning for production of the J34-WE-22 in their new plant in Kansas City. Development was to continue and by the year's end they would have contracts in hand for research to improve J34 operability and durability and for the development of two new engines of higher thrust, the J46 and J40.

This volume will concentrate on the development and production of the J46 family of engines. Because the J46 incorporated many of the developments and improvements resulting from the earlier J34 improvement efforts, these will be reviewed as a prelude to the J46 development effort. The full and lengthy story of the J34 will to be covered in a later volume.

Paul Lagasse, in his Master's thesis[1], discusses in great detail the shortcomings of the Westinghouse business model as it applied to their Steam Turbine and later Gas Turbine Divisions. His analysis shows that Westinghouse had many business, organizational, and industrial plant shortcomings that created internal obstacles to the successful development and production of gas turbines. Among these were the lack of production management skills, corporate finance support, internal machine shop capabilities, test facilities and a proper research group. These shortcomings were not addressed in a timely enough manner to affect the outcome of the various engine programs, resulting in the late delivery of under-developed and under-performing engines and, in the case of the J40, failure to deliver ANY of the higher power engines the Navy was counting on for their fighter programs (the J40-WE-10 and J40-WE-12). As a result of the Navy's withdrawal of orders and WAGT's failure to develop a broader customer base for its engines, Westinghouse withdrew from the aviation gas turbine business altogether in 1960.

The J46 program had a long development cycle and WAGT's planned growth program, based largely on Rolls-Royce technology, was intended to take the engine to the end of the 1950's, but in the end, the engine remained an orphan linked to only one airframe, the Chance Vought F7U-3 "Cutlass". That airframe had no future in the world of supersonic development and the J46 growth program was too little too late.

Neither Westinghouse nor the Navy could have foreseen the rapid changes in mission requirements that would be brought about by both advances in aeronautical knowledge and unexpected world events such as the Korean War, Russia obtaining nuclear weapons, and the rapid development of the need for supersonic flight. These were beginning to occur in the short period after the J46 (and its predecessor the J34-WE-38/J34-WE-32) development had begun. The developing mass of transonic and supersonic data collected from the Douglas Skystreaks and Skyrockets, the Bell series of X-1 research aircraft and National Advisory Committee for Aeronautics wind tunnel research began to indicate the need for far greater engine thrust. Even better fuel economy was needed than any of the contracted WAGT engines were expected to deliver.

The explosion of a nuclear weapon by the Soviet Union in 1949 brought sharp focus on the increased need for fast-climbing, supersonic interceptors with increased mission radius. In addition, the outbreak of war in Korea forcefully demonstrated the limitations of the current Navy jets, causing a sharp increase in accelerated need for follow-on equipment with enhanced performance. As these evolving requirements emerged, the Navy contracted with WAGT to deliver more powerful and efficient versions of the J40 in parallel with the original designs. WAGT responded with proposals that indicated the new requirements could be met using the same basic design approach and remain within the basic installation envelope. The proposed lengthened development cycle would cause a six month schedule slip in the J40 program and shift that program's focus to the higher power models, even as development of the initial models continued to facilitate early airframe development (the Douglas XF4D-1, XA3D-1, and McDonnell XF3H-1). These events occurred in parallel with the J46 program and are related since the engines used many common components and systems. The full J40 story is covered in the author's companion volume, *Westinghouse J40 Axial Turbojet Family*.

The J46 program, which started out with many airframe programs (McDonnell F-88, Lockheed F-90, Douglas MX-656 (the X-3) the Douglas F3D-2 and -3, Vought F7U-3 and a bit later the Convair XF2Y) identified for its use, began to lose its potential customers due to schedule slippages. Airframe designs were either changed to higher power engines to offset weight increases, or were cancelled outright. In the end, only the Vought XF7U-3 Cutlass program was left and it is in relation to that aircraft that the J46 engine is best known.

The little known J34-WE-32 engines used in the F7U-1 models in limited numbers are now largely forgotten. Yet the technical improvements planned for the J46 engine were all tested on the J34-WE-32 and -38 (non-afterburning version of the -32). Many of the design problems and performance short falls of the upcoming J46 design were first experienced in this earlier starting program.

As BuAer and the airframe companies began to integrate afterburner equipped engines into regular production airframes, the efforts to define the requirements for new engine control systems along with improved reliability will be discussed. The human factor issues related to afterburner use and control became a major discussion focus point all through J34-WE-32 and J46-WE-2 development. Even the pilot operated throttle quadrant was redesigned for commonality across all afterburner equipped Navy aircraft. Perhaps of all the projects Westinghouse undertook for BuAer, the development of fail-safe controls and human factors improvements had the most lasting impact on the Navy after Westinghouse had departed the gas turbine field.

The engine control system intended for the J46 and J40 programs was developed on the J34-WE-32/-38 engines. Far advanced for its time, this control system was a vacuum tube (thermionic valve) based system that ultimately was undone by the inability of the electronics to perform reliably and provide for full fail-safe operation. The primitive emergency afterburner fuel control system provided for in the initial design of this control was found to be unacceptable. By the time an accurate non-electrical back-up emergency fuel control capable of satisfactory control of the engine was developed, the performance difference between the electronic and mechanical controls narrowed to the point where the drawbacks of the electronic system made it unattractive, even if it did deliver better fuel economy and reliability through more precise temperature control. Much time and effort was spent on developing alternate controls for the engines, contributing greatly to schedule delays and to some extent their performance shortfall.

In addition to all the other design challenges and materials shortages during the Korean War, the military's transition from the previous "AN" technical specifications to the MIL-E type specifications proved to require significantly more work than anticipated. In some cases unplanned retesting had to be done to the new requirements. The introduction of newer fuel types (JP-3/JP-4) also caused many additional delays as materials and controls were researched and modified to handle the characteristics of the new fuels.

The marriage of the J46 to the F7U-3 Cutlass was a mismatch that caused much grief. While the lower power of the J40-WE-22 in the F3H-1N Demon found its difficulties mostly in airframe power to weight problems of (seemingly) unanticipated severity, the F7U-3 mismatch was of a different nature. The F7U-3 had pitch drag characteristics similar to a plain delta wing, in that small increases in pitch caused very large drag increases. Carrier landings with engines having long response times on short final were bound to cause severe problems and did. As a result, pilots were taught to fly the approach using the speed brake for rate of decent control, keeping engine rpm's and thrust high while holding the stick still in pitch but not in roll, a very unnatural way to control the airplane. The nickname "Gutless Cutlass" became attached to the airframe and a comparison of weight to thrust ratios of contemporary aircraft are explored in the final chapter. Was the nickname justified?

The technique of moving the throttles to full military thrust just before touchdown became a standard for aircraft landings that is still in use today.

The J46's technical development has never been described in detail. Being a classified project while underway, many details were hidden from view in the Navy's files. The classified security status was removed in 1965, but by then, interest in the J46 engines, already out of service, had waned. WAGT and most of the division's records were no longer in existence. As a result, the full technical details of most of the models of the J46 have never been published, and the few details in print are often wrong or applied to the wrong model. The bulk of the data published refers to the J46-WE-8(/A/B) used in the production F7U-3.

It is fortunate for the researcher that all of the engine models that entered development produced detailed model specifications that became part of the contractual record and these, along with quite a bit of the design change correspondence related to these specifications and the contracts were retained in the files. It has allowed the recreation of the development story in large part, even if the in-house WAGT engineering specifics are no longer available.

This volume describes the development and technical challenges as well as a technical description for each J46 model. Some development issues affected more than one model and while every effort has been made to reduce redundancy in the text, it could not be entirely eliminated.

This volume focuses on the chronological technical description of the engine development and the involvement of individuals has been largely ignored for both space reasons and the lack of comprehensive records in regard to those individuals. Paul Legasse's work covers some of the high level personnel story and the reader is encouraged to read his document.

It was not the intention of the author to rehash or question the decisions made by the parties involved in the engine family's development, only to present the history and information as they have emerged from the files. Viable engines of the desired models were very late in being delivered and a bit down on performance, but the record shows that all involved were working within their technical limits and organizational constraints to satisfy the Navy's requirements. If anything, it provides insight into both the development and the challenges of changing requirements in this transitional period in turbojet development, a time when the vast majority of the turbojet engines designed by industry failed technically or failed to meet the needs of the marketplace. The large research and plant investments made by the traditional aircraft engine industry giants began to allow them to produce successful designs well matched to the needs of the military and, a bit later, civilian aviation requirements. Westinghouse fell further and further behind and failed to find customers, military or civil, for their later designs.

Now relegated to history, numbers of J34 and J46-WE-8(B) engines are in museum collections, some as cutaways, allowing interesting comparison to contemporary products from other manufacturers.

---

[1] The Westinghouse Aviation Gas Turbine Division 1950-1960: A Case Study in the Role of Failure in Technology and Business, Paul D. Legasse, Master's Thesis for Master of Arts Degree, 1997, University of Maryland. A .pdf file of this thesis can be downloaded from the Aircraft Engine Historical Society web site, www.enginehistory.org.

Figure 1: Typical J46 diffuser seen from the exit end. It evenly split the airflow into three zones, the innermost and outermost flows used for cooling the annular combustion chamber and the middle zone went into the combustor section. Note the many spot welds needed to manufacture this component. Very small differences in weld locations, weld height, component placement or handling damage could create unpredictable flow patterns downstream and cause hot spots in the turbines. WAGT P48097, 11/14/1952. *Courtesy Hagley Museum and Library*

Chapter 1
Earlier Research and Development

In 1947, with the then current J34-WE-22 finally coming off the production lines in small quantities, the Bureau of Aeronautics command of the U.S. Navy issued a Letter of Intent (LOI) contract to Westinghouse for research to improve the J34 engine. The accumulating early experience with this and preceding models of the engine (as well as the earlier J30) indicated the need for significant on-going development to improve reliability, operability, and durability. This was increasingly imperative as the engine was already being used in different airframes, both as a prime driver or boost engine.

As will be seen, it was hoped that if possible, the improvements would increase the engine's thrust and improve its specific fuel consumption numbers.

The Westinghouse Aviation Gas Turbine Division (hereafter WAGT) was still operating as a sub-component within the Westinghouse Steam Turbine Division. They operated their research, development and production lines at the shared Essington, Pennsylvania factory site in a suburb of Philadelphia. Within this organization structure, WAGT lacked a dedicated research laboratory, machine shop or adequate floor space, sharing these facilities, such as they were, with the Steam Turbine Division on a low priority basis.

The WAGT chief engineer was Reinout Kroon. He and his engineering team were steeped in the methods and practices of the Steam Turbine Division. The turbines of the parent group were typically individual units for commercial customers and the basic method of the group was an experimental approach of learning from experience. In this approach, they commonly modified an individual turbine until it was satisfactory, regardless of how much time it took. The method did not incorporate working to delivery schedules or designing for production, each turbine being a singular project unto itself. Such turbine designs were not sized for future growth or for any allowance for improved efficiency over time. Utilization of this approach when designing and constructing gas turbines had already produced great delays and difficulties while bringing the earlier J30 to production readiness, even with the help of Pratt & Whitney. The J34 design was based on the J30, incorporating most of the basic mechanical design layout of the earlier engine scaled up to a 24 inch intake diameter. Even with the experience of the J30 behind them, the J34 prototypes had experienced delays. The establishment of a new production line in Kansas City was also a contributor to the delay as new workers were trained and equipment put in place, but the J34 was now considered past those early production issues and would soon be safely in production in quantity in Kansas City, leaving the development engineering effort in Essington.

A brief review of the J34-WE-22 being produced (WAGT model 24C4B) at this point is in order. It was a single shaft, three bearing design, with an eleven stage compressor with a 50% action/50% reaction blade/stator workload split design, an annular diffuser, a dual annular combustor (which allowed two rings comprising a total of 60 spray nozzles for fuel delivery), and a two stage turbine with a fixed nozzle. The J34-WE-22 did not have an afterburner and utilized a fixed exhaust nozzle with ground adjustment for the outlet size being done by tabs on the airframe itself. The compressor case and discs were made of aluminum alloy with steel blades and guide vanes (stators). The diffuser was aluminum and the combustor, a double annulus section, was constructed of high temperature steels welded to form the combustion basket and liner. The turbine discs were steel with blades of a Stellite alloy. Starting was accomplished by an electrical starter/generator mounted directly ahead of the shaft. The accessory gearbox was driven from the front of the compressor shaft by a bevel gear, the accessories themselves being grouped under the bottom of the engine on an aluminum gear case. The mounting of the engine controls and accessories allowed for easy modification as required for dual engine installations where the left and right engines had to have mirror image connections. In the case of the Douglas F3D-1/-2 SkyKnight, the installed engine gear cases were assembled rotated in opposite directions to accommodate the airframe pilot escape tunnel design. The later J46, as will be seen, would not offer such flexibility for such unique installation needs. The J34 engine offered a basic 3,000 pounds of static thrust at sea level normal day conditions with a specific fuel consumption (SFC - measured as pounds of fuel per hour per pound of thrust) of about 1.04 at military rpm and power. It was rated to 50,000 feet and normally operated at up to 12,500 rpm at anything above the minimum cruise power settings.

All WAGT performance guarantees were based on mathematical calculations of performance under varying conditions up to the maximum speed and altitude.

J34 Design Overview

The challenges of achieving a successful axial design that functioned well at all power settings, airspeeds, altitudes and ambient temperatures are best appreciated if such a layout's operational elements are reviewed.

The axial compressor uses rows of blades attached to spinning discs. Each row is called a stage, the first being the one with the largest blades. It may or may not have a row of fixed fins in front of it, these presenting the airflow to the spinning blades behind at the optimum angle to allow the blades to achieve maximum efficiency. The spinning blades push the air backwards at an angle, adding energy by accelerating the air and also compressing it to some degree. The accelerated air then encounters another row of fixed guides (now called "stators") where it both slows down, adding to the compression, and has its direction changed to again present the flowing air at the best angle for the next stage's row of spinning blades. The next stage has smaller blades as the air is now compressed to some degree and can achieve additional compression with less surface area. At each stage, the air is compressed further until it exits the compressor and flows into the diffuser, where the air flow is stabilized, slowed to some extent and the compressed air released in a smooth stream back into the combustor.

The challenges of an axial compressor are many. WAGT used a 50/50 action/reaction split on each stage of their compressors as the workload split between the spinning blades and stator, each contributing half of the compression achieved by the stage. Assuming the design point balance is achieved, the air flowing into the next stage is flowing into a smaller space at a higher temperature. The blades and stator of the next stage must be designed to this requirement. By the time the air is compressed through all of the stages to the design compression ratio (say 4.0) of the entire compressor, the air coming out of the rear of the compressor is now occupying approximately a quarter of the space of the uncompressed air of the first stage.

Starting such a compressor is a challenge, as at first the slowly spinning blades are not achieving much compression and most of the volume of compressed air simply cannot get out of the rear of the compressor. The later stages are contributing almost nothing to the compression as they are deeply stalled in the low pressure air that is traveling slowly. As the starter increases the RPM, each later stage begins to un-stall and begins to add to the compression effort. Only when a steady air supply at an adequate level of compression is coming out of the rear of the compressor can the engine combustion be started. Up to this point, the starter motor has been slowly increasing the speed of the compressor shaft against an increasing workload as the compressor begins to become more effective. Starter motors for such a compressor have a tough life. Start times can be up to a minute on each start before the engine achieves an idle speed where acceleration to higher RPMs can be accomplished unaided and the starter can disengage.

Other considerations that affect the detailed compressor design are: the trade-offs between intake size and RPM limits of the chosen compressor discs and blades to move the desired amount of air (at the time of the J46 design measured as pounds of air per second at military RPM and throttle setting), the air volume in excess of the air needed for combustion at full throttle needed to provide an adequate surge line, and the number of stages needed to accomplish the targets (the compression workload per stage). The early design of the J34 compressor showed that they required 11 stages to accomplish a 3.9:1 compression ratio.

Among other considerations is the overall efficiency of the engine. Higher compression ratios deliver higher fuel efficiency, but can add weight through additional stages and increased engine strength to handle the internal pressures, such extra weight being easier for a bomber to trade off against the lower fuel weight than in a fighter design.

Even at the low compressor ratios of the early WAGT engines (3.5 to about 4), designing a compressor was no easy thing.

The rest of the engine almost seems simple by comparison, but closer examination reveals additional challenges lurking there as well.

The diffuser must present a smooth and even airflow at even temperatures to the combustor. The outer blades of the compressor increase the airflow speed and pressure slightly more than the inner. The entry temperatures to the diffuser can be significantly higher along the outer edge of the diffuser than along the blade roots. All of this must be equalized in as short a distance as possible in the diffuser through mixers and guide vanes.

The combustor must split the incoming air into flows for cooling the outer and inner components of the combustor and for combustion. As combustion occurs, temperatures rise as the air flows toward the outlet of the combustor and additional cooling air is admitted in controlled streams to provide a film of air over the metals to keep them within their operational temperature limits. Hot spots of any kind must be avoided.

Combustion requires a steady supply of measured fuel that has been vaporized to the greatest extent possible at every possible throttle setting. The fuel control must balance the current throttle setting against the pressures and temperatures in the engine and only allow a safe amount of fuel to be delivered for the current thrust demand, temperatures and pressures. At no time can excess fuel be delivered that would raise the operating temperatures inside the combustor section or the turbine entry temperatures beyond the safe limit of the materials in those sections, regardless of the throttle setting. The combustor must provide even burning with little or no carbon buildup within the combustor or clogging of the spray nozzles (or fuel evaporators). Most importantly, the pressures in the combustion section must never be allowed to approach too closely the higher pressure of the incoming air supply coming out of the diffuser. If this occurs, the compressed air cannot flow out into the diffuser and the compressor blades stall as the airflow breaks down. For any given compressor operating speed, the pressure where compressor stall will occur is called the "surge" line. The control unit, to achieve the highest possible thrust and fuel efficiency, must operate the engine as close to the surge line as possible without ever going too close. The addition of afterburners on jet engines added extreme difficulties to the design of control units, as the lighting of an afterburner results in a large rise in back pressure at the rear of the engine at the moment of ignition.

The engine starting system generally uses ignitors (spark plugs) designed to live in the hot environment of the combustion chambers while also giving a long life. Getting the position of the ignitors correct for consistent ignition both on engine start and restarts in flight after a blowout (flameout) was a matter of trial and error. Finding ignitor designs that gave long life and reliable operation in service proved to be a seemingly never-ending search.

The turbine section operates within the swiftly escaping gas from the combustor, the blades (now called "buckets" in the USA) being spun by the gas, extracting only the amount of energy needed to achieve the required RPM of the engine (and accessories). The hot gas flows over guide vanes (nozzles) to aim it at the best angle to the spinning turbine blades. In the case of a two stage turbine, the gas again flows over a second stage of nozzles and then past the second stage blades. At no time can the turbine be allowed to speed beyond a designed maximum speed or be exposed to gas temperatures beyond the material's limits. Higher speeds and temperatures shorten the life of the turbine wheels and blades considerably and damage to this section of WAGT engines was generally the most frequent cause of rejection to overhaul before the normal overhaul time limit was reached.

The RPM and temperature operating limits of the engine have to be guarded by a control system that protects the engine from damage while delivering maximum performance under all conditions as automatically as possible. A fail-safe control and fuel delivery system must be provided to allow the pilot to maintain safe flight in the event of a major failure of a major control or fuel component to the greatest extent possible. The control and fuel system must function perfectly regardless of aircraft maneuvers including both positive/negative and zero g-loads in all directions, including a catapult launch from a carrier.

The above general description of the J34 axial design and its challenges give us a good basis to appreciate the Bureau of Aeronautics actions as they became more familiar with the engines being delivered and operated in service.

Improving the Breed

On June 26, 1947, BuAer awarded a sole source Letter of Intent contract to WAGT (contract NOa(s) 9051). The statement of work gives the reader good insight into the existing problems and shortcomings of the J34-WE-22. That BuAer was willing to undertake an expenditure of over two million dollars to improve the engine is an indication of strong support to WAGT. Hindsight shows us that this was a missed opportunity for Westinghouse to make an early investment in the fledgling division for the future by building adequate research, test, and machine shop capabilities to match BuAer's contribution. Certainly the scope of this contract makes it clear that a massive effort would be needed to produce much better engines.

Statement of Work for Contract NOa(s) 9051 and the Resulting Development Work[1]

This contract covered most of the functions of the engine and can be read as BuAer's areas of concern and desired improvements to the existing engine. This contract would later become the focal point for other issues with the engine that were identified after the contract was put in place.

While the contract does not specifically spell it out, the end result of much of the work was to be applied to two J34-WE-22 engines loaned to WAGT. These were to be torn down and the modified, upgraded parts and systems were to be utilized to build two new engines deemed the Model J34-WE-30. Some records refer to these engines as J34-WE-30a, but no specific reason for that change was identified.

With a contract in place, work began on all of the tasks listed. With parallel work and discussions going on during the duration of the contract across the engine Statement of Work (SOW), following the individual topics becomes difficult if a chronological sequence of events across the entire contract were to be followed. Here we will discuss each development topic from the SOW and follow it to its conclusion.

Task I, A.1 Develop Fuel System and Control

This task led to the development of the electronic control system utilized by the J34-WE-32/-38 and the J40-WE-6. It is possible this control system was also used on the two

J34-WE-30 engines at the end of this contract, but no specific build description of the latter engines was located.

The electronic system was an isochronous (constant speed) system utilizing a Woodward X-770 governor to maintain a consistent speed. The system operated the engine at a constant RPM above 60% power settings. It utilized circuits incorporating vacuum tubes and could operate an adjustable nozzle automatically. It had total control of the engine and offered linear throttle response to the pilot from idle to maximum thrust even on afterburner equipped engines. The afterburner was modulated through three "zones" of thrust, the last being maximum thrust. Versions of this system were tested on a house J34 test engine for use on the J34-WE-32/-38 and the J40-WE-6/-8. When in operation on the test stand, it performed as designed and allowed superior fuel control, over-temp protection and lowered pilot workload. The emergency fuel system designed for the J34 (see Task I, A.3 below) was fully integrated into the automatic system.

The electronic system was ultimately rejected by BuAer. Actual use of the system revealed design weaknesses. The vacuum tubes suffered from signal drift as they aged and they also physically failed with relatively short life-cycles in service use. Pre-aging of the tubes before service use along with the development of "ruggedized" versions of the tubes improved the situation. Engine vibration also contributed to tube failure and WAGT's attempts to be allowed to mount the control on the airframe itself were rejected by BuAer, it being considered that the standing requirement for all engine components was that they be mounted on the engine itself with no exceptions. In the end, other considerations drove BuAer's decision to require a different control on the J46 (and the later J40 production engines).

The speed control and the temperature circuits would occasionally override each other in a feedback loop that would result in the engine "hunting" and the thrust fluctuating slightly up and down. Also, the pilot could return the throttle to a prior position after an adjustment and not receive the same engine thrust even at the same altitude and speed of the aircraft. This made formation flying very tiring, with constant pilot throttle adjustments required to stay in position.

The main objections to the system were safety related. The control required electrical power from the aircraft electrical supply to operate. This violated BuAer's fail-safe requirement. Also, the emergency fuel control system, which was built into the fuel control, was totally hydraulic and, in the beginning, offered only limited pilot control. As the capabilities of the emergency system were improved, it moved closer to being able to provide fully adequate control at all times and it became the replacement main control, implemented in steps as electrically dependent components were replaced in stages. The J46 would utilize two of the mechanical systems, as will be described in the J46 development chapter.

The above description was taken from many memos and reports. WAGT filed a patent application for this system which gives the best description of the electronic and mechanical elements of the system, along with detailed circuit diagrams.[2]

This was ongoing work and no report was issued under this contract. The work was transferred to the X24C8 (J34-WE-32/-38) and X24C10 (J46) development contract and will be discussed further in those chapters.

Task I, A.2 Engineering Liaison on Control System Development

This was accomplished via regular meetings and memos to BuAer describing the results.[3] Meetings were held with Chandler Evans, Holley Carburetor Company, Simmons Accessories, and Bendix Products, all of whom were working on non-conventional control system approaches.

Task I, A.3 Emergency Fuel System Development

WAGT produced report A-600, January 23, 1948 that included their recommendations for an emergency fuel system. *(The report was not found.)* Other documents refer to Appendix A of the report and indicate it recommended that an emergency fuel pump be added to each engine, with either a pressure or speed sensor to be integrated into the engine control to automatically switch to the emergency pump in the event of a sensed failure. Whether the pilot would have the ability to manually switch over to the emergency pump is not clear. However, the design was tentatively approved at a conference on February 3, 1948 and WAGT quoted quantities, shipping schedules and prices in a memo on February 25. It is clear the intent was to install the pump and switch on all existing J34 engines in the field and on all new engines produced.[4] Pumps would begin to be delivered in October, 1948 if BuAer responded to the proposal immediately.

BuAer asked for a flight test of the system and WAGT proposed using an F2H-1 Banshee, which had two engines and was an excellent platform for such a test which required holding constant speeds at constant altitude.[5]

On April 1 and 6, WAGT was asked if it was possible to quickly develop kits for emergency fuel system pre-flight checks for the J34 that were also suitable for the Douglas D-558-II (Skyrocket) airplane. The response was that four kits could be developed for $3,037 with delivery of the first two kits within eight weeks of receipt of a commitment and the other two kits within an additional two weeks. However, other work would be delayed: development of an independent emergency fuel system (four weeks); improved fuel control (two weeks); and the J34 afterburner control and development (under contract NOa(s) 5382) (three weeks). Additional funds to cover the work would have to be allocated to one of the existing contracts.[6]

Only later did BuAer discover that WAGT was not planning to deliver certain elements of the emergency fuel system called out in the approved report (A-600). These were a selection system between the primary and emergency system, with a switch in the automatic position allowing automatic switchover with no pilot action needed in a failure of the primary system. The "Check" position would select the emergency system to prove it was working correctly. "Reset" would return the system to the "Automatic" function. The system had to be operable with an external voltage supply between 17-30 volts. In addition, an externally powered *(from the engine)* indicator light had to illuminate in the cockpit if the emergency system was activated. These features had to be included in any emergency system on the J34-WE-32/-38 engine or the J46-WE-2 engine.[7]

As a result of the discovery of the missing elements of the emergency fuel system in the planned control system, the specifications for the J34-WE-32/-38 and J46-WE-2 engines were modified to spell out the specific emergency fuel related functions and elements required in the system.[8]

BuAer approved the final design, covered in report A-869 on July 21, 1949.[9]

However, upon further review of report A-869 on the planned control system elements, BuAer made the following additional objection comments:[10]

1. There was insufficient detail relative to the endurance characteristics of the system, i.e. number of hours of operation and types of failures possible.
2. There were no provisions for indicating operation of the emergency pump and/or control to the pilot.
3. There were no circuit diagrams showing how failure of the electrical power supply accomplished automatic transfer to the emergency system.
4. The provided sketch lacked clarity of detail and used non-consistent nomenclature to the text of the report making the control operation difficult to understand.

At about this time, BuAer began pushing WAGT to develop what became known as the "Simple Control System", which, except for the automatic trim control, did not use any electrical power. The full story related to this control will be covered in the J46-WE-8 development chapter.

Task I, B. Unusual Flight Conditions Reliability

The WAGT Report A-869 that closed this subject was not found in the files. Most likely this was a dramatic improvement in both fuel delivery and lubrication related to high "G" and zero/negative "G" conditions frequently experienced by aircraft both in maneuvering and level flight as disturbances from turbulence were encountered. See the next related task.

Task I, C.1. Improve the Lubrication System for High Altitudes and Speed.
Task I, C.2. Improve Lubrication under Unusual Positions.

Both of these tasks were responded to in WAGT Report A-622. While this report was not found in the files, its basic recommendations can be derived from a contemporary patent application filed only weeks after the above report was forwarded to BuAer.

The patent description states that the best lubrication system for a gas turbine power plant minimizes the distance oil has to travel and the pressure drop oil has to experience, both to reduce external tubing/piping needed and to facilitate cold starting, especially in extreme conditions of high altitude.

The pump should incorporate both scavenge and pressure delivery elements in a common case. Such a pump should be mounted for easy access and minimal disruption of piping during removal and/or installation.

The filters and control valves should be serviceable without disassembly or removal of the pump from the engine.

Lastly, it stated that such a system should be installable on an existing engine without extensive alterations to the engine.[11]

In the patent application image, the oil scavenge and supply pump is at the bottom of the accessories case. The supply line goes to and around in a spiral on the inside of + the air intake (presumably for oil cooling), emerges to travel down the top of the engine to a coupling, there splitting off a supply line to the number one bearing, then continuing along the outside of the case to the middle number two bearing and on to the rear bearing. Return lines bring the oil back from the bearings to the sump.

This design description does not address negative or

zero "G" conditions, which require the scavenge system, oil supply pickups and air-oil separators to continue to operate normally even in unusual attitudes for some period of time and resume normal operation automatically if the oil supply is interrupted for a short period, all without pilot action or damage to the engine. Report A-622 may have addressed this.

Task I, D.1 Develop Better High Altitude Combustion and Airspeed

Overall engine performance at altitude suffered from issues related to the J34's fuel spray system. This carried over the J30–WE-5 approach and incorporated two hollow circular tubes with fuel spray nozzles screwed into one side of each ring, the nozzles of each ring then aligning with openings of the respective dual annular combustion chamber burner cage. There were 60 total spray nozzles. The hollow tube was kept full of fuel with the amount of fuel sprayed into the combustion chamber being controlled by a simple fuel line pressure regulator that balanced pilot demand and compressor output pressures. To adjust for minor differences in spray fuel ring pressure due to gravity, the spray nozzles' orifices were smaller on the bottom half of the rings than on the upper half.

NACA had tested a Model X19B (J30) which used a very similar system. McDonnell Aircraft Company issued a complaint that they were unable to achieve stable flight in their XF2D-1 aircraft at speeds slower than 500 mph above critical single engine altitude (approx. 35,000 feet) because of the very high idle speed (10,500 rpm or above) of the engine at that altitude. Such a high idle would have produced a significantly higher thrust creating the problem. They asked for a significant reduction in engine idle speed.

BuAer responded to the request with a memo describing the issues NACA reported with the J19B fuel system and recommended possible changes to the current J34 design as a result. They stated "the combustor tests on the X19B engine showed that the principle cause of deterioration of combustion efficiency with increasing altitude is the rapid deterioration of both spray characteristics and nozzle-to-nozzle distribution of fuel as the rate of fuel flow decreases and the static head between top and bottom nozzles of the fuel manifold becomes an appreciable percentage of the pressure head across the nozzles."

They went on to state "It has been clearly established, in several cases, that duplex nozzles are superior at low flow rates to the fixed orifice nozzles of the type used in the X19B *(J30)* and 24C *(J34)* engines."

"Recent tests at the Flight Propulsion Research Laboratory of the National Advisory Committee for Aeronautics have demonstrated that spring loaded nozzles, in conjunction with a flow divider, are superior to duplex nozzles at all flow rates between 8 and 70 per cent *(SIC)* of rated capacity of the duplex nozzles. Furthermore, fuel pressure requirements of the spring loaded nozzle are essentially constant of the entire flow range. These pressures are approximately 50 psi for the nozzle used in the NACA tests. It is believed the spray characteristics of the NACA nozzle are such that not more than 20, possibly as few as 16, would be required for the 24C engine."[12]

BuAer requested immediate action on the C24-4B engine and advised WAGT they could view a demonstration of the spring-loaded nozzles at the NACA laboratory in Cleveland at their earliest convenience.

It was noted that at a conference with WAGT to discuss the situation, WAGT responded to the suggestion of the possibility of using only 20 or 16 nozzles instead of the 60 currently used by stating it might not be practical with the current double annulus combustion chamber. WAGT advised three lines of investigation:

1. Utilize the present double annulus combustion chamber with a reduced number of nozzles (possibly 6 in inner annulus and 12 in outer).
2. Utilize the "Peabody" can type combustor with a single spring loaded nozzle for each can.
3. Redesign the present combustion chamber liner to provide a single annulus with a minimum number of nozzles.

"Of these three approaches the first and second appear to be the less extensive."[13]

The "Peabody" can-type combustor design referenced is known today as a can-annular design, where the combustion takes place in a series of cans inside the annular combustion area where some of the air coming out of the diffuser is routed into the individual cans for combustion and the rest routed around the cans to cool them. The design offers better combustion control in a limited space and potentially easier maintenance if parts are damaged, but overall tends to weigh more than a pure annular design.

The final report on this work was WAGT A-851, issued on June 7, 1949. It summarized the results:

"A generally improved double annulus combustor which will fit the 24C-8 engine for operation on AN-F-48 fuel has been developed. The principal performance gains over previous 24C combustors are in radial temperature distribution, combustion efficiency and altitude limits. It was found possible to incorporate the same improvements in the 24C-4D engine and this was done."

"The effect of a change in fuel from AN-F-48 *(100/130 avgas)* to AN-F-58 *(JP-3)* on 24C double annulus combustor performance was appraised for development purposes and a change in combustor configuration was made which nearly eliminated the coke formed by AN-F-58 fuel."

"Advance work was done on a single annulus combustor with spring-loaded nozzles for the 24C-10 engine."[14]

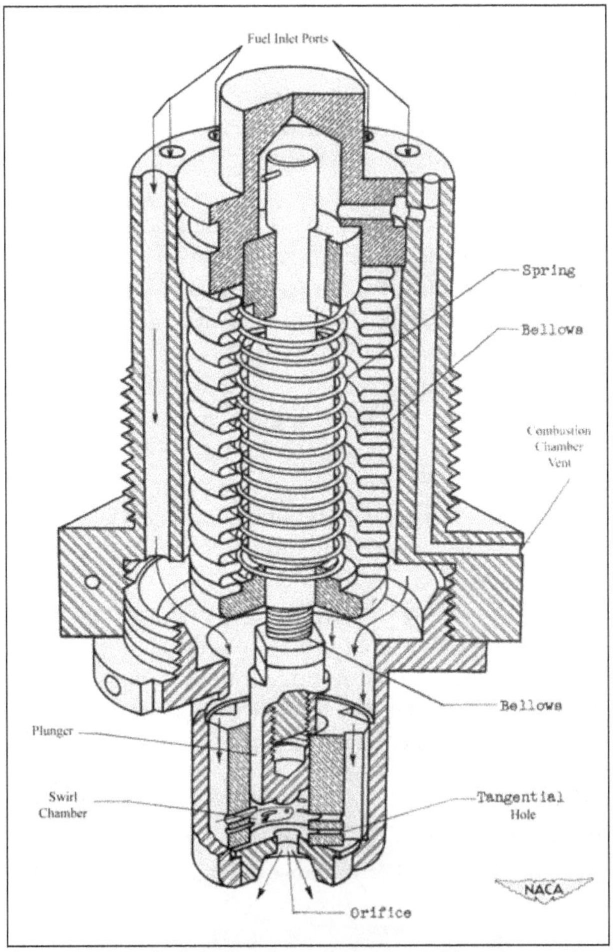

Figure 1 NACA designed spring pressure variable area nozzle. This nozzle design strongly resembles the WAGT fuel nozzle used on the J40-WE-6/8 engines. NACA Report E8D14.

The Westinghouse 24C8 engine was the J34-WE-32 engine, which was developed as the precursor to the J46 engine under the J46 development contract. The J34-WE-32/-38 engines are covered in the next chapter.

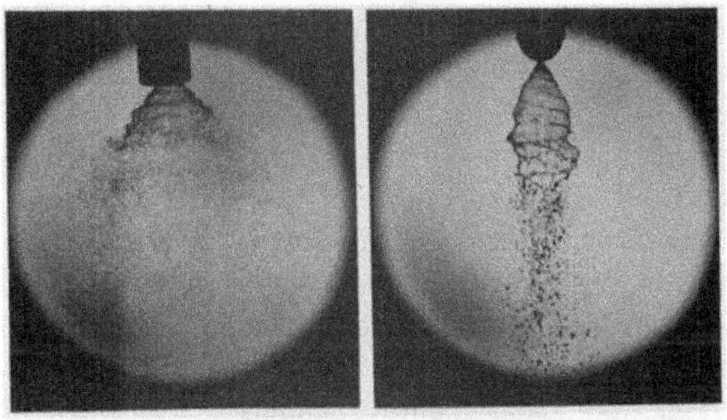

Figure 2 Spring-loaded nozzle (L) vs. WAGT type fixed orifice nozzle spray characteristics at 40 lb/hr using avgas. Both nozzles were comparable when the rate passed 70 lb/hr. NACA Report E8D14.

The detail of the report included the statement that during the investigation, a decision to operate the 24C8 engines only on gasoline type fuels had been reached. The anti-coking knowledge gained would be used in development of the 24C10 engines.

The detail on the third item shows the conclusion was a bit overstated in terms of accomplishments on the task. While test apparatus had been built and initial testing had begun, the letter of intent contract for the 24C10 model (J46) had been received and the development effort on a single annulus combustor transferred to that contract. Further discussion of combustor development will be covered in the J46 development chapters.

Task I, E. Develop High Altitude Starting and Ignition Shielding

High altitude starting was addressed in WAGT Report A-864 (*not found*). The approach basically covered restarting occurring automatically if the engine encountered a flame out (blow out). A flame probe was added to the combustor to sense the active ionized gas of the flame if combustion was occurring. When a blowout occurred, the lack of an ionized gas in the gap of the flame probe elements caused the ignition system to turn on and begin energizing the spark plugs. Once a flame was detected again, the starting circuit disengaged automatically. If damage occurred to the circuitry of the flame probe to the ignition circuit, the fail-safe action was for the ignition to automatically turn on and stay on until engine shut down. No damage would occur to the starter circuit or the spark plugs from continuous operation. Interference shielding was addressed in Report A-814.[15]

WAGT Report A-684 (not found) proposed a starting and ignition system design and A-814 reported on testing of the design. The test report states the primary objective was "to develop an ignition system so designed that although the fuel flow is interrupted during a 10 second period of negative acceleration. When the acceleration is removed, the engine will return to normal operation without further action by the pilot. The ignition system would also permit engine starting at all altitudes up to and including 40,000 feet and with flight speeds up to and including an equivalent ram pressure ratio of 1.4. It should comply with AN-E-30 paragraph D23e Performance, and AN-E-32 paragraph 7-2d(1)c Radio Interference."

The summary states that the ignition unit performed in continuous duty in a temperature range of -67 to 165°F and met the required AN-E-30 specification. Various spark plug modifications were investigated and tested to eliminate ceramic insulator failures. The tests failed to eliminate the failures, but a modification of the spark plug shell that extended the shell past the insulator provided much cooler spark plug operation and improved ignition ability.

Detailed findings showed the motor-breaker ignition systems could provide long life, but the two tested, from General Laboratory Associates, Inc. and Scintilla, although providing adequate power, were too heavy for engine applications.

Interrupter plugs produced high energy spark but the electrode durability was poor and had to be improved before being considered for duty use.

Finally, the ceramic cracking of insulators came as a result of the available insulator materials not being able to handle the temperatures and cooling rates to which they were exposed.

The present design, 14G510, was proving to give excellent durability with many lasting 150 hours of service without problems. The plug design is shown below.[16]

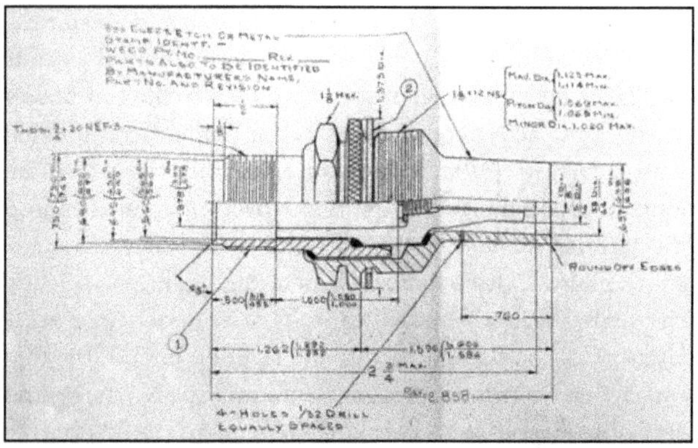

Figure 3: J34 Spark Plug Design. WAGT Drawing of Part 62G488, 3/18/1948

Task I, F. Develop an Integral Compressor Bleed Manifold

This work was reported in Report A-763, although the report was not located. It was approved by BuAer March 30, 1949.[17] The J46 and J40 engines incorporated an integral bleed manifold that allowed up to 10% of the compressor air flow to be extracted for cockpit and other compartment pressurization or de-icing. Such air would have to go through an air conditioner system to regulate its temperature and pressure to the desired setting. The air in raw form could be used as an anti-icing flow to engine compressor entry vanes if necessary.

Task I, G. Develop Quick Disconnect Flanges for the Air Intake and Turbine Exhaust Attachments.

This task was reported on in WAGT Reports A-598 and A-762 (neither was located). The clamp design was accepted by BuAer on January 12, 1948[18] after WAGT demonstrated that the clamps could withstand the bending and shear loads required in the maneuvering limits specification for the J34-WE-22. The clamps made installing and uninstalling the engines much quicker.

Task II, A. Develop the Necessary Gear Box for High Altitude Operation and Provide a Pad Drive, If Required, for the Variable Area Exhaust Nozzle.

The WAGT report, A-613, was not located. However, the recommendations in the report were accepted by BuAer on April 12, 1948 with two discussion points. First, in connection with high speed, hot day operation, the present oil cooler configuration would only cool the oil to 240°F. The rubber parts of the gear box had to be changed to AMS3226 specification materials to operate at the higher temperature. They pointed out that BuAer wanted an oil to fuel cooler *(interestingly, called a "fuel to oil" cooler in the memo, obviously in error – Auth.)* developed under Task I, sub-task C of the 9051 contract, presumably to lower oil temperatures.

They also stated that while WAGT's tests indicated that spline shaft oil lubrication would not be necessary at high altitudes, should it turn out to be necessary in service, WAGT's suggested solution of flooding the gearbox with oil was unacceptable and that some other solution would have to be found.[19]

Task II, B. Develop the Compressor to Provide Reliable and Efficient Performance at High Altitudes and Airspeeds.

WAGT submitted report A-769 dated October 28, 1948 (not found) to close this task. A-769 was a plan for development of the compressor and was accepted by BuAer on December 28, 1948.[20]

The plan was executed and resulted in two reports. The first was the final report, A-852, dated June 15, 1949. This indicated that the development program produced improved J34-WE-30 and J34-WE-34 compressors and a new design for the XJ34-WE-32 engine. The new compressors met the design points of 59.3 lb/sec airflow and a pressure ratio of 4.15:1 at 12,500 rpm. The compressor's efficiency was good at maximum engine rpm and further development work was being done under contract Noa(s) 9670 to make best use of the compressor in the J34-WE-32.[21]

Confusingly, an interim report on this task was issued (A-855, June 17, 1949) after the final report because the contract called for an interim as well as a final report. This report shows in detail the investigations undertaken to improve the compressor and achieve the results reported in A-852.[22] Using nine pressure probes per stage, the compressor had been tested from 5,830 to 13,500 rpm and up to a pressure ratio of 4.31 at 12,500 rpm (the stall point at full unlimited throttle). A full range of data for a total of 238 test points was taken. The tests showed a rough running point at 9,500 rpm. While the last eight stages of the compressor were operating at maximum efficiency, the first three stages we lightly loaded at full rpm and could be redesigned to take more of the compressor's workload. It noted that current J34's in service were operating successfully in the 9,500 rpm range without problems. The surge line of the compressor was tested at all rpm's and throttle settings and it was discovered that throttle movement and speed change in approaching the stall limit had very little effect on pressure ratio at the point of the stall. One major finding was that the use of higher compressor ratios than the current 4.15 redesign would require the variable area adjustable exhaust nozzle intended for the J34-WE-32 design.

Compressor development on the J34 continued to the end of the J34 program to address blade and disc failures and improved afterburner performance. A National Advisory Committee for Aeronautics (NACA) investigation of third stage J34 blade failures in 1953 showed that at certain rpm's the blade vibration was significantly above safe levels. Changes to the first two stages of stator vane angles removed the third stage blade excessive vibration and increased the airflow through the compressor by four percent at all speeds.[23]

Task II, C. Develop an Automatically Controlled Variable Area Exhaust Nozzle for the Model J34-WE-22 engine.

No final report on this task was referenced in the contract materials reviewed. WAGT did design a variable area nozzle and submit drawings to BuAer, but the drawings did not surface in the files.[24] It is known that BuAer had explored letting contracts for automatic nozzle control designs and asked WAGT to provide various engine data so the request for bids could be let.[25]

Design work on an automatically controlled variable area exhaust will be discussed in later chapters dealing with the J34-WE-32/-38 and the J46.

Task II, D. Develop a Reliable and Consistent Turbine Outlet Exhaust Gas Temperature Measuring Device.

Final WAGT report A-727 (July 23, 1948) was submitted to fulfill this task. It referenced report A-623 which contained drawings of the solution (report not found). The final report states that A-623 reported that only development of reliable thermocouples could adequately address the temperature measurement needs of a control system for the engine. The solution in A-727 was the follow-on work based on A-623's accepted proposed solution.

The first design utilized nine thermocouples (in lieu of the then current three being used) in the turbine exit airflow. They were wired in parallel to ensure all hot spots in the exhaust were captured in the average of the readings being sent to the cockpit indicator and the control system. This direct signal failed to produce the desired accuracy. The tested range of calculated turbine inlet temperatures was still too large to enable use of a single turbine setting on the engine that would ensure the cockpit indicator was accurate without adjustment regardless of which engine was installed in the airframe. The first design's results showed that the use of such a single setting would have either meant that many engines would not have met their thrust specification on final test if set too low, or would have allowed other engines to suffer hot spot damage if set too high. Practically speaking, the then current process of setting the engine indicators and control units to match each individual engine at installation on the airframe would have had to be continued.

Efforts were expended in testing both numbers and thermocouple locations and the best location found was along the exhaust nozzle wall. Even with careful adjustment, the best measurements still resulted in too wide a difference over many engines to achieve the goal of engine interchangeability without resetting indicators in the cockpit.

A balancing thermocouple circuit was designed and tested. It worked effectively in a 150 hour engine test and analysis proved that various failures of either the circuit or one of the engine thermocouples would have only a minor effect on actual indicator readings, hence no unsafe operating condition would result.[26]

Task II, E. Provide Engineering Assistance for Wind Tunnel and Flight Tests of the Model J34-WE-22 Engine.

This was accomplished through provisioning of WAGT personnel at the test sites, along with appropriate tools and parts.

Task II, F. Prove the Reliability of the Accessory Drives for Models J34-WE-7, -9, -24 and -26 Engines

Reports A-613 and A-837 were submitted on this task. Neither was located in the archives. Both were approved. The subject of the accessory gearbox location and overall design would be a major component of J46 development. The J34's gearbox allowed for major differences in the accessory locations for various airplanes. The McDonnell F-88 and Douglas F3D-1 and F3D-2 had mirror image gearboxes, the later aircraft having the engines also rotated slightly in their mounts to fit the airframe. Such flexibility would not be available in the J46 and caused problems.

Task II, G.
a. Make a Design Study of, and Submit Preliminary Design Drawings on, an Inlet Screen for the Model J34-WE-22

Reports A-704 and A-725 were submitted and approved. The screen was designed and tested for strength and report A-725 covered this test. [27,28]

b. Make a Design Study of and Submit Preliminary Design Drawings on an Anti-icing System for the Model J34-WE-22 engine, and modify one (1) Government-furnished Model XJ34 or J34 engine to include one (1) set of anti-icing equipment.

Reports A-705, A-802 and A-832 were submitted, but only A-832 was located. The tests at Mt. Washington utilized J34-WE-22 Serial WE002051, an engine previously allocated to the Aeronautical Engine Laboratory. The compressor was tapped and hot air piping was run from there to the starter motor dome and intake struts. Due to unusually clear weather, extreme difficulty was at first experienced in getting ice to form on the engine, even with the piping blanked off. The tests were delayed until better conditions existed and report A-832 resulted from the late testing. This reported that engine operation at high engine speeds in heavy icing conditions was dangerous, as chunks of ice could break off from the intake and flow back through the compressor with some violence. The piped hot air was effective in keeping icing to a minimum except in the most extreme icing conditions.[29]

Task III, Item 1 – Services and materials necessary to improve Government-furnished Model J34-WE-22 engines to give them longer life and higher performance including the design, manufacture and furnishing of three (3) sets of improved parts assembled on such engines; the preparations and furnishing of drawings, design analyses and related data, monthly progress reports thereon; and the following:

a. Develop and Apply New Engine Materials.

This was an ongoing task that analyzed using new materials; for the compressor case (magnesium alloy), blades (ceramic, stamped steel), compressor discs (titanium), and burner basket (ceramic lining). Details on this research as it was applied will be covered in the J46 development chapters.

b. Develop a Thrust Measuring Kit to Permit Flight Test Measurement of Engine Performance Without Special Engine Mounts.

WAGT initially recommended using a modified Brown Flight Test Recorder to capture the electrical pulses from low pressure sensors on the engine, subsequently reducing the data on the ground for calculating the thrust attained during the flight.[30] Report A-587 dated October 22, 1947 was submitted (not located). This approach was approved by BuAer and three thrust measurement kits were requested to be manufactured.[31] These were to be used on the F3D-1, F2H-1 and F6U-1.[32]

Drawings for the kits were not submitted to BuAer until August 2, 1948.[33]

They were designed to be interchangeable with existing engine components and to require no airframe changes, the engines with and without the kit being installationally interchangeable in the field. The outline of the approach to be used for kit development was covered in WAGT Report A-587,[34] accepted by BuAer on December 1, 1947.[35]

The final test report on the kit developed stated "The principle of operation of these kits is that the performance of the installation components may be determined from knowledge of the status and total pressure and its distribution at the compressor inlet ... and the temperature and temperature distribution at the turbine outlet."

Parts were modified to allow for pressure rake and temperature probes at the front and rear of the engine. These replaced the original parts with those modified to accommodate the various ports needed. Instruments were added to record the readings on a chart while the engine was operating. The temperature recorder was made by Brown Instrument Co. and the pressure recorder by Westinghouse and Shaevitz Engineering, Collingswood, NJ.

The kit did not give a direct thrust reading in the cockpit while the engine was in operation. The recordings had to be examined and the data reduced by hand, with adjustments for various elements, such as the difference between actual and measured airflow rates at different airspeeds.[36] One stated limitation was that the kit could not measure all compressor inlet and turbine outlet pressure values on a single run because that would have required 64 pressure transmitters and only 48 could be provided for at any one time. Because of that limitation, at least three separate engine runs had to be made to cover the entire range, meaning in this case three flights with the sensors being relocated as necessary between the flights. Overall accuracy of the kit was given as +/- 1.5% above 10,000 rpm and +/- 5% from idle (5,000 rpm) to 10,000 rpm.

Some technical details of the aerodynamic and thermodynamic limits of the then current J34 engine design are mentioned in this report. It states that the greatest pressure variation at the compressor inlet should be no more than five percent of the mean pressure. The T_5 (turbine in temperature) should not vary more than 200°R above the mean value. Curiously, on the test engine used to prove the kit and the supporting data reduction process, the actual variation on the engine T_5 was found to be 207°R, attributed to the fact the engine was extremely old and worn. The report showed that at 12,392 rpm, the adjusted thrust of the test engine was 2,860 pounds, below the specification of 3,000 pounds for a current J34. WAGT stated this was to be expected of an early, worn engine in the series.

The basic contract called for three of the developed kits to be produced. BuAer later asked for prices for three more kits, to be used on the F3D-1, F2H-1 and F6U-1 airplane models. The proposal presented was for three kits at a total price of $61, 182.[37] Apparently no order was forthcoming, as the subject is not mentioned again in the records.[38] Interestingly, WAGT did not get around to quoting prices until February 10, 1949, possibly because the focus had moved to the J46 and J40 programs by that time.[39]

Task IV, Item 1 – Test water injection on Government-furnished Model J34-WE-22 turbo jet engines to provide thrust augmentation for take-off power, and provide and furnish reports and data thereon.

BuAer pushed WAGT on these tests, a memo in January, 1948 reminding Westinghouse that they had asked in October for tests to be run in 60 days on the Chance-Vought water injection ring with report(s) to be filed on the results.[40]

The tests were finally run and a final report (A-848, August 31, 1949) submitted. The initial tests were with an unmodified Chance-Vought supplied ring. Thrust increases of 10-12% using a 70/30 water/alcohol mix were obtained, but extremely poor radial temperature distribution resulted in rapid blade lengthening. The test was of a ring mounted just in front of the intake, injecting a water-methanol mix at rates from 0-10 gallons per minute.

The ring was modified and they tried injecting 80 octane gasoline from the ring in the same position, and injecting water-methanol with the ring in the exhaust duct behind the second stage of the turbine.

All of the results showed modest thrust increases over the standard thrust and some radial temperature distribution differences ranging from modest to severe.

Although the contract did not call for it, BuAer requested at low priority[41] that attempts be made to modify the ring to operate in the diffuser. Problems encountered were gasket leakage, blowout and melting. After several attempts, the tests were discontinued. The conclusion was that water injection or some variation using gasoline or methanol only was impractical on the J34 using the current injection apparatus. It was recommended that tests be done trying compressor inter-stage and combustion chamber injection.[42]

BuAer approved the report, but requested the task be kept open and that testing of any approach to injection that appeared feasible be considered. Any method that might be applied to later ("modern" was the word used) WAGT engines or later models of the J34, including afterburner models, was to be included in the consideration of the approach taken.[43] A proposed program for further testing was requested.[44]

NACA during this period (1947-1950) published over 100 reports on various studies of water injection on turbojets of all kinds. None were specifically on the J34 engine, but these reports were available to Westinghouse during this period for their consideration.

Much of the work related to contract Noa(s) 9051 overlapped the development work of the 24C8 engine, which was designated the J34-WE-32 and J34-WE-38 without an afterburner installed. As such, the contract became a funding source for development on the later model engine.

At the completion of the contract, using parts from the three sets of parts created under the contract, two J34-WE-22 engines were taken apart, the new parts installed and the modified engines re-designated as J34-WE-38 engines, effectively becoming the early prototypes for that model. One of the new J34-WE-38's (Serial WE002060) was sent to the AEL for testing and the other (Serial WE002070) sent to NACA for testing in their facilities. Later, WE002060 was tested at the AEL with the new electronic power control system intended for the production J34-WE-32/-38 engines (as well as the J40-WE-6/-8 models).[45]

With all this development work either complete or continuing, the next chapter reviews the development to the next engine that directly influenced the J46 development, the J34-WE-32/-38.

While this volume will concentrate primarily on the technical development of the J46 engine and its service life issues, this review of the major challenges being worked on related to its predecessor the J34-WE-30 and leading up to the J34-WE-32/-38 designs hopefully have given the reader a basis to better understand the unresolved issues that were still a challenge in moving forward to the later J46 development. The full development history of the many other J34 variants will require a volume of its own.

Chapter 1 – Earlier Research and Development

[1] Bureau of Aeronautics contract Noa(s) 9051, dated June 26, 1947. U. S. Navy Bureau of Aeronautics Contract Correspondence, Container 256 **[B256]**, Volume 1 **[V1]** Record Group 72.3.2 **[RG72.3.2]**, National Archives College Park, Maryland **[NACP]**
[2] WAGT Patent Application WM 5234, October 11, 1949, Cyrus F. Wood, Subject: Jet Engine Power Regulator. **B258V7RG72.3.2NACP**
[3] BuAer Memo, July 30, 1948, Subject: Contract NOa(s) 9051 – Model J34=WE-22 Engine, Engineering Liaison and Assistance to Government Control Contractors. **B257V5RG72.3.2NACP**
[4] WAGT Memo, February 25, 1949, Subject: Independent Emergency Fuel System Suitable for J34 Engines, Our Reference N-92393. **B257V3RG72.3.2NACP**
[5] WAGT Memo, March 11, 1948, Subject: Contract NOa(s) 9051; Independent Emergency Fuel System, Request for Flight Test of. **B257V3RG72.3.2NACP**
[6] WAGT Memo, April 21, 1948, Subject: XJ-34 and J-34 Engines Emergency Fuel System Pre-Flight Check Kits, Douglas D-558-II Airplane, **B257V4RG72.3.2NACP**
[7] BuAer Memo, March 3, 1949, Subject: Contract Noa(s) 9051, 9433, 9791, 2670 and 9212, Model J34-WE-30a, -34, -32; XJ46-WE-2, and XJ40-WE-6, -8, and -10 Engines, Emergency Fuel System, Contractor Furnished Components. **B257V6RG72.3.2NACP**
[8] BuAer Memo, March 25, 1949, Subject: Westinghouse Models J34-WE-30, -30A, and -34 Engine Emergency Control System Requirements – Deviations from. **B257V6RG72.3.2NACP**
[9] BuAer Memo, July 21, 1949, Subject: Contract NOa(s) 9051 – Development of Emergency Fuel System. **B258V7RG72.3.2NACP**
[10] BuAer Memo, September 26, 1949, Subject: Contract NOa(s) 9051 – J34 Engine Control System Final Report on. **B258V7RG72.3.2NACP**
[11] WAGT Patent Application WM 7990, March 5, 1948, Lubrication Means. **B257V3RG72.3.2NACP**
[12] NACA Report E8D14, Gas Turbine Engine Operation with Variable Area Fuel Nozzles, Harold Gold, David M. Straight, July 8, 1948, Flight Propulsion Research Laboratory, Cleveland, Ohio
[13] BuAer memo, September 8, 1947, "Contract NOa(s)-9051– J34-WE Engines Unsatisfactory Idling at Altitude. **B256V2RG72.3.2NACP**
[14] WAGT Report A-851, Combustion Development, Final Report, Contract NOa(s) 9051, Task I4D **B257V6RG72.3.2NACP**
[15] WAGT Patent Application WM 5081, December 29, 1949, Earl O. Setterblade, Combustion Apparatus. **B258V8RG72.3.2NACP**
[16] WAGT Report A-814, March 31, 1949, Final Report on J34 Ignition System Development Tests, Contract NOa(s) 9051, Task I, Sub Task B & E. **B260VENCLRG72.3.2NACP**
[17] BuAer Memo, March 30, 1949, Contract Noa(s) 9051, Item I, Task I, Sub-task F – final Report on Integral Compressor air Bleed System for Improved 24C Engine, Report A-763. **B257V6RG72.3.2NACP**
[18] BuAer Memo, January 12, 1948, Contract Noa(s) 9051, Task I, Sub-task "G", Design Study on Quick Disconnect Flanges on J34 Engine. **B257V3RG72.3.2NACP**
[19] BuAer Memo, April 28, 1948, Contract Noa(s)-9051, Task II, Sub Tasks A and F, Design Study of Proposed Development on Accessory Drive Gearbox for J34 Engines, Report A-613, Acceptance of. **B257V3RG72.3.2NACP**
[20] BuAer Memo, December 28, 1948, Contract Noa(s) 9051, Item 1, Task II, Sub-Task B – Design Study of J34 Compressor Development, Approval of Report on. **B257V5RG72.3.2NACP**
[21] WAGT Report, A-852, Final Report on Compressor Development. Item 2, Task II Sub-Task B of Contract Noa(s) 9051. **B257V6RG72.3.2NACP**
[22] WAGT Report A-855, June 17, 1949, Interim Technical Report of J-34 Compressor Development Item 2, Task II, Sub-Task B of Contract NOa(s) 9051. **B257V6RG72.3.2NACP**
[23] NACA Report, RM E52116, May 18, 1953, Investigation of Blade Failures in a J34, Eleven-Stage Axial-Flow Compressor, Howard F. Calvert, André J. Meyer, Jr. and C. Robert Morse, **NACA digital reports database**.
[24] BuAer Memo, July 6, 1949, Contract NOa(s) 9051, Item 2, Task II, Sub-task C – Variable Area Exhaust Nozzle Development – Reproducible Drawings – Forwarding of. **B258V7RG72.3.2NACP**
[25] BuAer Memo, November 28, 1947, NOa(s)-9051, J34-WE Engine, Request for Internal Performance on. **B256V2RG72.3.2NACP**
[26] WAGT Report A-727, July 23, 1948, Final Report on Development of an Exhaust-Gas Temperature Measuring Device for the J-34 Turbo-Jet Engine, Contract NOa(s) 9051, Task II, Sub Task D. **B257V5RG72.3.2NACP**
[27] BuAer Memo, November 19, 1948, Contract NOa(s) 9051 – Item 2, Task II, Item G, Air Inlet Screen Design Study, Approval of Report on. **B257V5RG72.3.2NACP**
[28] BuAer Memo, September 24, 1948, Item 2, Task II, Sub-Task G, Air Inlet Screen Strength Test, 18 August 1948. **B257V5RG72.3.2NACP**
[29] WAGT Report, A-832, May 18, 1949, Additional Anti-Icing Tests of a J-34 Turbojet engine on Mt. Washington, Contract NOa(s) 9051, Item 2, Task II, Sub-Task G., Proj 644. **B257V6RG72.3.2NACP**
[30] BuAer Memo, October 27, 1947, Contract NOa(s)-9051 Item 3, Task III – Sub-Head B – Development of Thrust Measuring Kit for Flight Test Work. **B256V2RG72.3.2NACP**
[31] BuAer Memo, December 1, 1947, Contract NOa(s)-9051 – Task III, Sub-task "B" – Thrust Measuring Kit – Approval of Report on. **B256V2RG72.3.2NACP**
[32] BuAer Memo, September 27, 1948, No Subject, Request for thrust kit proposal. **B257V5RG72.3.2NACP**
[33] WAGT Memo, August 2, 1948, Contract NOa(s) 9051, Item 3, Task III, Sub-Task B. Reproducible Drawings, Forwarding of. **B257V5RG72.3.2NACP**
[34] WAGT Report A-587, October 22, 1947, Thrust Measuring Kit for Flight Test Work Item 3, Task III – Subhead B, Contract NOa(s) 9051. **B256V2RG72.3.2NACP**
[35] BuAer Memo, December 1, 1948, Contract NOa(s)-9051 – Task III, Sub-Task "B" – thrust Measurement Kit – Approval of Report on. **B256V2RG72.3.2NACP**
[36] WAGT Report A-726, July 6, 1948. Flight Test Measuring Kit. Curiously, the issue date on the report cover is "6-28-30". (A penciled note gives the date as "7/6/1948") **B260VENCLRG72.3.2NACP**

[37] WAGT Memo, February 10, 1949, Thrust Measuring Kit, Our Reference Neg. 92886. **B257V6RG72.3.2NACP**
[38] BuAer Memo to WAGT, September 27, 1948, No Subject. Memo covers request for prices of three thrust measuring kits similar to the ones produced under contract NOa(s) 9051. **B257V5RG72.3.2NACP**
[39] WAGT Memo, February 10, 1949, Subject: Thrust Measuring Kit, Our Reference Neg. 92886. **B257V6RG72.3.2NACP**
[40] BuAer Memo, January 28, 1948, Contract NOa(s)-9051 Thrust Augmentation Tests on the J34 Turbo-Jet Engine. **B257V3RG72.3.2NACP**
[41] BuAer Memo, June 22, 1948, Contract NOa(s) 9051 –Item 4, Task IV, - Water Injection Tests on J34 Turbo-Jet Engine. **B257V4RG72.3.2NACP**
[42] WAGT Report A-848, August 31, 1949, Final Report of Water Injection Tests on the J34 Turbojet Engine, using the Chance Vought Injection Ring. Contract NOa(s) 9051 – Item 4, Task IV. **B257V6RG72.3.2NACP**
[43] BuAer Memo, March 30, 1950, Contract NOa(s) 9051 – Task IV, Item 1 – Continuation of Investigations of Water Injection Methods in J34 Turbo-jet Engines. **B258V8RG72.3.2NACP**
[44] BuAer Memo, May 16, 1950, Contract NOa(s) 9051, Task IV, Item 1: continuation of investigations of water injection methods in J34 engines. **B258V8RG72.3.2NACP**
[45] AEL Report 1274, November 19, 1953, Report on Evaluation of Electronic Control for J34 Engine with Variable Area Exhaust Nozzle. **B77V98RG402NACP**

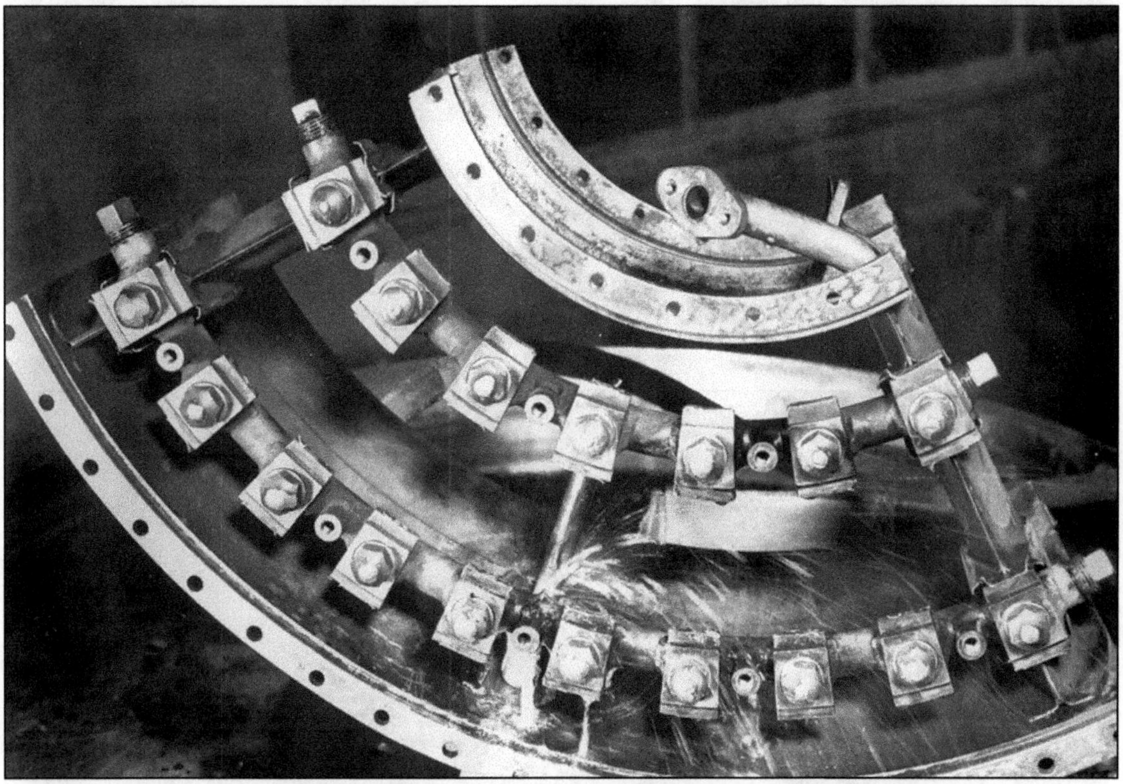

Figure 4: Quarter section J34 spray nozzle assembly being developed for the J34-WE-38. The section fit into the quarter section combustor test rig. WAGT P46126, 1/28/1952. *Courtesy Hagley Museum and Library*

Figure 5: J34-WE-38 compressor blade tip clearance measurement tools in place on test bed. WAGT P48105, 11/17/1952. ***Courtesy Hagley Museum and Library***

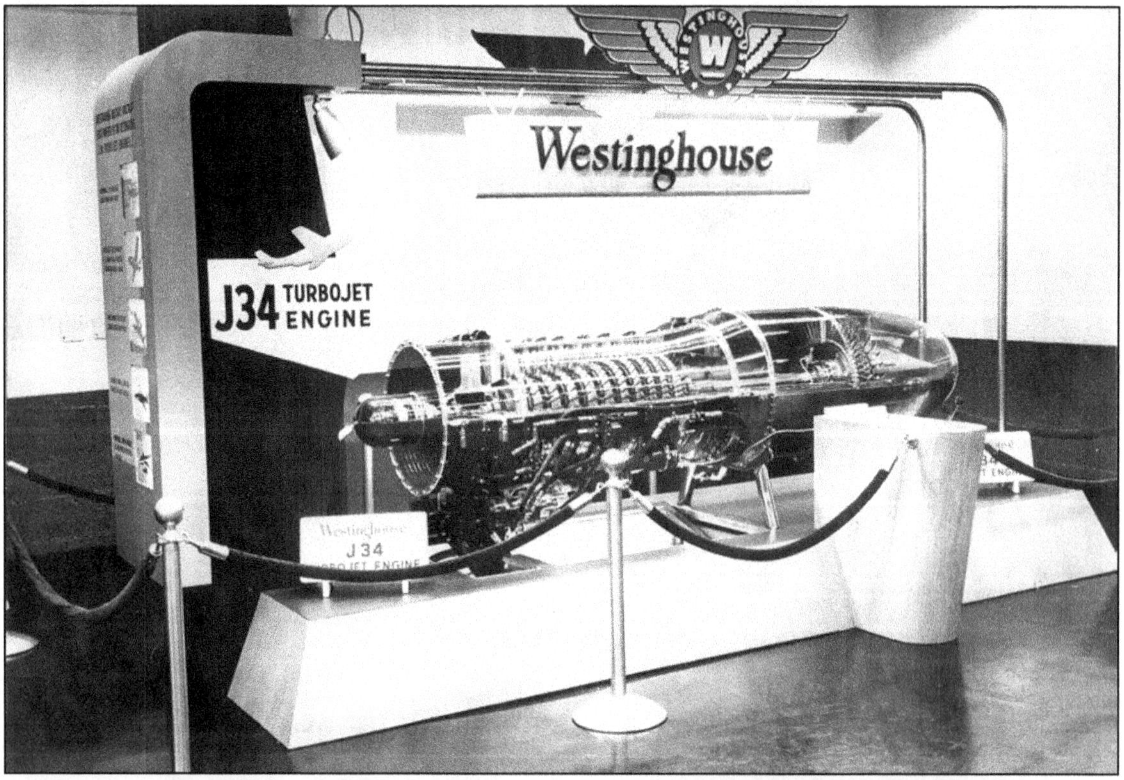

Figure 6: Westinghouse J34 display. Engine appears to be a J34-WE-34 or -36 cutaway. Photo card not captioned. WAGT P49911, 8/10/1953. ***Courtesy Hagley Museum and Library***

Chapter 2
XJ46-WE-2 and J34-WE-32/-38 Procurement & Initial Development

BuAer began to recognize that the J34 was not going to give any more additional thrust than the 3,000-3,200 pounds being delivered by the J34-WE-22, J34-WE-30 and J34-WE-34, at least in the short term. More thrust was needed to allow for more capable airframes than the early Navy fighters. All of the research into improving the J34 was coming together and it seemed reasonable to ask that a slightly more powerful version of the J34 be planned with an afterburner (A/B) that could still fit within the airframes already using or planning to use the earlier J34's. If the significantly more powerful engine could stay close to the diameter of the J34, a pair of them at about 3,500-4,000 pounds military thrust, and also having an afterburner, would allow development of a new interceptor with significantly improved overall performance. WAGT had informal meetings with BuAer representatives and acknowledged BuAer's interest in a memo of October 31 of that year informing them that a proposal for the new engine was being prepared and it would be sent to the Navy by the end of November.[1]

Procurement

The initial proposal was not actually sent to BuAer until December 30 and was then subsequently modified on January 27, 1948 upon further discussion with the Navy. The work was to be in two phases, the first including initial development of the new model through flight substantiation and a Phase II taking it through qualification for service use. The reason for the split was not mentioned, but likely it was for budgetary considerations.[2]

BuAer's stated intent for the new engine was to include features desired by BuAer and to incorporate the basic engine design improvements developed by Westinghouse as a result of all of the research development efforts to date.

The main focus was to be on the enlarged J34, which was given the WAGT internal model designation of X24C10. Such a designation was a break from prior internal model designations at WAGT and may have been done to hide the fact this was in reality a largely new engine. The maximum size of the new engine was based on the limits of airframes under consideration for the new engine, all of which at that time were being designed to the earlier J34 installation dimension envelope. This limited the combustion area of the engine to a 1.1 inch maximum diameter growth. Airflow was to be increased 20 percent over the earlier J34's and the pressure ratio increased to assist both in lowering fuel consumption and improving high altitude combustion. The weight including the A/B was estimated at 1,701.6 pounds.

The compressor would be a 12 stage unit using curvic clutch construction, three main bearings, a single annulus combustion chamber, separately removable fuel nozzles, a two stage turbine, a moveable exhaust nozzle, an afterburner, a revised accessory gear box (with a remote drive for airplane accessories) and integrated controls for both the basic engine and afterburner. The control system was to incorporate a fully automatic emergency fuel system, speed control using direct speed sensing and automatic controls similar to those being developed for the J34-WE-30 (24C4C) under contract NOa(s) 9051. Given all the major differences from the earlier engines, it seems a new major WAGT model number would have been logical, but the engine remained throughout its life in the earlier WAGT 24C model group. BuAer recognized the scope of the changes would result in a new engine by assigning it the J46 designation instead of a sub-model in the J34 series.

The basic engine at 10,100 rpm was to produce 4,200 pounds of sea level thrust at Take-off/Military rating, and 6,100 pounds with the afterburner operating. Specific fuel consumption ratings were to be .98/2.5 respectively using AN-F-48 grade (100/130) aviation fuel with a guaranteed minimum SFC of .82 in cruise. The engine would operate reliably to 60,000 feet and restart up to 45,000 feet.

The specification (WAGT-X24C10-2) had to be submitted and approved within 9 months, with Flight Substantiation Test reports within 24 months, the Qualification Test conducted within 28 months and the test reports within 28.5 months, all from contract start.

Phase I cost was to be $1,869,196 and Phase II $1,302,197.

With a two year delay in getting a fully qualified engine into production in the offing, WAGT had also proposed to develop a J34 engine model incorporating many of the development improvements, also having an afterburner and obtaining thrust ratings of 3,500/4,900 pounds. This engine was the WAGT model XC24C8. Aerodynamically it would have been identical to the J34-WE-30 under development and retain the main structural design of the earlier

J34 models, but being modified to integrate an afterburner with an adjustable nozzle. The new control was to allow rapid and positive changes in engine speed, afterburner and exhaust nozzle. BuAer designated it the XJ34-WE-32. This engine's schedule would be 7 months to specification delivery (Appendix A to WAGT-X24C10-2), 13.5 months to Flight Substantiation Test Reports and 16 months to Qualification Test, with the test reports to follow within 16.5 months, all from contract start.

The main airframe targets for the J34-WE-32 were listed as the McDonnell F2H-1 and F2H-3 and for the J46-WE-2 , the Douglas XF3D-1, D-558-II (Skyrocket), XS-3 and D-571 (later styled XF4D-1), Vought XF7U-1, McDonnell XP-85, XP-88, Lockheed XF-90 and Curtiss XP-87.[3,4]

The contract was immediately amended two days later to authorize the second phase of development and added one engine of each type as a deliverable for the Qualification Test and later delivery to BuAer.[5,6]

Development Begins

Development began immediately on both engines. Many configuration and performance problems encountered during development in large degree affected both engines, so the discussion here will follow both as they moved through development. Items related specifically to the J34-WE-32 (and variants) will be limited here to only those items affecting both engines, as space in this volume will not allow the full J34-WE-32 story to be explored.

With the program started, BuAer immediately raised a new requirement, asking that WAGT investigate the use of a solid propellant type starter. BuAer stated they understood that the use of such a starter would require significant development work, but it also appeared the present location of the electric starter on the rear of the accessory gearbox of the J34-WE-22 would not allow enough room for the larger size of a solid propellant starter. WAGT was asked to investigate moving the starter to the front of the gearbox, moving it to the rear engine exhaust nozzle cone or other variations of the current accessory locations. Further, BuAer noted that WAGT had indicated they were going to relocate the J46 electric starter to the No. 1 bearing support and that starters other than electrical would probably be used on that engine. Comments were requested.[7]

WAGT had earlier sent comments indicating that any redesign for a solid propellant starter required a way to duct hot gases to and away from the starter turbine. The new front bearing support would require a complete redesign. They anticipated a successful redesign of the struts of the front bearing supports for these gases would be possible, with the exhaust gases being passed through the engine air stream. A shut-off valve on the engine bleed would be necessary to prevent contamination of the bleed air until startup was fully accomplished and a certain engine RPM reached.[8]

BuAer's request for further comments resulted in WAGT submitting report A-724 on June 21. This report is an exhaustive review of the changes and challenges needed to modify both the engine and gearbox to accommodate a higher output type XII F generator drive, a higher output type XII J hydraulic pump, and a solid propellant starter. Analysis showed:

1. The existing XJ34-WE-32 and J34-WE-34 designs had moved the starter from the gearbox to the center of the engine air inlet duct.
2. It was possible to re-locate the starter pad and place it back on the accessory gearbox, although it was not practical to do so because of the resulting increases in engine envelope size, engine weight and development time (10 additional months).
3. The higher torque capacity type XII F generator pad and type XII J hydraulic pump could be accommodated but were not practical for the same reasons stated for item 2 above.
4. The present planned XJ34-WE-32 accessory gearbox design was the most practical in light of development time, engine envelope and engine weight.
5. The uprated accessory gearbox requirements needed to be incorporated in the XJ46-WE-2 gearbox design.
6. A solid propellant starter could probably be accommodated on the nose position if the entire nose intake and strut supports were redesigned. Without detailed information on such a starter, the layout effort, likely schedule delay and any weight implications could not be determined.

The conclusion that the requested changes should be considered for the J46 and not the J34-WE-32 was supported by many pages of detailed analysis for each change. Photos of an inverted wooden mockup of the possible arrangement for the -32 were included, but details are unlabeled.[9]

Figures 1 and 2: Wooden mockup of proposed J34-WE-32 accessories. WAGT Report A-724 (undated), P39257 and P39259.[9]

BuAer primary comments on the submitted draft specification were to change the fuel to AN-F-58 (JP-3) and to remind WAGT that the starter for the J46 would certainly be something other than electrical. For the J34-WE-32, the starter position would be discussed separately.[10]

The first mention of a possible non-afterburning version of the J34-WE-32 came in late June in a report covering evaluation of installation options for the J34-WE-38 (X24C7) in the F3D-1. The XF3D-1 J34-WE-22 engines had their accessory gearboxes turned 40 degrees off the normal center line position to accommodate the airframe escape tunnel. The engines were "handed" in that the gearboxes were assembled to the engine in off-longitudinal positions in opposite directions on this aircraft, resulting in the J34-WE-24 and -26 sub-models of the J34-WE-22. Minor accessory relocations were then made, such as moving the oil pump on the left-hand engine to the low side of the gearbox to ensure satisfactory scavenging operation. Some special fittings for piping were also supplied. Performance-wise the two modified engines were otherwise identical to the J34-WE-22.

On the new J34-WE-38, redesigning the basic engine to allow it to be installed with the engine rotated about its longitudinal axis in either direction or the use of accessory field conversion kits were looked into; the results creating subtypes of the planned J34-WE-38. The spare parts organization of BuAer had objected strongly in the past to having sub-types of engines, the practice causing them significantly increased tracking and stocking efforts to ensure the proper parts were maintained and shipped for each engine.

The current design of the J34-WE-38 had provisions for only a standard configuration with the accessory gearbox on the bottom centerline. Use of this engine in the F3D-1 would force the redesign of the fuselage with the engines mounted lower and slightly outboard, thus increasing the fuselage frontal area. The landing gear would not allow enough room for engine removal or installation in the new position. A crash skid would be needed to protect the engine fuel and oil connections in a crash. The lower engine accessory gear box would be the first rigid structure to contact the lowest cable on a carrier landing barrier crash. It was concluded that the type of modifications made to the J34-WE-22 for the XF3D-1 were a more desirable approach.

Ironically, the development changes included in the XJ34-WE-38 (mainly more thrust) that made it attractive for use in the F3D-1 made it difficult to incorporate alternate accessory configurations. WAGT acknowledged that retaining the current design was best in reducing the servicing load in general.

Two main avenues were then investigated: redesign the engine to allow rotation of the entire engine through the various angles on each side; or use field conversion kits to allow the front portion of the engine to be converted to the 40 degree of center configuration (similar to the -24, -26 but done in the field instead of on the production line).

The analysis of the first possible approach covered the problems of oil scavenging, fuel dumping, oil tank location, front and rear engine mounts location and strength, all of which presented major challenges to resolve without adding weight to the engine.

The second approach, the use of field modification kits, brought up other issues. The rotation of the entire front housing through 40 degrees would position the grouped accessories as a bulge to one side of the engine and require new bolt-hole patterns. Splitting the accessories in some way to eliminate the bulge would increase weight. Moving the bolt holes would preclude using the engine on experimental aircraft that planned to use an air-driven fuel pump and experimental afterburner or, preclude upgrading prior J34 engine models to use the new front section of the J34-WE-38. The front engine mount design would be significantly different from the current one and replacing the front bearing support in the field was not practical.

Oil scavenging would be a challenge, also making the installation of the parts needed to modify the engine beyond field modification installation capabilities. Left and right engines presented different problems and needs in this area.

Finally, the report summarized the two approaches and impacts as:

Proposal One – Various Longitudinal Rotations:

A. 10 Degree rotation. No delay to the engine program if a Letter of Intent was received by August 1, 1948.
B. 20 Degree rotation would delay the engine program as much as 24 months depending on a decision on the 20 degree roll requirement on the oil tank. The oil tank decision could handicap the engine for use in planned normal installations.
C. 30 Degree rotation. Extensive design changes would delay the engine program at least 18-24 months. The

impact on airframe designs currently proposing the use of the engine would be drastic.

Proposal Two – Use Field Conversion Kits:

A. Field conversion of standard engines would delay the standard engine program 2-3 months with subsequent delay to at least one production program being designed to use the standard engine. There would be a definite impact of the changes to standard normal installations *(not explained)*.
B. The 24 months required for design and production of the modification kits would probably delay the F3D-1 airplane production schedule.

In summary, WAGT noted that the 11 degree bent tail pipe design on the F3D-1 presented a problem for the J34-WE-38, which had a variable area exhaust nozzle, the actuating system for which was not designed for use with bent pipes. WAGT noted that redesign of the F3D-1 to use the standard J34-WE-38 might allow it to accommodate the standard variable exhaust nozzle. WAGT recommended that the F3D-1 be redesigned to use the standard J34-WE-38 engine as currently designed.[11]

Looking ahead, the J46 was slated to be used in the Douglas XF3D-3 variant and the accessories mounting issues for that aircraft were to be just as relevant to the planning of the J46 layout.

It should be apparent that BuAer's proposed desire to use a non-electrical (solid propellant, pneumatic, etc.) starter in addition to the accessory arrangement redesign considerations of the coming engines must have caused many long and difficult discussions not only with WAGT but between BuAer and the airframe companies during this period. The challenges of designing the accessories package for any given engine installation are exposed in these reports and memos.

The Mock-Up Board (MUB) for the XJ34-WE-32 was held near the end of October and made the following comments regarding the accessories. (No mock-up of the XJ34-WE-38 was at the meeting, the Board's assumption was that only minor changes to that engine would be made relative to removal of the afterburner and various A/B related elements.) Paragraph 4 in the introduction is of interest, and shows that the debates over engine accessory layout were still ongoing while not providing any sign of final decisions.

"The major points of discussion by the mock-up board concerned the general arrangement of accessories on the engine with regard to servicing, removal and replacement. Since the engine is designed specifically for installation in already existing aircraft, major changes affecting the engine envelope were not possible. Engine accessories such as the three booster coils and the power regulator which are necessary for engine operation and which should be mounted on the engine could not be so mounted due to their size and obvious interference in most aircraft installations. The generator installation clearances will not permit removal or installation of a generator with cooling air inlet in other than the downward position. Since generator clearance involves the design of several features of the engine in the accessory section, the mock-up board agreed to require no change pending receipt of information from the contractor on what generators can be satisfactorily installed in the present space. In general, it is desired to point out, that the policy of tailoring the XJ34-WE-32 engine to fit in certain aircraft already designed, has seriously compromised the maintenance serviceability and accessibility of the engine and accessories. Many undesirable features of this engine, although recognized by the mock-up board, were accepted since satisfactory correction would have necessitated interference in certain installations."[12]

The balance of the mock-up board's comments related primarily to adding quick disconnects to the accessories, the study of which was tied to a similar requirement for the J40 engines and the reply to the board was linked to the J40 study completion. On balance, the engine's layout and controls were deemed acceptable with minor changes.

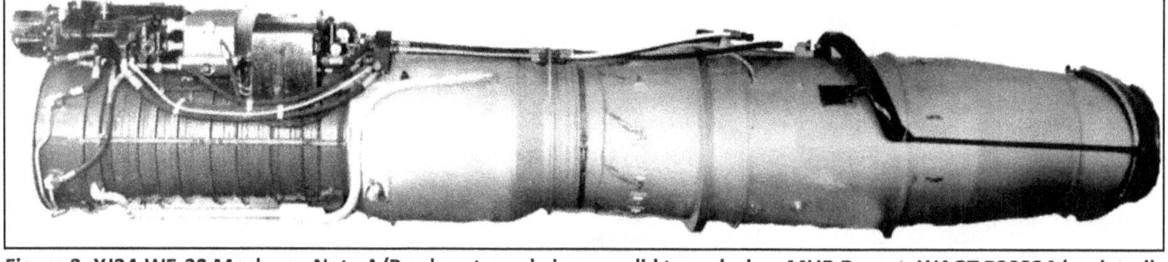

Figure 3: XJ34-WE-32 Mock up. Note A/B exhaust nozzle is an eyelid type design. MUB Report, WAGT P39824 (undated).

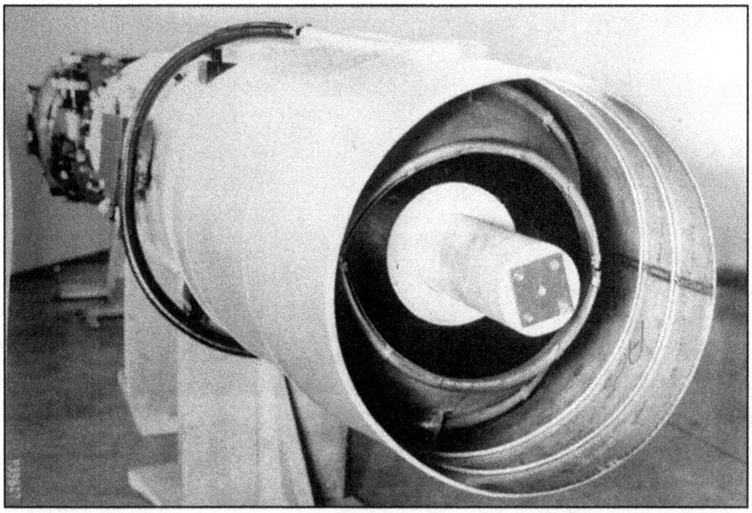

Figure 4: XJ34-WE-32 mockup of exhaust nozzle with nominal shroud in place. (White dowel in center not part of engine, used for lifting and rotating mockup.) MUB Report, WAGT P39827 (undated).

Backing up a bit, in July, BuAer had looked ahead and tried to advance the anti-icing testing ("Project Summit" on Mt. Washington) for both the XJ34-WE-32 and XJ46-WE-2 engines. Arrangements were made to move up the availability of a 40 day test period on Test Stand No. 1. Subsequently, five additional 20 day test periods would be available to both Westinghouse and General Electric which they could schedule as appropriate. BuAer also suggested WAGT consider other anti-icing designs including the reduction of inlet guide vanes. An experimental unit with reduced inlet guide vanes should be installable on a J34-WE-22 for testing.[13]

WAGT responded in August that the disassembly effort required to replace a complete assembly on the front of a J34-WE-22 was impractical in the weather conditions at the top of Mt. Washington. It might be possible to fabricate a test section of alternate anti-icing units that could be tested on an adapter on the front of the engine. They suggested Test Stand No. 2 would be more suitable for such a testing approach. They thanked BuAer for their efforts to arrange opportunities for anti-icing testing.[14]

A verbal request from BuAer to investigate speeding up delivery of the qualification test of the XJ34-WE-32 and a subsequent improvement in production initiation was made. WAGT responded with a memo stating that a review of the work to be accomplished showed that the original schedule in the proposal and subsequent contract was realistic. The detailed initial production would be:

J34-WE-32 Initial Production Program	
Month	**No. of Engines**
*November 1949	2
*December 1949	4
*January 1950	8
*February 1950	16
*March 1950	20
April 1950	20
May 1950	30
June 1950	40

*Fifty hour flight substantiation specification engines

The schedule and development assumed the accessories section would remain on the centerline bottom of the engine. The elements of the engine requiring extensive development prior to production were:

1. Variable area exhaust nozzle controls (fuel, hydraulic, and electrical)
2. Lubrication system
3. Compressor
4. Turbine
5. Anti-icing system
6. Afterburner

The memo went on to detail the work needed in each of those areas as a means of supporting WAGT's claim of their inability to speed up production availability.[15]

Within the same time frame, WAGT forwarded their report A-741 to BuAer covering a design study of the accessory arrangement for the XJ46-WE-2. The accessories to be accommodated were:

- Generator
- Main Fuel Pump
- Hydraulic Pump
- Tachometer
- Governor
- Booster Fuel Pump
- Afterburner Fuel Pump
- Oil Pump
- Starter

It was noted that the pads for the booster fuel pump, afterburner fuel pump and oil pump had special requirements and would not meet the requirements of the standard AN-E-30 specification.

The requirement for magneto and vacuum pump drives was deviated from since such accessories were not used on the engine and including them in the comparative study would have resulted in an enlarged gear box and added unnecessary gearing. This deviation had been approved earlier for 24C and the X40E type engines and it was assumed it would be approved for the XJ46.

Many arrangements were studied, most being subvariants of two main ones: starter external on gearbox and starter internal in the inlet duct. The weight from the inlet flange of the compressor to the forward-most part of the engine and the oil reservoir weights had to be analyzed due to their impacts on the design of the front bearing support structure which had to accommodate the various stresses.

It was assumed all the accessories themselves weighed the same in each arrangement. Drawings of both arrangements in forward and side views were included in the report. The analysis identified the following findings for each layout:

Starter Location Analysis	
External	**Internal**
Pros: - No accessory in the air inlet duct - All accessories easily accessible for servicing - Engine oil reservoir in the gearbox	Pros: - Engine oil reservoir in gearbox - Similar layout to 24C engines - Weight was 146 pounds - Smaller gearbox - More adaptable to airframe manufacture
Cons: - Weight was 174 pounds - Gearbox was considerably larger (3 inches deeper and 2 inches wider)	Cons: - Starter not as accessible and had a cowling to be removed - Solid propellant starter installation complex

WAGT recommended the internal design be adopted, with the gearbox and intake casing being one casting.[16]

The next memo from BuAer on the starter location subject is an example of how the Navy's desire to incorporate new starter technology was pressed during development of the J34-WE-32/-38 and the J46. In a long discussion, the following points were:

1. Comments regarding accessory temperatures were not valid, as whether installed in the intake or on the gearbox, the atmospheric temperatures would be the same.
2. The discussion of solid propellant starter exhaust entering the bleed air was not correct, since the starting process would be over before the bleed air would enter the aircraft.
3. In view of the long development delay that would be imposed if the starter was relocated on the J34-WE-32 gearbox, the recommendation to retain the starter on the intake was approved.
4. BuAer was moving forward to identify two sources of starters independent of the engine manufacturer to insure an uninterrupted supply in war time emergencies. Westinghouse, as the engine supplier, was not considered a logical source of supply for the starter.
5. BuAer had under development a 40HP solid propellant type starter with the Eclipse-Pioneer Division of Bendix Aviation Corporation. *(The technical details of which were provided further along in the memo – Auth.)*
6. A 35HP pneumatic type starter had been developed by the AiResearch Manufacturing Company and small scale production was underway. A smaller 20HP version was under development. An auxiliary power unit designed to provide a self-contained air source for airplanes using a pneumatic starter had been developed by AiResearch, having a capacity to drive the 35HP starter. *(Technical details of both the 20HP starter and the auxiliary power unit were listed.)*
7. In view of the large scheduled procurement of J34 engines, it was imperative that provisions had to be made to accommodate both types of starters on those engines.
8. It was recognized that it was impractical to incorporate a starter of any type on the gearbox on the J34-WE-32/-38. BuAer understood that the No. 1 bearing support on those engines was being changed to a six strut arrangement to accommodate anti-icing provisions. It was requested that provision be made to handle both the solid propel-

lant and pneumatic type starters on the No.1 bearing support location.

9. It was BuAer's intent to use either a solid propellant or pneumatic type starter on the J46 but it was not clear what provisions in the accessory design on that engine had been made to accommodate such starter types. They presumed the starter solution arrived at for the J34-WE-32/-38 would be reflected in the design of the XJ46 starting system.
10. On a matter unrelated to the starter situation, they recognized that the maximum ram air temperature of the J46 would be 215ºF, higher than electrical accessories at that time could operate reliably, and that it was more practical to develop accessories that could tolerate these higher temperatures than to provide air-conditioning cooling for the accessories. They felt moving to a turbine type starter would help relieve the accessory overheating problem.[17]

Behind the scenes, a new development schedule was seen at WAGT during meetings and BuAer asked the Bureau Area Representative (BAR) to have a copy of the chart sent to HQ.[18] Almost the same day, the WAGT signed copy of Amendment 1 was received back at BuAer. The schedules within the amendment were unchanged.

The monthly progress report arrived on September 27 and the requested schedule was attached. The report showed the mock-up of the XJ46-WE-2 would be available by January 1, 1949.

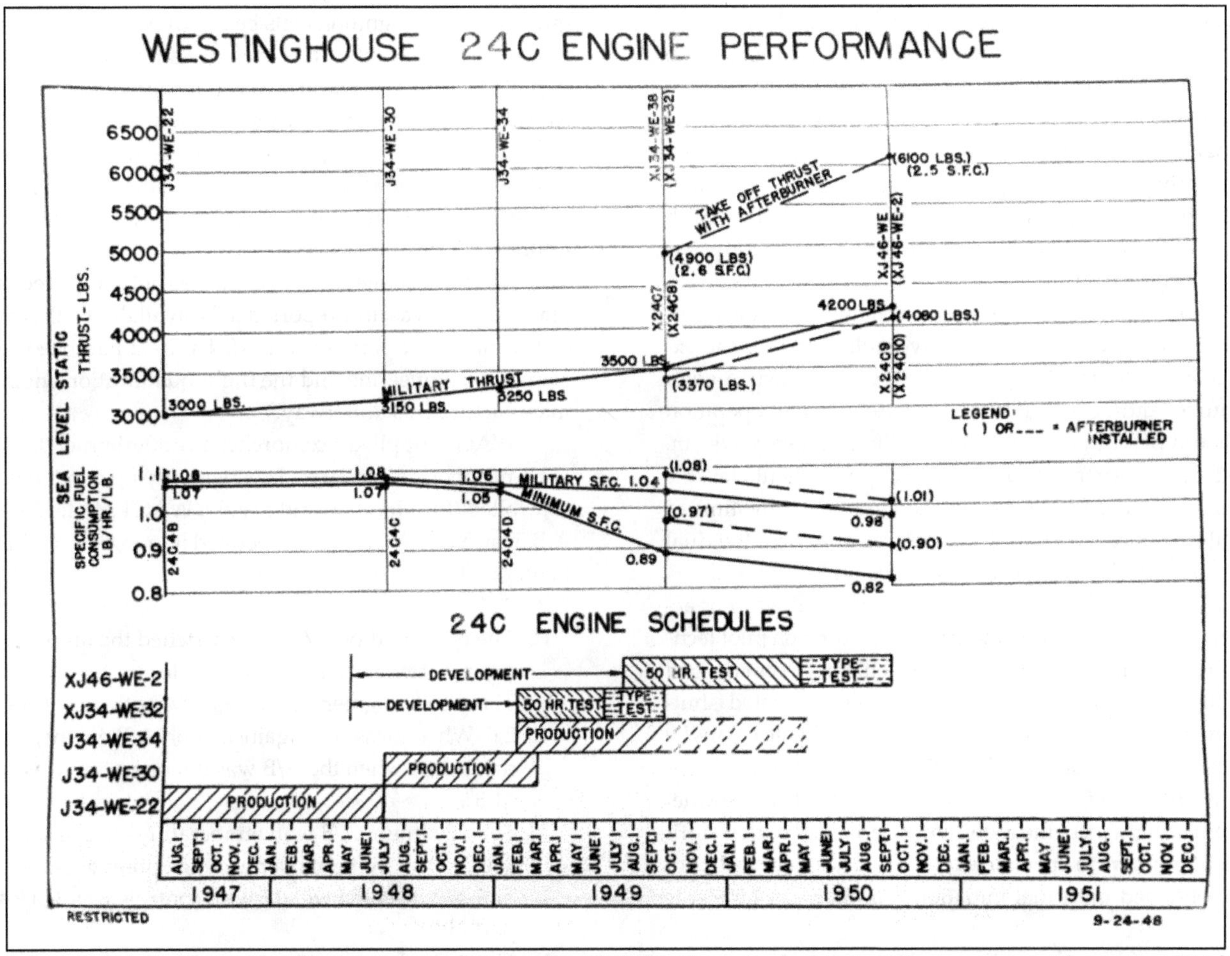

Engine schedule for the 24C family of engines as of September 24, 1948. Chart attached to August progress report for contract NOa(s) 9670. Note that performance part of chart does not relate to schedule portion below. For WAGT, this is a detail schedule.

A WAGT memo in early October stated that while the goal was to ultimately mount all the electrical devices on the engine, the Power Regulator was a problem due to its dimensions exceeding the planned installation envelope of the engine. It would fit on the XJ40 series of engines, but it could not be scaled for the smaller engines and would have to be mounted on the airframe (the cool cockpit was suggested).[19] A complete, annotated copy of this memo was located in the NOa(s) 9212 contract file (XJ40 development) and a written note on it says that

WAGT was informed that it was the policy of BuAer that all electrical components necessary to operate the engine had to be mounted on the engine. WAGT had noted in their memo that currently the electrical accessories could only operate reliably at temperatures of 165°F or less. BuAer told them that they had to run reliably at the 215°F ram temperatures guaranteed by the specification. BuAer's position on cooling vs. upgrading the accessories was noted above.

Another requirement surfaced regarding the proposed use of the J46 in the Douglas F3D airplane. Douglas wanted a 6.5 degree bend approximately 29 inches from the tail and the afterburner shroud. BuAer commented to WAGT that they wanted to know what would be involved in tests, costs and engine availability. They also stated the bend (axial deviation) should be where the extension pipe would join the afterburner and not in the A/B itself.[20]

BuAer review of the challenges of emergency control of a variable area exhaust nozzle indicated that it was impractical to provide a system to automatically move the nozzle to the military rated position in the event of both the primary and emergency controls failing. For non-afterburning engines they accepted WAGT's design which locked the nozzle in the position it was in when the hydraulic line failed.

For an afterburner equipped engine, they now wanted dual power supplies and hydraulic lines for the exhaust nozzle. In the event of failure of the primary power source and/or the hydraulic lines, normal operation should proceed automatically on the emergency system. If both the primary and emergency system should fail, the nozzle should lock in the position existing at the time of failure. Once the control system was qualified, the dual power supply and hydraulic lines could be removed.

Automatic starting as WAGT planned was not acceptable as it was dependent to some extent on pilot technique and in combat situations the pilot might be engaged with other matters. The mechanically activated shutdown valve should be retained and an alternate means be provided to actuate it other than moving the throttle below idle cut-off. They suggested a variable discharge fuel pump might meet the requirement, adjusted so that sufficient pressure to open the variable fuel nozzles would not be obtained until fuel initiation rpm on the engine was reached.[21]

In mid-December BuAer clarified the Douglas special requirements for the F3D, stating that the airplane would not use afterburning, but the extension pipe bend would be required for both the J34-WE-38 and the non-afterburner version of the J46-WE-2. They asked that WAGT advise as soon as practical whether the Douglas requirements could be met. This is the first time that a non-afterburning version of the J46 is mentioned. This engine would be designated the J46-WE-4 as development moved forward.[22]

The monthly activity report BAR summary covering March concerned itself with the status of the MUB action items for the J34-WE-32 and it stated the XJ46 mock-up would not be available on the contractual required date of January 5 but would slip to late February or early March.[23]

Contract Amendment 2 was created December 29 to add a requirement for piping, outline and installation drawings of the WAGT model 24C7, designated the J34-WE-38, a non-afterburning version of the J34-WE-32, to be used in the F3D aircraft to supplant the J34-WE-36 then planned to be used in that aircraft. The specification for the new model would be included as an appendix to the XJ34-WE-32 specification. Cost was $148,263.00.[24]

In response to the J34-WE-32 MUB Item 5 which requested that the ignition coils be mounted on the engine in a location acceptable to at least one airframe and more if possible, WAGT found a location acceptable to the XF-88 and BuAer accepted the location and closed the MUB item. Interestingly, the XF-88 is also written as "XF88" within the same memo, showing the shift from the old "P" nomenclature to the new "F-" system was not uniform everywhere as yet.[25,26]

As expected the XJ46 mockup was not completed by January 5 but was now reported to be available by the end of February. All parts for the XJ34-WE-32 had been released by engineering and the flight qualification engine manufacture was underway.[27]

WAGT supplied a comprehensive afterburner development report in January. They were flying a manually controlled (fuel, ignition and nozzle) WAGT Model No. 3 A/B in an XF-80A-2 and had operated it to 35,000 feet. The main points in the report were:

1. The measured performance matched the test stand figures after accounting for duct losses.
2. The thrust augmentation was 30% with an SFC of 2.6. When measured against a non-A/B engine, the thrust loss when the A/B was not operating was 11.5%.
3. The maximum A/B wall temperatures decreased 300°F when in flight over ground running temps.
4. The A/B installed weight with controls was 340.9 pounds.
5. The A/B relit up to 20 thousand feet with no special pilot actions. Juggling fuel, engine rpm and eyelid nozzle operation obtained relights up to 30 thousand feet.
6. The spark plug was not needed at higher altitudes, A/B ignition being accomplished through accelerating the engine.

7. Full power A/B runs up to 35 thousand feet were made with no blowout tendency.
8. On the last flight, the eyelids were accidentally left closed and some damage to the A/B resulted when the turbine blades burned out. The pilot was not injured and the aircraft was undamaged on landing.
9. The location of the hot streak ignition injector was found to be incorrect, with no flame being observed making it through the turbine. Tests showed moving the injector to just past the compressor outlet produced flames five feet long out of the turbine.
10. The candle type spark plug worked satisfactorily.
11. No progress had been made in reducing basic thrust loss in non-A/B operation.
12. Testing of the new AN-F-58 (JP-3) fuel showed that it reached its maximum burning efficiency in a lean air-fuel ratio that precluded the high burning temperatures required for high A/B thrust augmentation.
13. Eight Model No. 3 A/B's were under construction with two scheduled for completion by February 4. The other six units would go to the Air Force MC812 program (XF-88).[28]

Proposal vs. Model Specification WAGT-24C8-2A Weight		
Change Area	**J34-WE-32**	**J34-WE-38**
Proposal Total (lb)	**1,539**	**1,376**
Weight changes		
Anti-icing provisions	43	43
Addition of electrical junction box and cable	12	12
Improved exhaust nozzle control	20	20
Dual fuel pumps	14	14
Speed control gear box redesigned	3	3
Deletion of starting components and addition of electrical leads	0	0
A/B ignition system (missed on bid)	9	0
Improved A/B regulator	9	0
A/B strengthening	62	0
Fuel filter housing	0	3
Total Approved (to date)	**172**	**95**
Not yet approved	25	25
New Proj. Adl. Total	**197**	**120**
New Engine Wt.	**1,736**	**1,496**

It must have raised concern at BuAer and the Power Plant Division to see that the tested A/B design was producing a thrust increase far below the relative percentage rate guaranteed for the J34-WE-32 (40%) and the J46-WE-2 (45%) and that the thrust loss of the base engine when operating without the A/B operating was so large. Shortfalls in thrust would plague both engines in the months and years ahead.

The battle over weight increases brought a detailed explanation of the causes of the weight increases between the proposals and the current model specifications for the J34-WE-32 and J34-WE-38.

The memo ends with a reminder to BuAer that at the time of the coordination of the Bid Specification, BuAer requested that WAGT remove as much margin as possible from the engine weight.[29]

Only two weeks later, BuAer approved some of the solutions to changes requested by the MUB for the J34-WE-32 and these added 12 pounds to the engine weight. It is not clear if these were in the 25 pounds of weight included in the "Not yet approved" category above or not.[30]

One of these changes was to incorporate a means of indexing the power control lever on the engine in order to rig it during installation of the engine. The proposed WAGT solution was to incorporate a pointer keyed to the power control lever shaft with scribe lines on the A/B fuel control body designating the static idle speed and maximum speed positions of the pointer. BuAer said the approach was approved as long as the indexing plate contained markings for ranges of travel for cut-off, idle, and military thrust and the point of normal thrust above the scribe lines. The scribe lines were to be marked in two degree increments.[31]

A verbal discussion started in January regarding the XJ46-WE-2 gearbox and the need to have it support two AND 20002 Type XII F generator pads. WAGT responded stating that the gearbox could be changed to accommodate this in two field interchangeable ways. The first, at no increase in charge or delivery delay, would increase the engine weight by 40 pounds, ~36 pounds for the extra AND 20002 pad and ~5 pounds for gears to change the pad speed from 7,000 to 6,000 rpm. The second approach would change the gearbox with one AND 20006 Type XVI pad replacing the two AND 20002 Type XII F generator pads. That approach would cost $28,105.00 which would include verification testing; such testing to be completed in six months after engine qualification but not to interfere with the qualification testing of the XJ46.

The two arrangements would be intra-convertible in the field by replacement of the gearbox cover and minor gear re-arrangement.[32]

Ultimately in May this resulted in Amendment 4 to the contract being issued, adding items 4(a), 8(a) and 9(a) for the development and testing of both gearbox arrangements and the supplying of a conversion kit for field conversion between arrangements. The cost had come down a bit to $27,092.00.[33]

(Note: Amendment 3 concerned a modification to the language in the contract defining which costs were to be allowable under the contract.)

The status report stated the assembly of the XJ34-WE-32 50-hour test engine would begin during April. The mock-up of the XJ46-WE-2 would be complete "on or about February 21, 1949". [34]

WAGT officially updated the specification for the J34-WE-32/-38, adding 12 pounds to both engines to reflect the mock-up changes that were approved.[35]

The requirement for a special tailpipe extension to fit the Douglas F3D-1 caused WAGT to meet with Douglas and review the F3D-1 detailed design. WAGT informed BuAer that their stress analysis confirmed it was possible to cantilever the special tail pipe on the J34-WE-38. It was believed this would be true of the XJ46 without an afterburner as well. The pipe would require a vertical tilt of the fixed portion of the exhaust nozzle of the J34-WE-38 of approximately three degrees and another bend of approximately eight degrees at a point twelve inches aft of the turbine housing rear flange. The nozzle would be about 35 inches long (six inches longer than the straight nozzle), the extra length being necessary for possible future use of a non-afterburning version of the XJ46-WE-2 engine.

The exhaust nozzle would incorporate the bend and the special nozzles furnished as part of the engine. The balance of the nozzle appeared to remain unchanged, but that needed to be verified in detail design. There would definitely be changes in the fixed portion of the nozzle, the two emergency control push rods, and a few hydraulic lines. There might be an increase in weight and reduction in performance of the currently specified J34-WE-38 engine. A handwritten note (not dated) on the back of the memo's circulation control sheet indicated that Douglas would use standard engines and supply the exhaust extension and longer nozzle actuating rods for the J34 engine. This seems to negate the analysis WAGT did and/or reflect a change on Douglas's part regarding the detail installation plan for the engines.[36]

The weight increases of the XJ46-WE-2 were subjected to the same analysis as the prior J34-WE-32 study.

Proposal vs. Current Specification Weight	
Change Area	**J46-WE-2**
Proposal total (lb)	**1,686**
Weight Changes	
Anti-icing provisions	40
Addition of electrical junction box and cable	12
Improved exhaust nozzle control	20
Dual fuel pumps	12
Speed control gear box redesigned	3
Deletion of starting components and addition of electrical leads	-4
A/B ignition system (missed on bid)	9
Improved A/B regulator	9
A/B strengthening	62
Addition of generator pad and pad speed change	40
Total Approved (to date)	**200**
Not yet approved – alternate gearbox generator layout	18
Not yet approved – J34-WE-32 MUB carryovers	25
New Proj. Addl. Total	**243**
New Engine Wt.	**1,929**[37]

The XJ46-WE-2 Mock-Up Board was scheduled for March 14-16, 1949 to be comprised of 14 BuAer and Power Plant Division members.[38]

A conference was held with WAGT during early February regarding the WAGT report A-800 "Design Study of the Independent Emergency Fuel System". It was discovered that the planned control's emergency fuel system did not supply certain components. (*Note: The planned control system was to be used on the J34-WE-30a, -34, -32; XJ46-WE-2; and XJ40-WE-6, -8, and -10 engines. This is a good example of how design issues could affect multiple engine models under contract. Full copies of some of the documentation on these types of issues were only retained in one relevant contract file, seemingly at random. Researching these topics was a challenge!*)

The missing or incomplete component or operational features were:

Preflight Check System - The components should operate on 17-30 volts with a three position switch offering three operating modes:

a. Automatic – The automatic emergency switch-over system is in effect with system running on the Primary system.
b. Check – The system switches to the emergency system and is locked there with an emergency indicating light illuminated.
c. Reset – The system can select the Primary system even if it has failed. Normally, selecting Reset would return the system from Emergency to Primary and put it back in Automatic operation.

It was noted that the emergency fuel system described in report A-600 had been approved as a result of work under contract NOa(s) 9051 and that deviations from the approved design in that report had to be specifically requested by the bureau.[39]

Anti-icing Provisions - WAGT noted that when design started, little information on the icing problem was to be had. Which parts to keep free of ice, the actual atmospheric icing conditions to be guarded against and the deterioration in engine performance during operation under icing conditions were unknown. Given the uncertainties, the XJ34-WE-32 mock-up included five hot air heaters in the struts of the inlet in addition to the heated inlet guide vanes.

Recent revisions to the AN-E-30 specification had included design conditions for anti-icing systems and these now included specifications for the allowable thrust loss when operating in icing conditions. Recent information based on measurements taken on a J34 engine on Mt. Washington was also now available.

The testing data indicated that the formation of ice on unheated guide vanes would practically close the inlet to the engine before appreciable ice built up on the struts or starter cowl on the engine inlet. It was believed ice in those locations would break off and pass through the engine without damage.

It now appeared that the complexity of the de-icing system could be reduced and a small weight saving realized by heating only the inlet guide vanes.[40]

J46-WE-2 Installation Features

WAGT released report A-813 "Engine Installation Features and Accessory Arrangement of the XJ46-WE-2 Engine" on March 8, 1949. This was the first general description of the engine since the submission of the original bid.

The summary of this report stated "The XJ46-WE-2 engine is designed so that (the) shape of its clearance envelope is the same as its predecessor the XJ34-WE-32. This engine embodies the following new features affecting installation which are not incorporated in the XJ34-WE-32 engine: (1) increased thrust, (2) reduced specific fuel rate, (3) increased thrust per unit frontal area, (4) two generator pads, (5) oil reservoir integral with gear box, (6) accessory drive gear box integral with front bearing support, (7) engine driven afterburner fuel pump, (8) externally removable fuel nozzles, (9) external fuel manifolds, (10) revised rear engine mounts, (11) single annulus combustion chamber (12) quick disconnects at both the inlet and turbine flanges and (13) ability to use AN-F-58 (JP-3) fuel." It was noted the A/B was the same as that of the XJ34-WE-32 with a larger diffuser.

Figure 5: XJ46-WE-2 Mockup Board Exhibit Engine. WAGT P40602A, 12/8/1949. *Courtesy Hagley Museum and Library*

J46-WE-2 Installation Features

A. Installation

a. The inclusion of two generator pads allowed for the generation of more electrical power for airframe components and allowed use of two different types of generators.
b. The oil system was changed to dry sump with a hot reservoir integral with the accessory gear box. This reduced foaming and decreased space requirements.
c. The single casting of the inlet and the gear box reduced the height of the engine and eliminated some parts.
d. The engine driven A/B fuel pump was used instead of an air-driven pump to allow the pump to run at a constant speed regardless of load, increasing reliability by avoiding light load/high rpm operation.
e. The single annulus design of the combustion chamber permitted a fuel nozzle design that allowed the nozzles to be individually removed through the diffuser struts without engine disassembly.
f. External fuel manifolds were now used to feed the new fuel nozzles.
g. The rear engine mounts were re-located on the combustion chamber outer casing to maintain the same distance of earlier engines.
h. The single annulus chamber allowed a reduced combustion chamber diameter.
i. Quick disconnect clamps at the inlet and between engine and A/B allowed for disassembly when the top of the engine was not easily accessible.
j. The use of AN-F-58 fuel met the latest BuAer requirements.

B. Accessory Arrangement

a. The bottom arrangement for the gearbox was selected to allow the engine to fit into the existing J34 airframes.
b. The hour-glass shape (XJ40) gearbox design was rejected because the extra length of the accessories mounted ahead of such a design. Compared to the relatively small increase in engine height when accessories were mounted underneath, the bottom location was deemed the better design choice.

C. Installation Dimensions

Installation Dimensions Comparison (Inches)		
Dimension	**XJ34-WE-32**	**XJ46-WE-2**
Length	184.0	191.5
Basic Diameter	27.0	29.0
Vertical Height	38.0	37.25
Horizontal Width	30.5	30.5

D. Installation Connections, Location and Requirements

a. Engine Mounts – A single mount on the top of the front bearing support and two pads on the sides of the combustion chamber outer casing. The A/B had one mount on the top of the unit to take vertical and side load only. Axial load went through the rear engine mounts of the combustion chamber outer mounts. The maximum load was 20,400 pounds.
b. Air Connections
   i. Inlet duct flange (quick disconnect)
   ii. Compressor air bleed (optional)
   iii. Anti-icing exhaust
   iv. Oil reservoir vent
   v. Governor alternator blast tube
   vi. Power regulator blast tube (cooling)
c. Fuel Connections
   i. Fuel booster pump inlet
   ii. A/B fuel pump inlet
   iii. Fuel pump seal drains
   iv. Combustion chamber drains
   v. Dump valve drain
   vi. Afterburner drain
   vii. Overspeed relay seal drain
   viii. Primary fuel regulator vent
   ix. Emergency fuel regulator vent
   x. Fuel filter drain
d. Oil Connections
   i. Oil filler port (use 1 of 4)
   ii. Oil reservoir drain

E. Accessory Drive Pads

a. Various (11 provided)

F. Control Connections

a. Power control lever

G. Electrical Connections (to Junction Box)

a. DC Supply
b. Anti-icing control switch
c. A/B ignition coil connection
d. Emergency fuel system indicator and selector
e. Thermocouple indicator connection

H. Electrical Connections (outside Junction Box)

a. DC supply from the junction box to engine ignition coils
b. High tension leads from the engine ignition coils to engine spark plugs (2)
c. DC supply from the junction box to A/B ignition coil
d. High tension lead from A/B ignition coil to afterburner spark plug
e. Junction box to power regulator
f. Junction box to voltage regulator

I. Engine Instrumentation

Engine RPM, Turbine Outlet Temperature, No. 1 Bearing Temperature, Bearing Oil Inlet Temperature, Emergency Fuel Pump Discharge Pressure, Emergency Fuel Control Pressure, A/B Fuel Pressure, Booster Fuel Pump Inlet Pressure, Emergency Control Indicator

J. Engine Control System and Operating Requirements

a. Covered in WAGT Report A-615
b. Essentially the same as that of the XJ34-WE-32. Fuel entered the booster pump from the airframe booster pumps. On exit, it went to the main pump and then to the oil cooler and fuel regulators. One outlet from the fuel regulator was a bypass back to the main pump inlet, other outlet routes were to the splitter valve (to the two fuel manifolds) and then to the dual flow fuel nozzles; another was to the emergency fuel pump and emergency fuel regulator to the fuel nozzles with internal changeover for both; the third was from the airframe booster pumps to the A/B fuel pump, then to the A/B fuel control and then to three outlets to supply the fuel rings (3) in the A/B.

J46 Control Elements

WAGT report A-615 described the power control system in detail. At this point, the planned control was the electronic design. The following describes the main elements and their function(s):

Primary Control – Consisted of a governor alternator regulator, power regulator, power scheduler, fuel regulator, exhaust nozzle actuator, and turbine out thermocouples. The control synchronized the fuel flow and the exhaust nozzle area to the pilot's power control lever position, providing power at the desired engine rpm and thrust, optimizing the specific fuel consumption and preventing the exceeding of the maximum allowable turbine out temperature.

On the Primary system, the power regulator received signals from the pilot's lever, the governor alternator speed and the turbine out thermocouples. The circuits within the control took those inputs to compute the required fuel flow and exhaust nozzle area and transmit signals to the fuel control valve and exhaust nozzle actuator regulator. These adjusted the fuel flow and exhaust area.

The detailed description of the internal functions of the control were covered in WAGT report A-688 section VI.

Emergency Control – A completely separate manually controlled system which, without movement of the pilot's control lever, took over control of the engine in the event of failure of the governor alternator output voltage. In the event of partial or progressive failure in the power control, the emergency control could be manually selected by the pilot. In emergency, the fuel flow to the engine was accomplished by the emergency fuel regulator while control of the exhaust nozzle area was accomplished by direct mechanical connection from the power lever to the exhaust nozzle actuator regulator. In case of failure of the hydraulic supply to the exhaust nozzle actuator the nozzle was locked in the position it was in when the failure occurred.

Dump Valve – This drained the fuel manifolds when the engine was shut down. Approximately one quart of fuel was dumped each time. A tank had to be provided to catch this fuel on the ground or deck. It could be blown overboard in flight or dumped into a safe container on the ground. A flame trap in the tank vent line was recommended.

Fuel Shut-off Valve – There were three positive, manually operated shut-off valves. The main valve was in the primary fuel regulator and was cam operated from the power lever. It remained closed until the power regulator was moved off a flat spot on the cam opening the valve. The A/B valve was toggled so that it remained seated until the power lever was advanced into A/B range. The emergency fuel regulator / fuel shut-off valve was manually opened and closed from the power lever. The shut-off valves prevented the flow of fuel if the engine was windmilling.

Fuel Booster Pump – This pump received fuel from the airframe booster pumps and supplied it under pressure to the gear type dual fuel pump.

Dual Fuel Pump – This pump combined the main and emergency fuel pumps in a single housing. Failure of the main pump selected the emergency pump automatically. The emergency pump was capable of supplying take-off fuel volumes. Selection of the emergency pump by the pilot could be manually accomplished. If the A/B fuel pump failed, the main and emergency fuel pumps supplied fuel to the A/B at take-off volumes automatically.

Afterburner Control – Advancing the control lever into the A/B range unlocked the A/B positive fuel shut-off, energized the A/B ignition coil, and successively admitted fuel to the three afterburner fuel rings *(providing fully modulated thrust in A/B – Auth.)*. Modulation of the A/B exhaust nozzle eyelids was automatic.

The A/B fuel pressure regulator sensed compressor discharge total pressure (air mass) and automatically scheduled the fuel pressure to establish the proper fuel flow. The schedule was based on A/B operation at 10,100 engine rpm with the exhaust nozzle area modulated to maintain maximum allowable turbine out temperature.

Afterburner Installation Provisions

A quick disconnect clamp was provided between the turbine housing and afterburner diffuser to allow installation of those components separately.

The A/B diffuser contained an expansion bellows to allow for differential expansion.

A single mount was to be provided that allowed for vertical and lateral loads only.

Engine Cooling Considerations

A fire seal was provided on the combustion chamber outer casing to the rear of the engine mounts. The fire seal was not designed to support airframe loads and it was recommended that the airframe fire seal designs be coordinated with WAGT.

The engine compartment isolated cooling air on both sides of the fire seal. Aft of the fire seal adequate cooling air was available for the airframe structure and cooling the A/B ejector.

In front of the fire seal, the accessories in general, including the fuel and oil lines, were expected to operate satisfactorily up to 150°F. The power regulator, ignition coils and leads had to be held to not more than 165°F.

The installation should not restrict A/B ejector cooling airflow, as reliability of the aft end of the A/B could be affected. The required cooling flow was 4 lb/sec between the engine and the ejector.

Weight

Basic dry engine weight plus unattached components was 1,622 pounds. The weight did not include the starter, generator, tachometer generator or hydraulic pump, these being Government Furnished Equipment (GFE).

Airframe attached accessory weights (lb) were: Alternator Regulator – 5.0; Power Regulator – 29.0; Engine Ignition Coils and Leads (2) – 13.0, and A/B Ignition Coil and Lead – 6.0. The total was 53.0.

Afterburner weight was 211 pounds.

Total weight was 1,886 pounds.

Starting and Ignition

Ignition - Ignition coils, high tension leads and spark plugs were provided to ignite the fuel in the combustion chamber and afterburner. The A/B coil was energized by a switch in the A/B fuel control when A/B was selected.

The three coils were to be mounted in the airframe near the engine with the maximum length of the high tension leads to the spark plugs being the maximum distance from the coils to the spark plugs. Airframe power should be supplied within 9 to 29 volts DC.

Electrical Starting System – No provisions were made for electrical starter cables in the front bearing support struts. If electrical starters were used, the cables had to come in through the aircraft inlet duct system, design of which would have to be coordinated with WAGT.

Air Starting System – A pneumatic or solid propellant starter could be located on the starter pad if separate ducts were incorporated into the aircraft inlet duct system. A thickness to cord ratio of between .11 and .26 was specified for such struts to minimize duct loss. A single strut was not allowable to avoid blade impulses in the compressor.[41]

February's monthly activity report showed the XJ34-WE-32 50-hour flight substantiation test engine was in assembly. The XJ46-WE-2 wooden mockup was completed and the mock-up board meeting was held on March 14-16. (Details below)[42]

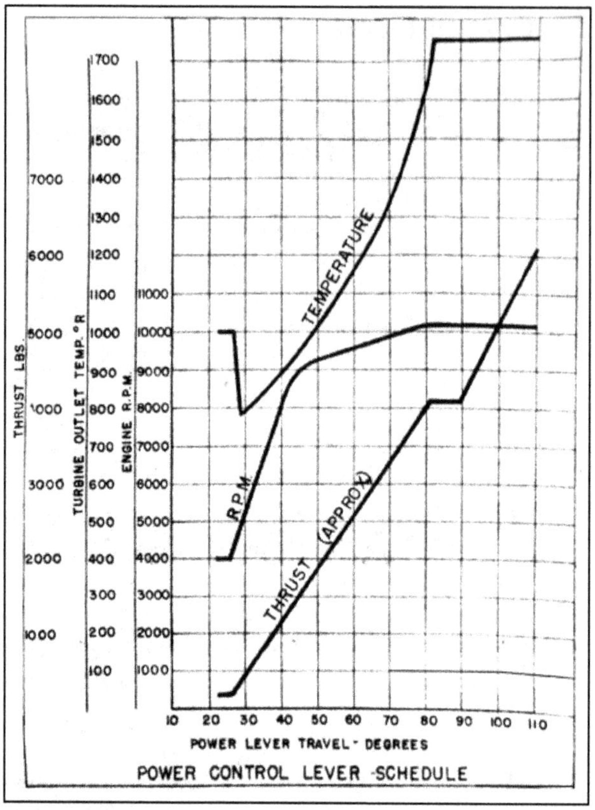

Figure 6: XJ46-WE-2 power schedule from WAGT report A-813. Note A/B thrust is to be fully modulated, not on/off.

The XJ46-WE-2 Mock-Up Board report was not issued until April 25, 1949, but it is appropriate for the details to be covered here, as activities related to the action items issued by the board were started immediately after the board finished meeting back in March.

XJ46-WE-2 Mock-Up Board Report[43]
March 14-16, 1949
Essington, Pennsylvania

The basic description of the engine was given, noting it was slightly larger than the J34 series preceding it and that it would be available for installation in later versions of aircraft presently using the earlier J34 series.

"The major points of discussion by the mock-up board concerned the control system, the variable area nozzle actuating system, and the lubrication system. Concerning the control system, the presently proposed system includes electronic components which will operate satisfactorily at temperatures up to 165°F only. In addition, the electronic components are so configured as to be unsuitable for engine mounting. The Bureau of Aeronautics' philosophy on control systems expressed at the subject conference is that control components should be engine mountable if such can be achieved. Location of the control components in the nacelle or engine compartment will be necessary, of course, if considerations of space, size, or environment so dictate. However, it is considered that development should be toward the end of engine mountability (*SIC*) and toward making the control components withstand the same extremes of environment as the basic engine itself. The extreme of locating the control components in the air conditioned portion of the airplane is considered undesirable and to be avoided if at all practicable. To this end, item 10 ......requesting investigation of the control mounting and environment limitation problem, was approved. This is considered to be a long range project and consideration of practical limitations will dictate that the control be at least initially built as presently conceived."

A fuel control system simplification was proposed by WAGT and discussed at the MUB and study was continuing. *(This report was never issued specifically for the J46, but appeared as "Proposed Simplification of the XJ40-WE-8 Control System", dated April 13, 1949.)* The fuel system related simplification was stated as Item 1. "The fuel metering components are simplified by combining the primary fuel regulator, and the emergency fuel regulator into a single assembly. The combined assembly will be functionally the same as the present design except for the elimination of the hydraulic acceleration limiter. The change makes it necessary for the pilot to control the acceleration of the engine during emergency operation. A topping governor is included providing the same speed control as the present system under emergency control as well as extending positive overspeed protection during primary control operation. The proposed system employs a simple throttle valve for emergency fuel control in the range below military rpm and the topping governor controls rpm at military. Thrust at military rpm is controlled by the variable area exhaust nozzle. As with the present design the proposed emergency system takes over control either automatically after alternator voltage failure or as a result of manual selection by the pilot." A table compared the behavior of both approaches when component failures occurred. The proposed change reduced power lever torque, reduced weight and would eliminate over 100 parts.[44]

XJ46-WE-2 Mock-Up Board Daily Decisions

Item	Change	Action Taken
1	Provide 30 second negative "G" oil supply	Provided
2	Relocate the oil level dip stick indicator next to the top of reservoir	Provided
3	Delete #1 bearing thermocouple and add a temperature bulb	Change to be made. Initial engines will retain thermocouple as a flight safety provision.
4	Remove the ejector shroud and make it an airframe item	WAGT to supply limits on air flow, skin temp, heat rejection and effect of weight to airframe manufacturers.
5	Show quick disconnect bond on engine air inlet	WAGT to provide data.
6	Improve routing of wiring harness on RH side of engine	To be done.
7	Provide bottom jack points	WAGT to provide data.
8	Provide adequate clearance for connections at both lower air-bleed ports	To be done.
9	Provide set of main engine mounting pads near turbine housing and eliminate flex joint	No. WAGT to submit data found in research on flex bellows joint.
10	Make power regulator, alternator voltage regulator, ignition coils engine mountable. Raise temperature limits	WAGT to study.
11	Provide ability to replace control components without need for factory calibration of new parts	WAGT to provide.
12	Prevent mechanism for locked hydraulic actuator or lines to avoid rupture from temperature rise	WAGT to provide.
13	Install short flex hose in aft section of tubing leading from oil pump to eyelid actuator	WAGT to provide.
14	Eliminate the mechanical control	No. To be studied by BuAer control engineers.
15	Make attachment provisions to the power control that can be routed to upper side of engine	WAGT to provide longer studs in appropriate places on both sides.
16a	Decrease power lever torque below 50 lb	WAGT to provide.
16b	Reduce power lever torque to 25 lb	WAGT to provide. Actual numbers to be coordinated later.

17	Make electrical control connections non-interchangeable	WAGT to provide.
18	Indicate by paint color portions of emergency system to be deleted if proposal is accepted	WAGT to show on mock-up.
19	Provide a "cruise" band on the power control for optimum engine cruise regardless of conditions.	Not approved.
20	Revise A/B operating and starting altitudes to 60,000 ft	Not approved. Referred to BuAer.
21	Engine should be operable at maximum output under all conditions of flight	Not approved.
22	Add ability for insulation blanket for afterburner shell be applied by WAGT or airframe manufacturer	WAGT to provide. Study what parts of A/B aft of compressor flange can be blanketed.
23	Study the simplified emergency fuel system and provide a suitable specification	Yes. BuAer to study and communicate decision.
24	Clarify allowable temperatures and heat rejection at that temperature at aft engine components or accessories to which cooling air need be supplied	WAGT to provide.
25	Make provisions for emergency eyelid operation and prevent loss of engine oil in the event of actuator or line rupture	WAGT to provide.
26	Study gearbox to provide: a. Sufficient drives for all engine accessories b. Provide two 400 HP power take-off drives, one on face and one 90 degrees to the face c. Add provisions to be able to mount gearbox at any of four locations around engine	Not approved.
27	Provide adequate clearance for install of hydraulic pump and tachometer generator	WAGT to provide.
28	Change all accessories to be quick removal type	Not approved. Proposal requested for weight and height of changes for quick disconnects.
29	Study to ensure all bolted connections are accessible with standard wrenches	WAGT to study. BuAer to provide list of standard tools.
30	Remove starters from engine and allow use of hand held ground starters. Eliminate dead weight of air starter in flight.	Not approved.
31	Supply starter and fairing with the engine	Not approved.
32	Improve clearance between duct attachment clamp and gearbox	WAGT to provide.
33	Study failure effect of engine driven booster pump on fuel system	WAGT to study.
34	Revise oil filling provisions to the normal one of oil rising up the neck to signify full	Not approved.
35	Define filtration requirements for all systems.	WAGT to provide in specification.
36	Investigate effects of moving fuel filter downstream of oil cooler on: a. raising fuel temp before filter and b. pressure drop through cooler. Provide provisions for alcohol injection into fuel filter	WAGT will study the first part, a. and b. WAGT to obtain information on necessary provisions from AMC.
37	Fuel dump valve should provide less than 10cc's drainage on starting	Not approved. WAGT to specify amount in installation manual.
38	Engine ratings be based on AN-F-58 fuel with heating value of 18,400 Btu/lb	WAGT to provide. Heating value to be established for engine specification.
39	Fuel system components to test for leakage to new AN-E-30 specification	Not approved.

40	Provide minimum number of straight thread self-locking oil drain valves to drain oil completely in +15/-20° nose up/down horizontal position	WAGT to study how to provide.
41	Increase maximum allowable fuel booster inlet pressure from 35 to 50 psi	Not approved.
42	Define engine boost pump and afterburner minimum inlet pressure with 100°F fuel between seal level and 8,000 ft	WAGT to provide and supply to airframe contractors.
43	Oil system should not damage system in a cold start	WAGT to provide. Was already a current specification requirement.
44	Test to ensure whether oil level indicator works properly from -65°F to +165°F	WAGT to test.
45	Study sludging characteristics of centrifugal oil air separator	Not approved.
46	Oil reservoir should allow for 15% of tank capacity	Already provided.
47	Add bleed air capacity of 6% of compressor flow to specifications	Not approved as written. Specification to state bleed data on one port and multiple ports.
48	Increase amount of allowable bleed air to 10% above 70% rpm	Not approved. See Item 47.
49	Rotate customer connections on upper two compressor bleed ports so face is tilted forward and closer to compressor case	Not approved.
50	Show method for servicing starter cowl	Done. Same as J34-WE-32.
51	Provide capacity and tap-off for anti-ice air	WAGT to provide if anti-icing of struts is found to be required.
52	Provide grommets where lines pierce support panels	WAGT to provide.
53	Move engine firewall (seal) aft to same vertical plane as the C-8 engines	WAGT to provide.
54	Design study for interchangeable parts between A/B and non-A/B engine	WAGT to conduct.

55	Reduce diameter of nozzle control yoke, and eliminate can in fairing	Not approved.
56	Reduce eyelid yoke envelope	Not approved.
57	Spark plugs should not require replacement during qualification test	Not approved. Specification coordination item.
58	Starting control should automatically bring engine to sustaining operation in 30 seconds	Not approved. Specification coordination item.
59	Move power control lever box as close to compressor as possible	WAGT to provide to extent possible.
60	Bore struts on engine and use an electrical starter	Not approved.
61	Provide stop on bellows between engine and A/B to restrict max length growth	WAGT to provide.
62	Provide air inlet protective screens	Not approved.
63	Don't use pipe threads except for permanent closures	WAGT to provide.
64	Provide separate cannon plug for thermocouple leads	Not approved. Present cannon plugs have chromel-aluminal pins.
65	Explain why non-standard turbine out thermocouples are being used	Not approved. Standard circuit resistance is provided.
66	Investigate emergency provisions to use main fuel pump even when the emergency control is selected	Not approved. Would be considered as part of fuel system simplification study.
67	Make main engine mount provisions for lower 45 degree mounting	Not approved. Would be considered in connection with special provisions of MX-656 engines.
68	Expedite BuAer action on Item 14	No action. Comment only.
69	Bring installation drawing up to date	Not a mock-up item.
70	Provide starting at -67°F without use of special lubrication and drive power needed at this temperature	First part not approved. Second part to be reviewed in light of AMC J34 cold tests.

71	Oil cooler and oil temperature regulating valve to be subjected to special tests	Not approved. To be considered by BuAer in specification coordination. May require contract mod.
72	Request weight reduction campaign on engine	Not approved. Normal process between WAGT and BuAer.
73	Revise components to prevent overheating after shutdown	WAGT to consider as a design problem.

The parallel development of the XJ34-WE-32 returned its focus to the oil system to determine what changes needed to be made to support inverted flight (for longer periods than the 10 seconds of the J34-WE-22/-30 engines).

WAGT determined certain changes needed to be made and some testing done to determine whatever changes might have to be made to support longer duration inverted flight:

1. The 2nd scavenge line in the No. 1 Bearing Support Cavity had to be relocated to provide scavenging when the engine was inverted.
2. A swivel had to be added to the oil intake in the oil reservoir to permit drawing oil from the reservoir during inverted flight.
3. Tests of the Nos. 2 and 3 bearing houses had to be made to determine the magnitude of oil leakage when the housings were inverted.
4. Tests had to be run to determine how much air would become entrained *(trapped)* in the oil during inverted operation. The exhaust nozzle actuator could be affected by this condition.

It was estimated that the investigations would take four months.[45]

In March, BuAer approved an expansion joint change on the XJ34-WE-32 that WAGT had proposed back in January in response to a MUB request. The change reduced the outside diameter of the joint from 28.25 inches to 27.75 inches and added two safety links across the joint at the horizontal centerline.[46]

BuAer requested the specifications for that engine be modified to include testing language specific to emergency system cutovers and operation.[47] The language and test requirements were later added to the XJ46-WE-2 specifications as well.

WAGT quoted the price for the alternate XJ46 accessory gearbox arrangement as $27,092.00, a $1,013,00 reduction in price from the original estimate.[48]

Comments on two earlier reports (A-681 on oil systems design and A-748 on engine installation features of the J34-WE-32) were received from BuAer. On the oil system, they noted that the design for measuring oil quantity in the reservoir could allow oil to spill on the deck when the oil was being checked and that inaccurate oil level indication would probably occur in cold weather. A modified and more acceptable type of oil level indicator was described for the XJ46-WE-2 and they requested that design be incorporated in the XJ34-WE-32 as well.

In addition, the ignition and starting systems had been changed from the continuous operation type to operating only during the starting cycle. The continuous operation type had previously been described as the simplest and lightest type of ignition system, but the proof testing of such a system had not been accomplished and the intermittent type substituted.

Both changes were approved.[49]

The monthly status report summary for March 1949 states the XJ34-WE-32 engine would begin running on or about April 15, with some parts having been borrowed from a J34-WE-34. If early testing went well, the flight substantiation test could start around the end of June, as some accessories would not be on hand until then.

The changes to the XJ46-WE-2 mockup were underway as a result of the MUB requests.[50]

The XJ34-WE-32 MUB Item 28 asked for a study of the possibility of adding a temperature override in the afterburner fuel control to prevent engine damage if the engine was operated with the A/B nozzle locked in the closed or near-closed position as a result of failure. WAGT report A-818 summarized the analysis of the request and recommended that such an override would add little in the way of additional safety and would add cost and weight.

The analysis showed that when the engine was on primary control, operation of the engine in A/B with the eyelids closed would result in the engine operating in stall. This would create a lot of vibration that would warn the pilot to reduce the throttle out of the A/B position. J34's had been operated in deep compressor stall for extended periods of time without damage.

If the engine was in emergency, damage could occur if the engine was operated in A/B with the nozzles closed, but two separate failures (both airframe power supply plus some element of the primary fuel system and a hydraulic or mechanical failure within the exhaust nozzle actuating mechanism) had to occur for the situation to be dangerous. Operation of the A/B in such a situation for 15 seconds or longer would destroy the turbine. Since the emergency control required the pilot to monitor turbine temperatures during both acceleration and steady state at or near military rpm, it was presumed that the high temperature risk would be covered by manual pilot operation of the power control to avoid it.[51] BuAer approved WAGT's recommendation on April 20.[52]

The proposed fuel system simplification was reviewed and BuAer reported the following:

1. The omission of the interconnection between fuel pumps, the connection that provided limited A/B operation if the A/B fuel pump failed, was not acceptable, since total failure of the A/B could occur if either the airframe fuel boost pump or the A/B fuel pump failed. Airplanes being designed to use the XJ46-WE-2 (and J40 models) required the A/B for takeoff. A review of the pumping arrangement appeared to be in order.
2. The balance of the proposed changes was acceptable.

WAGT was reminded that the changes did not affect the control system requirements. A written note attached to this memo states that dual hydraulic power sources and lines for qualification tests were required. The engines would be delivered without the emergency power source, but they could be installed later if required.[53]

The anti-icing layout proposed by WAGT in response to MUB Items 48, 49 and 51[54] was approved as long as the anti-icing system retained sufficient flexibility to incorporate additional anti-icing provisions for the starter cowl and struts. Their reasoning was stated:

1. BuAer did not feel it was yet possible to draw firm conclusions regarding the anti-icing requirements of struts and the starter dome. Simply heating only the inlet guide vanes might allow excessive accumulations of ice on the struts and starter dome.
2. It appeared reasonable to conclude the inlet guide vanes could withstand ice impacts from upstream ice breaking off. A recent report of a J47 failing after a piece of ice passed through and struck a rotor blade would require further consideration of this assumption.
3. The present planned application of the J34-WE-32 in airplanes not intended to operate for extended periods in icing conditions would tend to guard the engines from serious ice accumulation. The only exception (the Douglas F3D) would require reexamination of the anti-icing provisions if the airframe was modified for extended icing condition operation.[55]

It would seem the result of this approach was to agree with WAGT and then not agree and require full anti-icing provisions even if only half of them were ever actually utilized. The WAGT proposal had reduced weight by eight pounds, but BuAer's approach would put the weight back.

BuAer had to ask for clarification on the planned exhaust nozzle actuator bracket location, as two different drawings for the XJ34-WE-32 showed it at different locations. The location might impact the planned use in the F3D-1 airplane.[56]

In May, WAGT forwarded the updated specification WAGT-24C8-2D covering the J34-WE-32, with Appendix A covering the J34-WE-38.[57]

The closure of J46 MUB items began with WAGT submitting a report showing the air-bleed porting for the engine, which was based on AN-E-30 requirements:

XJ46-WE-2 Air Bleed Summary		
	Allowable Bleed % of Total Flow	
Number of Ports	Each Port (max.)	Total (max.)
1	4.0	4.0
2	3.0	6.0
3	2.6	8.0
4	2.5	10.0

The maximum air bleeds were for engine speeds above 89% of maximum and at altitudes up to 60,000 feet.[58]

The WAGT approach to providing accessory quick disconnects was forwarded in early May covering the J34, J46 and J40 engine mock-up board items for the relative engines.[59]

Almost lost in the flurry of activity, the monthly progress report summary for April noted that the first runs of the XJ34-WE-32 had been accomplished. The thrust at top speed (12,500 rpm) had been low. The engine was running with cast compressor guide vanes except for the first stage in place of the normal rolled vanes. The inlet guide vanes were opened up to increase air flow, which resulted in a slight increase in Military thrust (3,610 lb). (These ratings were above specification (3,370 lb) but would drop when the A/B was attached because of pressure losses in the A/B duct.) The turbine inlet temperature was 1,525°F. The engine had run for 21 hours satisfactorily configured with fixed tails and partial accessories.

The mock-up changes on the J46 were progressing satisfactorily. Procurement and manufacturing on the prototype engine were also underway.[60]

BuAer comments on the XJ34-WE-32 Specification WAGT-24C8-2D came back and the main changes were:

1. The power control radiated limits (radio interference transmission) were accepted for prototype development but were unsatisfactory for service aircraft. Additional development was mandatory to reduce the radiation.

2. The emergency power control requirements were not satisfactory but were under negotiation. They requested expedition on closing the requirement.
3. Shipping container requirements should be deleted and the container must conform to BuAer Experimental Specification XAK-22.
4. Change Take-Off to "Military with Afterburner" on the A/B altitude performance curves for the engine. (This was a 10-minute rating.)
5. The current 10-second negative "G" rating needed to be changed to a 30-second rating to allow service aircraft to perform their functions. It was requested that WAGT provide advice to the bureau relative to production delays, etc. if the XJ34-WE-32 and -38 were modified to incorporate this 30-second negative "G" requirement.[61]

A serious problem regarding engine performance data distribution came to light and became the subject of a long memo from BuAer. The situation was reviewed in detail, but the core issue was that preliminary, unapproved, or unsupported data of various types, formats and syntax was being given to airframe manufacturers and airframe companies who were in turn using the data in airframe proposals to BuAer. All such preliminary unapproved data sheets and performance curves were to be withdrawn immediately. Prior authorizations for distribution of data in the subject contracts was withdrawn. Henceforth, performance data would be distributed by BuAer or when specific authorization for data charts was given. Generally, the only authorizations that would be considered would be for data that was intended to be included in the approved specification for the engine. The contractor was encouraged to obtain approved specifications at the earliest practical date.[62]

WAGT acknowledged and agreed to the instructions in their response, also stating that informal distribution of data was never intended to interfere with BuAer operations but was done to assist the airframe companies to understand the latest information available in their own planning. They included a proposed information distribution matrix based on discussions with the airframe companies, with definitions of all types of information and asked that BuAer review and approve it.[63] Approval was granted within a few weeks, with distribution to go ahead on a trial basis and with the expectation that modifications would likely be needed based on experience moving forward. Additionally, they reminded WAGT that the approved specification was the ruling authority and no data was to be distributed that did not agree with the current specification. Early submission and approval of specifications were again emphasized.[64]

WAGT submitted an updated report on A/B development progress (A-838) in May, but a copy was not located in the archives.[65]

In spite of WAGT's submission of report A-836 (also not found) on the temperature limits of the electrical component of the J34-WE-32/-38 and J46 and which stated testing was underway to determine if the limits could safely be raised [66], BuAer continued to push for an expanded electrical component operating temperature envelope. They noted that the last specification for the XJ40-WE-6 stated the electrical component limits were 165°F in continuous operation and 200°F for a five minute operating period for the engine. Since the electrical components on the J34-WE-32 and -38 were essentially the same as the XJ40-WE-6, they asked that the specification on the J34-WE-32/-38 be modified by adding the 200°F five minute operating period limit as well.[67]

Afterburner pump inlet pressure requirements for the XJ46-WE-2 and possible solutions were reviewed by BuAer in June. They noted that the requirement was that the main and emergency fuel pumps had to sustain maximum afterburner fuel flow at 1.1 ram and 100°F fuel temperature even if the aircraft booster pump failed. If the A/B fuel pump inlet pressure requirement was too high and the airframe boost pump failed, the A/B fuel pump would cavitate and stop pumping fuel, resulting in A/B failure. During a take-off this could be disastrous. BuAer asked that the A/B fuel pump inlet pressure be lowered to avoid an A/B failure possibility and to reduce the airframe pump flow requirement, thus assisting in easing the airframe fuel supply solution.[68]

Having already made their revisions comments on the latest XJ34-WE-32/-38 specification regarding the oil supply for inverted flight, BuAer discovered after their review that the requirement for inverted flight was entirely missing from the current specification. In fact, they found it had been removed entirely from the prior version of the specification with no notice of the change. They requested that the paragraph covering the requirement as it appeared in the WAGT-24C8-2B revision be inserted in the final issue of the final specification.[69] It was a further oversight on BuAer's part that the paragraph also needed to be modified to call for a 30-second negative "G" capability as stated in their earlier correspondence.

BuAer requested WAGT complete a chart on engine characteristics to be provided for future airframe consideration to the airframe companies. It was returned completed (below) by WAGT a bit later.[70]

Westinghouse Turbojet Engine Dry Weights (Estimated) (lb)						
Engine Components	**J34-WE-32**	**J34-WE-38**	**J40-WE-6**	**X40E8A (J40-WE-10)**	**X40E7A (J40-WE-12)**	**J46-WE-2**
Engine	1,890.0	1,440.0	2,975	3,968	3,598	1,886
Compressor	457.0	457.0	1,130	1,566	1,566	498
Combustion Chamber incl. diffuser	254.6	254.6	368	395	395	273
Turbine 1st Stage Disc, Blades, Nozzle	95.4	95.4	241	301	301	128
2nd Stage Disc, Blades, Nozzle	76.7	76.7	254	289	289	108
Shaft, Turbine Housing, Bolts, etc.	60.3	60.3	375*	409*	409*	62
Exhaust System Nozzle	----	86.0	62	----	82	----
Inlet Assembly	57.0	57.0	108	111	111	200**
Accessory Drive and Attached Engine Components	419.0	353.0	459	529	465	355
Afterburner, exhaust collector & nozzle	270.0	----	----	568	----	287

* Included turbine bearing
** Included integral accessory drive and oil reservoir

A simplified control system had been proposed for the XJ46 and BuAer had reviewed the proposal. They asked detailed questions related to possible failure modes and made other suggestions related to the proposal. At this point, the primary control was still considered acceptable pending test verification, but the back-up emergency control, particularly as it related to the A/B and its interrelationship with the power controls in both primary and emergency modes, was undergoing continuing scrutiny. In response to a memo covering BuAer's comments and suggestions from a conference that was held, WAGT's response was:

1. BuAer – "What is the effect on the engine control if one of the teeter valve coils should fail? Would the main control valve be closed and cause the engine to shut down?"
   WAGT Response – "Failure of either coil in the teeter valve would cause a momentary motion in the direction controlled by the unfailed coil. As soon as there is any motion from the desired set point, the feed-back circuit operates to de-energize the unfailed coil causing the main control to stop in some mid-position. Further study has revealed that if a light spring is placed so as to urge the main control valve in(to) the open position any failure of the teeter valve coil controlling the opening of this valve will result in no change in fuel flow or pressure.
   The only effect will be to reduce the available power for moving the main control valve. Any failure of the other teeter valve coil will cause the main control valve to open wide and control of the engine will be maintained by the emergency overspeed valve.
   It is believed that the addition of the opening spring to the main control valve will satisfactorily prevent closing of the main control valve upon electrical failure of either teeter valve coil."
2. BuAer – "Is it possible to start the engine in the air using only the emergency power control system?"
   WAGT Response – "The design of the power control system is such that air starts can be made using only the emergency power system."
3. BuAer – "Would it be desirable to provide a pressure pickup downstream of the main control valve to sense pressure failure and change over to the emergency system?"
   WAGT Response – "Study had revealed that the emergency pump was automatically selected upon failure of the primary pump as indicated by a pressure failure. There were no valves downstream of the pumps that could stick or malfunction to lower pressure below that needed for minimum throttle flow setting. Inclusion of a pressure pickup would serve no purpose."
4. BuAer – "Why not use the three position switch for the preflight check system to standardize the cockpit

equipment in all aircraft using emergency systems? During the conference, reference (a), the discussion indicated that a reset mechanism would be required and that, when the emergency system is selected either manually or automatically, it should be locked in to prevent cycling between the primary and emergency systems in case the primary system should recover. (Such as the case of a bad electrical connection to the solenoid valve which continues to make and break the circuit)."

WAGT Response – "Further study of this problem revealed that the three position switch is more desirable than the original proposed three pole single throw switch. Accordingly, this Contractor will redesign the system to include provisions for using the pre-flight system as described in reference (d). The addition of these provisions to the simplified power control system will increase the weight of the system and consequently the engine by approximately three pounds (3) and will increase the volume by approximately 27 in^3 or a package 3"x3"x3". This preflight check system will be applied to engines in accordance with the earlier references."[71]

The challenge of integrating the electronic control and manual emergency back-up control with an operating A/B was one that would require Westinghouse to make many future changes, including the reduction of the electronic components and increasing the hydraulic component in an effort to satisfy additional emerging service requirements.

Only the BAR's cover letter for the May activity report has survived, but it states that the XJ34-WE-32 testing through 25.76 hours without its A/B had continued through May. Various configurations of the engine had been tested, with the 7th and 8th mentioned specifically. The 8th configuration consisted of:

1. Rolled compressor vanes in all stages except the first. Cast vanes for the first stage. The rolled vanes had been opened up 3 degrees
2. Compressor inlet guide vanes opened 2.8%
3. First state turbine nozzles closed 17%
4. Combustion screens were included
5. Fixed tails

Turbine blade growth was reported as .0027 inches average on the first stage and a .0023 average on the second stage. Total Military operating time during the period was 3.97 hours. Total engine running time to date was 39.13 hours, with total military time to date of 5.70 hours. The A/B would be installed and operated during the next series of tests.

Configuration Results		
	7th Configuration	8th Configuration
RPM	12,500	12,500
Thrust, lb	3,710	3,700
SFC (lb/hr/lb)	1.00	1.00
Airflow, lb/sec	60.0	60.0
Comp. Ratio	-----	4.38

The XJ46-WE-2 purchasing activity began during the month of May.[72]

WAGT surveyed the MUB Item 15 request for provisions for control rod brackets on the regulators and provision of excess stud length where necessary. They asked that the airframe companies mark up four different provided drawings of the existing design so they could learn what was required and then they would prepare a response.[73]

A six page memo received from BuAer in July contained the results of BuAer's review of the latest XJ46WE-2 specification, WAGT-24C10-2C. The memo swept up many of the emerging requirements from the MUB. It requested a standardization of the elements of the specification to those for the J34-WE-32/-38 and J40 engines including the common elements in those engines which had been pre-agreed to with BuAer. That also applied to the J46. It requested a rewrite of the specification to fully comply with AN-E-31a, -30a, -32a and-33a and took exception to the specification weight, added new requirements, and asked for further investigation of certain specification aspects. The specific elements of interest (other than simple typographical corrections or document organization) were:

1. After approval of the mock-up, all installation envelope changes to the engine had to be reviewed and approved by BuAer before airframe companies could be informed of the change.
2. The specification for AN-F-58 fuel stated that it could contain as much as 25% aromatic content. WAGT's specification that called for the fuel to contain between 10 and 20% aromatic content had to be removed. BuAer could not assure that the engine might have to operate with the maximum allowable aromatic content.
3. The engine control tolerances were further defined to state:
   a. At Military and Normal rated thrust positions, RPM should be maintained within ±1%.
   b. At cruise positions, RPM should be maintained within ±2%.
   c. At Idle position, RPM should be maintained within ±0 - 10%.
   d. At Military thrust position, the indicated temperature should be maintained within ±0 – 40°F of Mili-

tary maximum allowable measured gas temperature.

e. At Normal thrust position, the indicated gas temperature should be maintained within ±0 to 50°F of Normal maximum allowable measured gas temperature.
f. There should be no appreciable hunting with normal instrument damping.
g. Drift in RPM was acceptable within the above tolerances.

4. No major part of the engine which could not be disassembled in the field for inspection and adjustments could be disassembled for the qualification test. Specifically adjustment of compressor and turbine clearances would not be permitted.
5. A requirement was added to the specifications that the A/B fuel pump be tested to the same specifications as the primary and the emergency pump was to be added to the specification. It had to operate at full capacity only one quarter of the time required for the primary pump.
6. The initial run schedule of one and a half hours was defined in detail and had to be added to the specification. The new schedule included tests of the emergency control.
7. Radio interference transmissions were not in compliance with all ranges covered by AN/I-27a, but would be acceptable for qualification testing. The transmissions were not acceptable for service use and had to be corrected prior to production proceeding.
8. The specification weight was 1,886 pounds. The original approved weight was 1,710 pounds and only 40 pounds of the increase were approved by BuAer for changes to the engine gearbox. WAGT was asked to justify in detail the weight increases.
9. A new paragraph was added: "E-7a. The engine shall function satisfactorily for at least 10 seconds during inverted flight although the fuel flow to the engine may be interrupted during this maneuver. Satisfactory engine functioning during this condition shall be considered achieved when the engine will return to normal operation within the operating limits shown on curve 333152 without further action of the pilot when the negative acceleration is removed."
10. BuAer added the maximum permissible shear, axial and overhung moment loads on the inlet, and that A/B quick disconnect clamps were to be supplied with the engine.
11. It requested a statement be added stating that the maximum gas temperatures listed in the tables would be updated after final acceptance test.
12. WAGT was to add specific heat rejection, maximum operating temperatures and cooling requirements for each engine zone.
13. Stated that the emergency control requirements were still in negotiation and would be confirmed later.
14. Required a statement be added that provisions had been made to prevent loss of engine oil in the event of breakage or leakage of the exhaust nozzle actuator control line.
15. Added several paragraphs dealing with both the A/B and normal fuel pump emergency provisions at relative ram pressures and air temperatures.
16. Added a statement that the engine had to be able to operate normally after up to a 10 second oil interruption and the oil pump would re-prime itself after such an interruption.
17. Changed the oil pressure connection to 1-1/16 – 12, E-Z internal straight threat with room for a pressure transmitter.
18. Added oil cooler specifications: Fuel inlet temperature of 110°F ±0 – 5 degrees in all conditions. For cold temperatures the cold fuel temperature should be -65°F.
19. An oil temperature bulb should be installed for measuring oil-out temperature.
20. The starter drive should be able to handle 1,000 starts with the maximum torque required for a 15-second start. The starter pad should be upgraded to a 3-tooth type, strong enough to handle the requirement.
21. The spline adapters for generators had to be specified.
22. The air test description designed to prove there were no leaks in the fuel components was to be added to the specification in compliance with AN-E-32a.
23. WAGT was asked to confirm that the bleed air statement of six percent mentioned in a WAGT installation table they had shared with airframe personnel and discussed at the MUB was permissible. It again raised the issue of unauthorized distribution of information, this being a good example.

The closing paragraph of this memo states that the majority of changes requested where no explanation was given were included to "conform with previous agreements with Westinghouse on variable area nozzle engines."[74]

It is obvious from that statement that the specification control process was flawed. With multiple engines under development, there were many meetings occurring simultaneously and decisions taken at any one of those might or might not apply to other engine projects, even if the engineers for those projects were not in the meetings. At a minimum, the administrative control of the various specification documentation involved was struggling to keep up with such a diverse and uncontrolled decision-making process.

The monthly progress report for June stated the 50-hour flight substantiation XJ34-WE-32 engine would not be available at the end of June. Assembly of the engine would start at the end of July and the engine on test in August or early September. This represented a schedule slip of at least 90 days.

The initial XJ34-WE-32 engine was still in test and some of the testing was with a prototype A/B attached. The balance was with fixed tails (as opposed to the variable area exhaust nozzle intended). The control governor fuel system was that of a J34-WE-34. Total run time was 51 hours. Engine with A/B run time was 1.2 hours of the total of 10 hours for the month. Total engine military time was 11 hours.

The attained testing results of the A/B performance was 4,600 pounds with an SFC of 2.3 vs. the guaranteed 4,900 pounds at an SFC of 2.6.

A new design of the combustion chamber housing with a relocated anti-icing pad was being checked. The new accessory gearbox had been received and would be assembled into the second engine. The first engine was being tested with a J34-WE-34 gearbox assembly with modified oil piping and using an XJ34-WE-38 oil pump and oil cooler. The new oil reservoir was 70% complete. The A/B was being tuned to utilize the gas pressure and velocity distribution of the engine more efficiently. The flame holder was being revised to provide increased durability.

A new flexible joint to give improved installation clearances in the Chance-Vought F7U-3 airframe was in development. It would have a smaller outside diameter than the prior design.

The electronic governor (fuel control) system testing was being done on a house engine with 30 hours of testing satisfactorily accomplished. The second high pressure fuel pump example was rejected for excessive internal leakage found to be caused by a crack in its casting.

The fuel booster pump had undergone approximately 75 hours of testing, including tests to determine the effect of fluctuating fuel inlet pressures to the pumping system. The pressure regulator valve successfully dampened out fluctuations from six cycles per minute to six cycles per second.

The five fuel regulators had all been tested satisfactorily in 30 hours of running on a house engine. The first emergency fuel regulator had been delivered and was about to be bench tested. Three more emergency regulators were scheduled for July delivery.

The A/B fuel control drawings were complete and released for manufacture. A bread-board type hot-streak ignition valve was built and tested, with its layout drawings completed. The Pesco A/B fuel pump had bearing and seal modifications designed to improve durability. A meeting with the Thompson Products company was held to review the status of the air-turbine fuel pump for XJ34-WE-32 application. A sample unit was to be delivered to WAGT in July. Skinner Purifiers, Inc. completed delivery of their first sample 10 micron high pressure filter.

The Bendix-Scintilla ignition coils ran a total of 367 hours of continuous operation before one failed due to excessive bearing wear. Spark plug testing to increase ceramic durability was continuing.

The exhaust nozzle parts were being tested for durability and the valve settings for stability. An improved assembly was released for production with July delivery.

The emergency schedulers and shafts were being manufactured. The power control lever box would be released by early July for procurement.

The power regulators were to be rewired following the establishment of the tuning requirements. The power scheduler example was expected to be delivered in July. Production tooling for the governor alternator was in process with several minor changes to be made.

The electronic control cable and junction box were in the design process. BuAer changes to include a pressure switch and electric control relay assembly would affect delivery and weight of the assembly.

The radio interference reduction approach for each electric component was discussed in detail. The main impact was that inclusion of filters, etc. would slightly increase the overall weight of the electric assembly in general.

For the XJ46-WE-2, the engine was in early manufacture with much work to be done on the various parts. The compressor discs had been finish turned and were ready for broaching (the disc end slots for the blade roots) and clutching (cutting the clutch teeth on each side of the disc). The dies for blade forgings would be finished in early July. Manufacture of rolls for the vane sections was underway, with completion in August. The diffuser struts were having their spinning forms and dies made. The combustion chamber liner was in redesign, halting combustion tests.

The turbine housing layout would be ready for detailing in July. One turbine disc was finish turned with the other to be done by early July. Sample castings had been received for both rotating blades and the first stage turbine nozzle. Samples of the second stage nozzle were to be available the first week in July. The supplier drawings of the No. 2 and No. 3 roller bearings had been received.

The drawings for the cover and housing casting of the gearbox and No. 1 bearing support were released for manufacture. The remaining drawings would be available in July.

The following engine accessories were basically the same as those for the XJ34-WE-32 engine:

1. Dual high pressure fuel pump
2. Fuel booster pump

3. Primary fuel regulator
4. Emergency fuel regulator
5. A/B fuel regulator
6. Ignition system
7. Exhaust nozzle control
8. Power regulator
9. Governor Alternator

The A/B fuel pump layouts were complete and sample pumps would be ordered in early July.

The location of the thermocouple balance box and junction box were still to be accurately determined. When that was done, the layouts of the harness and brackets could be started.

Thirty dual orifice fuel nozzles designed by Ex-Cell-O had been ordered. A proposal had been received from the Delavan Manufacturing Company to make dual orifice nozzles and was being studied. Earlier test samples from that company had been tested and proved satisfactory.

The dump valve element design was complete and work on the splitter fuel valve was under way.[75]

It is instructive that the projected first run of the XJ46-WE-2 is not included in the progress report or any other scheduling information except for specific items within the report. Not all of those had projected completion dates either. With the lack of hard scheduling information, it is difficult to understand how BuAer was able to coordinate activities with the airframe companies.

BuAer asked how much weight could be eliminated from the J34-WE-32 and WAGT responded that the simplified anti-icing system had reduced the weight by eight pounds and a further four pound reduction was possible if an air starter was used instead of an electrical starter thus eliminating the need for the electrical cables to the starter. They commented that they appreciated BuAer's concern regarding the weight problem and that WAGT was "taking continued steps (not detailed) to keep the weight to a minimum consistent with the durability demanded by the engine."[76]

WAGT had responded to the XJ34-WE-32 MUB Item 18 back in January, but for some reason the response was not forwarded to BuAer by the BAR until May. The MUB had asked that WAGT investigate if the exhaust nozzle actuator could be moved further aft to assist one airframe company with design of the actuator extension design. WAGT's response said it was not possible without complete redesign of the A/B to handle the additional strength requirement and thermal stresses. The redesign would add weight to the engine.[77] A conference was held with Douglas and in July BuAer accepted the recommendation provided they were "compatible" with the agreements reached during the conference.[78]

A good example of the type of design issue that potentially affected weight, cost and installation envelope involved provisions for a "fire seal". This was Item 1 of the XJ34-WE-32 MUB. WAGT proposed a fire seal solution at a weight of 9.5 pounds and a cost of $2,356.00. It would require a contract amendment if approved. Their long March memo described all of the reasons WAGT recommended the solution.[79]

BuAer's response on July 27 analyzed WAGT's suggestion and discussed several issues. First, the reasons to design the fire shield on the combustion chamber outer casing rather than on the combustion chamber forward flange had been pointed out. Second, a review of the AN-E-30 paragraph D-17 revealed the statement: "Provisions shall be made on the engine for attachment of a liquid tight, temperature resistant fire shield." The WAGT Bid Specification WAGT-24C10-2B stated "A fire shield may be mounted on the engine at aft side of the forward flange of combustion chamber outer case." Specification WAGT-24C8-2A stated "A fire shield adapter shall be provided on the engine combustion chamber outer casing."

Since the proposed design was only meeting the specification, there should be no additional cost to BuAer. Also, it was considered the weight quoted was excessive and should be part of the (existing) dry weight of the engine, not an additional weight.

The proposal design was accepted but the cost and weight increase was not approved.[80] How WAGT was to implement the design without adding weight was not explained. WAGT later requested re-consideration of the decision over weight and cost. The memo reviewed the entire history of the fire seal provisions from the preliminary proposal to post MUB final solution. The pencil notes on WAGT's review show BuAer's reviewer was not convinced by the argument.[81] Two weeks after this memo, WAGT submitted a design change to the fire seal reducing its weight from 9.5 pounds to 7.5 pounds.[82]

WAGT's report A-836 on Accessory Temperature Limits was approved, but BuAer reminded WAGT that they wanted the XJ34-WE-32 temperature limits raised to 200°F continuous with a 250°F five minute over temperature limit rating. On the XJ46-WE-2, they wanted the maximum temperatures of the engine components to be raised to be compatible with the ram temperature limits of the basic engine. WAGT had stated they were investigating methods to improve the limits and verify current components and BuAer wanted the report on this work forwarded as they became available. The earlier information in A-836 should be incorporated in the installation manuals.[83]

The standard aircraft emergency three position switch was to use the governor alternator as a power source and BuAer approved this design element in early August.[84]

WAGT got their hand slapped for "Unexecuted Deliveries on Contractual Items" by BuAer and responded that they were aware of their contractual obligations and were reviewing their process to ensure BuAer was notified of any expected schedule delivery delays in the future.[85]

Back in June, WAGT had submitted their analysis of the effect of a failure of the engine driven booster pump on the fuel system. They covered the two items: the bellows-operated booster pump discharge pressure regulator and the booster pump itself.

They found in the first condition that a bellows failure on the booster pump would fail-safe, as the poppet valve would assume a position of minimum fuel flow restriction. No danger from fuel leakage into the engine compartment would occur, as the air side of the bellows vented to the atmosphere (outside the airframe) through the booster pump drain ports. The excessive fuel bypass could disturb the schedule of fuel flow (thrust) vs. the throttle position when operating at high altitude on the emergency fuel system, but even with this double failure condition, an engine overspeed could not occur.

In the second condition, they reported that if the pump failed, fuel would continue to flow through it to the dual fuel pump and then maximum engine fuel flow to the engine could be attained as long as the airframe booster pump was operating and was supplying a pressure of at least 34 inches of mercury. This was defined in Appendix B of the XJ40-WE-8 mock-up report and was equally applicable to the XJ46-WE-2.[86]

BuAer accepted the report in August.[87]

Later that month, BuAer approved a weight increase of three pounds on the XJ34-WE-32 to cover the requested changes in the ground handling provisions in the engine requested by the MUB.[88]

WAGT surveyed the removal of the forward (of two) combustion chamber drain valves from the XJ46-WE-2, as review showed the engine tended to always drain to the rear because the combustion chamber was sloped two degrees down at the rear, unlike prior engines. They understood that the engines would not be started in a "kneeling position." (Some early U.S. Navy fighters had provisions for the nose gear to partially retract to an intermediate position, lowering the nose and allowing more aircraft to be parked in tight hanger space.) None of the aircraft in which the J46 was planned to be used had the kneeling feature.[89]

The requested rewrite of the specifications to the Mil-E standards triggered a WAGT response that stated that in view of the careful analysis required of the new MIL-E-5007, -5008, -5009, and -5010 specifications which replaced the AN-E-30a series, there would be some delay in providing the requested revisions. The rewritten specifications would be submitted the week of September 12. Any changes to the contracts required would be submitted at that time as well.[90]

It was brought to the attention of BuAer that if WAGT's earlier proposal to eliminate the hydraulic fuses *(check valve)* in the hydraulic system and replace them with a shuttle valve was approved, the change would reduce the weight of the XJ34-WE-32 and -38 by four pounds. They requested approval.[91] The reference was to a long memo from July 29 that presented analysis of the recommendations from XJ40-WE-6 MUB for Items 4 and 21 regarding provisions for providing hydraulic lines and prevention of leaks in those lines. This memo also applied in full to the XJ34-WE-32, XJ46-WE-2, and XJ40-WE-8, all of which utilized the same basic A/B exhaust nozzle control. The basic analysis conclusions on the providing of duplicate hydraulic lines with fuses on the engines were that up to a year delay to develop the system would be needed and, depending on which particular solution BuAer chose, it would also add 20-40 pounds to a given engine.[92]

They also apologized for not explaining why they had deleted the reference specification regarding the oil interruption requirements. They explained their design of the oil system used a flop-over tube to avoid interruption of engine lubrication. Since they, the contractor, assumed full responsibility for the lubricating system, it appeared more logical to establish performance requirements for the system rather than to detail certain special tests which did not necessarily reflect operating conditions of the engine. Such special tests would require extensive design changes to safely accommodate their particular requirements. They proposed the following language for the specification:

"E-24g(2) The engine will mechanically withstand, with no detrimental effects, any oil interruption which may be imposed upon it by zero g flight maneuvers for a period of 10 seconds. This characteristic need not be demonstrated by the contractor."[93] BuAer accepted the proposed language.[94]

WAGT prepared Appendix B to specification WAGT-24C8-2D in July to cover modified testing requirements of interim YJ34-WE-32 engines to be built prior to the start of production. These were to be tested to the 50-hour flight demonstration level test requirements of AN-E-50 with some allowances for inspections during the test. These were to be allowed since some of the components might not be to the final production standard.[95] The appendix was dated July 25 and BuAer responded on August 24 with a few requested changes related to the new MIL-E-5009 specification covering engine testing. They asked a rewritten appendix be submitted for approval.[96] Looking ahead, WAGT responded on September 16, stating that preparing the engine to a level capable of meeting the new MIL-E-5009 specifications would delay the 50-hour test by about four months and would cost approximately $150,000 (and by implication need a contract amendment). The testing requirement for the anti-icing provisions could not be met because the anti-icing provisions on the engine would not be complete until just prior to the 150-hour qualification test of the J34-WE-32. The balance of the changes was incorporated in the Appendix. They requested approval of the initial submitted Appendix language with the acceptable incorporated language included.[97]

As a result of the exchange of letters, BuAer convened a conference to work out the final acceptable test levels of the de-rated engine. The conference was held on September 21, 1949, at which time WAGT presented the performance levels of the engine they could guarantee for the YJ34-WE-32 as:

Proposed YJ34-WE-32 Performance Guarantee
Ratings at Standard Sea Level Conditions

Rating	Thrust (lb)	RPM*	SFC(lb/hr/lb)**
Take-off (5 mins.) w-A/B	4,700	12,500	2.60
Military (10 mins.) w-A/B	4,700	12,500	2.60
Military (30 mins.) w/o-A/B	3,150	12,500	1.20
Normal w/o-A/B	3,020	12,500	1.12
Cruise 1 w/o-A/B	2,730	12,400	1.075
Cruise 2 w/o-A/B	2,260	12,000	1.04
Idle w/o-A/B	205	5,000	5.70

*Design values subject to revision as recommended by contractor and approved by procuring service, prior to qualification test.

**Heat of combustion assumed to be comparable with presently available fuel specification AN-F-48 grade 100/130 namely, an average of 18,800 BTU/lb liquid fuel at 77°F.

Two alternative 50-hour test cycles were proposed, and BuAer selected the "five cycles of 10 hours each" option. Production schedules for at least the first three months would stay the same, with YJ34-WE-32 performance engines shipping in lieu of production performance level engines. The de-rated engine was to be operated at a peak (turbine) inlet temperature of 1,525°F corresponding to the measured (turbine outlet) gas temperature of 1,380°F during takeoff and military conditions. The Power Plant Division of BuAer wanted to wait for engines until they could meet the full J34-WE-32 performance guarantees, but the Fighter Branch had airframe schedules in play and they required engines without delay. So, BuAer was forced into a situation of accepting not less than ten and possibly as many as 26 de-rated engines to the YJ34-WE-32 performance ratings. If additional engines beyond the maximum number agreed upon were ordered, a second 50-hour test would be run. After the XJ34-WE-32 passed its 150-hour Qualification Test, the YJ34-WE-32 engines were to be brought up to full WAGT-24C8-2D standard (modified as necessary from the results of the 150-hour test) at no expense to the government.[98] With some minor changes to the proposed test schedule selected, BuAer accepted the 50-hour test plan for the de-rated engines.[99]

With all the discussion on testing and performance standards resulting in agreement going on, it is appropriate to return to the ongoing development issues, of which there were many. Westinghouse was very busy with the following engines under simultaneous development, all using the same basic control system:

YJ34-WE-32
XJ34-WE-32
XJ34-WE-38
XJ40-WE-6
XJ40-WE-8
XJ46-WE-2

Given that a definitive starter and control system had not been defined as yet, BuAer submitted their comments back to WAGT later in August but had to delay their delivery of qualification test requirements for the electrical elements of the power control. These were to be submitted at an "early" date.[100]

Testing status of the XJ34-WE-32 was reported in July. The first engine was still in development testing. The A/B testing was up to 5.1 hours, with 19.52 military power time and a total engine time of 78.44 hours. The second engine (designated as the 50-hour test engine) went through a trial run, a temperature traverse and then a "green run". (The green run was WAGT language for the initial run of an engine through all of its power settings, with the operator adjusting the fuel controls and power controls to obtain the proper temperature read-outs. BuAer attempted to get

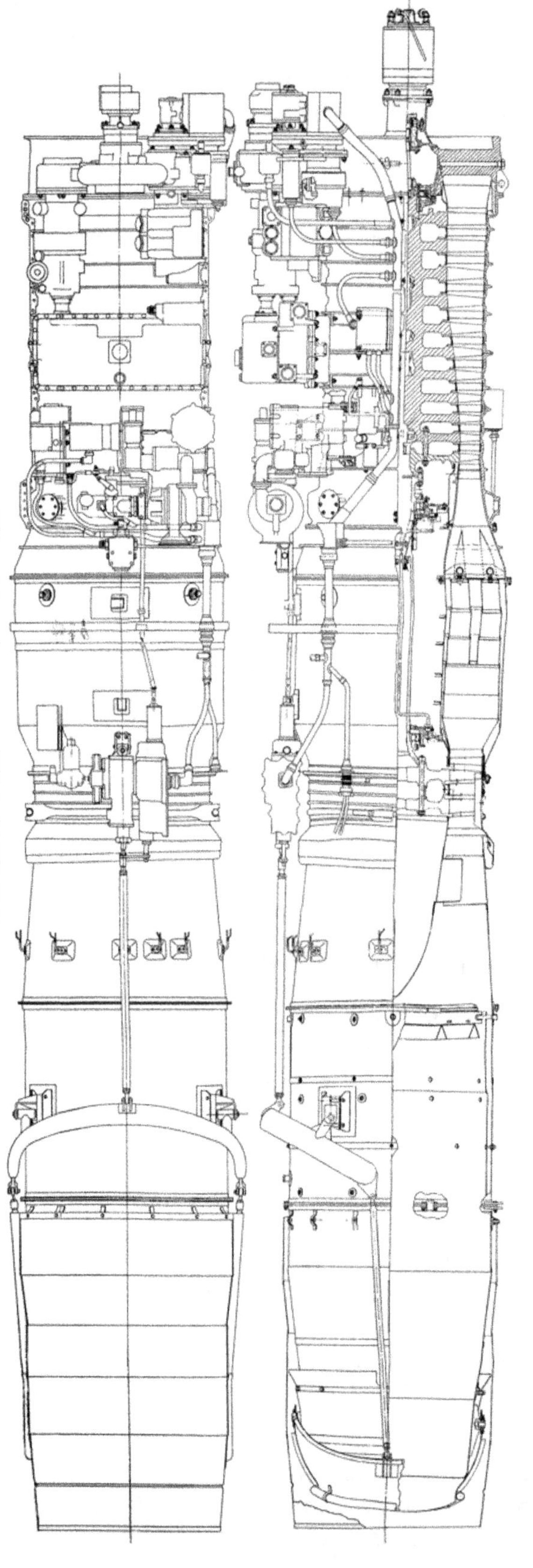

Westinghouse J34-WE-32

Assembly Drawing 58J55 May 17, 1950 (Modified)

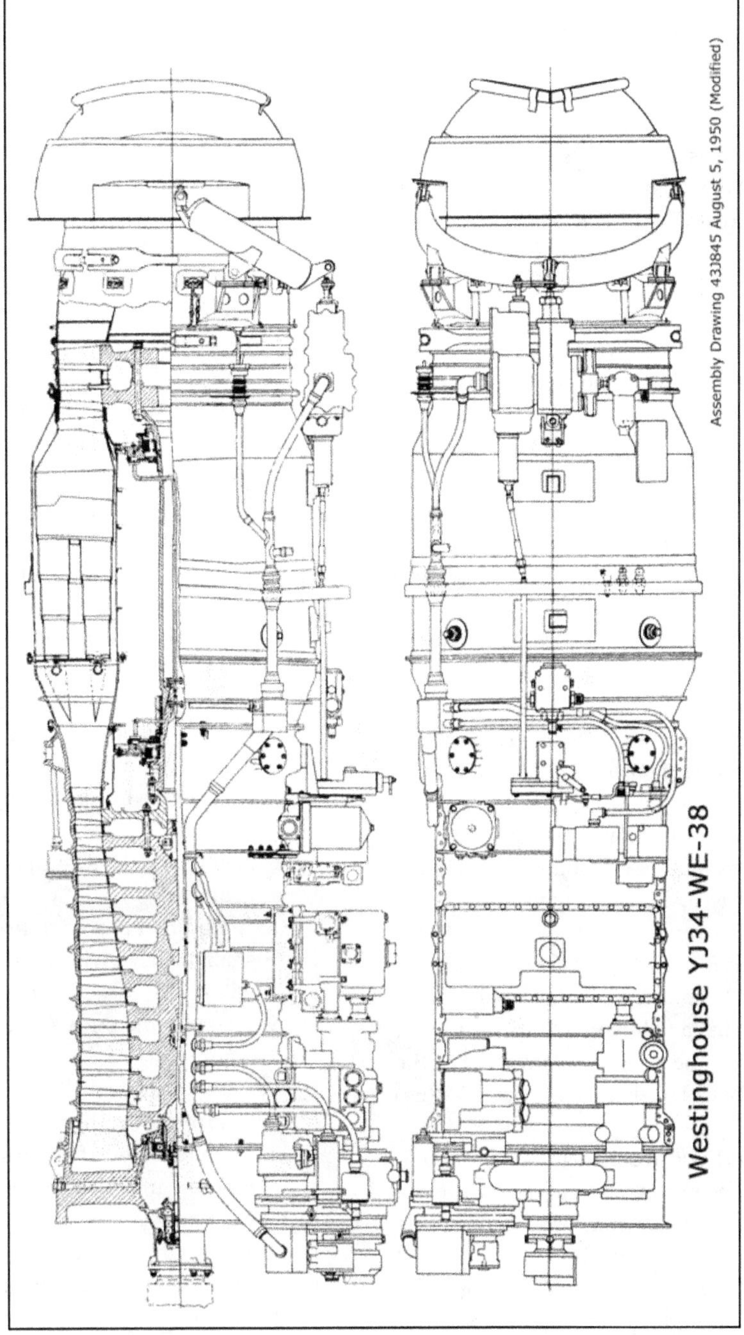

Assembly Drawing 433845 August 5, 1950 (Modified)
Westinghouse YJ34-WE-38

WAGT to call it the "break-in" run, but WAGT persisted in using "Green Run" for all of their engines.) During the green run, there was a sudden increase in oil consumption and the engine was stopped. Investigation showed the inner wall of the turbine shaft housing had collapsed, with the inner wall pulled in and distorted. The No. 3 bearing seal and No. 3 bearing were damaged. The shaft housing was being reinforced with four stiffening rings. The engine had resumed testing to determine the reason for poor combustion distribution.

As almost a footnote, the report stated that XJ46 activity had been stepped up.[101]

In early September BuAer asked for better drawings on the XJ34-WE-32 generator clearance.[102] Six days later WAGT sent over the requested drawing.[103]

The MUB for the XJ46-WE-2 had requested the A/B shroud (also called the Cooling Ejector and "Cooling Ejector and Shroud") be deleted as an engine part and be provided by the airframe manufacturer. The same request had been made by the XJ40-WE-8 MUB and ultimately that request was rejected on the grounds that the shroud required extensive development more easily accomplished by the engine contractor, so the shroud remained part of the engine.

Examination of the XJ46-WE-2 MUB's similar request resulted in the same decision being reached, with the shroud remaining part of the engine.[104]

WAGT responded to XJ46 MUB Item 36 that requested the fuel and oil coolers be moved and that provisions for alcohol injection be provided. The first two items were left where they were, as moving the fuel filter meant that unfiltered fuel would pass through the emergency valves if moved and moving the oil cooler would create an objectionable pressure drop during emergency operation with a failed A/B fuel pump.

A provision for alcohol was provided in the form of a .25 inch port into the filter housing on the upstream (unfiltered) side of the micronic filter element.[105] BuAer approved the recommendation and change on September 6.[106]

Analysis of certain failures of the simplified power control were presented to BuAer in a memo in early September. BuAer's concern related to failure of electric and electronic components of the power control system during take-off, when a sudden loss of power could be disastrous. WAGT's theoretical analysis (not supported yet by actual testing on the test bed with the engine running) showed that at military power settings, it was "unlikely" that there would be little to no thrust loss of the engine. Below military, the probability of a rapid reduction of thrust was possible. The characteristics of the expected behavior of the engine had resulted from designing the power control so that it was unlikely that a failure would cause a rapid reduction in thrust when the power lever was set in the military position.[107] This analysis covered the J40-WE-6 and -8 engines as well.

As an "FYI", WAGT notified BuAer that they had finalized the location of the exhaust nozzle actuator to the combustion chamber housing, which had moved the emergency scheduler shaft and the oil lines to the exhaust nozzle actuator closer to the engine, thus reducing its installation envelope "considerably."[108] BuAer approved this change.[109]

BuAer had requested WAGT explain the differences between the XJ34-WE-32 and XJ34-WE-38 in detail and give the weight on the XJ34-WE-38 in light of those differences. While the -38 version of the J34-WE-32 would play only a minor role down the road, the differences are presented here for completeness of the development record.[110]

The testing of the second XJ34-WE-32, (X24C8-2) designated to be the 50-hour test engine, continued. It was in its eleventh configuration. The high oil consumption had been cleared up. In the current test configuration, the compressor was exhibiting a high degree of stalling in the 11,000 – 11,500 rpm range.

During an oil consumption check run, the compressor had a third stage blade failure in the ball root sector. The blade had a fatigue failure thought to be due to a nick in the extreme corner of the ball root section where the failure started.

Conversion of a J34-WE-32 to a J34-WE-38

Remove:

1. The A/B at the turbine housing aft flange.
2. The fuel and control lines to the A/B including the A/B ignition system.
3. The A/B control (w/built-in filter) and its brackets.
4. The A/B fuel pump and air valve.
5. The lead from the electric control cable which is connected to both the A/B control and the fuel regulator.
6. The power lever control gear box and its bracket.
7. The bulkhead at the forward turbine housing flange supporting the fuel supply lines to the A/B and the thermocouple connection.

Add:

1. The exhaust collector and quick disconnect clamp.
2. The variable-area exhaust nozzle assembly.
3. The exhaust nozzle yoke, linkage, and hardware.
4. A fuel filter, fuel filter bracket and connecting hose assemblies.
5. A new lead from the electric control cable to the fuel regulator. (A connector was available on the control cable.)
6. A cover for the air valve boss on the air bleed manifold.
7. A new bracket to support the power control lever gearbox.
8. A modified power control lever gearbox (cover, shaft, Geneva wheel removed and new cover added).
9. Plugs for the three A/B fuel line holes in the firewall.
10. A bulkhead at the forward turbine housing flange to support the thermocouple connectors.
11. Change the relief valve setting in the oil pump to limit the maximum pressure of the high pressure element.

The dry weight in Specification WAGT-24C8-2D was 1,448 pounds. The WAGT model number for the -38 engine was 24C7.

Figure 7 & 8: J34-WE-38 Mockup Left and Right Views. MUB Report. WAGT P40732 and P40733, 9/30/1949.

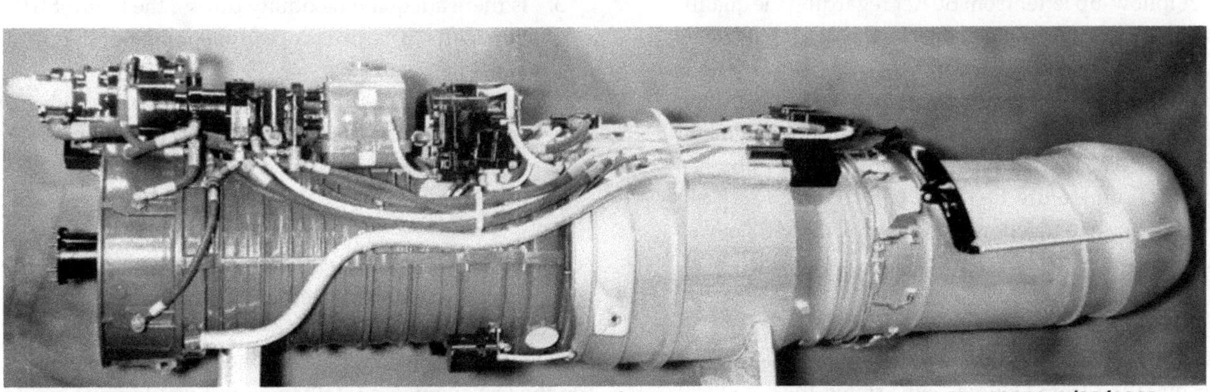

Figure 9: J34-WE-38 Mockup with proposed XF3D-2 bent extension. MUB Report. WAGT P40851, 9/30/1949.

No A/B testing on this engine had occurred as yet. The first engine (X24C8-1) was still showing large non-A/B thrust losses which needed to be improved. It had a total of 85 run hours on the engine at this point.

The second engine was to have all the compressor blades on the third stage and the compressor spindle replaced before testing resumed on or about September 23. The 14% closed turbine nozzles were to be checked to try to improve the engine stall characteristics.[111]

WAGT notified BuAer that the specification delivery for the XJ46-WE-2 was delayed again, from September 12 to October 3.[112]

The monthly progress report for August updated the earlier information. The first -32 engine now had 125 hours of total test time with 5.74 hours of A/B time and 23 hours at military settings. The electronic control was full of problems, there were cracks in the diffuser bellows, and other things breaking. The thrust losses in the A/B and high fuel consumption had not been cleared up. (Remember the YJ34-WE-32 being negotiated reflects the thrust and SFC issues.)

The second engine test status was as reported earlier. The BAR added that the October 6 completion date for the 50-hour test would not be met. No new date was suggested.

The XJ46-WE-2 engine program was being stepped up, with layout and design being pushed to completion and manufacture of parts accelerated.[113]

Convair sent in a special request for an electrical signal on the YJ34-WE-32 that would go to a relay that would trigger a two speed airframe fuel boost pump and asked if it could be provided without affecting interchangeability of the engine. Convair had discovered that their current boost pump lost pressure when the A/B fuel pump began delivering fuel to the A/B and were looking at changing to a two speed pump.[114] This turned out to be anything but a simple change. Providing the connector would create electrical radio interference and delay production of electrical control modules and BuAer disapproved the change. WAGT sent out a survey to all the airframe companies to see if they wanted an A/B operation signal from the engine control.[115,116]

The "official" survey of moving the exhaust nozzle actuator on the J34-WE-32 and -38 was sent out in mid-September.[117]

The last day of September BuAer approved specification WAGT-24C8-2D subject to the inclusion of the paragraph pertaining to oil interruption that had been previously agreed upon. Appendix B regarding the YJ34-WE-32 would be handled as a separate matter.[118]

A follow-up letter from BuAer regarding the qualification test of the hydro-mechanical control elements states they were provided in a letter in July (by reference). It went on to add a new requirement for the shaking of the electrical and electronic components continuously from between 75 to 225 cycles per second during the Endurance Test, Low Temperature Test and High Temperature Test. The frequency had to follow along approximately in phase with the power control lever motion from Idle to Military.[119]

The survey on moving the exhaust nozzle actuator to a new location and redesigning the flexible joint brought a major objection from Douglas, who stated the new location caused a major interference within the wing of the MX-656 (X-3). The flexible joint relocation however was acceptable.[120] Lockheed's response came in a bit later. It praised the actuator relocation, saying it would provide a real improvement from the standpoint of disconnect speed, disconnect cleanliness and interchangeability between afterburning and non-afterburning installations. However, they noted in the potential application in the F-90 airplane, the change would make it more difficult if not impossible to install or remove the lower half of the tailpipe clamp in the space available. If the actuator could be swung down and forward after disconnecting the push rod to the yoke, it would improve things. They asked that "backfire" loads in the tailpipe be actually tested by WAGT on the flexible joint and clamp, and add a quick disconnect clamp on the joint. If the installation space around the joint was approximately the same dimensions as the Pratt and Whitney joint and seal, it would be adequate.[121]

Carrying the discussion forward, Air Material Command next weighed in, stating "....(WAGT) proposes to change from a bellows type afterburner flexible joint to a non-expanding spherical friction joint. The Air Material Command is dubious as to whether such a friction joint can be made tight enough to seal off leakage of fuel and combustible gases during starting and hot exhaust gas thereafter. If clearances are such as to prevent such leakage, this would tend to increase the friction in the joint to an extent to jeopardize its flexibility."[122]

Douglas later accepted the new actuator location and redesigned flexible joint. No reason for the change of position was noted.[123]

BuAer followed up on the Air Material Command's comments by issuing a request to WAGT to provide information on the proposed flexible joint:[124]

1. How the joint operates to take care of expansion of the afterburner longitudinally and radially.
2. How the joint remains gas tight during all conditions.
3. Is there adequate flexibility during the highest friction loads to assure proper functioning of the joint?
4. An enlarged sketch of a cross section of the joint with pertinent notes and without dimensions would enable a more rapid diagnosis of the operation of the joint.

It was not possible to determine exactly which flexible joint design was being tested on the 50-hour flight certification engines during this design process. The author believes it was the new design, since if the old design was on the test engines, a totally new test would have had to be run incorporating the new design.

WAGT responded to the information request in the later part of December. They explained the design and attached a new drawing for BuAer to study.[125] BuAer released the design subject to satisfactory test and operation.[126]

Returning to the project in the mid-October time frame, the progress report for September arrived and started with the X24C8 testing status. The controls were still giving problems and until they were resolved the 50-hour test would not be run. The total test times were:

X24C8-1 Total 145.28, Military 32.70
X24C8-2 Total 33.57, Military 4.60

The blade failure had been reported previously, but no new information was added.

The progress on the XJ46-WE-2 was gradually being stepped up. The BAR added that WAGT was making every effort to meet their schedule.[127] So, for the month of September, not much progress on either the XJ34-WE-32 or XJ46-WE-2 appears to have been accomplished.

In late October, BuAer sent a memo to WAGT saying they knew WAGT recommended quick disconnect fittings on the dual fuel pump, booster fuel pump, governor alternator and overspeed relay, but that no money was available so they could not approve the proposal.[128] Some investigation must have occurred, because at the end of November a memo on this subject was sent to WAGT following a Navy Cost Inspector's cost review. The inspector determined the WAGT engineering and drafting rate should be lowered from $3.45 to $3.23/hour in calculating the cost of the proposal. This indicates that WAGT had submitted a proposal along with their recommendation for the quick disconnects.[129]

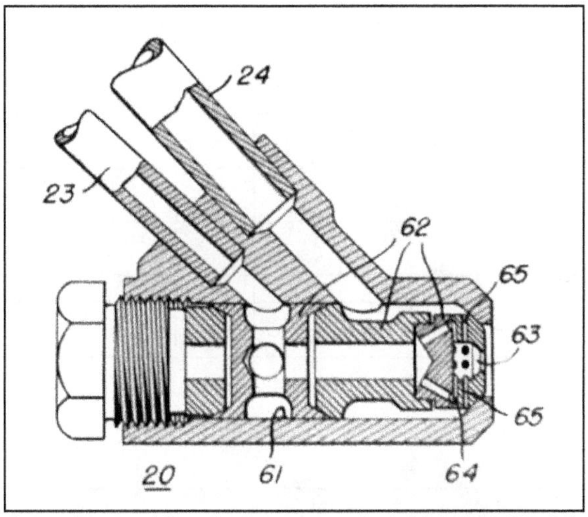

Figure 10: Duel flow fuel spray nozzle. Initial XJ46-WE-2 fuel injection system design. WAGT invention drawing WM5212, 9/6/1949.

BuAer went back to an earlier WAGT response regarding provisions for electric and non-electric starters and cover plates for the machined pad for the electric starter radio noise filter when an electric starter is not used on the J34-WE-32. BuAer requested that a cover plate be provided if the engine was to be shipped without an electric starter.[130]

Trying to cover all the bases on the combustion chamber drain valve issue, BuAer approved removing the forward valve but required the opening be kept and fitted with a plug just in case a requirement surfaced later for a nose down engine start capability.[131]

A few days later, they pre-approved the design clearances for the generator drive on the assumption that no "serious" airframe clearance issues would surface from the survey in progress. They also reminded WAGT that the generator clearance allowed for was based on the dimensions of the General Electric GM76 400 ampere rated generator, but that BuAer had no control over the generator dimensions in the future and that larger generators might be installed to the maximum allowed under the AMD10305 specification dimensions. They asked WAGT to determine if the oil tank pocket could be increased to allow for a 6.5 inch diameter generator at a later date.[132] Douglas responded to the survey saying they had no issues with the redesign.[133]

BuAer approved specification WAGT-24C8-2E for models J34-WE-32, J34-WE-38, and YJ34-WE-32, -38 engines. This approval would allow WAGT to produce a final build standard for those models. The main body covered the J34-WE-32; Appendix A the J34-WE-38; Appendix B the YJ34-WE-32; and Appendix C the YJ34-WE-38.[134]

WAGT got around to surveying the addition of an A/B operation signal on the junction box. Such a connector would be energized from the airframe DC supply voltage through a micro switch in the power scheduler. Movement of the throttle into the afterburner operation range would close the switch turning the indicator on. On the J46-WE-2 a single pin could be added to the control cable assembly junction box. On the J34-WE-32, given its advanced design and parts release status, they proposed either changing the alternate ignition connector to be the A/B signal line, or providing a two pin split of that line to provide both signals.[135]

WAGT suggested qualifying the main fuel pump at maximum fuel flow and pressure with the engine operating at -10°C and a ram pressure ratio of 1.35:1, thereafter determining the actual life of the pump at the fuel flow and pressure required for (engine) operation at -10°C with a ram pressure ratio of 1.5:1.[136] BuAer accepted the proposal in mid-November.[137]

The October activity report for the XJ34-WE-32 testing shows the first engine up to 191 hours running time and the designated 50-hour test second engine approximately 33 hours. Power control issues on the latter engine were preventing the start of the 50-hour test. The controls were still under development on the first engine. The probability of starting the 50-hour test in November was considered small. The XJ46-WE-2 was progressing. Some development tests of component parts and accessories had been going on continuously, but the report had not included a specific assembly start date.[138]

Installation envelope issues with the XJ46-WE-2 MUB Item 10 were addressed with a proposal by WAGT in late November. They advised BuAer that the power regulator and alternator voltage regulator in the present configuration could not be mounted within the XJ46-WE-2 engine envelope due to volumetric considerations. There was insufficient volume and time before the start of the 150-hour qualification test within which to redesign them. In addition, such a redesign was estimated to cost $150,000.

The current ignition coils could be redesigned to a "flashlight" shape (by Scintilla Co.) to fit the envelope for a total estimated cost of $35,742.00. These redesigned coils could be available fourteen months after receipt of authority to proceed, but not before the 150-hour test.[139]

A qualification and production schedule for the J34-WE-32 and -38 engines was presented by WAGT in November and accepted by BuAer later in the month.[140,141]

Current vs. Proposed J34-WE-32 and -38 Qualification and Production Schedule

Contract Noa(s)	Engine Model	Action	1949 S	1949 O	1949 N	1949 D	1950 J	1950 F	1950 M	1950 A	1950 M	1950 J	1950 J	1950 A	1950 S
Present															
9670	XJ34-WE-32	50-Hour QT*	-------	--›10/6											
	XJ34-WE-32	150-Hour QT	-------	-------	-------	-›12/31									
	XJ34-WE-38	50-Hour VT**					No firm date in time to meet January deliveries.								
9791	YJ34-WE-38	150-Hour VT								No firm date in time to meet April deliveries.					
	YJ34-WE-32	50-Hour Prod.			2	4	4	16	20						
	YJ34-WE-38	50-Hour Prod.					4								
	J34-WE-32	150-Hour Prod.								12	22	25	31	40	
	J34-WE-38	150-Hour Prod.								16	18	22	30	35	
Proposed															
9670	XJ34-WE-32	50-Hour QT*	-------	-------	-›11/26										
	XJ34-WE-32	150-Hour QT						‹------	-------	------›					
	XJ34-WE-38	50-Hour VT**				‹--	-------	------›							
9791	XJ34-WE-32	50-Hour DR-Prod.***			2	2	2#	0	4#				Hold to a minimum.		
	YJ34-WE-32	50-Hour Prod.								16	20		Hold to a minimum.		
	YJ34-WE-38	50-Hour Prod.						4							
	J34-WE-32	150-Hour Prod.										17	31	40	40
	J34-WE-38	150-Hour Prod.								16	18	22	30	35	

* QT – Qualification Test
** VT – Verification Test
*** DR-Prod – De-rated Production
\# - Not to be shipped unless BuAer requested delivery 30 days ahead of indicated schedule.

WAGT surveyed a final location for the XJ46-WE-2 exhaust nozzle actuator and A/B flexible joint in mid-November. The actuator attachment point was moved from the A/B diffuser to the aft end of the combustion chamber. The A/B flexible joint was changed from a bellows type to a friction type with the capability of deflecting up to two (2) degrees in any direction about the center of rotation of the joint.[142] These design changes mirrored the changes on the J34-WE-32 for the same components.

The BAR had to write the BuAer Chief in early December to notify him that the first prototype XJ34-WE-32 had been wrecked at about the 200 hour point when a leak in the oil cooler allowed gasoline to mix with the lubricating oil and contaminated the No. 2 and No. 3 bearings. The turbine shaft and turbine were badly damaged when the stub-shaft sheared as a result.

The second XJ34-WE-32 would have to be used to develop the controls. As a result, the start of the 50-hour test on the XJ34-WE-32 would be delayed until at least the latter part of December and the XJ34-WE-38 schedule slipped similarly. If the control issues were not resolved, the schedule would slip further.[143]

In response to BuAer's earlier direction to re-write the Specification for the XJ46-WE-2 to the new "Mil-Specs" MIL-E-5007 through MIL-E-5010, WAGT responded with a revised development program based upon Contract Noa(s) 9670. The additional cost would be $694,503.00 with a six (6) month extension to the delivery schedule of the 50-hour engine and a six (6) month slip in the production delivery as well.

The primary reason for the cost and time was the change in the severity of the qualification test requirements.

The high temperature limited life components of the engine would be subjected to two and a half times as much operation at the limiting temperatures. Accelerations which were apt to cause severe distortions were increased four times over the prior "AN" testing requirements.

They stated that there were two possible development approaches to build longer life into the engine. The first was to reduce stress levels throughout the engine on all parts having limited life. All parts designed on a creep basis (such as the hot turbine parts) would increase in weight. The second approach was to obtain increased life by refining the original design through experience (repeated redesign of failed parts) until the increased life was obtained in the engine. They stated this could be done without "detriment to engine performance and weight". The approach would require additional engines, engine test time and additional engineering time for analysis of testing and redesign of parts.

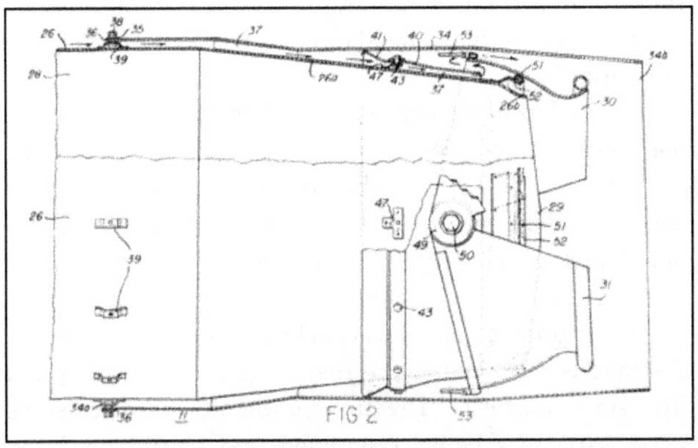

Figure 11: J34-WE-32 Exhaust nozzle design layout. WAGT invention drawing WM5092, 9/23/1949.

Finally, WAGT stated the reliability experience obtained from service operation of the early J34's and J30's served to greatly lessen the cost of the proposed XJ46 program. (Exactly how was not specified.) And lastly, such experience would not benefit certain phases of development of high temperature parts and control. *(So, the prior experience just stated does not assist in the development of improvements in the high temperature areas, and WAGT has neatly introduced the control development issue indirectly.- Auth.)*[144]

The simplified emergency fuel systems for the XJ46-WE-2 and XJ40-WE-8 engines could be developed for an estimated $170,685 on the former and $29,495.00 on the latter according to a WAGT proposal in early December.[145]

In a follow-up to BuAer's decision to disapprove the proposed YJ34-WE-32 method of obtaining an A/B operating signal, WAGT now stated that they had sent out a survey asking for the need for such a signal in early November and that they would take no further action unless requested by BuAer.[146]

The issue of generator drive clearance on the J34-WE-32 and -38 were finally resolved with a BuAer memo stating that the generator on the J34-WE-32 for the F7U-1 would not seriously interfere with the scavenge oil line. On the Douglas F3D-1, the adapter should be retained on the J34-WE-38 engine, (but removed on the J34-WE-32). They requested WAGT submit revised drawings of the -38 engine generator clearance envelope showing the effect of the one (1) inch lowering of the oil scavenge line on the F3D-1 clearance.[147]

The eventual investigation of the reservoir to provide for generators of 6.5 inch diameter showed that the oil capacity would be reduced to the point where the oil system would be marginal in operation. Redesign to recover the capacity would be extensive and most development testing would need to be repeated, a total process that could take as long as six months. Pattern and casting changes on the compressor inlet and outlet housings would have to be made as well. WAGT, in an effort to partially support the larger generator requirement, had reduced the generator mounting flange from its 2.0 and 2.25 inch cut-back dimensions to 2.0 and 1.25 inches respectively. BuAer could also change the requirement for a fully rotatable blast tube (air spout) to limit it to generators of a length of 13 inches or more as shorter generators could be removed without touching the oil reservoir.[148]

BuAer approved specification WAGT-24C8-2E for the J34-WE-32 and -38, requesting only that the specification engine weights be changed to read 1,689 pounds for the -32 and 1,439 pounds for the -38. The changes reflected earlier approvals of weight additions and subtractions resulting in a net minus nine (-9) pound change in both engines.[149]

The responses related to the A/B operation signal started coming in and pointed out that more discussion was needed. Uniformly, the airframe contractors thought the signal should indicate that the A/B was ACTUALLY operating and by how much (remember, the A/B fuel control on the J34-WE-32 and J46-WE-2 was to be a fully modulated capability). The survey proposed a signal that only showed that the A/B SHOULD be operating and not by how much.[150,151] Vought was OK with the signal as proposed, but wanted the connector put elsewhere as hooking up to the proposed connector location was a problem for them on the F7U-1 and F7U-2. (This goes back to efforts to provide thrust measuring kits on the early J34 engines.)

In mid-December, WAGT submitted a proposal to develop a short afterburner model X23C10M of the J46 for $780,032. The bid assumed the proposals for MIL-spec development and development of the simplified emergency fuel system would be accepted and that the use of machine tools and special tooling currently being used in the basic XJ46 development program would be used.[152] No indication of the source of the requirement was given.

XJ34-WE-32 50-Hour Verification Test

The wrecked first XJ34-WE-32 engine was rebuilt and prepared to undergo the 50-Hour Flight Verification Test. A deviations list from the planned first production engine had been compiled before the engine was wrecked and it would have to be updated prior to the test, however the changes were expected to be minor. The first list had noted 22 differences from the production engine.

WAGT stated, "It is the contractor's opinion that the above listed differences are minor only and that the durability to be expected of the test engine is similar to or less than that which may be expected of the production units. As detailed in reference (a), the engine test will be under slightly de-rated conditions and will be applicable to the first ten production engine units."[153] In other words, the unit would be tested to the YJ34-WE-32 de-rated specifications. The first XJ34-WE-32 (X24C8-1) would be used for the test and the second for continued controls testing.

BuAer was now under pressure to supply engines to Vought to initiate ground and possibly limited flight testing. The BAR sent a wire to BuAer HQ asking permission to accept two XJ34-WE-32 engines before the 50-Hour test on the engine was completed. (They would have had to be to the YJ34-WE-32 performance specifications at that point.) The written notes on the distribution sheet show that the Power Plant Division (PPD) objected, stating that design deficiencies still existed in the engine and that they were to be installed in actual F7U airframes. They recommended the request be denied. The PPD commander communicated the denial to the BAR by telephone and no written denial was issued.[154]

A conference was held in early December and at that conference WAGT informed BuAer that the J34-WE-32 could only meet its specification performance with an afterburner designed with an internal shroud or liner and without an ejector cooling shroud. It was stated the guarantees could not be met when an ejector shroud was required.

(The reader is directed to NACA Research Memorandum RM E51J4 dated December 17, 1951, Full-Scale Investigation of Cooling Shroud and Ejector Nozzle for a Turbojet Engine – Afterburner Installation, Lewis E. Wallner and Emmert T. Jansen. The report shows how little prior research had been done in the subject area prior to the report. The illustrations clearly show a J34 of unidentified model attached to a clam-shell type afterburner, but the engine and afterburner manufacturer details are not given.)

BuAer asked that since an earlier J34 model had met its performance specifications with an ejector shroud, they wanted a detailed explanation of why it was not possible in this instance.[155]

WAGT responded a month later, stating "The engine designed with an afterburner on Contract Noa(s) 5382, XJ34-WE-11, was guaranteed to have a thrust that was 30% higher than the basic engine, the J34-WE-30, to which it was added. The J34-WE-32 engine is guaranteed to have a 40% boost over the thrust of the basic engine. The afterburners are quite similar. It was possible to obtain the 30% boost on the -11 afterburner but has not yet been possible to achieve the 40% required on the -32 afterburner." At a conference, the development efforts on the -32 afterburner were reviewed and it was agreed that WAGT would advise BuAer by January 25, 1950 of the performance that the -32 afterburner configuration could produce.[156]

The monthly progress report stated the reassembled XJ34-WE-32 could start the 50-hour test by December 30 if the control system problems were resolved. The 500-hour sea level qualification test on the dual fuel pump, booster pump and fuel filter were completed but teardown showed a spline tooth failure in the dual pump (gear box side). A spline of higher core strength pump was being assembled and testing repeated. The XJ46-WE-2 was "progressing satisfactorily."[157]

A mid-January wire was sent from the BAR updating the J34-WE-32 and -38 testing status. As it turned out, the second XJ34-WE-32 originally intended for the 50-hour test and which had been testing the control system, had a turbine failure on January 7. It was caused by a faulty outer nozzle ring which was oversized and allowed the nozzle outer shroud to rotate, thus causing the vanes to drop out. Damage was limited to the first stage turbine blades and first stage nozzles. The first production engine (serial WE023001) was being assembled and would be called 24C8-3 and be used for the 50-hour test. Use of the latter engine would delay deliveries of production engines on contract Noa(s) 9791.[158]

A memo addressing a serious design control process disconnect between the WAGT design engineers and BuAer was sent to WAGT on the subject of status of a proposal for a simplified fuel system development. In summary, it stated that while BuAer agreed with the recommendations on ways to simplify the XJ46-WE-2 and XJ40-WE-8 fuel systems, no funding was available for such development work. It had come to the attention of BuAer that current development work on the fuel systems of those engines was based on assumed future acceptance of the proposed simplification changes and that when such assumptions came to light, WAGT engineers and representatives had been cautioned not to assume the bureau could or would negotiate a procurement. They recommended WAGT should review their designs for the subject engines to determine which components depended on the simplified fuel system.

BuAer's failure to accept WAGT's proposal on the simplified fuel system was not considered justification for delays in the respective engine developments.[159]

The December activity report only added the information that the 50-hour and 150-hour tests on the XJ34-WE-32 and XJ34-WE-38 had not begun during December as expected. The -32 was experiencing vibration and the turbine was found to be out of balance. Control problems were preventing further testing of the XJ34-WE-38. They expected the -38 test to start during the week of January 23rd. Qualification testing on the -32 and -38 engine accessories was underway. The ignition system was qualified with the exception of the moisture test, which would be conducted during the 150-hour test. The (modified) fuel pumping system was undergoing the altitude test. The XJ46-WE-2 was expected to begin testing in March.[160]

WAGT surveyed the location of the drain ports for the Afterburner Air Turbine Fuel Pump Shroud on the YJ34-WE-32. They would be located 60 degrees either side of the horizontal centerline of the pump on a 3.25 radius whose center was on the horizontal centerline of the pump and 11/16 inches aft of the vertical centerline of the Fuel Inlet Flange. The shroud could be rotated and if rotated to a position other than that shown on the referenced drawing, a(n) (airframe) drain which falls at the lowest point should be utilized.[161]

Closing the XJ46 MUB Item 53, BuAer approved the fire seal installation plan along with the two pound weight increase for the fire seal itself.[162]

BuAer notified WAGT that they accepted the stated deviations (January 23, 1950 memo) on the J34-WE-38 engine for the 150-hour test with two exceptions: 1) any engines shipped with an earlier fuel scheduler that could not meet the fuel consumption guarantee at the No. 2 cruise point would have to be replaced by WAGT later at no cost to the Government; and 2) the one and a half pound overweight condition of the test engine would not be accepted on production engines.[163]

At BuAer's request, WAGT had reviewed a list of possible generators for use on the J34-WE-32 and -38. They sent their comments to BuAer and reminded them that the generators to be used on the engines would be running at continuously high speeds due to the high continuous operating speed of the model engines in question. That speed was 2/3 of the engine speed of 12,500 rpm.[164]

XJ34-WE-38 150-Hour Qualification Test

The XJ34-WE-38 started its 150-hour qualification test on January 30. The test was conducted with AN-F-48 grade (91/98) avgas and AM-O-9 Grade 1010 oil. At the start, the engine had accumulated 20 hours of operating time. The speed governor head was leaking oil at the seal and a check showed the splined shaft was too long. A check of the drawing showed the spline had been machined to the incorrect length. The shaft was ground off and reinstalled. No leak was observed on the test stand after the repair but after installation on the engine, the leak continued, so a new governor was installed. Several preliminary starts and accelerations were made before the test was started with the engine time at 20:50.

Through February 8, the engine had completed seven of the fifteen ten-hour periods of the test. Mechanically, the engine had performed satisfactorily. It was operating at a peak turbine inlet temperature of 1,525°F ±40°F at the military rated position.

The regulating screws on the power regulator had to be adjusted to keep the military speed within the model specification requirements and prevent the turbine outlet temperature from drifting. The BAR anticipated more work on the power regulator was still going to be needed to eliminate temperature and engine speed issues.

At the 70 hour point, the engine was shut down for test cell adjustments, at which point engine oil smoking was noticed. That would require an investigation as to cause.

The pre-test radio interference test had been run and was satisfactory. The test would be repeated at the end of the 150-hour test.

At the 100 hour point (on February 10), the engine was mechanically sound. Everything was operating satisfactorily except for the power regulator. It continued to require occasional adjustment to prevent the turbine outlet temperature from drifting.

The minor leaks in the governor and the alternator drive gearbox were thought to be due to out of specification parts and loose fittings respectively.

Very low oil consumption was observed, being "negligible". Thrust and SFC's were within the guarantees.

The engine oil smoking was found to have been caused by the mixing of AN-F-58 and AN-F-48 fuels. Such smoking had been experienced in previous testing with AN-F-58 fuel. *(This curious explanation begs many questions, such as how did mixed fuel and/or AN-F-58 fuel get into the engine during this test? If it did, were the test results valid? These were not addressed by the BAR in his report.-Auth.)*

At 130 hours (February 15), engine operation mechanically was still satisfactory. Oil consumption continued to be very low with thrust and SFC still meeting guarantees except for the No. 2 cruise point (which was known to be a problem prior to test). Acceleration time from idle to military speed (rpm) was averaging three to six seconds and idle to military thrust thirteen to sixteen seconds.

(I'll leave it to the pilots reading this to think about a throttle slam on approach to landing and having to wait for up to sixteen seconds for the power to arrive. – Auth.)

But there were problems. At the 125th hour the electric control relay assembly refused to switch over to emergency. The relay was a hermetically sealed unit but upon opening the box loose components were observed. The vendor had been contacted about the problem and the needed improvements. The relay would have to be re-qualified and a re-test had already been started.

On February 15 the engine completed the 150-hour test. The engine had 182.81 running hours on it at the completion. Mechanically the engine was satisfactory. The engine thrust and SFC guarantees were all met. Radio interference checks at the end of the test showed 10% of the tested frequency points did not meet specifications. A total of 84 engine starts were accomplished vs. the 75 required.

Both the power regulator and electrical control would need to be re-qualified. On tear-down, other noted failures observed were:

1. The combustion basket had two spot welds broken loose on the inner annulus and a small crack on the inner side of the outer annulus.
2. The spark plugs showed deterioration. Metal tubes were partially burned off and the porcelain tip broken off on one plug.
3. The ignition leads showed deterioration. The porcelain filament had broken off and part of the insulation remained in the plug.
4. A small crack was visible in the area of the cooling air holes in the combustion liner.
5. The mounting trunnion in the exhaust collector was cracked.
6. The exhaust nozzle was cracked in several places.
7. The hinge-point on the eyelet nozzle was cracked.
8. The first turbine stage nozzle (50 vanes) was cracked and bent, with some vanes showing erosion and pitting.
9. First stage turbine blades – two blades were cracked on the leading edge and another blade was cracked on the trailing edge.
10. Second stage turbine blades – four blades showed shrinkage in the vicinity where the casing was "gated".
11. The dual fuel pump spline shaft was badly chewed up.
12. The second stage compressor diaphragm showed minor cracking on the outer shroud.
13. Accessories were all in good condition.

The BAR considered the test successful. [165,166,167,168]

Later in March, WAGT submitted report A-949 covering the details of the test and the results. Seven action areas were detailed with correction plans for each. They were:

1. First stage (turbine) nozzle vane cracking was to be addressed by a program to lighten the gas loads on the first stage vanes.
2. The variable area exhaust nozzle displayed issues on disassembly and was redesigned. It would have to be requalified and introduced with engine WE024005.
3. Oil leaks at the starter pad were found to be caused by the "O" ring, which was to be relocated to prevent future leaks.
4. Excessive wear on the spline shaft of the dual fuel pump would be corrected by using a harder and stronger material for the shaft.
5. The electric control relay failure was traced to vibration wear of the relay armature at its pivot points. Harder materials and a spring would be added to hold the armature in position securely at all times.
6. Slight high temperature drift of about 50°F of the electronic power regulator occurred as well as improper tail area scheduling. An investigation was underway as to the cause(s).
7. The spark plugs and leads which were used were the same as those used on the J34-WE-30 and -34 without experiencing the failures exhibited on the test engine. An investigation was underway to determine the cause and attempt to duplicate it.

WAGT recommended the engine be accepted for production.[169] It was produced under contract Noa(s) 9791. We will leave off the J34-WE-38 engine at this point, as the minor changes to the J34-WE-32 from that point forward that would affect the J46-WE-2 are the only ones of interest herein.

While the J34-WE-38 test was underway, discussions on the XJ46-WE-2 power regulator were continuing. BuAer, noted that even if the voltage regulator and ignition coils were redesigned as WAGT proposed, they would still have to be mounted on the airframe. Based on that they rejected the proposal.[170]

Revised performance guarantees for the early J34-WE-32 engines (to be called YJ34-WE-32) were sent over in February.[171]

Revised J34-WE-32 Performance Guarantee Ratings at Standard Sea Level Conditions				
Rating	Thrust (lb)*	J34-WE-32 SFC** Guarantee	YJ34-WE-32 SFC** 9/30/49	YJ34-WE-32 SFC** 1/25/50
Take-off (5 mins.) (A/B)	4,900	2.60		
Military (10 mins.)	4,700		2.60	
	4,800			2.60
Military (30 mins.)	3,370	1.08		
	3,150		1.20	
	3,270			1.135
Normal	3,020	1.03	1.12	1.08
Cruise 1	2,730	1.00	1.075	1.05
Cruise 2	2,260	.97	1.04	1.03
Idle	205	5.70	5.70	5.70

*Design values subject to revision as recommended by contractor and approved by procuring service, prior to qualification test.

**Heat of combustion assumed to be comparable with presently available fuel specification AN-F-48 grade 100/130 namely an average of 18,800 BTU/lb liquid fuel at 77°F.

A survey was sent out in February to incorporate main fuel regulator drains on the XJ34-WE-32, J34-WE-38 and XJ46-WE-2. These drains would be plugged on shipment. The design functioned properly if the back pressure at the drain was always two pounds per square inch lower than the inlet total pressure. Normally, no fuel would flow from the drains, but up to ten (10) drops per minute was allowable. Preferably the drains should be piped overboard or else a collection tank could be used if the back pressure requirements could be met. Drainage after shutdown was estimated to be approximately .25 of a pint. The location of the drains on each engine model was described. (This is the first memo where copies were sent to the Air Force Air Material Command, reflecting the fact that procurement of an Air Force version of the XJ46-WE- 2 was being negotiated. This would be designated the XJ46-WE-1 and is described in Chapter 8.)[172]

The survey responses regarding providing an A/B operation signal were reviewed by BuAer and they withdrew the requirement in late February.[173]

Unexpectedly, BuAer issued a letter notifying WAGT officially that funds allocated to the contract would be expended by approximately April 15,1950. They requested an estimate be forwarded of the total expenditures required to complete the contract.[174]

The Second XJ34-WE-32 50-Hour Test

The engine completed the test on April 1, 1950. During the test:

1. The compressor radial retaining screws that held the compressor diaphragms (stator segments) in position broke off repeatedly.
2. The power regulator had to be adjusted on three occasions due to temperature and rpm drift (also encountered on the XJ34-WE-38 150-hour test).
3. Three air turbine pumps had to be replaced. A new design by Thompson Products was about to start testing on or about April 5.
4. At 42.5 hours, the boost pump failed due to a cracked bellows. Although this pump design had performed satisfactorily in over 800 hours of lab testing, plans were made to redesign it with a shorter bellows in an attempt to reduce the chance of such failures.
5. The emergency scheduler had to be cleaned at 42.5 hours as the tail *(clam)* shells would not close properly. A small piece of rubber was found blocking the orifice opening. After removal, the tail operated correctly again. (Nothing was said about the source of the rubber or how it could have gotten into the scheduler.)

6. The A/B fuel line quick disconnects were replaced due to leakage. The cause was found to be the "O" rings taking a permanent set which destroyed their sealing qualities if the quick disconnects were opened and then reconnected.
7. At the post test calibration, it was found the engine could not be calibrated when the emergency control was operative because the nozzle eyelids would not close.
8. The eyelid seal became distorted and warped as the test progressed because hot gases were leaking past the seal. The leaks affected the SFC's and the non-A/B Military rating could not be met. Pre- and post-testing calibrations showed all other SFC's performance were met except the non-A/B Military rate. The deviations from the de-rated YJ34-WE-32 performance projects were slight.
9. Other than the problems listed, the engine was mechanically sound and performed satisfactorily.

The fire seals were further discussed based on questions the Air Force raised about the proposed final design. The questions regarded:

1. The adequacy of the seals between the engine and the fire seal under fluctuating temperatures,
2. The ability of the clamp to be properly torqued and remain torqued under fluctuating temperatures,
3. The ability of the seal to withstand fore and aft structural and aerodynamic forces without slipping on the combustion chamber, and
4. The likelihood of "wicking" occurring in the asbestos impregnated metal mesh seal thus causing an additional fire hazard.

BuAer considered the biggest question to be the third item as it depended on the loads the airframe placed on the seal. The MX-656 project (the Douglas X-3) was expected to produce the highest such loads. Repositioning of the fire seals on the engine design was not anticipated to be required. Any service issues would be addressed as they were uncovered.[175]

In early April, BuAer approved the relocation of the exhaust nozzle actuator and new flexible joint redesign for the XJ46-WE-2.[176]

Back in December WAGT had surveyed a change in the engine clearance. They wanted to change the turbine housing and afterburner diffuser on the XJ46-WE-2 engine to utilize the air-cooled turbine housing feature used successfully on the XJ40 engine design. The change would allow closer tip clearance control of the turbine blade tips. The relocation of the exhaust nozzle regulator on the combustion chamber was included in the change and the spherical friction joint used was similar to the one used on the XJ34-WE-32 engine.[177] BuAer approved the change in April provided it had no effect on engine cost, weight or development schedule.[178]

WAGT requested and received approval to transfer charges for two development projects started under contract NOa(s) 9051 to the NOa(s) 9670 contract in April. The projects were related to fuel system development with a variable exhaust nozzle and control system development.[179]

In response to a WAGT survey (Engine Change No. 56) of a redesign of the A/B ignitor valve for the YJ34-WE-32, the Douglas response was a bit confusing, as it stated the change in design, "as applicable to the XJ34-WE-17 engine, will not affect this Contractor's current design". This appears to be a typo in the text, as the XF3D-1 was initially planned to be flown with modified model J34-WE-11 engines called the J34-WE-17 and J34-WE-19. It was actually flown with the J34-WE-24 and J34-WE-26 models of the J34-WE-22. These had their accessory sections rotated in mirror image directions to clear airframe obstructions. The author presumes the writer meant to say "the replacement for the XJ34-WE-17".[180]

The March progress report stated the XJ34-WE-32 engine was in build-up and would be ready for test on May 4, 1950. (Thus demonstrating the importance of timeliness of the monthly progress reporting process, where the March report was not received by BuAer until April 21, making it useless for management decision making. The test was long over then and the results reported via other vehicles.)

Various component qualification tests had been completed on the dual fuel pump, which showed spline shaft wear. BuAer temporarily accepted the pumps until an improved shaft model was available. The booster pump showed similar wear after 1,100 hours of testing. The new design bellows on the booster pump needed to go through a 500 hour test which was to start immediately.[181]

XJ46-WE-2 Testing Begins

The March progress report stated the initial XJ46-WE-2 had been built and run. After two hours, the engine was damaged when one of the first stage radial retaining screws broke and went through the engine. Seventy five percent of the compressor blades had been damaged. Screws in the 6th and 7th stages had also broken but did not go back through the engine. The screw size was to be increased from No. 10 (.190 inch diameter) to .250 inch diameter size. It was planned that the engine would resume testing the first week of May.

The survey on the main fuel regulator drains was modified after objections to the high allowable back pressure. WAGT modified the allowable back pressure on the drains by lowering it to 1.5 pounds per square inch and the survey was sent back out.[182]

BuAer felt the need to clarify their requirements for fuel system drains in general. It stated there were two types of drainage provisions for engines:

1. Drainage from which known relatively large quantities of fuel were normally expected to drain at some point in the operating sequence. The fuel manifold dump valve fell into this category. Drainage from this valve at engine shut down was not considered indicative of malfunction.
2. Drainage provisions for which relatively insignificant quantities of fuel were expected during normal operation. The fuel pump seal drains, overspeed relay seal drain, and fuel regulator fell into that category. Drainage of significant quantities of fuel from those drain provisions would be considered indicative of a malfunction of seal provisions.

Dump tanks should be used for category one type drain provisions. Overboard drains would be satisfactory for the category two type drains.

Afterburner drains were difficult to place in either category as the drains might discharge large quantities of fuel after a false or aborted start. It was not clear that running a "dry" (no fuel) start cycle would clear away most if not all of the surplus fuel. The early experience with A/B equipped J34's indicated a need for drain valves. The Naval Air Test Center was reporting the drain on their test J34-WE-34 was not needed to clear fuel, but was useful in draining excess oil coming from the oil drainage system after shutdown.

One airframe manufacturer had questioned the possible fire hazard of the drain piping through the fire seal to the dump tank, connecting to a tank that might be full of fuel and vapor. The contract recommended the drain be made to drain overboard directly.

In view of the concerns, BuAer stated the following combustion chamber and A/B provisions were now required:

1. Provision should be made for manually draining the combustion chamber and afterburner drains by the ground crew after false or aborted start attempts or to check for liquid accumulations after prolonged shut down. Suitable manual drain valves and access provisions were required in the airframe installation.
2. Deletion from the engine of automatic combustion chamber drain valves and replacement with a plugged boss. The proper boss was to be selected by the airframe contractor according to the ground angle of their particular installation.
3. Deletion of the requirement to connect the combustion chamber and afterburner drains to the airframe dump tank.[183]

The design of the safety locks on the rear trunnions of the engines showed that the current WAGT design was not acceptable to Douglas, as the locking stud had to be in place when the engine was installed on the Douglas airframes or the stud projecting out would block the installation. Review of Chance Vought showed no issues with the design and based on which airframe company was to use which engines, it was determined only the J34-WE-32 engines needed a new design.[184]

Taking up work that was started back on earlier versions of the J34-WE-22, at BuAer request, WAGT produced a proposal to convert the J34-WE-32 and -38 engines to use the "standard" throttle arrangement developed earlier. The J46 and J40 models would all enter production using the standard arrangement. The J34-WE-32 and -38 would require modification to use the standard arrangement. Four specific changes had to be made:

1. The power scheduler would be changed so that:
    a. The primary power control and the automatic ignition control switch were closed in the range between approximately 9° and 13° from the "off" position.
    b. The connection from the aircraft DC bus to the power control should be closed, or opened, depending upon the direction of the power lever movement, in the range between approximately 11° to 15° from the "off" position.
2. The electric control relay assembly was to be changed so that the ignition relay was normally open instead of normally closed.
3. The power regulator was to be changed so that the ignition control relay contact arrangement would match that of item 2 above.
4. The electric control cable would be changed by the addition of a jumper in the main junction box to provide an alternate DC input connection directly to the power control, by-passing the switch in the power scheduler.

The modified throttle quadrant arrangement was proposed to be incorporated on the 47th J34-WE-32 and approximately the 75th J34-WE-38. The development cost was $20,267.00 with no additional production cost.[185]

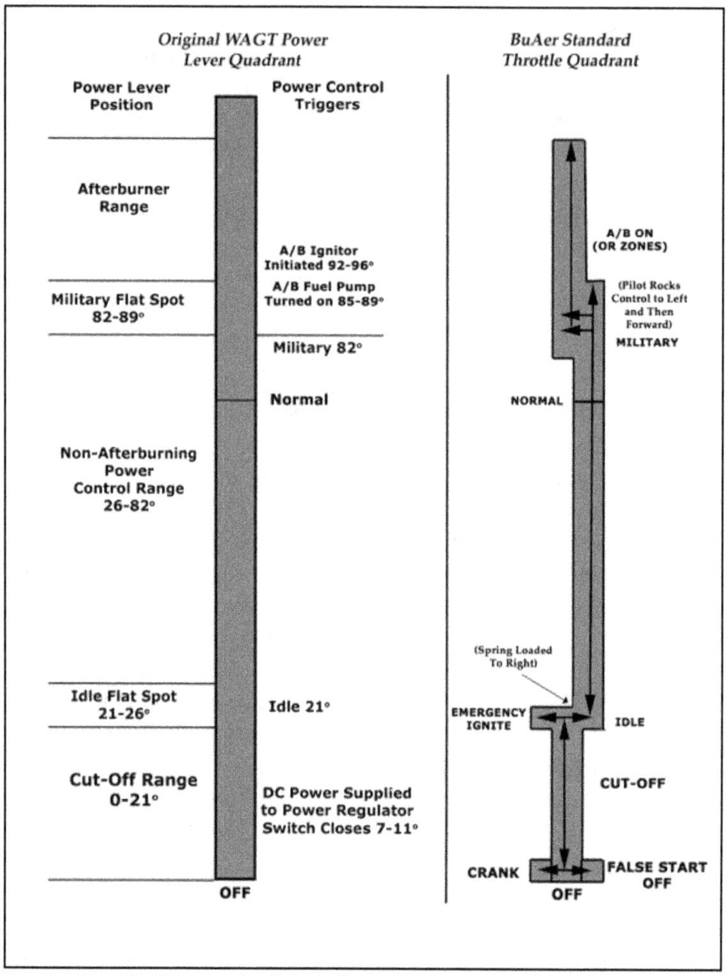

New BuAer standard throttle and power control arrangement based on WAGT analysis. WAGT Memo May 1, 1950. Figures 1 and 2 (combined by author)

In an attempt to clear up all the surveys related to drains of all kinds, WAGT sent over a memo with the following status reporting on all open surveys related to the subject:

1. For the XJ46-WE-2, since it was concluded no weight change would accrue from removing the front fuel drain and leaving a plugged boss, WAGT elected to leave the front drain in place and not generate any drawing changes. BuAer was asked to disapprove the original survey.
2. The survey covering the relocation of the exhaust nozzle regulator and the afterburner flex joint was approved by BuAer on April 12.
3. The survey on changing the main fuel regulator drains was approved by BuAer on April 12. The survey had been issued after the fact since the drains change had been included on the mock-up review and was technically not needed. The survey had been issued to determine any back pressure requirements and estimated drainage quantities. Based on the feedback received, (reviewed with BuAer previously), approval of WAGT's proposed provisions was requested.
4. The drain ports on the afterburner air turbine pump shroud were surveyed primarily to satisfy a request from Douglas. Since a new pump design was in the works, the drain ports design would be surveyed again with the Change in Design for the new pump.
5. A pressure operated dump valve was recommended as essential for safe operation of the engine. Approval by BuAer was requested.
6. The survey covering changes to the A/B ignition system was approved by BuAer on November 18, 1949.
7. The survey to add two jack pads to the No. 2 bearing housing and the resulting recommendations

were submitted to BuAer September 16, 1949. WAGT records showed no response from BuAer had yet been received and approval action was requested.[186]

It is clear that attempts were being made to get all the paperwork involved in the approvals process up to date.

By mid-May BuAer was trying to get production underway for the XJ34-WE-32 and sent a memo to WAGT stating that there was no requirement for YJ34-WE-32's to be built under the production contract NOa(s) 9791. Instead, eight XJ34-WE-32 engines could be delivered with the agreement that WAGT would later rebuild those engines to the J34-WE-32 standard established by the 150-Hour Qualification Test. The XJ34-WE-32 engines would have operating restrictions placed on them and Vought was so informed. The satisfactory completion of the first 50 hours of the 150-Hour Qualification test (determined after a teardown of the engine at that point) would be considered completion of the contract requirement for the 50-Hour test. BuAer asked WAGT to give them an estimate of the cost savings of having eliminated the separate 50-Hour flight verification test.[187]

On May 31, 1950, Amendment 5 to the contract was signed. It changed the General and Administrative cost rate in the contract from 2.5% to 2.787%.[188]

In a follow-up to a BuAer request for improved boost pump requirements data for the J46-WE-2 and J40-WE-8, WAGT responded saying the J40-WE-8 requirement was double that of the J46-WE-2, which was very similar to the J34-WE-32. The J46-WE-2 was going to use a pump almost identical in performance to the J34-WE-32 and it was assumed the current in-tank boost pumps available to airframe manufacturers would be adequate. When service experience with the J34-WE-32 was available, the issue would be re-examined. WAGT's plan for this item was considered satisfactory by BuAer.[189,190]

Drains were back on the table again in June when BuAer asked the cost of replacing both the combustion chamber drain valves with bosses.[191] Weeks later WAGT changed the boss threads to a non-straight design.[192]

On June 5, BuAer initiated contract Amendment No. 6 to pay for the development of the J32-WE-32 and -38 to use the standard power quadrant.[193]

WAGT continued to struggle with control system problems all through June and July on the J34-WE-32. The 150-hour qualification test was started on August 7. It was stopped after 31 hours due to failure of the oil feed line to the Nos. 2 and 3 bearings. The failure was at the weld point which permitted oil to flow freely causing an oil fire at the No. 3 bearing which wrecked the No. 3 bearing seal and damaged the first stage turbine nozzle vanes. Until that failure, the engine had operated satisfactorily. Afterburner thrust was low by 50 pounds below the 4,900 pound thrust specification. The low cruise SFC was above specification, a known factor still not resolved.

However, the next statement of the BAR report backtracks on its "satisfactory performance" comment and reported that while on primary control, going from non-A/B into A/B, the engine had a "stall" that was considered very serious and WAGT had been so informed. No power regulator drift had been observed. It was considered that the addition of "shock" mounting of the tubes in the control unit which control the temperature and speed was the reason for the improved behavior of that part of the control.

The engine had accumulated 80 hours of running prior to the start of the test, having accomplished 340 starts at that point. In green and acceptance testing, some elements had to be replaced due to poor performance: ignitor valve (leakage), A/B shut-off valve (leakage), booster fuel pump (bellows failure), power regulator (drift of speed, temperature and fuel hunting), alternator drive gearbox (oil leak in alternator), main fuel valve and actuator (twice-erratic performance due to armature failure). The test engine had the old style armature installed which was not causing trouble. A new armature design was under parallel test. Oil scavenging was insufficient, with oil building up in the gearbox (a known problem). A production engine (WE024017) was being used to clear up that problem.

All elements which needed further verification based on the results of the XJ34-WE-38 type test (except ignition wire leads and plugs) were incorporated into the engine.

It was planned to restart the test using the controls from the present engine on a new engine in the latter part of August.[194]

The progress came to a halt with the receipt of test problems reported by NACA. The NACA (National Advisory Committee for Aeronautics) sent over results from their high altitude wind tunnel testing of the XJ34-WE-32. These reported that the engine suffered from adverse temperature distribution in the turbine section when the engine was operated above 25,000 feet. The inversion problem was severe enough to shorten the life of the turbine blades and lower performance at altitude. In the NACA report (not issued until November 29, 1951) it was stated "the maldistribution of turbine temperatures was primarily attributed to adverse radial-velocity gradients at the compressor outlet (combustor inlet). In order to improve the compressor-outlet velocity profiles, the engine manufacturer designed and fabricated a stage of stationary mixer vanes that was installed directly downstream from the last stage of straightening vanes at the compressor outlet." The "installation of the mixer vanes produced a more favorable compressor-outlet velocity gradient which, coupled with increased turbulence, greatly improved the turbine radial temperature distribution and, unlike the engine without the mixer, the distribution was not appreciably affected by changes in flight condition, corrected engine speed, or temperature level." Further down in the report, it was noted that the location of the thermocouples that reported temperature to the power controller was in the wrong location when the temperature inversion occurred, so the power controller was limiting the thrust of the engine due to the incorrect temperature signals being received.[195]

It is clear that WAGT's lack of an in-house high altitude test cell prevented them from discovering the problem earlier. In the meantime, while they developed the solution and had the fix tested, BuAer de-rated the XJ34-WE-32 and YJ34-WE-38 engines in service use as follows:

- Takeoff on a standard day or warmer – the maximum exhaust gas temperature (T5) was lowered by 35°F.
- Takeoff at below standard day temperature – the maximum exhaust gas temperature (T5) was lowered by 125°F.
- For all flight operation, operation limited to -125°F below log book maximum exhaust gas temperature (T5)

These changes were the approximate reduction in thrust of 2.1 pounds for each Fahrenheit degree reduction in T5.[196]

The August activity report stated that WAGT was still awaiting a new design of the power regulator, power scheduler, electric control relay and fuel valve motor actuator, all proving troublesome during the YJ34-WE-38 endurance test. The XJ34-WE-32 type test was now expected no earlier than the first week of October and possibly later pending the incorporation of the redesigned components.[197] The one serious omission was the fact that the de-rating of the engine due to the temperature inversion problem was not mentioned and any delays while a solution was found were not considered in the outlook.

In early October, BuAer rejected WAGT's statement that "an adverse compressor outlet velocity profile shift was, alone, responsible for the observed undesired shift in radial distribution of turbine outlet temperature." It stated that WAGT efforts on resolving the problem through changes only to the velocity were inadequate and that an active effort needed to be exerted to improve combustor performance.[198]

WAGT's response to the BuAer concern stated the first mixer vane design produced a velocity near the hub lower than desired and a second design was being procured. Tests of an XJ34-WE-38 engine taking both a turbine inlet and outlet traverse showed inconsistent data results. The reason(s) were being investigated further. Tests had shown that the two different sized combustor annuli *(holes to us laymen)* had different blowout limits. A changing balance between these two annuli would produce a changing balance of temperature and the effect would be most pronounced at high altitudes and low ram pressure ratios. When the velocity profile was stabilized the conditions of operation of the annuli would be determined.[199]

The oil scavenging issue in the J34-WE-32/-38 was addressed with a survey of a proposed change to the scavenge oil hoses. They were to be reversed where they connected to the gearbox in order to obtain a reversal of oil flow through the gearbox thereby utilizing the "pumping" effect of the internal gears and materially aiding scavenging, particularly at high altitudes.[200]

In mid-October, WAGT sent BuAer its test plan to qualify the electronic control elements.[201]

The 150-hour qualification test started on October 17. The test engine (X24C8-7) was assembled to the latest production build standard except for the following differences:

1. A Thompson A/B air turbine fuel pump was installed instead of the normal Pesco pump. The Pesco pump had completed a 50 hour lab test but had suffered seal leakage due to excessive seal wear. An identical pump was flying in the XF-90 without problems. The Thompson pump development was started later than the Pesco pump and the qualification test was being used as a pump test to speed development.
2. A new oil reservoir designed for simpler manufacture and planned for incorporation on the 21st J34-WE-32 and 31st J34-WE-38 engines.
3. A new oil cooler required to be used with the new oil reservoir. It was identical to the one used on the J34-WE-38 except for the location of one boss.
4. A revised carbon seal on the bevel gear package to reduce oil leakage into the starter drive cavity. The new seal was to be incorporated starting with the 47th J34-WE-32 and any engine incorporating the older seal would be updated at overhaul.
5. The acceleration control rod seal in the emergency fuel control might leak more than the specified 10 drops per minute. Production parts were to be altered for a tighter fit. The actual rod being used on the test engine had exhibited no tendency to leak.
6. The quick disconnect clamp between the turbine housing and the afterburner utilized two bolts at

each joint instead of one. This was done to reduce turbine housing distortion due to lower radial pressure exerted on the machined flanges.

7. The A/B exhaust collector tail area was slightly greater than the initial production A/B's. The enlarged A/B's were to be incorporated on production engines starting with the 21st J34-WE-32 or sooner if possible.
8. The emergency scheduler shaft was the same used in the -38 type test. It showed excessive pin wear on the -38 engine and was being redesigned with a harder material, not yet available.
9. The anti-icing valve and hot gas problem were installed on the combustion chamber housing. In lieu of the non-available mixed gas conduit tube, blank-off plates were installed on the valve discharge ports. The de-icing assembly was installed to determine its resistance to the high temperatures and vibration of the engine.
10. Long thermocouples were incorporated to obtain more accurate temperature data but were not expected to survive the test.

XJ34-WE-32 150-Hour Qualification Calibration Results Prior to Test Start

Rating	Thrust (lb)	SFC Guarantee	SFC Primary	SFC Emergency	Proposed Guarantee*
Take-off	4,900	2.60	2.33	---	2.50
Military	3,370	1.08	1.035	1.038	1.08
Normal	3,020	1.03	1.00	.998	1.03
90% Normal	2,730	1.00	0.98	.974	1.00
75% Normal	2,260	0.97	0.97	.962	1.00

*WAGT proposed that the guarantees for the A/B SFC rating be lowered and the 75% Normal cruise SFC rating be raised.

The engine had 14.83 hours of running time from green, acceptance and calibration testing. The qualification test was begun using the AN-E-33 specifications. Several stops were necessary during the first 30 hours of the test. At 11 hours, the long thermocouples which had been used to get a more accurate picture of turbine temperature, broke and shorted out the harness. They were replaced with standard thermocouples. On restart, a burned out tube in the power regulator caused fuel "hunt". The tube was replaced. Also at 11 hours, the booster fuel pump leaked and investigation showed the bellows had failed. The pump was replaced with one of the same design. At 21 hours, the second booster pump had to be replaced, having developed a bellows failure as well. (A new pump bellows with double the number of folds to reduce the stress on the folds was under development.)[202]

The September activity report (which covered up to October 25) stated the qualification test run was up to 40 hours, but added no additional information on the test status or results. Another section reported the qualification test of the emergency and primary controls was underway and that the qualification testing for the J34-WE-32 A/B afterburner control was also underway.[203]

A special test status report covering the first 50 hours of the test added that to date the performance had been good, maintaining turbine inlet temperature at peak conditions, 1,525 +0-40°F. Switch-overs between primary and emergency controls were accomplished with no incidents. Oil consumption was negligible, but evidence that the gearbox was loading up with oil was investigated to cause, whereupon it was discovered the scavenge arrangement was not up to the latest planned configuration currently being surveyed. This was corrected and the problem cleared up.

At 21 hours the A/B shroud had to be repaired by welding due to a severe crack, thought to be caused by resonance with operating in A/B. The send booster pump had a bellows failure and the redesign with double the number of convolutions in the bellows cracked after 5,000 cycles of operations. Another was under test with thicker gauge bellows material.

At 30 hours the engine would not go above idle thrust. This was caused by a thyratron tube burning out. At 40 hours the A/B shroud had to be replaced due to another severe crack developing. At 45 hours the power regulator was replaced owing to erratic operation. At 46 hours a generator (used to load the gearbox) failure occurred due to brush failure. The generator loading was not resumed, but the overhung weight was retained. Also at 46 hours, on emergency operation, the exhaust tail operation was erratic and would not close properly. The emergency scheduler was removed and cleaned out. It was found that bits of "O" ring material were blocking the orifice. The damaged "O" rings were replaced.[204]

The A/B fuel control test plan for the J34-WE-32 was approved by BuAer November 6, 1950.[205]

Picking up the test progress for the next 50 hours, the next report states that at 57 hours the A/B ejector shroud had to be welded again, only to be replaced at the 100 hour point as the cracks continued to progress in spite of additional brackets added for strength. It was not known for sure why it was cracking, possibly either from resonance caused by the exhaust muffler or the A/B when it was in action. Three additional fuel booster pumps failed in the bellows component. Testing of new bellows designs continued. At 80 hours the bevel gear package had to be removed due to a manufacturing error which allowed high pressure oil to enter the starter compartment, causing the starter to function erratically. A tapped hole had been drilled completely through and not detected. At 90 hours, rubbing was detected in the turbine section. After inspection, the second stage turbine vane assembly was rotated, but whether the vane assembly was distorted or warped would not be known until teardown assembly. The A/B fuel regulator had to be adjusted on A/B runs shortly after 50 hours owing to the fuel nozzles "coking up". The A/B section was removed and the A/B nozzles cleaned before testing resumed.[206]

A BuAer memo stating the J34-WE-38 Qualification Test that ended September 23 was unacceptable was received on November 10. Before covering this, the final test reports on the XJ34-WE-32 will be described to conclude that effort.

The last 50 hours of the XJ34-WE-32 test were reported on November 22. The engine performed more satisfactorily during the last 50 hours of the test. The booster pump bellows failed once again (6 failures in total). Engine calibration after the test ended produced results very close to the initial calibration and well within the percentage limits of AN-E-32.

The teardown results were:

1. The A/B squeal baffle was distorted and buckled.
2. The exhaust collector had cracks.
3. The outer shroud had cracks (three shrouds had been used during the test).
4. The A/B flame holder was distorted and had some minor cracks.
5. The A/B fuel nozzle rings were coked up, mostly confined to the center (inner) fuel ring. No coking was experienced on the middle ring and only slight coking on the outer ring.
6. The exhaust nozzles had some cracks around the weld areas. (WAGT stated some cracking in A/B sheet metal fabricated parts was to be expected.)
7. The spark plugs showed some deterioration but were still serviceable.
8. The combustion basket had minor cracks (less extensive than had been experienced on the XJ34-WE-38 test).
9. One section of the 5th stage compressor diaphragm was cracked.
10. The vanes in the first stage turbine nozzle diaphragm were cracked and some vanes bent.
11. The second stage turbine disc showed rubs on the seal lands.
12. The balance of the engine was in good condition.

The report added that WAGT was developing fixes to the components that developed trouble during the test.[207]

The parallel qualification testing results on the YJ34-WE-38 and the recent NACA altitude wind tunnel testing caused BuAer to withhold approval for service use of the engine. During the 150-Hour Qualification test, problems were encountered with:

1. Electric control relay.
2. Power regulator.
3. Power scheduler.
4. Main fuel regulator.
5. Emergency scheduler linkage.
6. Power lever torque values.
7. Power lever position tolerance.
8. Engine weight.

The engine would only be acceptable for service use if the eight items above and the high altitude performance deficiency (temperature inversion) were all corrected and the specified component qualification tests completed.

However, BuAer did consider that the subject test was a satisfactory basis to accept a limited number of engines to replace the YJ34-WE-38 engines previously accepted for experimental flight use and service evaluation. The BAR was authorized to accept and ship sixteen (16) ZJ34-WE-38 engines as replacements for sixteen (16) YJ34-WE-38 engines previously accepted and subsequently found to be unsatisfactory. BAR acceptance of engines beyond the sixteenth (16th) ZJ34-WE-38 was NOT authorized.

The ZJ34-WE-38 engines had to conform to the parts list of the engine that had completed the qualification test, with some exceptions:

1. The engines must be equipped with "fixes" to the power regulator, power scheduler, electric control relay, fuel valve actuator assembly and emergency scheduler shaft.

2. The engines had to have deep-immersion thermocouples and have their control systems re-adjusted to restrict engine operation to safe turbine temperature limits.

The YJ34-WE-38 engines previously accepted were to be returned to WAGT for scrapping and for re-use of mechanically sound parts reusable in production engines. WAGT had agreed to incorporate any mechanical changes to the ZJ34-WE-38 engines as changes were indicated and verified.

BuAer stated they would consider that the development and verification requirement of Item 8(c) of contract NOa(s) 9670 would be satisfactorily completed upon either (a) satisfactory completion of qualification testing of the XJ34-WE-32 engine incorporating "fixes" for deficiencies cited (in this memo) or (b) waiver, by separate BuAer correspondence, of the power lever torque and/or weight deficiencies. By direction, BuAer requested that no further effort be directed specifically toward J34-WE-38 development.[208]

The October status report stated the component qualification testing on the XJ34-WE-32 and -38 was progressing satisfactorily.

The Power Level Torque (XJ46-WE-2 MUB Item 16) item was reduced to below the 25 pound-inches BuAer requested as the maximum and WAGT asked for approval of this change (and for the J34-WE-32 /-38) in late October.[209]

The A/B fuel controls were approved November 6.[210]

On November 8, WAGT notified BuAer that coordination with Chance Vought had resulted in a mutually agreed configuration of the ejector rear-end shape of the A/B on the -32. The new shape allowed the contractor to add his own extension piece having the desired ejector shape to fit a particular airplane as well as provide additional cooling air flow. Installation involved the airframe manufacturer unbolting the standard ejector and bolting on the airframe manufacturer's extension. It was requested the new ejector configuration be incorporated beginning with engine WE23021 and retroactive to all previous engines. For acceptance testing, a short ejector extension to duplicate the original aerodynamic shape would be provided by WAGT.[211]

WAGT forwarded their official qualification test report (A-1061 – not found in the files) as required by the contract on November 17.[212]

On December 14, 1950, WAGT surveyed a new shock mount design in the power regulator. These were expected to provide better vibration damping and operate satisfactorily at higher temperature than the current mounts.[213]

Contract Amendment 8 dated December 21 was signed to add $495,529.00 with no additional fixed fee to cover anticipated expenditures.[214]

On the basis of the results of the qualification test of the Thompson Products Model TT70000-1 A/B fuel pump, it was considered qualified according to AN-E-32 and approved for use on the J34-WE-32 engine.[215]

On December 27, BuAer asked that the revised power control system qualification test requirements be included in the next revision of the specification.[216]

The fuel pumping system with the exception of the bellows was considered qualified for the J34-WE-32/-38 engines. Any unqualified bellows on pumps delivered before the final approved bellows was qualified were to be replaced at no cost to the government.[217]

The revised sea level static performance guarantees proposed were approved in early January 1951 and BuAer asked WAGT to update the specification with the changes and all other approved changes since the current version of the specification had been approved.[218]

Convair contacted WAGT concerning use of both the J34-WE-32 and J46-WE-2 engines on their "Y2-2" airplanes (later XF2Y-1) in early January. WAGT sent a letter of concern to BuAer reminding them of past challenges faced in the MX-656 (Douglas X-3 "Stiletto") when trying to tailor the airframe to be able to use these two engines with their different installation requirements alternatively in the same airframe.[219] Given how stretched WAGT was at that time with work on the J34, J46 and J40 series, this might have been a subtle way of trying to dissuade BuAer from going down the path to yet another one-off derivation with little chance of production occurring. This was happening with the J46 (see next chapter) and was taking up a lot of engineering time.

BuAer finally released a provisional qualification test approval on January 15, 1951. This was to allow its use in a limited quantity of service type aircraft. The approval was contingent upon correction and satisfactory verification of the following deficiencies:

1. "Engine-driven fuel boost pump bellows: (may be verified by satisfactory accumulation of 50-hours miscellaneous operation (including afterburner operation) on house engines of the -32 model).
2. Turbine temperature inversion (may be verified by completion of either a 150-hour endurance test of corrective design on a -32 engine; or by successful verification of the already designed "mixer" on a -36 engine, plus accumulation of 7-1/2 hours afterburner time on -32 engines, as outlined."

The approval memo then went on to state: "Normally the model would not be qualified before satisfactory completion of certain component tests and demonstrations of durability which had not been accomplished to date. The power regulator would require considerable redesign to ensure consistent and dependable performance over a complete 150-hour endurance test and in production engines. Distortion and/or cracking of afterburner sheet metal components, flame-holders and fuel rings would have to be eliminated. Further development of the -32 model through satisfactory completion of development of the J34-WE-38 model, as set forth (by reference) would be required. However, due to the fact that airplane applications of these engines appear to be very limited, and in view of the extreme importance of added emphasis on XJ40 and XJ46 engine developments, it is considered that the above further work should not be pursued except where it might contribute to the XJ40 or XJ46 programs. Accordingly, action is being initiated for termination, for convenience of the government, of items 5a, 8b and 8c of the subject contract.

"The following deficiencies, demonstrated by the subject test, should be corrected at the earliest practicable date by normal production change in design procedure:

1. Failures of afterburner ejector support brackets;
2. Buckling of the first row compressor outlet straightening vanes;
3. Cracking between header holes at support points of the combustion chamber liner;
4. Oil leakage into the starter cavity of the power take-off gearbox;
5. Fatigue failure of oil reservoir tube support bracket;
6. Sticking of air shut-off valve;
7. Hazard of clogging of emergency scheduler orifice (jet);
8. Distortion of turbine housings;
9. Inconsistent operation of power regulators of the same configuration.

"Further, effort should be devoted to the significant reduction of distortion and/or cracking in afterburner sheet metal components, flame holders, and fuel rings.

"Due to the urgent requirements of the F7U airplane program, the Bureau of Aeronautics perforce must accept a certain number of J34-WE-32 engines, under contract NOa(s) 9791, prior to complete verification or correction of the above-listed deficiencies. The conditions of such acceptance will be established by separate correspondence. However, it is known to be probable that the "mixers" mentioned ...... above will not be available for all such engines. Accordingly any engines accepted without mixers shall be equipped with reinforced deep immersion control thermocouples. These thermocouples must be verified for flight by successful completion of 7-1/2 hours of afterburning time plus at least 50 hours of accumulated service in operation of house engines."[220]

The NOa(s) 9670 development contract termination notice (TN-1) was sent out January 23, 1951, terminating development of the XJ34-WE-32 and -38 on Contract NOa(s) 9670, leaving the XJ46-WE-2 and -4 development tasks in place.[221]

Since many of the J34-WE-32 component problems, particularly related to the control system were common to the J46 and J40 programs, the net effect of the termination was to end development charges for those items to the J34-WE-32/-38 contract items, moving the charges to the J46.

From this point, the J34-WE-32 and -38 were to move into the background as they entered production and initial limited service use in the Chance Vought F7U-1, Consolidated XF2Y-1 and McDonnell F2H-4. Various sub-models were produced under contract NOa(s) 9791. They had already made their contributions to the J46 by identifying and helping to resolve some of the design and service challenges of afterburner equipped engines with variable exhaust nozzles. The identified design flaws and the lessons being learned were continuously fed back into the J46-WE-2 and J40 programs.

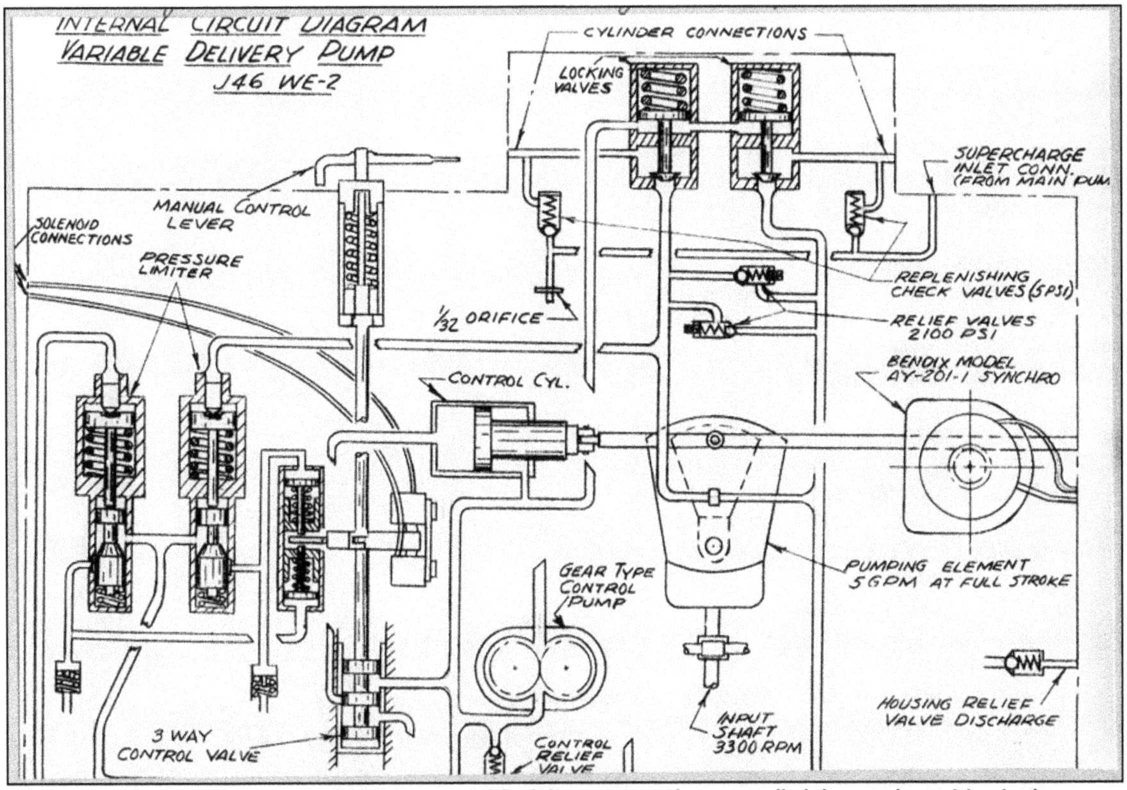

Figure 12: Diagram of circuits in the Vickers variable delivery pump that controlled the nozzle position in the SHC and hydro-mechanical control systems. WAGT P49258, 4/24/1953. *Courtesy Hagley Museum and Library*

Chapter 2 – J46-WE-2 and J34-WE-32/-38 Procurement & Initial Development

[1] WAGT Memo, October 31, 1947, "Further Development of 24 (Inch) Frame Turbo-Jet Engines." **B556VC1RG72.3.2NACP**
[2] WAGT Memo, January 27, 1948, "Development of X24C-8 (XJ34-WE) and X24C-10 Turbo-jet Engines, Our Reference N-91965", **B557ENCLRG72.3.2NACP**
[3] Specification WAGT-X24C10-2 (Proposal Draft), December 23, 1947, **B557ENCLRG72.3.2NACP**
[4] BuAer Request for Authority to Contract, Clearance No. Air 6710, NOa(s) 9670, Cost-Plus-Fixed Fee contract, Approved April 14, 1948, **B556VC1RG72.3.2NACP**
[5] Contract Noa(s) 9670, Amendment 1, April 16, 1948, **B552V1RG72.3.2NACP**
[6] Specification WAGT-X24C10-2A, March 24, 1948, **B552V2RG72.3.2NACP**
[7] BuAer Memo, June 1, 1948, "Contract NOa(s)-9670, XJ34-WE-32 Turbo-Jet Engine, Location on." **B552V2RG72.3.2NACP**
[8] WAGT Memo, May 14, 1948, "Solid Propellant Starters for Turbo-Jet Engines", **B552V2RG72.3.2NACP**
[9] WAGT Report A-724, June 21, 1948, Design Study of XJ34-WE-32 Engine Accessory Gear Box Arrangements Contract NOa(s) 9670, **B552V2RG72.3.2NACP**
[10] BuAer Memo, June 2, 1948, "Contract NOa(s) 9670, XJ34-WE-32 and XJ46-WE-2 Bid Specification WAGT-X24C10-2B dated 7 April 1948, Comments Concerning." **B552V2RG72.3.2NACP**
[11] WAGT Report (no "A" number assigned), June 29, 1948, Design Study of J34-WE-38 Engine Suitability For The Douglas F3D-1 Airplane, **B552V2RG72.3.2NACP**
[12] Mock-Up Board Report for Model XJ34-WE-32 Engine Contract NOa(s) 9670, Approved November 22, 1948, **B552V3RG72.3.2NACP**
[13] BuAer Memo, July 29, 1948, "Contract NOa(s) 9670 – Models XJ34-WE-32 and XJ46-WE-2 Engines, Anti-Icing Provisions." **B552V2RG72.3.2NACP**
[14] WAGT Memo, August 12, 1948, "Contract NOa(s) 9670 – Models XJ34-WE-32 and XJ46-WE-2 Engines – Anti-icing Provisions", **B552V2RG72.3.2NACP**
[15] WAGT Memo, August 3, 1948, "XJ34-WE-32 (X24C8) Turbo-Jet Engine Time Required for Development and Initiating Production Contract NOa(s)-9670", **B552V2RG72.3.2NACP**
[16] WAGT Report A-741, August 6, 1948, Accessory Arrangement Design Study Turbo Jet Engine XJ46-WE-2 Contract NOa(s) 9670 Item 1(a), **B552V3RG72.3.2NACP**
[17] BuAer Memo, September 3, 1948, "Contract NOa(s)-9670, XJ34-WE-32 and XJ46-WE-2 Turbo-Jet Engines; Location of Starters on", **B552V3RG72.3.2NACP**
[18] Naval Speedletter, September 23, 1948, "Noa(s) 9670 XJ34 and XJ46 Development Program Outline, Request for", **B552V3RG72.3.2NACP**
[19] WAGT Memo, October 5, 1948, "Contract NOa(s) 9212 XJ40WE-2,-4,-6. Contract NOa(s) 9670 XJ34WE-32 and XJ46WE-2. Location of the Electrical Control Components." **B552V3RG72.3.2NACP**
[20] BuAer Memo, October 14, 1948, "Contract NOa(s)9670, XJ46 Engine Configuration Variation – Requirement for." **B552V3RG72.3.2NACP**
[21] BuAer Memo, December 2, 1948, "Contract NOa(s) 9212, XJ40-WE-2 Engine; and Contract NOa(s) 9670, XJ46-WE-2 Engine – Design Study of Control System for.", **B552V3RG72.3.2NACP**
[22] BuAer Memo, December 16, 1948, "Contract NOa(s) 9670, XJ46 Engine; Configuration Variation for.", **B552V3RG72.3.2NACP**
[23] BuAer Memo, December 23, 1948, *"Contract NOa(s) 9670 – Monthly Progress Report (November) – Forwarding of.", **B552V3RG72.3.2NACP**
[24] BuAer Contract Noa(s) 9670 Amendment 2, December 29, 1948. **B552V1RG72.3.2NACP**
[25] WAGT Memo, December 6, 1948, "Contract NOa(s) 9670 – XJ34-WE-32 Engine, Mounting of Ignition Coils on", **B552V4RG72.3.2NACP**
[26] BuAer Memo, January 14, 1949, "Contract NOa(s) 9670, "XJ34-WE-32 Engine – Mounting of Ignition Coils on", **B552V4RG72.3.2NACP**
[27] BuAer Memo, January 14, 1949, "Contract NOa(s) 9670 – Monthly Progress Report (Summary cover letter) (Month of December 1948) – Forwarding of." **B552V4RG72.3.2NACP**
[28] WAGT Report A-796, January 20, 1949, Afterburner Progress Report, Contract NOa(s) 9670 Item 7. **B552V4RG72.3.2NACP**
[29] WAGT Memo, January 14, 1949, "Weight Differences between Bid Specification and Model Specification of J34WE-32 and -38 Engines." **B552V4RG72.3.2NACP**
[30] BuAer Memo, January 26, 1949, "Contract NOa(s) 9670, XJ34-WE-32 Turbo-Jet Engine, Mock-Up changes, Approval of", **B552V4RG72.3.2NACP**
[31] WAGT Memo, January 29, 1949, "Contract NOa(s) 9670, XJ34-WE-32 Turbo-Jet Engine, Mock-Up Changes, Approval of", **B552V4RG72.3.2NACP**
[32] WAGT Memo, February 2, 1949, "XJ46 Turbo-Jet Engine Gearbox Configuration", **B556VC1RG72.3.2NACP**
[33] BuAer Contract Amendment 4 to Contract NOa(s) 9670, May 25, 1949. **B556VC1RG72.3.2NACP**
[34] BuAer Memo, February 15, 1949, "Contract NOa(s) 9670 – Monthly Progress Report (Month of January 1949) – Forwarding of." **B552V4RG72.3.2NACP**
[35] WAGT Memo, February 18, 1949, "Contract NOa(s) 9670 – Westinghouse Specification WAGT-24C8-2A dated 24 December 1948 – Weight Revision to." **B552V4RG72.3.2NACP**
[36] WAGT Memo, February 15, 1949, "Contract NOa(s)-9670 XJ46 Engine Configuration Variation for, Our Reference WG-60999", **B552V4RG72.3.2NACP**
[37] WAGT Memo, February 24, 1949, "Weight change on XJ46-WE-2 Engine Since Submission of Bid Specification", **B552V4RG72.3.2NACP**
[38] BuAer Memo, February 25, 1949, "Contract NOa(s) 9670, Westinghouse XJ46-WE-2 Engine Mock-Up Inspection, Scheduling of", **B552V4RG72.3.2NACP**
[39] BuAer Memo, March 3, 1949, "Contract NOa(s) 9051, 9433, 9791, 9670 and 9212, Model J34-WE-30a, -34, -32; XJ46-WE-2, and XJ40-WE-6, -8, -10 Engines, Emergency Fuel System, Contractor Furnished Components", **B257V6RG72.3.2NACP**

[40] WAGT Memo, February 23, 1949, "Contract NOa(s) 9670 – XJ34-WE-32 Engine, Anti-Icing of. Item 2, Task II, Sub-Task G, Contract NOa(s) 9051, Anti-Icing Test of a J34 Turbo Jet Engine on Mt. Washington, Project #644.", **B552V4RG72.3.2NACP**
[41] WAGT Report A-813, March 8, 1949, Engine Installation Features and Accessory Arrangement of the XJ46-WE-2 Engine, Contract NOa(s) 9670 Item 1(a), **B552V4RG72.3.2NACP**
[42] BuAer Memo, March 22, 1949, "Contract NOa(s) 9670 – Monthly Progress Report (Month of February 1949) – Forwarding of.", **B552V4RG72.3.2NACP**
[43] BuAer Report, April 25, 1949, Mock-Up Board Report for Model XJ46-WE-2 Aircraft Engine, Contract NOa(s) 9670, **B552V5RG72.3.2NACP**
[44] WAGT Report (not numbered), April 13, 1949, Proposed Simplification of XJ40-WE-8 Power Control System, April 13, 1949, **B553V9RG72.3.2NACP**
[45] WAGT Memo, March 7, 1949, "Contract NOa(s) 9670 Turbo-Jet Engine. Characteristics of Oil System.", **B552V4RG72.3.2NACP**
[46] BuAer Memo, March 24, 1949, "Contract NOa(s) 9670 – XJ34-WE-32 Engine Afterburner Expansion Joint", **B552V4RG72.3.2NACP**
[47] BuAer Memo, March 25, 1949, "Westinghouse Models J34-WE-30, -30A and -34 Engine Emergency Control System Requirements – Deviations from.", **B552V4RG72.3.2NACP**
[48] WAGT Memo, April 4, 1949, "Contract NOa(s) 9670 XJ46-WE-2 Turbo Jet Engine Gear Box." **B552V4RG72.3.2NACP**
[49] BuAer Memo, April 6, 1949, "Contract NOa(s) 9670 – Report No. A-748 "Engine Installation Features and Accessory Arrangement of the XJ34-WE-32 Engine, approval of." **B552V4RG72.3.2NACP**
[50] BuAer Memo, April 11, 1949, "Contract NOa(s) 9670 – Monthly Progress Report (Month of March 1949) – Forwarding of.", **B552V5RG72.3.2NACP**
[51] WAGT Report, A-818, XJ34-WE-32 Turbo-Jet Engine Design Study of Turbine Outlet Temperature Override for Afterburner Fuel Control – Contract NOa(s) 9670, **B552V4RG72.3.2NACP**
[52] BuAer Memo, April 20, 1949, "Contract Noa(s) 9670 – XJ34-WE-32 Engine Design Study of Turbine Outlet Temperature Override for Afterburner Fuel Control", **B552V4RG72.3.2NACP**
[53] BuAer Memo, April 12, 1949, "Contracts NOa(s) 9212, XJ40-WE-2 Engine; Contract NOa(s) 9670, XJ46-WE-2 Engine, and XJ34-WE-32 Engine – Revised Design of Control System for", **B552V4RG72.3.2NACP** *(See contract NOa(s) 9212 for a full copy of this memo)*
[54] WAGT Memo, March 23, 1949, "Contract NOa(s) 9670 – XJ34-WE-32 Anti-Icing System Design", **B552V4RG72.3.2NACP**
[55] BuAer Memo, April 19, 1949, "Contract NOa(s) 9670 – XJ34-WE-32 Engine Anti-Icing System Design – Release of.", **B552V4RG72.3.2NACP**
[56] BuAer Memo, April 26, 1949, "Contract NOa(s) 9670 – XJ34-WE-32 Engine – Attachment Point of Exhaust Nozzle Actuator", **B552V5RG72.3.2NACP**
[57] BuAer Memo, May 3, 1949, "Contract NOa(s) 9670, Item 3(b) – Model Specification WAGT-24C8-2D dated 5 April 1949 for Engine J34-WE-32 – Forwarding of.", **B552V5RG72.3.2NACP**
[58] WAGT Memo, April 27, 1949, "Contract NOa(s) 9670 – XJ46-WE-2 Compressor Air Bleed Limits.", **B552V5RG72.3.2NACP**
[59] WAGT Memo, May 3, 1949, "Contracts NOa(s)-9212 and 9670 XJ34-WE-32, XJ46-WE-2 and XJ40-WE-6 Engine Mock-Up Board Changes", **B552V5RG72.3.2NACP**
[60] BuAer Memo, May 16, 1949, "Contract NOa(s) 9670 – Monthly Progress Report (Month of April 1949)", **B552V5RG72.3.2NACP**
[61] BuAer Memo, May 23, 1949, "Contract NOa(s) 9670, Westinghouse Model Specification WAGT-24C8-2D dated 5 April 1949 Covering the Models J34-WE-32 and -38 Engines – Approval of", **B552V5RG72.3.2NACP**
[62] BuAer Memo, May 24, 1949, "Contracts NOa(s) 9670, 9212, 10067 and 10114, Engine Data – Incorrect Distribution of.", **B552V5RG72.3.2NACP**
[63] WAGT Memo, June 7, 1949, "Contracts NOa(s) 9670, NOa(s) 9212 – Noa(s) 10067 and NOa(s) 10114 Engine Data – Distribution of", **B552V5RG72.3.2NACP**
[64] BuAer Memo, June 29, 1949, "Contracts NOa(s) 9670, 9212, 10067, and 10114, Engine Data; Distribution of.", **B552V5RG72.3.2NACP**
[65] BuAer Memo, May 24, 1949, "Contract NOa(s) 9670 – Afterburner Progress Report", **B552V5RG72.3.2NACP**
[66] WAGT Cover memo to Report A-836, (June, 1949), June 7, 1949), "Contract NOa(s) 9670 – XJ34WE-WE-32 and XJ46-WE-2 – Accessory Temperature Limits.", **B552V5RG72.3.2NACP**
[67] BuAer Memo, June 16, 1949, "Contract NOa(s) 9670, Westinghouse Specification WAGT-24C8-2B Covering the Model J34-WE-32 and -38 Engines – Electrical Component Temperature Limits for.", **B552V5RG72.3.2NACP**
[68] BuAer Memo, June 8, 1949, "Contracts NOa(s) 9670 and 10114 XJ46-WE-2 and XJ40-WE-8 Turbo-Jet Engines, Fuel Pump Inlet Requirements, Establishment of:", **B552V5RG72.3.2NACP**
[69] BuAer Memo, June 13, 1949, "Contract NOa(s)-9670 Westinghouse Specification WAGT-24C-2D Dated 5 April 1949 Covering the Models J34-WE-32 and -38 Engines – Oil Interruption Requirements for.", **B552V5RG72.3.2NACP**
[70] BuAer Memo, June 14, 1949, "Contracts NOa(s) 9212 and NOa(s) 9670, Turbo-jet Engine Research and Development Programs.", **B552V5RG72.3.2NACP**
[71] WAGT Memo, June 9, 1949, "Contract NOa(s) 9670 and 10114; XJ46-WE-2 and XJ40-WE-8 engines, Simplified Power Control System for,", **B552V5RG72.3.2NACP**
[72] BuAer Memo, June 24, 1949, "Contract NOa(s) 9670 – Monthly Progress Report (Month of May 1949) – Forwarding of.", **B552V5RG72.3.2NACP**
[73] WAGT Memo, June 29, 1949, "Contract NOa(s) 9670; WEC Turbo-Jet Engine XJ46-WE-2; Make Attachment Provisions for Control Rod Brackets on Regulators; Survey of.", **B552V5RG72.3.2NACP**
[74] BuAer Memo, July 8, 1949, "Contract NOa(s) 9670 – Westinghouse Specification WAGT-X24C10-2C, dated 9 May 1949, Covering the Model XJ46-WE-2 Engine", **B552V5RG72.3.2NACP**
[75] WAGT Monthly Progress Report for June 1949, Contract NOa(s) 9670, Report A-871, July 8, 1949, **B552V5RG72.3.2NACP**

[76] WAGT Memo, July 7, 1949, "Contract Noa(s) 9670 Westinghouse Model XJ34-WE-32 Engine, Reduction of Dry Weight of.", **B552V5RG72.3.2NACP**
[77] WAGT Memo, January 3, 1949, "Contract NOa(s) 9670 – XJ34-WE-32 Engine – Attachment Point of Exhaust Nozzle Actuator.", **B553V6RG72.3.2NACP**
[78] BuAer Memo, July 21, 1949, "Contract NOa(s) 9670, XJ34-WE-32 Engine, Attachment Point of Exhaust Nozzle Actuator", **B553V6RG72.3.2NACP**
[79] WAGT Memo, March 10, 1949, "Contract NOa(s)-9670 XJ34-WE-32 Engine Fire Shield Our Reference Neg. 92804", **B553V6RG72.3.2NACP**
[80] BuAer Memo, July 27, 1949, "Contract NOa(s) 9670-XJ34-WE-32 Engine Fire Shield, Approval of." **B553V6RG72.3.2NACP**
[81] WAGT Memo, August 17, 1949, "Contract Noa(s) 9670 – XJ34-WE-32 Engine – Fire Shield", **B553V6RG72.3.2NACP**
[82] WAGT Memo, August 25, 1949, "Contract Noa(s) 9670-XJ34WE-32 Engine Fire Shield Adapter.", **B553V6RG72.3.2NACP**
[83] BuAer Memo, August 1, 1949, "Contract NOa(s) 9670-XJ34-WE-32 and XJ46-WE-2 Engines – Accessory Temperature Limits", **B553V6RG72.3.2NACP**
[84] BuAer Memo, August 2, 1949, "Contracts NOa(s) 9670 and NOa(s) 10114 Model J34-WE-32, XJ46-WE-2 and XJ40-WE-8 Engines; Emergency Fuel System, Contract Furnished Components", **B553V6RG72.3.2NACP**
[85] WAGT Memo, August 2, 1949, "Contract Noa(s) – 9212 and 9670; Unexecuted Deliveries on Contractual Items; Request for Clarification of", **B553V6RG72.3.2NACP**
[86] WAGT Memo, June 30, 1949, "Contract Noa(s) 9670 – XJ46-WE-2 Engine – Effect of Failure of Engine Driven Booster Pump on Operation of the Fuel System.", **B553V6RG72.3.2NACP**
[87] BuAer Memo, August 12, 1949, "Contract Noa(s) 9670 – XJ46WE-2 Engine – Effect of Failure of Engine Driven Booster Pump on Operation of the Fuel System", **B553V6RG72.3.2NACP**
[88] BuAer Memo, August 15, 1949, "Contract Noa(s) 9670, XJ34-WE-32 Engine Mock Up Changes, Approval of", **B553V6RG72.3.2NACP**
[89] WAGT Memo, August 11, 1949, "Contract Noa(s) 9670 – XJ46WE-2 Engine – Forward Combustion chamber Drain Valve – Elimination of.", **B553V6RG72.3.2NACP**
[90] WAGT Memo, August 12, 1949, "Contract Noa(s) 9670 – Specification WAGTo-X24C10-2C dated 9 May 1949 Describing XJ46-WE-2 Engine.", **B553V6RG72.3.2NACP**
[91] WAGT Memo, August 16, 1949, "Contract Noa(s) 9670 XJ34-WE-32 and -38 Engines, Elimination of Hydraulic Fuse.", **B553V6RG72.3.2NACP**
[92] WAGT Memo, July 29, 1949, "Contract NOa(s) 9212 – XJ40-WE-6 Engine Mock-Up Board Decisions, Items 21 and 4, and Contract NOa(s) 9670 – XJ46-WE-2 Engine, and Contract Noa(s) 9670 – XJ34-WE-32 Engine, and Contract NOa(s)10114 – XJ40-WE-8 Engine.", **B304V7RG72.3.2NACP**
[93] WAGT Memo, August 12, 1949, "Contract Noa(s) 9670 Specification WAGT-24C8-2D dated April 5, 1949, Oil Interruption Requirements", **B553V6RG72.3.2NACP**
[94] BuAer Memo, August 24, 1949, "Contract Noa(s)-9670, Westinghouse Specification WAGT-24C8-2D Dated 5 April 1949 Covering the Models J34-WE-32 and -38 Engines – Oil Interruption Requirements for.", **B553V6RG72.3.2NACP**
[95] WAGT Proposed Appendix B, July 25, 1949, Specification WAGT-24C8-2D, **B553V6RG72.3.2NACP**
[96] BuAer Memo, August 24, 1949, "Contract Noa(s)-9670 – Westinghouse Specification WAGT-24C8-32, Appendix B Covering Requirements for the Model YJ34-WE-32 Engine.", **B553V6RG72.3.2NACP**
[97] WAGT Memo, September 16, 1949, "Contract Noa(s)-9670 Westinghouse Specification WAGT-24C8-2D, Appendix B Covering Requirements for the Model YJ34-WE-32 Engine", **B553V7RG72.3.2NACP**
[98] WAGT Memo, September 30, 1949, "Contract Noa(s) 9670 – 50 Hour Tests of YJ34WE-32 Engine.", **B553V7RG72.3.2NACP**
[99] BuAer Naval Speed Letter, October 12, 1949, "Contract Noa(s)-9760 – fifty-hour test on Westinghouse XJ34-WE-32." *In spite of the subject line referencing the XJ34-WE-32, the body of the speed letter refers to the engine as the YJ34-WE-32.* **B553V7RG72.3.2NACP**
[100] BuAer Memo, August 26, 1949, "Contract Noa(s) 9670 – Westinghouse Specification WAGT-X24C10-2C, Dated 9 May 1949, Covering the Model XJ46-WE-2 Engine – Ignition System Qualification Test Requirements for.", **B553V6RG72.3.2NACP**
[101] BuAer Memo, August 18, 1949, "Contract Noa(s) 9670, Monthly Progress Report (Month of July); Forwarding of.", **B553V6RG72.3.2NACP**
[102] BuAer Memo, September 2, 1949, "Contract Noa(s) 9670 – Westinghouse Model XJ34-WE-32 Engine Generator Drive Clearance Envelope, ", **B553V6RG72.3.2NACP**
[103] WAGT Memo, September 1949, "Contract Noa(s) 9670 – Generator Installation Clearance on the XJ34-WE-32 and -38 Engines.", **B553V6RG72.3.2NACP**
[104] BuAer Memo, September 6, 1949, "Contract Noa(s) 9670 – XJ46-WE-2 Turbo-Jet Engine, Afterburner Cooling Ejector, Furnishing of", **B553V6RG72.3.2NACP**
[105] WAGT Memo, July 14, 1949, "Contract Noa(s) 9670 – XJ46-WE-2 Fuel Filter – location of, and provision for alcohol injection.", **B553V6RG72.3.2NACP**
[106] BuAer Memo, September 6, 1949, "Contract Noa(s) 9670, XJ46-WE-2 Fuel Filters, Location of and Provision for Alcohol Injection", **B553V6RG72.3.2NACP**
[107] WAGT Memo, September 1, 1949, "Contracts Noa(s) 9670 and Noa(s) 10114, Analysis of the effects of certain failures in the simplified power control.", **B553V6RG72.3.2NACP**
[108] WAGT Memo, September 2, 1949, "Contract Noa(s) 9670 – Location of Emergency Scheduler Shaft and Oil Lines to the Exhaust Nozzle Actuator.", **B553V6RG72.3.2NACP**
[109] BuAer Memo, September 14, 1949, "Contract Noa(s) 9670 – Location of the Exhaust Nozzle Actuator", **B553V6RG72.3.2NACP**
[110] WAGT Memo, August 30, 1949, "Contract Noa(s) 9670 – Difference between the XJ34-WE-32 and the XJ34-WE-38 Engines.", **B553V7RG72.3.2NACP**
[111] BuAer Naval Speedletter, September 21, 1949, "Contract Noa(s) 9670-Item 5a-XJ34-WE-32, Contractor's Number (X24C8-2) – 50-Hour Engine – Information on." **B553V7RG72.3.2NACP**

[112] WAGT Memo, September 19, 1949, "Contract Noa(s) 9670 – Specification WAGT-X24CV10-2D dated 9 May 1949 Describing XJ46-WE-2 Engine.", **B553V7RG72.3.2NACP**
[113] BuAer Memo, September 24, 1949, "Contract Noa(s) 9670, Monthly Progress Report for August 1949, - Forwarding of.", **B553V7RG72.3.2NACP**
[114] BuAer SpeedLetter, September 28, 1949, No subject., **B553V7RG72.3.2NACP**
[115] BuAer Memo, October 13, 1949, "Contract Noa(s) 9937; YJ34-WE-32 afterburner operation electrical signal", **B553V7RG72.3.2NACP**
[116] WAGT Memo, September 26, 1949, "Contract Noa(s) 9670 W.E.C. Turbo-Jet Engine Model YJ34-WE-32 Afterburner Operation Electrical Signal Chance Vought Aircraft Request For.", **B553V7RG72.3.2NACP**
[117] BuAer Memo, September 13, 1949, "Contract Noa(s) 9670; WEC Turbo-Jet Engine Model XJ34-WE-32 and -38. Relocation of Exhaust Nozzle Actuator and Afterburner Flexible Joint, Survey of.", **B553V7RG72.3.2NACP**
[118] BuAer Memo, September 30, 1949, "Contract Noa(s)-9670, Westinghouse Specification WAGT-24C8-2D, dated 5 April, 1949, Including Appendix A, Covering the Models J34-WE-32 and -38 Engines; Approval of.", **B553V7RG72.3.2NACP**
[119] BuAer Memo, October 11, 1949, "Contract Noa(s)-9670; Westinghouse Specification WAGT-X24C10-2C of 9 May 1949 Covering the Model XJ46-WE-2 engine – power control qualification test requirements for", **B553V7RG72.3.2NACP**
[120] Douglas Aircraft Company, Inc Memo, October 21, 1949, "Contract Noa(s) 9670; WEC Turbo-Jet Engine Model XJ34-WE-32 and -38. Relocation of Exhaust Nozzle Actuator and Afterburner Flexible Joint, Survey of.", **B553V7RG72.3.2NACP**
[121] Lockheed Aircraft Corporation Memo, November 3, 1949, "Contract Noa(s) 9670; WEC Turbo-Jet Engine Model XJ34-WE-32 and -38; Relocation of Exhaust Nozzle Actuator and Afterburner Flexible Joint, Survey of.", **B553V7RG72.3.2NACP**
[122] BuAer BAR Memo, November 18, 1949, ""Contract Noa(s) 9670; WEC Turbo-Jet Engine Model XJ34-WE-32 and -38; Relocation of Exhaust Nozzle Actuator and Afterburner Flexible Joint, Survey of.", **B553V8RG72.3.2NACP**
[123] Douglas Memo, November 30, 1949, "Contract Noa(s) 9770 (sic), Model F3D-1 Airplane – WEC Turbo-Jet Engine Model XJ34-WE-32 and -38. Relocation of Exhaust Nozzle Actuator and Afterburner Flexible Joint, Survey of", **B553V8RG72.3.2NACP**
[124] BuAer Memo, December 1, 1949, "Contract Noa(s) 9670 – XJ34-WE-32 engine afterburner flexible connections; comments on", **B553V8RG72.3.2NACP**
[125] WAGT Memo, December 21, 1949, "Contract Noa(s) 9670 – XJ34-WE-32 Engine Afterburner Flexible Connection.", **B553V8RG72.3.2NACP**
[126] BuAer Memo, January 13, 1950, "Contract Noa(s) 9670 – XJ34-WE-32 engine-afterburner flexible connection design", **B553V9RG72.3.2NACP**
[127] BuAer Memo, October 24, 1949, "Contract Noa(s) 9670 – Monthly Progress Report for September 1949 – Forwarding of.", **B553V7RG72.3.2NACP**
[128] BuAer Memo, October 25, 1949, "Contract Noa(s) 9670; engine accessory quick disconnect mountings for the XJ46-WE-2 engine", **B553V7RG72.3.2NACP**
[129] BuAer Memo, November 30, 1949, "Contract Noa(s) 9670 – Engine Accessory Quick Disconnect Mountings for the XJ46-WE-2 Engine.", **B553V8RG72.3.2NACP**
[130] BuAer Memo, October 25, 1949, "Contract Noa(s) 9670 XJ34-WE-32 engine provisions of electric starter power supply; comments on", **B553V7RG72.3.2NACP**
[131] BuAer Memo, October 28, 1949, "Contract Noa(s) 9670, XJ46-WE-2 engine, forward combustion chamber drain valve; elimination of", **B553V7RG72.3.2NACP**
[132] BuAer Memo, October 31, 1949, "Contract Noa(s)-9670, Westinghouse Model XJ34-WE-32 engines; generator drive clearance envelope", **B553V7RG72.3.2NACP**
[133] Douglas Memo, November 17, 1949, "Contract Noa(s) 9670 WEC Turbo-jet Engine Models J34-WE-32 and J34-WE-38, Gear Box Scavenge Oil Line Rework for Generator Clearance; Survey of.", **B553V8RG72.3.2NACP**
[134] BuAer Memo, October 31, 1949, "Contract Noa(s) 9670 Specification WAGT-24C8-2E dated 21 October 1949, Including Appendices A, B, and C.", **B553V7RG72.3.2NACP**
[135] WAGT Memo, November 1, 1949, "Contract Noa(s) 9670; WEC Turbo-Jet Engine Models J34-WE-32 and J46-WE-2, and Contract Noa(s) 10114; WEC Turbo-Jet Engine Model J40-WE-8; Desirability of Afterburner Operation Electrical Signal, Survey of.", **B553V7RG72.3.2NACP**
[136] WAGT Memo, October 25, 1949, "Contract Noa(s) 9670 and Noa(s) 9791 J34-WE-32 and -38 Turbo-Jet Engines Fuel Pump Qualification Test.", **B553V8RG72.3.2NACP**
[137] BuAer Speedletter, November 16, 1949, "Contract NOa(s) 9670 and NOa(s) 9791; J34-WE-32 and -38 turbo-jet engines fuel pump qualification test", **B553V8RG72.3.2NACP**
[138] BuAer Memo, November 18, 1949, "Contract Noa(s) 9670 – Monthly Progress Report for October 1949", **B553V8RG72.3.2NACP**
[139] WAGT Memo, November 17, 1949, "Contract Noa(s)-9670 – Engine Mounting of Power Regulator, Alternator Voltage Regulator and Ignition Coils on XJ46-WE-2 Engine – Item No. 10 of Mock-Up Board Report – Our Reference Neg. 92986", **B553V10RG72.3.2NACP**
[140] WAGT Memo (attachment), November 9, 1949, "Contract Noa(s)-9670 Our Reference WG-60999 qualification of J34WE-32 and -38 Engine Models", **B553V8RG72.3.2NACP**
[141] BuAer Memo, November 25, 1949, "Contract Noa(s) 9670, delivery and qualification of models XJ34-WE-32 and -38 engines; approval of", **B553V8RG72.3.2NACP**
[142] WAGT Memo, November 17, 1949, "Contract Noa(s) 9670; WEC Turbo-Jet Engine Model XJ46-WE-2; Relocation of Exhaust Nozzle Actuator and Afterburner Flexible Joint; Survey of.", **B553V8RG72.3.2NACP**
[143] BuAer Memo, December 2, 1949, "Contract Noa(s) 9670 – Qualification of J34-WE-32 and -38 Engine Models.", **B553V8RG72.3.2NACP**
[144] WAGT Memo, December 2, 1949, "Contract Noa(s) – 9670 Revised Proposal for Development of the XJ46-We-2 Turbo-Jet Engine Our Reference WG-60999", **B553V8RG72.3.2NACP**

[145] WAGT Memo, December 1, 1949, "Contracts Noa(s)-9670 and 10114 Proposal for the Simplification of the Emergency Fuel Systems on the XJ46-WE-2 and XJ40-WE-8 Engines, Our References WG-60999 and WG-62360", **B553V8RG72.3.2NACP**
[146] WAGT Memo, December 1, 1949, "Contract Noa(s) 9670, 9791, and 10114; WEC Turbo-Jet Engine Models YJ34-WE-32, XJ46-WE-2 and XJ40-WE-8; Afterburner Operation Electrical Signal.", **B553V8RG72.3.2NACP**
[147] BuAer Memo, December 8, 1949, "Contract Noa(s)-9670 – Westinghouse Models XJ34-WE-32 and -38 engines; generator drive clearance envelope", **B553V8RG72.3.2NACP**
[148] WAGT Memo, December 30, 1949, "Contract Noa(s) 9670 – XJ34-WE-32 Engine Generator Drive Clearance, WECO Drawing 62F28.", **B553V9RG72.3.2NACP**
[149] BuAer Memo, December 9, 1949, "Contract Noa(s)-9670 – Westinghouse Specification WAGT-24C8-2E dated 21 October 1949 covering the Models J34-WE-32, -38 engines", **B553V8RG72.3.2NACP**
[150] Bell Aircraft Corporation Memo, December 9, 1949, "Contract NOa(s)-9670, WEC Turbo-jet Engine Models J34-WE-32 and J46-WE-2 and Contract Noa(s)-10114, WEC Turbojet Engine Model XJ40-WE-8, desirability of afterburner operation electrical signal, survey of", **B553V8RG72.3.2NACP**
[151] Douglas Memo, December 16, 1949, "Contract NOa(s)-9670, WEC Turbo-jet Engine Models J34-WE-32 and J46-WE-2 and Contract Noa(s)-10114, WEC Turbojet Engine Model XJ40-WE-8, desirability of afterburner operation electrical signal, survey of", **B553V8RG72.3.2NACP**
[152] WAGT Memo, December 16, 1949, "Development of XJ46WE (X24C10M) Engine Our Reference N-93313", "Contract NOa(s)-9670, WEC Turbo-jet Engine Models J34-WE-32 and J46-WE-2 and Contract Noa(s)-10114, WEC Turbojet Engine Model XJ40-WE-8, desirability of afterburner operation electrical signal, survey of", **B553V8RG72.3.2NACP** (The actual proposal was not in the file)
[153] BuAer Memo, December 19, 1949, "Contract Noa(s) 9670 – XJ34-WE-32, 50-Hour Flight Verification Test Engine –Deviation from First Production Engine.", "Contract NOa(s)-9670, WEC Turbo-jet Engine Models J34-WE-32 and J46-WE-2 and Contract NOa(s)-10114, WEC Turbojet Engine Model XJ40-WE-8, desirability of afterburner operation electrical signal, survey of", **B553V8RG72.3.2NACP**
[154] BuAer BAR Teletype, December 19, 1949, No Subject, **B553V8RG72.3.2NACP**
[155] BuAer Memo, December 20, 1949, "Contract Noa(s) 9670; request for information on afterburner design", "Contract NOa(s)-9670, WEC Turbo-jet Engine Models J34-WE-32 and J46-WE-2 and Contract Noa(s)-10114, WEC Turbojet Engine Model XJ40-WE-8, desirability of afterburner operation electrical signal, survey of", **B553V8RG72.3.2NACP**
[156] WAGT Memo, January 16, 1950, "Contract Noa(s) 9670 Performance of Afterburner used on J34-WE-32 Engine", **B553V9RG72.3.2NACP**
[157] BuAer Memo, December 23, 1949, "Contract Noa(s) 9670 – Monthly Progress Report for November 1949 – Forwarding of.", **B553V8RG72.3.2NACP**
[158] BuAer Speedletter, January 11, 1950, "Contract Noa(s) 9670 – 50-Hour Qualification Test on XJ34-WE-32 (X24C8-2) Engine – Information on.", **B553V9RG72.3.2NACP**
[159] BuAer Memo, January 20, 1950, "Westinghouse XJ46-WE-2 and XJ40-WE-8 turbo-jet engines, simplified fuel systems for; status of action on Westinghouse proposal", **B553V9RG72.3.2NACP**
[160] BuAer Memo, January 20, 1950, "Contract Noa(s) 9670 – Monthly Progress Report for December 1949.", **B553V9RG72.3.2NACP**
[161] WAGT Memo, January 18, 1950, "Contract Noa(s) 9670 and 9791; WEC Turbo-Jet Engine Model YJ34-WE-32, Proposed Fuel Drain Ports on the Afterburner Air Turbine Fuel Pump Shroud, Survey of.", **B553V9RG72.3.2NACP**
[162] BuAer Memo, January 25, 1950, "Contract Noa(s) 9670 – location of XJ46-WE-2 fire seal; approval of", **B553V9RG72.3.2NACP**
[163] BuAer Memo, January 30, 1950, "Contract Noa(s) 9670, XJ34-WE-38 150 hour type test engine, deviations from first production engine assembly.", **B553V9RG72.3.2NACP**
[164] WAGT Memo, January 31, 1950, "Contract Noa(s) 9670 Westinghouse Model XJ34-WE-32 and -38 engines, Generator Drive Clearance Envelope.", **B553V10RG72.3.2NACP**
[165] BuAer Memo, February 8, 1950, "Contract Noa(s) 9670, item 8c, XJ34-WE-38, 150 hour verification test in accordance with AN-E-32 and as modified by model specification WAGT-24C8-2E of 21 Oct 49.", **B553V10RG72.3.2NACP**
[166] BuAer Memo, February 10, 1950, "Contract Noa(s) 9670, item 8c, XJ34-WE-38, 150 hour verification test in accordance with AN-E-32 and as modified by model specification WAGT-24C8-2E of 21 Oct 49.", **B553V10RG72.3.2NACP**
[167] BuAer Memo, February 15, 1950, "Contract Noa(s) 9670, item 8c, XJ34-WE-38, 150 hour verification test in accordance with AN-E-32 and as modified by model specification WAGT-24C8-2E of 21 Oct 49.", **B553V10RG72.3.2NACP**
[168] BuAer Memo, March 2, 1950, "Contract Noa(s) 9670, item 8c, XJ34-WE-38, 150 hour verification test in accordance with AN-E-32 and as modified by model specification WAGT-24C8-2E of 21 Oct 49.", **B553V10RG72.3.2NACP**
[169] WAGT Report A-949, March 28, 1950, Qualification Test J34-WE-38 Engine, Contract Noa(s) 9670, Engine X24C7-1, **B554V11RG72.3.2NACP**
[170] BuAer Memo, February 13, 1950, "Contract NOa(s) 9670; engine mounting of power regulator; alternator voltage regulator and ignitions coils on the XJ46-WE-2 engine", **B553V10RG72.3.2NACP**
[171] WAGT Memo, February 10, 1950, "Contract Noa(s) 9670 and NOa(s) 9791 Performance Ratings of J34-WE-32 Engines.", **B553V10RG72.3.2NACP**
[172] WAGT Memo, February 10, 1950, "Contract NOa(s) 9670, 9791, 10385 and 10114; WEC Turbo-Jet Engine Model Numbers YJ34-WE-32, J34-WE-38, XJ46-WE-2, J40-WE-6 and XJ40-WE-8; Main Fuel Regulator Drains, Survey of.", **B553V10RG72.3.2NACP**
[173] BuAer Memo, February 24, 1950, "Contracts Noa(s) 9670, 9212, and 10114; XJ40-WE-6 and -8, XJ34-WE-32 and XJ46-WE-2 engines; field survey, approval of", **B553V10RG72.3.2NACP**

[174] BuAer Memo, February 24, 1950, "Contract Noa(s) 9670 – depletion of funds; inadequate information concerning", **B553V10RG72.3.2NACP**
[175] BuAer Memo, April 10, 1950, "Contract Noa(s) XJ34-WE-32 and XJ46-WE-2 engines; fireseal attachment provisions", **B554V11RG72.3.2NACP**
[176] BuAer Memo, April 12, 1950, "Contract Noa(s) 9670; Model XJ46-WE-2 engine; relocation of exhaust nozzle actuator and afterburner flexible joint; approval of", **B554V11RG72.3.2NACP**
[177] WAGT Memo, December 6, 1949, "Contract Noa(s) 9670 – XJ46-WE-2 Engine Request for approval to changes to engine Clearance Envelope.", **B554V11RG72.3.2NACP**
[178] BuAer Memo, April 13, 1950, "Contract Noa(s) 9670; XJ46-WE-2 engine; changes to engine clearance envelope; approval of", **B554V11RG72.3.2NACP**
[179] BuAer Memo, April 14, 1950, "Contracts NOa(s) 9051 and NOa(s) 9670; request for transfer of charges", **B554V11RG72.3.2NACP**
[180] Douglas Memo, April 14, 1950, "Contract Noa(s) 9670 and 9791, WEC Turbo-jet Engine Model YJ34-WE-32, Proposed Change in Design of Afterburner Ignitor Valve; Survey of, Engine Change No. 56", **B554V11RG72.3.2NACP**
[181] BuAer Memo, April 21, 1950, "Contract NOa(s) 9760, item 7, monthly progress report (month of March); forwarding of.", **B554V12RG72.3.2NACP**
[182] WAGE Memo, April 19, 1950, "Contract NOa(s) 9670, 9791, 10385, and 10114 WEC Turbo-Jet Engine Models YJ34-WE-32, J34-WE-38, XJ46-WE-2, J40-WE-6, XJ40-WE-8; Back Pressure Requirements for Main Fuel Regulator Drains, Survey of. Engine Change Nos. 66 for YJ-34-WE-32, 7 for J34-WE-38, 79 for XJ46-WE-2, 59 for J40-WE-6 and 69 for XJ40-WE-8.", **B554V12RG72.3.2NACP**
[183] BuAer Memo, April 25, 1950, "Westinghouse XJ34-WE-32 and -38, XJ46 and XJ40 engines; fuel system drain provisions, comments concerning", **B554V12RG72.3.2NACP**
[184] WAGT Memo, April 25, 1950, "Contract NOa(s) 9670 and 9791; WEC Turbo-Jet Engine Model YJ34-WE-32; Airframe Safety Locking of Rear Trunnions.", **B554V12RG72.3.2NACP**
[185] WAGT Memo, May 1, 1950, "Contracts NOa(s)-9670, 9791, 10114 and 9212; Proposal for Modification of Controls on J34-WE-32, J34-WE-38, J46-WE-2, J40-WE-6 and J40-WE-8 Engines ", **B556VC2RG72.3.2NACP**
[186] WAGT Memo, May 3, 1950, "Contract NOa(s) 9670, 9791, 10385, and 10114; WEC Turbo-Jet Engine Models YJ34WE-32, J34WE-38, XJ46WE-2, J40WE-6 and XJ40WE-8. Survey Status Report.", **B554V12RG72.3.2NACP**
[187] BuAer Memo, May 17, 1950, "Contract NOa(s) 9670; J34-WE-32 engine qualification tests", **B554V12RG72.3.2NACP**
[188] Contract Amendment 5, May 31, 1950, **B552V1RG72.3.2NACP**
[189] WAGT Memo, May 1, 1950, "Contract NOa(s) 9670; WEC Turbo-Jet Engine Model XJ46-WE-2, Fuel Pump Inlet Requirements, Establishment of:, **B554V12RG72.3.2NACP**
[190] BuAer Memo, June 2, 1949, "Contract NOa(s) 9670; Model XJ46-WE-2 engine fuel pump inlet requirements", **B554V12RG72.3.2NACP**
[191] BuAer Memo, June 2, 1950, "Contracts NOa(s) 9670, 9791, 10385, and 10114; Westinghouse Electric Corporation Turbo-Jet Engine Models YJ34-WE-32, J34-WE-38, XJ46-WE-2, J40-WE-6 and XJ40-WE-8; survey status report", **B554V12RG72.3.2NACP**
[192] WAGT Memo, June 7, 1950, "Contract NOa(s) 9670 XJ46-WE-2 Engine Special Control Linkage Attachment Provisions.", **B554V12RG72.3.2NACP**
[193] BuAer Contract Amendment 6, June 5, 1950, **B552V1RG72.3.2NACP**
[194] BuAer Naval Speedletter, August 14, 1950, "Contract NOa(s) 9670, item 8b, 150 hour qualification test XJ34-WE-32 engine model; information on.", **B554V1RG72.3.2NACP**
[195] NACA Research Memorandum, RM E51K06, November 29, 1951, Investigation of Turbine-Outlet Temperature Distribution of XJ34-WE-32 Turbojet Engine, W.R. Prince and J.T. Wintler, Lewis Flight Propulsion Laboratory Cleveland, Ohio
[196] BuAer Telegraph, August 31, 1950, "XJ34-WE-32, YJ34-WE-38 Engine NACA Altitude Wind Tunnel Tests", **B554V1RG72.3.2NACP**
[197] BuAer Memo, September 28, 1950, "Contract NOa(s) 9670, item 7, monthly progress report (August 1950); forwarding of.", **B554V1RG72.3.2NACP**
[198] BuAer Naval Speedletter, October 3, 1950, "Contract NOa(s) 9670; data from NACA altitude wind tunnel tests of XJ34-WE-32 engine", **B554V1RG72.3.2NACP**
[199] WAGT Memo, October 23, 1950, "Contract NOa(s) 9670 – XJ34-We-32 Altitude Wind Tunnel Testing", **B554V1RG72.3.2NACP**
[200] WAGT Memo, October 6, 1950, "Contracts NOa(s) 9670 and 9791, WEC turbo-jet engine models YJ34-WE-32 and J34-WE-38, revised scavenge oil line rerouting, survey of, engine change No. 9 for J34-WE-38 and engine change No. 68 for J34-WE-32.", **B554V1RG72.3.2NACP**
[201] WAGT Memo, October 16, 1950, "Contract NOa(s) 9791, Electric Controls System, Qualification Testing of:", **B554V1RG72.3.2NACP**
[202] BuAer Memo, October 25, 1950, "Contract NOa(s) 9670, XJ34-WE-32 engine; 150 hour qualification test.", **B554V1RG72.3.2NACP**
[203] BuAer Memo, October 26, 1950, "Contract NOa(s) 9670, Item 7 – Monthly Progress Report (September 1950) – Forwarding of.", **B554V1RG72.3.2NACP**
[204] BuAer Memo, October 30, 1950, "Contract NOa(s) 9670, item 8b, XJ34-WE-32 150 hour qualification test; information on", **B554V1RG72.3.2NACP**
[205] BuAer Memo, November 8, 1950, "Contract NOa(s) 9670; J34-WE-32 engines; afterburner fuel controls, qualification testing.", **B554V1RG72.3.2NACP**
[206] BuAer Memo, November 7, 1950, "XJ34-WE-32 150 Hour Qualification Test (2nd 50 hours)", **B554V1RG72.3.2NACP**
[207] BuAer Memo, November 22, 1950, ""XJ34-WE-32 150 Hour Qualification Test (Last 50 hours)", **B554V1RG72.3.2NACP**
[208] BuAer Memo, November 10, 1950, "YJ34-WE-38 engine qualification test ended 23 Sep 1950", **B554V1RG72.3.2NACP**
[209] WAGT Memo, October 23, 1950, "Contract NOa(s) 9670 – XJ46-WE-2 Decrease Power Lever Torque.", **B554V1RG72.3.2NACP**

[210] BuAer Memo, November 6, 1950, "Contract NOa(s) 9670; J34-WE-32 engines, afterburner fuel controls, qualification testing of", **B554V1RG72.3.2NACP**
[211] WAGT Memo, November 8, 1950, "Contracts NOa(s) 9670 and NOa(s) 9791 J34-WE-32 Engines; Afterburner Shroud Attachment Provisions for F7U-1 Airplane.", **B554V1RG72.3.2NACP**
[212] WAGT Report A-1061, November 17, 1950, Qualification Test J34-WE-32 Engine Contract NOa(s) 9670, **B555V2RG72.3.2NACP**
[213] WAGT Memo, December 14, 1950, "Contracts NOa(s) 9670 and 9791. WEC Turbo-Jet Engine Models J34WE-32, -38, J46WE-2 and -4. Proposed Power Regulator Shock Mounts, Survey of. Engine Change No. 69 for -32 engines; No. 10 for -38 engines; No. 87 for -2 engines and No. 3 for -4 engines.", **B555V2RG72.3.2NACP**
[214] BuAer Contract NOa(s) Amendment 8, December 21, 1950, **B556VC2RG72.3.2NACP**
[215] BuAer Memo, December 26, 1950, "Contract NOa(s) 9670; J34-WE-32 engine afterburner fuel pump (WEC P/N 60E953; component qualification test", **B555V2RG72.3.2NACP**
[216] BuAer Memo, December 27, 1950, "Contract NOa(s); Westinghouse Specification WAGT-24C8-2E dated 21 Oct 1949 covering the Models J34-WE-32 and -38 engines; control qualification test requirements for", **B555V2RG72.3.2NACP**
[217] BuAer Memo, January 2, 1951, "Contract NOa(s) 9670, J34-WE-32 and -38 engines, fuel pumping system component; qualification tests", **B555V2RG72.3.2NACP**
[218] BuAer Memo, January 10, 1951, "Contract NOA(S) 9670 Westinghouse Specification WAGT-24C8-2E covering the Model J34-WE-32 and 38 engines; revision of", **B555V2RG72.3.2NACP**
[219] WAGT Memo, January 8, 1951, "Contract NOa(s) 9791 and 9670; WEC Turbo Jet Engines, Models J34WE32 and J46WE2, Use of in Y2-2.", **B555V2RG72.3.2NACP**
[220] BuAer Memo, January 15, 1951, "Contract NOa(s) 9670, XJ34-WE-32 turbojet engine; qualification test of", **B555V2RG72.3.2NACP**
[221] BuAer Contract Termination Notice, NOa(s) 9670, TN-1, January 21, 1951. **B555V2RG72.3.2NACP**

Chapter 3
XJ46-WE-2 (and -4) Continued Development

With the development focus now primarily on the XJ46-WE-2, we find that the May 1950 activity report is a contradictory statement on the engine test progress. On one hand, the BAR stated that "the progress on the XJ46-WE-2 has improved immensely over the last month". But the evidence suggests otherwise. The first test engine had only eight hours total run time on it. It was suffering from compressor stalling between 8,500 and 9,500 rpm. It had been reassembled with a different configuration first stage turbine nozzle and was back on test. The second engine which was scheduled for June assembly was delayed as parts kept being removed to keep the first engine running. And, as a backdrop, the intended XJ34-WE-32 150-hour engine was suffering from power regulator trouble, afterburner shut-off valve leakage and other minor control problems. The J34-WE-38 production engines were found to be suffering from bearing failure problems resulting from fuel leaking past the seals and washing the lubricants out of the bearings. All of these components were shared with or were almost identical in design to the ones being used on the XJ46-WE-2.[1]

Development in June 1950 continued towards the contractual target date of completion of the 150-hour Qualification Test in September. The various items left from the MUB were being resolved and approvals sought to enable a build standard for the engine to be defined.

The wiring harness arrangement changes asked for in MUB Item 6 were approved June 14.[2] The MUB Item 17 was cleared days later when the decision to allow the electrical connections at the junction box to be non-interchangeable between the engine types was approved.[3]

BuAer finally got around to finishing their internal cross-department review of the conditionally approved engine data distribution plan that WAGT had proposed. Only two modifications were made and full approval was granted.[4]

Continuing the MUB item clearances, BuAer approved Items 24, 32 and 63 in late June on the understanding that the changes proposed involved no weight changes to implement.[5]

After receiving further comments on drains, particularly in light of the memo sent out stating the general requirements, BuAer revised their requirements. The combustion chamber and afterburner drain provisions were changed to be:[6]

1. Retention on the engine of automatic combustion chamber and afterburner drain valves.
2. The connection of the combustion chamber drain valve or valves (according to engine ground attitude) should be to lines draining directly overboard. The requirement for manual valves was deleted. (They could be added if experience so dictated.) The afterburner drain valve was not to be used unless experience proved it necessary.

The new requirements kept arriving. An alternate engine starting provision was requested by Chance Vought in early June. This approach would utilize the single 30 kva alternator which was to be driven by a Vickers hydraulic pump/motor combination (driven from either engine) as a starter by applying an external 400 cycle power source to the alternator. The hydraulic coupling would then be controlled to apply torque for starting through the alternator drive pad and power takeoff system.

WAGT responded in mid-June that the limiting factors on engine starting were the ability of the alternator pad to absorb the torque load and for the start process to obtain a specification start to idle within 40 seconds. The normal pad limit was set at 6,600 lb-in with 1,500 lb-in continuous. WAGT was willing to accept a modification to raise the 1,500 lb-in to 2,500 lb-in below 2,500 engine rpm, as long as the combined torque of both the alternator and DC generator did not exceed 1,500 inches above 2,500 engine rpm.

They stated that if BuAer approved the proposed specification change, WAGT would be willing to negotiate the specification change and that Chance Vought would be informed.[7]

BuAer processed Amendment 7 to the contract to pay for the additional costs to switch from the AN specifications to the MIL-spec standard. The cost was $669,958.00.[8]

The oil drain provisions for the -2 were approved and Item 40 of the MUB was closed mid-way through July. Since the engine would successfully start at temperatures of -65°F, WAGT had pointed out that providing special oil drainage provisions beyond the normal gearbox drain was unnecessary. [9]

BuAer got around to responding to the Chance Vought starter proposal with a memo to the Dallas BAR

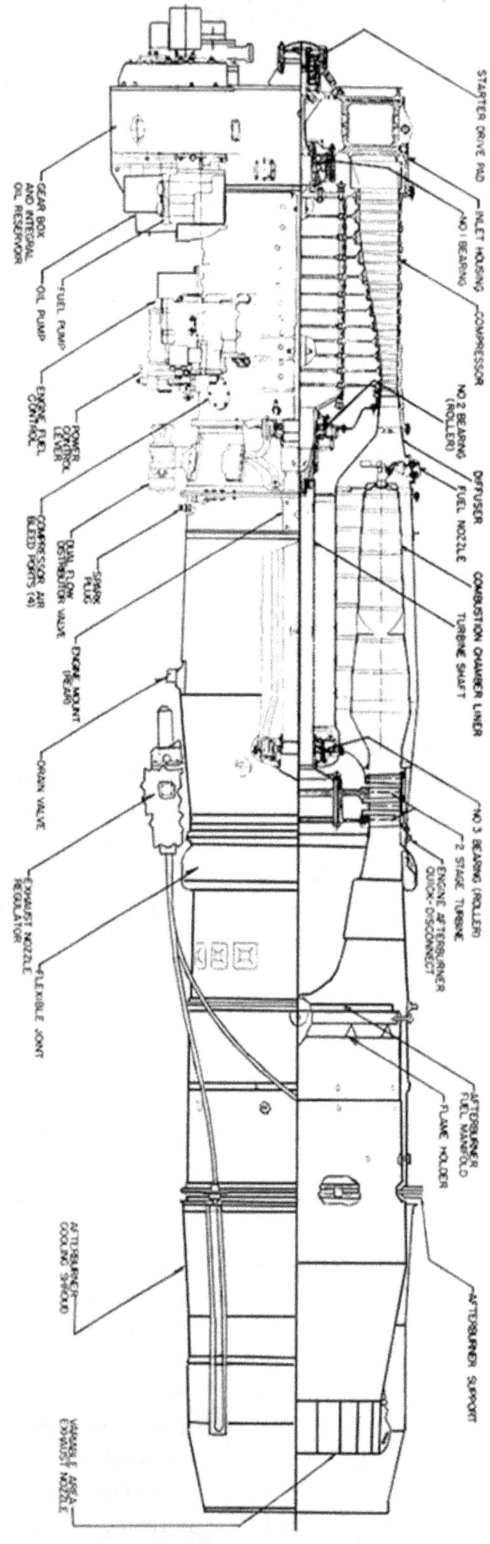

Westinghouse XJ46-WE-2
March 28, 1950

WAGT P42491 Courtesy Hagley Museum and Library

(assigned to the F7U-3 contract) in late July. They reminded him that the starter load curves provided by WAGT were estimates made before an engine had actually run and had not been checked yet based on actual starting tests of the engine. The requirements that had to be met for a new starter system were:

1. The system had to be able to start the engine with the time specified by Figure 1 of Specification MIL-E-5007.
2. The torque-speed output characteristics as well as the electrical power output characteristics when used as a starter had to be furnished to BuAer so the starting performance and requirements of the system could be calculated.
3. The unit as an alternator had to meet the requirements of Specification MIL-S-6099. Proof that it met the specification had to be provided to BuAer prior to acceptance of the aircraft.
4. A detailed drawing of the starter layout and installation had to be provided to BuAer and approved prior to initiating any development of the starter-generator.

From this we can gather that Chance Vought was to pick up the tab for this development effort, although exactly how they would do that without a J46 engine example to use in testing was not covered.[10]

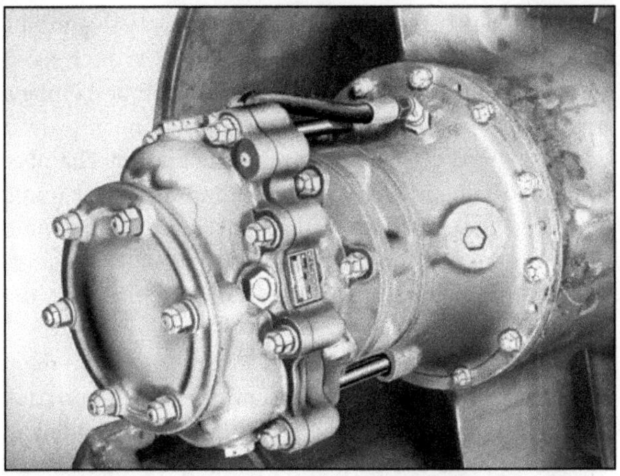

Figure 1: Chance Vought Starter Gearbox. WAGT 50676-1A, 12/18/1953. *Courtesy Hagley Museum and Library*

On the first day of August, BuAer approved the torque limit increase on the generator pad. This is the first memo in the contracts file that references an XJ46-WE-4 engine. The XJ46-WE-4 was to be a non-afterburner version of the -2.[11]

A wire from BuAer in August ordered the -2 mock-up be returned to WAGT from Chance Vought Aircraft in the configuration received (except for length).[12]

Backtracking a bit, the combined June/July activity report summary stated that the -2 testing was up to 47 hours of test time and the engine was still stalling in the 6,000 rpm range. New compressor blade angles were to be tried. A second engine was expected to be on test by mid-September. The second engine would utilize "atomized" combustion (walking stick evaporators instead of spray nozzles). A third engine would also utilize vaporized combustion and the outlook was to build it in late 1950 or early 1951.[13]

At this point, Douglas informed BuAer that the cooling air pressure supplied by the airframe would exceed the 2 psi and 5 psi limits of the power regulator and governor alternator set by WAGT's installation manuals. They asked that the pressure limits be raised to 12 psi. This would: 1) provide a unidirectional air flow and 2) prevent possible explosive mixtures from the accessory department from being drawn into either of the components and ignited. WAGT began a test program to determine the cooling requirements of the components.[14]

The move to the MIL-E specifications meant that WAGT now had two versions of the XJ46-WE-2 under development, one to the old AN specifications and the other to the new, tougher, specifications. The second engine was now designated the XJ46-WE-2A by BuAer. WAGT proposed a new schedule for the -2A engine and also modified the contract guarantee schedule for the XJ46-WE-2 and the BAR reported the schedules looked reasonable. [15,16]

XJ46-WE-2A:

- Model Specifications by August 30, 1950 (no change)
- Engine Operating Recommendation by August 30, 1952
- Development and test including accessory drive gearbox with a single AND20006 Type XVI generator pad, through a modified 150-Hour qualification test, test report delivery, and engine parts list by April 30, 1952
- Final design drawings and specifications for Model XJ34-WE-2A by May 31, 1952

XJ46-WE-2 New Dates (both 9 month slips):

- XJ34-WE-2 50-Hour Flight Substantiation Test March 31, 1951
- XJ34-WE-2 Delivery of 150-Hour Qualification Test Engine after successfully completing the test June 30, 1951

On top of the schedule slip news, the NACA (National Advisory Committee for Aeronautics) sent over results from their high altitude wind tunnel testing of the XJ34-WE-32, reporting that the engine suffered from adverse temperature distribution in the turbine section when the engine was operated above 25,000 feet. A mixer vane stage was inserted at the exit from the diffuser and this solved the problem.

The mixer vane stage was also to be incorporated into the J46 and J40 series of engines.

Figure 2: Example of mixer vanes (from J40). P50333-14, 10/12/1953. *Courtesy Hagley Museum and Library*

The monthly progress report for August states test engine No. 1 had 48.67 hours of test time and that temperature traverse testing was currently underway. The No. 2 engine would be available to begin testing the first week of October. Apparently the first engine had switched to walking stick vaporization for fuel combustion, as the report refers to the No. 1 engine using it the same as the up-coming No. 2 engine. The first two engines had different compressors and evaluation of them would dictate which design went into the No. 3 test engine.[17]

The September activity report stated the second XJ46-WE-2 engine had been assembled and submitted for test. After one hour of testing, it was reassembled with different compressor blade angles to try to resolve compressor stalling that had been observed. The No. 1 engine was up to 55 hours of testing and was being assembled for turbine tests in the research laboratory.[18]

In late October, BuAer sent a memo to the WAGT BAR related to the projected use of titanium in engines, in this case primarily the J40-WE-10/-12. The memo expressed concern over the availability of the material and possible related engine delivery delays, stating such delays would be unacceptable. The J46-WE-2 was also considering the use of the metal in the compressor, but this was not specifically addressed in the memo.[19] The primary reason to use titanium in the engines was to reduce weight.

WAGT responded a few days later stating they had a firm allocation schedule for titanium metal from DuPont, the amount of metal sufficient for the engines already under contract (one each XJ46-WE-2 and -2A, 167 J46-WE-2, 41 J46-WE-4).[20] This must not have been of much comfort to BuAer, who intended to order far larger quantities of engines than the few hundred of various types then under contract.

The October status report discussed the compressor stall issue on the No. 2 engine. After reassembly, stall was still experienced and after one hour the second stage compressor disc failed and the cause of the failure was under investigation. Engine No. 1 was due to go to the Aviation Gas Turbine Laboratory (AEL) for cold running tests that did not require that a complete compressor be installed. Accordingly, the second stage disc was removed off and the compressor from the No. 2 engine was to be installed on the No. 1 engine to allow it to go to AEL and the engine No. 1 compressor was to be installed in the No. 2 engine to allow it to resume testing.[21]

The No. 3 engine parts were being diverted to the first two engines to keep their testing progressing. The necessary tooling for this engine was overdue. This last engine was now scheduled to be assembled by January 31, 1951.

WAGT surveyed relocating the exhaust nozzle control and aft combustion chamber drain valve in late October. The nozzle control pivot was to be moved forward so the control would not extend aft of the plane of the quick-disconnect clamp, thus providing improved access to that clamp. The rear combustion chamber drain had to be moved as a result and a track or guide added to the A/B diffuser bottom to reduce buckling loads on the lengthened push-rod system. The change would add three pounds to the -2 and one pound to the -4.[22]

In mid-November, BuAer added the requirement that all active specifications state the maximum fuel consumption of the engine. This was being added to aid the airframe companies in planning for their aircraft fuel systems.[23]

The Power Level Torque (MUB Item 16) item was reduced to below the 25 pound-inches BuAer requested as the maximum and WAGT asked for approval of this change (and for the J34-WE-32 /-38) in late October.[24]

The relocation of the oil cooler was surveyed in early November. At the time the preliminary drawing of the oil cooler had been produced, the magnitude of hydraulic force that would be needed to move the iris A/B nozzle and the plug type nozzle in the -4 were unknown. It was now known on the -2 to be approximately 4,500 lbs. On the -2, an additional high pressure lube pump would be located on the aft face of the current lube pump. The new pump would interfere with the current oil cooler location. The new location would also improve the cooler's performance as well. On the -4, the extra pump would not be needed, but for parts commonality, the oil cooler was being relocated to the same location as the -2.[25]

MUB Item 41 for the -2 was approved on the grounds that the maximum engine fuel inlet pressure called out in specification WAGT-X24C10-2E was in compliance with MIL-E-5007.[26]

Contract Amendment 7 was approved covering the additional costs of switching over to the MIL-E specifications with their tougher and more extensive testing requirements.[27]

Figure 3: New evaporator fuel system with inner combustion chamber liner. Note early long versions of walking sticks. Later vapor discharge tube length was shortened by about half. WAGT P43750, 1/3/1953. ***Courtesy Hagley Museum and Library***

Figure 4: Inner and outer combustor liners for the annular combustion chamber. WAGT P43747, 1/3/1951. ***Courtesy Hagley Museum and Library***

On November 17, WAGT forwarded report A-1048 that included the starting requirements on an XJ46-WE-2 engine. The curves included were for any starter that could achieve acceleration from stop to an idle speed of 4,000 rpm within 30 seconds (against the specification time limit of 40 seconds). The Chance Vought hydraulic starter was tested and started the engine in 38 seconds.[28]

Chance Vought notified BuAer that the power regulator cooling ducts of the F7U-3 airplane would provide adequate cooling airflow from airspeeds 150 knots and above, no airflow from 0-150 knots and reversed airflow during ground operation. In a terminal velocity dive the regulators would be pressurized to 2.8 psi against the specification of 2.0 psi. BuAer forwarded the information to WAGT.[29]

Douglas responded to the oil cooler relocation survey with a large concern over interference with the flap mechanism in the new oil cooler position because the drip pan would have to be moved and there was no room to do so. Also, the new cooler location would interfere with the access doors to the drip pan. The drip pan, flap mechanism and wing structure had already been fabricated. They suggested moving the oil cooler as close as possible to the bottom of the engine and rotating it upwards to clear the drip pan and access doors by at least three quarters of an inch.[30]

Contract Amendment 8 was signed in December adding $495,529.00 to the contract with no increase in the fixed fee to cover anticipated expenditures.[31]

Redesign of the shock mounts for the power regulator was surveyed in mid-December. They noted the preferred mounting of the shock mounts was with the long shaft extension pointing forward and with the axis of the shock mount pins mounted within 5 degrees of vertical with respect to level flight attitude.[32]

Chance Vought responded to the oil cooler survey identifying a problem of interference with their dump tank design. WAGT had also found interference between the dump tank and the A/B fuel lines. They noted the location of the hydraulic lines to the nozzle control and the routing of the harness leads were not firm yet and might pose a problem with the dump tank. It was suggested Chance Vought lower the dump tank and provide special brackets to pick up the engine flange.[33] WAGT had to admit the original drawings sent with the survey were flawed in that they did not include all the piping and other wiring that would go below the oil cooler, leaving the impression there was more room in that area than there actually was.[34]

BuAer asked WAGT why the center of gravity of the XJ46-WE-2 had shifted as much as it had. It did not appear the change from the "atomizing" (spray nozzles) combustor type to the vaporizing type could completely account for the shift.[35] WAGT did not respond until April. This informative letter states the 1.7 inch change was caused by the following:

1. Replacing the atomizing combustion chamber with the vaporizing type,
2. The eyelid exhaust nozzle was replaced with an iris type exhaust nozzle,
3. The 7th through 12th stage steel compressor discs were replaced with titanium discs, and
4. A total weight reduction of 65.8 lbs. (new engine weight 1,849 pounds)

A chart was included in the response that showed all the new moment arm (weight multiplied by the distance from a fixed point) distance calculations followed by the new center of gravity calculation that resulted.[36]

The movement of the oil cooler required the jack lugs be moved as well and WAGT surveyed this change in early January.[37] Later WAGT had to explain the change did not apply to the J46-WE-1 because the jack lugs on that engine were not in the same location.[38]

The November status report stated the No. 2 engine had a second stage compressor failure and was repaired. The stator vanes on stages one through six were re-twisted and reinstalled. The No. 2 engine was to start turbine efficiency experiments.[39] The December report stated the No. 2 engine now had 40 hours of testing completed, with stalling still occurring and the compressor and turbine still not matched correctly (as of January 12). A different first stage nozzle was being installed and the vaporizing combustion chamber installed as well. It would be back on test January 15. The No. 1 engine was still in cold testing. A diffuser section had failed and was repaired and the testing was continuing. The report ends with the caution that considerable work needed to be accomplished before the 50-Hour flight substantiation test could be completed by the end of March.[40]

In mid-February, WAGT asked for information on whatever starters BuAer planned to use on the production engines. They also commented that they had three starters on hand but these would not be enough for all the demands of the engine program. Any information that could be provided would allow WAGT to procure adequate starters for the testing program.[41]

A request for windmilling information on the J46 brought a WAGT response that the characteristics were very similar to the other J34 model engines. Since the fuel pump used fuel for lubrication and cooling, the fuel control inlet should remain in the "on" position to ensure the pump was bathed in fuel while windmilling. BuAer's request was related to plans to fly the F7U-3 for range in cruise with one engine windmilling.[42]

The reasons for engine development lagging were obvious. The pressure on WAGT related to getting the J34-WE-32, -38 and J40 models into production was diverting resources. There must have been intense pressure to get the J46 out the door as well. WAGT made plans to accelerate the development at BuAer's request and sent over a memo that raised more questions than concrete, measureable answers about how the development would be compressed.

The plan explained that up to then, (to paraphrase) a trial and error, then redesign and repeat process was now to be modified to allow testing to occur in parallel with designing. In particular this would apply to the compressor design. Component testing was to be "much more vigorous".

The A/B performance, endurance and reliability needed to be improved and the engine control needed more development.

Manpower and facilities were being increased to speed the design and testing. Discussions with WAGT's suppliers were to be used for additional component testing at their facilities where feasible.

Then, after stating the obvious (above) and offering no supporting details on how the changes would allow them to meet the schedule, WAGT made a suggestion that a de-rated engine be considered for the target qualification testing date. The 50-hour flight qualification test was to be changed to a 150-hour qualification test to the AN-E-32 standard (instead of the Mil-E standard) of a "reduced performance" engine. A 150-hour test of a full performance engine to the AN-E-32 standard and then a full performance 150-hour test to MIL-E-5009 standard would be accomplished as development allowed, finally achieving the full specification guarantees. This would allow engines to begin production for limited service use on production airframes as soon as possible.

The expense to accomplish the proposed "changes" would require an additional $2,620,817.00 in funds that included a fee of $76,334.00. The fee was considered appropriate because the work to be accomplished was considerably increased. It was requested that the contract be amended accordingly if the changes were acceptable.

The distribution cover sheet has the comment on it that "if the scope of the work requested herein is the same as that originally called for in the contract, it is recommended that no fee be allowed."[43]

Before BuAer responded, WAGT sent over an amended additional cost breakdown that lowered the price to $2,619,889.00 with a fee of $76,307.00.[44]

The development progress status of the J46 was firmly in the background of the request for more money and the new approach to getting an engine into the soon to be produced airframes. The January status report arrived in mid-February and it stated the compressor stalling problem was not resolved. Because of this, A/B testing had not begun. The 50-hour flight qualification test date of March 31 could not possibly be met. The No. 1 engine involved in the cold testing for turbine evaluation had not been run for five weeks, awaiting data evaluation. The No. 2 engine had now accomplished 86 hours of testing, 36 of which were with the new vaporizer combustion system.

Almost as an afterthought, the report included the comment that the power regulator and combined fuel control system for the J34-WE-32 and -38, a version of which was to be used on the J46, had gone through 400 hours of testing. Fuel valve instability in the form of a high frequency "hunt" was still occurring.[45]

Materials shortages were beginning to occur at this time and would increase during the Korean War. One early example was the high temperature magnet wire used in the solenoids of the control system. The manufacturer (Sprague Electric Company) could not ship WAGT's order for the wire until September. However, the Bureau of Ships had returned a shipment to the manufacturer and it was still allocated to contract Nob(s)-46383. WAGT asked the allocation be moved to any of the engine contracts. If the change was effected immediately, control element deliveries would remain on schedule.[46]

At this point WAGT sent over the following suggested new contract deliverables due dates:

NOa(s) 9670 Suggested Revisions February 23, 1951		
Contract Item	**Old Date**	**Rev. Date**
8(a) Development and Test of XJ46-WE-2 through 150-hour Qualification Test and Delivery of Reports	June 30, 1951	June 30, 1952
9(a) Delivery of XJ46-WE-2 Qualification Tested Engine	June 30, 1951	June 30, 1952
10(a) Final Design Drawings	March 31, 1952	March 31, 1953
11 Final Design Summary Report	March 31, 1952	March 31, 1953

The Power Plant Division was against paying any additional fixed fees to Westinghouse, as the actual scope of the work had not increased. They argued the only change was to the delivery schedule dates and that the extra work was being generated by Westinghouse efforts to accelerate the development process to meet whatever new dates were to be agreed upon. Basically, it was WAGT's problem.[47]

The type of exhaust nozzle on the XJ46-WE-4 was defined in March. It would be a moveable plug exhaust nozzle using an experimental actuator. In WAGT "speak", this meant the control system would be able to move the plug back and forth to give control over the nozzle size area while the engine was in operation. In this case, the familiar bell crank and control rod system would be used to move the nozzle. Some J34's had used plug nozzles that were only manually adjustable during production acceptance testing to allow the tester to trim the turbine out gas temperature to obtain the guarantee thrusts and SFC's at the lowest gas temperature. Unfortunately, the design drawings for the preliminary -4 exhaust nozzle layout and adjustor were not retained in the files.[48]

Figure 5: Combined Fuel Regulator. WAGT P45044, 8/9/1951. *Courtesy Hagley Museum and Library*

The development continued while the schedule and costs were being worked out. While the February 1950 status report states that satisfactory development progress on the XJ46-WE-2 could be seen, considerable work needed to be done. The compressor, burner, turbine and lube system were under investigation. Three engine configurations (E-11, -12 and -13) had been tested with the E-13 apparently giving the best results, the severe stalling of the earlier configurations not being experienced. Teardowns had shown that mechanical redesign was necessary in some areas and was underway:

- The radial retaining screws had failed in fatigue,
- The brackets retaining the diffuser splitter fairings were cracking,
- The "candy canes" in the vaporizer were working loose allowing upstream leakage of fuel, and
- Compressor discs Nos. 1 and 2 showed fatigue cracks in blade grooves.

A new configuration (E-14) was in test using a variable area 1st stage turbine nozzle. The No. 3 engine was awaiting parts which were expected soon as a result of WAGT's accelerated work efforts.[49]

The March progress report states that after testing six compressor designs in an attempt to achieve the design values, aerodynamic problems still existed. One of the earlier compressor designs (which had shown stall) had been modified for research purposes and it was currently under test. If stall was found to be acceptably reduced the design would become the interim compressor. If this compressor proved successful, the number four engine would be assembled with complete accessories and some endurance running attempted.

In an effort to improve the fail-safe provisions for the A/B, particularly during takeoff, WAGT proposed adding a third transfer valve to the fuel system that would allow the main fuel pump to supply fuel to the A/B if the A/B fuel pump failed. The weight increase would be about two pounds. BuAer approved this change in mid-April for the -2 engine, stating that since the Air Force's -1 engine requirements were not detailed yet, they could not approve it for that model.[50]

Combining the voltage regulator and electric control relay assembly into a new unit to be called the Auxiliary Electric Control was proposed in a survey during January. The unit would also house the circuit for positioning the exhaust nozzle during emergency operation and potentiometers for adjusting the minimum and maximum exhaust nozzle areas for emergency operation. A written note on the distribution sheet states that the Power Plant Division had stated that possible airframe locations for the control had not been surveyed and that BuAer still did not want the unit mounted on the airframe, as they had explained in a long memo in February.[51,52]

WAGT surveyed the airframe companies for help in identifying possible locations to engine mount the auxiliary electric control in mid-April.[53] Two weeks later, while waiting for the responses to come back, WAGT stated to BuAer that an engine mounted installation did not appear possible. "Due to the geometry of the auxiliary electric control.....it appears doubtful as to whether any particular installation will be able to tolerate an engine mounted location. Since the XJ46WE-2 and -4 model specifications state that the voltage regulator shall be airframe mounted and since the voltage regulator comprises a substantial amount of the space required by the auxiliary electric control, it is this contractors (*SIC*) belief that from the installation standpoint this unit can best be mounted in the airframe." Letters from the airframe companies in response to the original survey were attached to support WAGT's position with the responses from the new survey to be forwarded as they arrived.[54]

The question of starter types to be used on the production airframes was answered by BuAer in April. The F3D-3 would use an AiResearch A7S20-3 air turbine type starter. The F7U-3 would use a combined AC starter-generator to turn a hydraulic constant-speed drive in the reverse direction to start the engine. The installation drawings were not available yet, nor were the torque and horsepower output curves for the F7U-3 system.[55]

The anti-icing design of the XJ46-WE-2 was approved, including the lack of provisions for the capacity and tap off for hot air for the starter fairing struts on the engine. WAGT had stated it was not necessary for the inlet housing struts and would not be necessary for the starter fairing struts either.[56]

A change in engine mount design was surveyed in mid-April. This changed the diameter of the reamed hole in the mount bushings from 3/4 to 5/8th of an inch. The change increased the clearance between mounting fixtures and the bottom of the groove between the mount lugs. The survey also corrected an error in the earlier drawings.[57]

A change in the ignition system from high tension to low tension high energy surface gap type was surveyed in April. The new type would reduce the engine weight, improve high altitude ignition and cold starting. The new units would replace the three coils.[58]

The April progress report picked up the development of the interim compressor. Five different engine configurations were tested. The testing changes made to configuration No. 19 based on results of the prior combinations resulted in the engine showing no trace of stall. However, the turbine inlet radial temperature distribution indicated very high turbine hub temperatures. The cause was under investigation.

Teardowns were still revealing mechanical defects and failures. Excessive oil consumption was traced to the No. 2 bearing rear seal which had broken from severe rubbing with the spline coupling. Excessive distortion and cracking of the vaporizing plate from overheating was found; loose "candy cane" vaporizing tubes were still being found; and the vaporizing plate was showing distortion and cracking from overheating.

The Nos. 2 and 3 engines had been rebuilt to the same configuration (No. 19) that showed no stalling characteristics except that the outlet guide vanes had been replaced by the mixer. No. 2 was back on test and No. 3 was being fully assembled with its A/B and controls.[59]

BuAer alerted the BAR in Dallas that based on conversations with WAGT as of May 7, the expected de-rated thrust of the early production -2 engines would be:

J46-WE-2 (de-rated)
Early Production
Expected Ratings as of May 7, 1951[60]

		Maximum		Military	
Altitude	Flight Speed (kts)	Thrust (lb)	SFC (lb/hr/lb)	Thrust (lb)	SFC (lb/hr/lb)
SL Guar	0	5,800	2.50	3,960	1.10
SL	500	6,635	2.74	3,695	1.39
35,000 ft	500	2,609	2.298	1,635	1.207
35,000 ft	574	2,934	2.25	1,752	1.222
35,000 ft	650	3,335	2.21	1,193	1.236

A survey was sent out in May to gather responses on removal of the afterburner drain boss and afterburner lifting lug and bumper.[61]

A set of performance graphs for the proposed de-rated -2 and -4 engines was forwarded to BuAer on May 11, 1951. These curves were not found, but a telling note on the distribution sheet reflects some of BuAer's reaction to the new curves: "Consider afterburner minimums, flight speed limits and starting limits at high altitudes as shown in enclosure very detrimental. Also, note great reduction in afterburner thrust at high airspeeds on sea level curve."[62]

BuAer accepted WAGT's solution of using a relief valve in the exhaust nozzle actuator oil lines to prevent line rupture from occurring in the event of system failure due to temperature rise on May 21, 1951. This was MUB Item 12.[63]

WAGT had found that the new iris type exhaust nozzle required higher oil pressure to operate the nozzle actuator and that raised the oil temperatures too high for safe engine operation. An alternate design using a variable displacement pump and an electrical control unit operated the nozzle within safe oil temperatures. The primary power regulators could easily be adapted to operate with the variable displacement pump, but the emergency exhaust nozzle position circuits could not. It was found the use of a magnetic amplifier to control the pump would be satisfactory. The amplified current during primary control operation would be coming from that supplied from the engine governor alternator. When operating in the emergency mode, an additional AC power requirement on the airframe bus would be needed as well as the already existing drain for exhaust nozzle operation in emergency. WAGT surveyed the airframe manufacturers on whether they could supply the current and also state if they were using a common AC/DC ground connection.[64]

Amendment 9 was signed May 24, 1951, changing the fixed overhead rate to 2.34%. It was to be applied to total cost incurred under the contract exclusive of cost of facilities and major plant alterations and rearrangements and General and Administrative overheads.[65]

In June, BuAer asked for an update and details on the fuel system simplification survey response status.[66]

Specification update WAGT-X24C10-2G that covered the -2, -2A and -4 engines was approved June 15, 1951. BuAer took exception to the contaminated fuel requirements, component maximum temperature limits and the power lever torque in the afterburning range, stating that although they were unacceptable for service engines, they would be acceptable for experimental engine development. They considered the specification an interim one.

They also requested a separate specification for the de-rated 5,800 pound thrust XJ46-WE-2 and 4,150 pound thrust XJ46-WE-4. These two engines would be designated the XJ46-WE-8 and -10 respectively.[67]

A survey was done in early June covering proposed changes to the ejector. The changes were: 1) shortening the ejector by four inches in length; for interim engines the cooling needs had been raised to 4.0% of engine airflow (from (3.6%) but were expected to rise further with higher temperature A/Bs; 2) inclusion of an attaching frame on the ejector to allow for extensions if the airframe manufacturers required one; 3) redesign of the ejector inlet bell mouth connection to provide more clearance between the engine afterburner mounting flange and the bell mouth lip; 4) "rigidify" the two ejector cowls that housed the exhaust nozzle push rods. This was done to eliminate any cracking of the cowls due to thermal expansion.[68]

WAGT had requested use of an AEL test cell for their accelerated testing. BuAer responded in late June stating they supported the WAGT efforts, but the request lacked sufficient data for them to justify the test cell being diverted for WAGT use.

While stating that a project office was being set up to coordinate WAGT work at AEL, BuAer asked that WAGT supply the number of hours by engine model that would be accumulated per month to be used for AEL planning purposes.

The next paragraph in the BuAer memo gives us good insight into some of the weaknesses in WAGT's work planning and BuAer's awareness:

"By virtue of the request for use of an AEL test cell, and, specifically, on the basis of (request in letter), Westinghouse obviously estimates that the work at AEL will be considerable and will contribute significantly to engine development. Accordingly, this bureau considers that it is inconsistent for Westinghouse: (a) to state in paragraph 7 of reference (a) that no improvement to engine development schedules can be quoted, and (b) to make no mention in reference (a) as to how much the total costs of the several engine development contracts might be reduced. Reference (a) satisfactorily details the manner in which the Westinghouse organization can be aided toward meeting its contractual obligations to effect various engine designs. However, reference (a) bears evidence of having been prepared without sufficient estimation effort on the side of the issue which is most important from this bureau's standpoint; namely in return for effecting the proposed arrangement, how will the government benefit in terms of expedition of deliveries under, and reduction of costs of, the several engine development contracts?"

BuAer stated that WAGT was obligated to satisfactorily explain this point since none of the contracts were based on assistance from AEL.[69]

No official WAGT response to this memo was found in any of the contract files, but the use of the test cell was granted and both J40 models and J46 models were tested there under lease, with WAGT paying for the fuel burned. AEL did not produce any reports on the results of these tests. It was found that the AEL test results for a given specific engine did not match the WAGT test results and work was done to determine why this was so but the cause of the differences was never definitively determined.

Development continued with detail refinements. WAGT notified BuAer that MUB Item 8 was resolved by replacing the Emergency Fuel Regulator with the new combined Fuel Regulator. This change now provided sufficient clearance to both of the lower air bleed ports.[70]

In late June, a further simplification of the A/B fuel system was proposed. WAGT noted that the specification called for the emergency A/B fuel system to function satisfactorily up to a ram pressure ratio of 1.15:1 at a fuel pressure of 2.0 psi below ambient pressure up to 6,000 feet altitude. To meet this requirement, WAGT had:

- Enlarged the capacity of both primary and emergency fuel pumps,
- Designed the engine-driven fuel booster pump with a capacity equivalent to both elements of the dual engine pump, and
- Added a shuttle valve in the pump and afterburner control to sense an afterburner pump failure resulting from loss of its inlet pressure and then route fuel to the afterburner from the dual engine pump.

They then argued that there was no reason to guard against pump over-speeds (possible with an air-turbine driven fuel pump), since the -2 engine pump used engine oil-lubricated bearings, could not overspeed and had no metal to metal rubbing parts. The pump being used had already demonstrated its reliability.

As a result, WAGT wanted to remove the emergency valves in the existing afterburner control as an unnecessary complication. Doing so would cause the emergency features to no longer meet the specification, since the A/B fuel pump could not operate with 100°F temperature fuel at fuel inlet pressures much lower than 50 inches of mercury absolute and because the engine fuel booster pump was in series with only the dual engine pump.

To restore full specifications emergency compliance, WAGT proposed additionally to insert a "T" valve with a simple swing check valve inside at the inlet of the afterburner fuel pump. A 1.25 inch fuel line would run from the discharge of the engine fuel booster pump to the new T-valve. In the event of loss of airframe fuel pressure, the A/B fuel pump, using positive fuel pressure from the existing engine fuel pump, would automatically meet the 1.15:1 ram pressure ratio requirement. If the booster pump failed, fuel would flow through the check valve at a reduced pressure of 0.1 psi.

The weight increase on the engine from the change would be about two pounds.[71] WAGT informed BuAer two weeks later that based on the survey responses and subject to BuAer and Air Force approval, they were going ahead with the simplified design and that a layout of the proposed piping was now available and being sent out to the interested parties.[72]

The above survey was in progress when BuAer complained that the J40-WE-6 emergency system provided "greater flexibility than the standard arrangement" required by the specification, in that the emergency fuel pump and emergency fuel control could be selected independently. They stated that if standard cockpit arrangements were to be maintained, changes to the standard could be tolerated. The increased flexibility which the J40-WE-6 emergency system provided was considered by BuAer to be a disadvantage rather than an advantage. This memo applied to the J46 as well as the J40, as it had the same capabilities built into its emergency controls.[73]

With all the increased focus on accelerating development of the J46, the May progress report was not encouraging. The compressor design which had been found to operate without stalling in the normal operating range was a poor performer overall. It developed only a 4.65 pressure ratio, had a low compressor efficiency of 80.3% and achieved a low turbine inlet temperature of 1,150°F, all well below their design targets.

The third and fourth engines had been built with the third having begun running on May 26. It was stopped and on June 3 the compressor was modified by twisting the tips of the 11th and 12th stages to increase air flow at the hub. The compressor was now found to be in stall up to 8,600 rpm. On this engine, the eyelid actuators proved to need redesign and no variable nozzle running was accomplished, a fixed iris nozzle being used and operated manually with the emergency control. Accelerations in this mode had to be carefully watched to avoid over-temping the turbine.

Mechanical failures were experienced. The vertical drive shaft driving the accessory gearbox failed. The cause was a seized shaft on the A/B fuel pump impeller. A large quantity of oil was found in the A/B fuel pump indicating excessive leakage past the seal. A second vertical shaft failed, this time due to improper assembly of the afterburner pump. A total of only 12 minutes of A/B testing was accomplished and that at low fuel flow rates. The compressor failed on June 10 while running at 9,750 rpm. A fatigue fracture of a 2nd stage compressor groove allowed the blade to pull away from the disc. The entire blade path of the compressor had to be replaced along with the compressor housing. This was the sixth compressor disc to be found with fatigue cracks in the blade groove channels.

No running had been accomplished at the design turbine inlet temperature of 1,525°F.[74]

The struggle to get the engine to run properly was absorbing a lot of resources. It could not have been good news that BuAer rejected the AN-E-32 cold exhaust nozzle test results on the J45-WE-32/-38 actuator, stating that WAGT failed to maintain the actuator oil temperature at -55°F ±5°F and instead tested with the oil temperature range of -52 to +93°F. Damage to the "O" ring and actuator piston were found at teardown. The testing results were incomplete and lacked detail, such as resistances and emergency schedule torques which had not been provided in the test report.

BuAer required that the cold actuator tests be repeated, skipping the hot and room temperature tests.[75] The J46 had to pass the even tougher MIL-E cold testing specifications on its actuator which would require it to produce more force to operate an iris nozzle than the J34-WE-32/-38 design.

In acknowledging the BuAer request that separate specifications be written for the -8 and -10 de-rated engines a rare reference to the XJ46-WE-6 was found. This engine was to be a J46-WE-2 with an afterburner that could operate to a 55K foot ceiling, sometimes called the "high altitude" afterburner. The project was cancelled to divert funds to complete the J46 development. No evidence was found that any hardware was built or an actual specification written for this model.[76]

The continuing problems with the electronic control systems being used on the J40-WE-6 and J34-WE-32/-38 and in test for the XJ40-WE-8 were reviewed at a conference on July 26. The reliability in flight was the major concern. The controls suffered from:

- Vacuum tube failures (both physically due to vibration and electrical),
- Circuit signal drift affecting the ability to control both turbine inlet temperature and rpm properly,
- An inability to maintain power settings or return to the same power settings during flight, and
- Inconsistent performance from one engine to the next.

These factors were combining to shorten overhaul life, increase maintenance time, and increase pilot workload. The performance concerns were long running and at the conference WAGT discussed a new "simple hydraulic control" (SHC) for the engines.

The proposed replacement design appeared to be quite acceptable at the conference and WAGT offered four alternate proposals.

1. Qualifying the J40-WE-8 and J46-WE-8 and J46-WE-10 (replacing the J46-WE-4) engines with the electronic control system,
2. Qualifying the above engines with the emergency portion of the present electric control systems,
3. Qualifying the above engines and the J40-WE-6 engine with the SHC, or
4. Qualifying the respective control systems with a 50-hour test preliminary to additional later qualifications in order to gain time.

Every effort was to be made to avoid interim combinations and concentrate on qualifying engines with the SHC.

The following schedules were reviewed:

- J40-WE-8 (Current w/electronic control) October 1951, w/SHC January 1952
- J46-WE-8 (Current w/electronic control) September 1951, w/SHC February 1952
- J46-WE-10 (Current w/electronic control) October 1951, w/SHC January 1952
- J40-WE-8 – Keep the present electronic system (No Date)

BuAer studied the proposed schedules and issued a Speedletter to the BAR on August 6. They had come to the conclusion that the emphasis must be placed immediately on development of a simplified control system and that the XJ40-WE-8 and XJ46-WE-2/4 (de-rated) engines should be qualified with such systems at the earliest practicable dates.

With WAGT's concurrence, the letter stated that:

1. The shift in technical effort could be accomplished under the present development contract items, with no change to the current estimate of total contract cost.
2. The specifications would be revised to reflect the simplified control system.
3. The design and performance features of the SHC would be defined in detail at an early date, with revisions to meet BuAer's minimum requirements as necessary.
4. WAGT would make every effort to complete the qualification tests with the simplified controls before the dates quoted in the WAGT conference notes memorandum.
5. Development of the complete temperature-sensing electronic control would be continued by WAGT to the extent possible under the existing contractual terms, but no effort would be made to verify it for production engines until reasonable service reliability had been demonstrated.
6. The decisions on the XJ40-WE-10/-12 were deferred until preliminary results of the simplified control for the XJ40-WE-8 became known.[77]

And so the electronic control, which had been looked to as a big step forward in providing almost total automatic engine management along with gains in ease of pilot control, improved SFC's and longer engine life was to be pushed aside for something with improved reliability and better A/B failsafe operation. It is a puzzle why it took BuAer so long to finally reject this system, which struggled from the very first to operate reliably and deliver the expected improvements. Now the race was on to deliver a mechanical power control that could meet the specifications.

Almost immediately, BuAer began trying to understand the impacts of changing to the SHC relative to the existing emergency control provisions. A request was made to WAGT to confirm that the emergency warning signal and preflight check system would comply with the specification requirement.

WAGT covered the five possible solutions, depending on the control system installed:

1. For J46, J40-WE-8 and J40-WE-10/-12 engines with the electronic primary control system, with the cockpit switch in the "Automatic" position, the emergency fuel system took over automatically in the event of an engine electrical system failure. If either the primary or emergency fuel pump failed, the remaining pump was selected automatically.
   a. If automatic selection occurred due to pump failure, the cockpit warning light circuit was closed (i.e., the cockpit light came on.)
   b. In the J40-WE-10/-12, since both pumps operated in parallel at low altitudes, the light came on if either pump failed.
2. If the cockpit switch was placed in the "check" position, the emergency fuel pump on the J46 and J40-WE-8 and the emergency fuel control were both manually actuated and locked-in. The light in the cockpit came on.
3. In the "Reset" position, the primary control was selected even if components of the primary system had failed. A failed fuel pump would not be

reactivated in this case, the cockpit light remaining on.

4. If the SHC was used, there was no longer an emergency fuel control to select either automatically or manually. The failure of a fuel pump turned on the light. On the J40-WE-10/-12 with their parallel fuel pumping system, there was no need for a switch, since there was nothing to be manually checked.
5. At the July 26 conference, a tentative approval was given to mount an emergency exhaust nozzle over-ride lever in the cockpit to permit emergency movement of the exhaust nozzle to either the full open or closed position in the event of a power failure. It was recommended that for the A/B versions with the two-position (clamshell type) exhaust nozzle, a cockpit switch should be provided as an emergency override for the "Micro-jet" or "Aerotec" switches until they were proven to be reliable.[78]

The June activity report is not available, but the July report shows that progress was being made. Most testing was now focused on the controls and A/B evaluation using the No. 4 engine. On August 14 the compressor on that engine was badly damaged when the engine was FOD'ed by an unknown object. The damage would slow development until a fifth development engine was assembled, expected in September.

The No. 2 engine was lacking compressor diaphragms. The No. 1 engine was to be used for turbine performance tests in the laboratory (assumed to be WAGT's). The No. 3 engine was being used for a pre-provisioning conference where the parts inventory would be determined.

Teardown of engines in July revealed:

- Excessive distortion and cracking of the vaporizing plate.
- Increased difficulty in removing and installing the 2nd stage turbine shoes due to shrinkage of the 2nd stage nozzle.
- Numerous axial cracks found on the aft section of the outer combustion liner.
- Numerous cracks evident in the engine diffuser fairing reinforcing brackets.
- Thermal cracks in the trailing edges of both the first and second stage turbine nozzle vanes.
- Buckling of the outer combustor liner.

Most of the distortion, cracking and buckling was traced to the faulty temperature distribution which had now been corrected.[79]

In late September, WAGT sent over an updated memo re-calculating the additional funds required to complete the contract. The updated listing reflected the change in scope, materials cost increases, and new costing rates since the earlier determination. The amounts were now $2,737,272.00 which included WAGT's $79,726.00 fee.

The new scope items were:

- The qualification of the XJ46-WE-8 to the AN-E-32 specifications by February 1952,
- The qualification of the XJ46-WE-2 to the AN-E-32 specifications by June 1952, and
- Qualification of the XJ46-WE-2A to the MIL-E-5009 specifications by February 1953.

The memo included several other topics. The reduced thrust -2 would be officially called the -8 and the specification to cover the -8 would be the WAGT-X24C10-6. The change to the SHC caused a rescheduling of the -8 (see above). The balance of the contract remained unchanged.[80]

MUB items were still being cleared, with Item 52 finally being approved on October 29, 1951. WAGT had changed the design to eliminate the use of grommets at bulkhead openings and replaced them with fittings and universal unions to prevent wear at those pass-through locations. The delay had apparently been caused by the lack of resources at the Power Plant Division to oversee WAGT development in detail. The written note on the distribution sheet says "There has long been a dearth of PP-2 personnel for monitoring the Westinghouse development contracts in as much detail as is apparently desired by Power Plant Division. In view of Westinghouse acceleration of effort for the past year or so, several relatively insignificant letters such as this have been extensively delayed by continual receipt of work of higher priority."[81]

MUB Item 70 was approved as well, the data provided on -67°F lubricants, low temperature starting power and speed requirements being deemed satisfactory.[82]

The last approval in October was for MUB Item 54 which requested a design study to show the extent of interchangeability between A/B and non-A/B versions of the engine. The approval specifically covered the electronic control version of the engine and informed WAGT that if the SHC was later installed, a new MUB might be required for an inspection.[83]

The August activity report was not located. The September report stated that the development program had stopped on August 14 due to lack of engine parts. Testing had resumed again September 25 using the No. 2 engine. After 3.87 hours of testing, the engine was wrecked because of fatigue failure of the 2nd stage compressor disc. Testing was again delayed until October 6 when the No. 5 engine became available. This engine had the new 1st and 2nd compressor stages. After 4.5 hours of running, the engine's turbine assembly was wrecked due to failure of the lock washer design in the vaporizing plate assembly. The bolts became loose and one of them passed through the assembly. The No. 4 engine was in assembly and a test start the week of October 15 appeared possible. The No. 5 engine was built to the XJ46-WE-8 configuration and was scheduled for the unofficial 150-hour type test.[84]

Figure 6: XJ46-WE-2 #2 second stage compressor disc failure. WAGT P45343, 9/25/1951. *Courtesy Hagley Museum and Library*

The survey on changing the oil cooler installation location was still outstanding. WAGT wrote to BuAer stating that they had resolved the installation issue with Douglas by way of personal discussions with the Douglas engineers regarding putting an indent in the drip pan. They asked for BuAer approval. (This memo is the first to include the XJ46-WE-3 designation, which initially was the Air Force equivalent to the XJ46-WE-8.)[85]

The story of the XJ46-WE-2 fades into the background at this point, the narrative shifting to the effort to get the de-rated XJ46-WE-8 and its sub-types' production ready. The contract requirement to deliver the XJ46-WE-2 remained until the contract ended in its entirety, but the engine never attained its performance guarantees in its original form. To meet the original guaranteed performance, its design would have to undergo major design changes and emerge as the J46-WE-18 intended for the ground attack version of the F7U, the A2U-1.

Figure 7: Rigid fuel lines on XJ46-WE-2 #5, soon rejected for hoses due to BuAer objections as a result of fracture experience on other engines. WAGT P47627, 9/14/1952. *Courtesy Hagley Museum and Library*

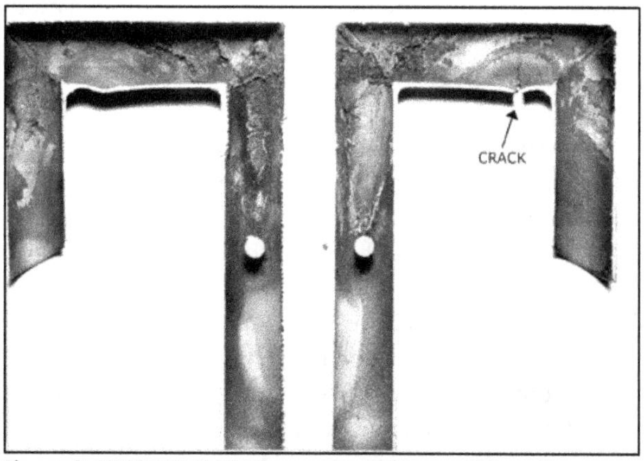

Figure 8: XJ46-WE-2 #1 Evaporator tube condition after testing on engine. The material thickness was increased as a result. Note the now shorter length of the discharge end. WAGT P46181, 8/30/1951. *Courtesy Hagley Museum and Library*

Figure 9: First XJ46-WE-2 on the test bed. WAGT P42536, 4/7/1950. *Courtesy Hagley Museum and Library*

Chapter 3 – XJ46-WE-2 (and -4) Continued Development

[1] BuAer Memo, June 20, 1950, "Contract NOa(s) 9670, Item 7, monthly progress report for May; forwarding of.", **B554V12RG72.3.2NACP**
[2] BuAer Memo, June 14, 1950, "Contract NOa(s) 9670; XJ46-WE-2 wiring harness routing", **B554V12RG72.3.2NACP**
[3] BuAer Memo, June 15, 1950, "Contract NOa(s) 9670; XJ46-WE-2 interchangeability of electric connections to junction box.", **B554V12RG72.3.2NACP**
[4] BuAer Memo, June 15, 1950, "Westinghouse engine data; schedule for distribution of", **B554V12RG72.3.2NACP**
[5] BuAer Memo, June 27, 1950, "contract NOa(s) 9670; XJ46-WE-2 engine, mock-up change proposals; approval of", **B554V12RG72.3.2NACP**
[6] BuAer Memo, June 27, 1950, "Westinghouse XJ34-WE-32 and -38, XJ46 and XJ40 engines; fuel system drain provisions.", **B554V12RG72.3.2NACP**
[7] WAGT Memo, June 16, 1950, "Contract NOa(s) 9670; WEC Turbo-Jet Engine Model J46WE-2. Use of Alternator Generator to Start Engine in Chance Vought F7U-3.", **B554V12RG72.3.2NACP**
[8] BuAer Contract Amendment 7, August 14, 1950, **B552V1RG72.3.2NACP**
[9] BuAer Memo, July 21, 1950, "Contract NOa(s) 9670; Model XJ46-WE-2 engine drain valves", **B554V12RG72.3.2NACP**
[10] BuAer Memo, July 28, 1950, "Contract NOa(s) 9937; model F7U-3 aircraft, engine starting.", **B554V12RG72.3.2NACP**
[11] BuAer Memo, August 1, 1950, "Contract NOa(s)-9670, Westinghouse Specification WAGTX24C10-2E dated 25 April 1950 covering the Models XJ46-WE-2, -4 engines; starter drive requirements for", **B554V1RG72.3.2NACP**
[12] BuAer Teletype, August 2, 1950, No subject, **B554V1RG72.3.2NACP**
[13] BuAer Memo, August 16, 1950, "Contract NOa(s) 9670, item 7, monthly progress report (June and July 1950); forwarding of", **B554V1RG72.3.2NACP**
[14] WAGT Memo, August 17, 1950, "Contracts NOa(s) 9791, NOa(s) 9670 and NOa(s) 10385, J34-WE-32 and -38 Engines, J46 Engines and J40 Engines; Pressure Limitations for the Power Regulator and Governor Alternator", **B554V12RG72.3.2NACP**
[15] WAGT Memo, August 25, 1950, "NOa(s) 9670 Revised Proposal for Development of the XJ46-WE-2A turbo-Jet Engine Our Reference WG-60999", **B554V1RG72.3.2NACP**
[16] BuAer Memo, August 30, 1950, "Contract NOa(s) 9670 Revised Proposal for Development of the XJ46-WE-2A Turbo Jet Engine", **B554V1RG72.3.2NACP**
[17] BuAer Memo, September 28, 1950, "Contract NOa(s) 9670, item 7, monthly progress report (August 1950); forwarding of.", **B554V1RG72.3.2NACP**
[18] BuAer Memo, October 26, 1950, "Contract NOa(s) 9670, Item 7 – Monthly Progress Report (September 1950) – Forwarding of.", **B554V1RG72.3.2NACP**
[19] BuAer Memo, October 23, 1950, "Contracts NOa(s) 9670, 10067, and 10114; use of titanium in XJ46 and XJ40 engine designs", **B554V12RG72.3.2NACP**
[20] WAGT Memo, November 21, 1950, "Contracts NOa(s)-9670, 10825, 10114, 10067 and 10385 – Use of Titanium in J46 and J40 Engines", **B554V12RG72.3.2NACP**
[21] BuAer Memo, November 22, 1950, "Contract NOa(s) 9670, item 7, monthly progress report (October 1950); forwarding of", **B555V2RG72.3.2NACP**
[22] WAGT Memo, October 19, 1950, "Contract NOa(s) 9670 WEC Turbo-Jet Engine Models XJ46WE-2 and XJ46WE-4. Relocation of Exhaust Nozzle Control and Aft Combustion Chamber Drain Valve, Survey of. Engine Change Nos. 85 for J46WE-2 and 1 for J46WE-4.", **B554V1RG72.3.2NACP**
[23] WAGT Memo, November 17, 1950, "Revision of Engine Model Specifications to show maximum fuel consumption with Primary Control Contracts NOa(s) 10385, 10067, 10943, 11028 and 9670", **B554V12RG72.3.2NACP**
[24] WAGT Memo, October 23, 1950, "Contract NOa(s) 9670 – XJ46-WE-2 Decrease Power Lever Torque.", **B554V1RG72.3.2NACP**
[25] WAGT Memo, October 24, 1950, "Contract NOa(s) 9670 WEC Turbo-Jet Engine Models XJ46WE-2 and XJ46WE-4. Relocation of Oil Cooler, Survey of. Engine Change Nos. 86 for XJ46WE-2 and 2 for XJ46WE-4.", **B554V1RG72.3.2NACP**
[26] BuAer Memo, November 14, 1950, "Contract NOa(s) 9670 – Westinghouse Model XJ46-WE-2 engine; fuel inlet pressure requirements for", **B554V1RG72.3.2NACP**
[27] BuAer Contract Amendment 7, November 17, 1950, **B556VC2RG72.3.2NACP**
[28] WAGT Report A-1048, November 17, 1950, Final Report on Starting Requirements of XJ46-WE-2 Engine; Contract NOa(s) 9670, **B554V1RG72.3.2NACP**
[29] BuAer Memo, December 11, 1950, "Contract NOa(s) 9670; Model J46-WE-2 engine, power regulator cooling", **B555V2RG72.3.2NACP**
[30] Douglas Memo, November 30, 1950, "Contract Noa(s) 9670, WEC Turbo-Jet Engine Models XJ46-WE-2 and XJ46-WE-4, Relocation of Oil Cooler; survey of engine Nos. 86 for XJ46-WE-2 and 2 for XJ46-WE-4.", **B554V2RG72.3.2NACP**
[31] Contract NOa(s) 9670 Amendment 8, December 21, 1950, **B556VC2RG72.3.2NACP**
[32] WAGT Memo, December 14, 1950, "Contracts NOa(s) 9670 and 9791. WEC Turbo Jet Engine Models J34WE-32, -38, J46WE-2 and -4. Proposed Power Regulator Shock Mounts, Survey of. Engine Change No. 69 for -32 engines; No 10 for -38 engines, No. 87 for -2 engines and No. 3 for -4 engines.", **B554V2RG72.3.2NACP**
[33] WAGT Memo, January 6, 1951, "Contract NOa(s) 9670; WEC Turbo-Jet Engine Model J34-WE-2; F7U-3 Airplane Manifold Dump Tank Installation, Engine Mounting Provisions for.", **B554V2RG72.3.2NACP**
[34] WAGT Memo, January 17, 1951, "Contract NOa(s) 9670, Turbo Jet engine Model J46WE2, Oil Cooler Relocation Survey.", **B554V13RG72.3.2NACP**
[35] BuAer Memo, January 11, 1951, "Contract NOa(s) 9670; XJ46-WE-2 center of gravity location", **B554V2RG72.3.2NACP**
[36] WAGT Memo, April 14, 1951, "Contract NOa(s) 9670; XJ46-WE-2 Center of Gravity Location.", **B555V3RG72.3.2NACP**
[37] WAGT Memo, January 10, 1951, "Contract NOa(s) 9670 WEC Turbo Jet Engine Models XJ46WE-2 and -4. Jack Lug Relocation,

Survey of. Engine Change No. 88 for XJ46WE-2 and No. 4 for XJ46WE-4 Engines.", **B554V2RG72.3.2NACP**

[38] WAGT Memo, January 16, 1951, "Contract NOa(s) 9670 WEC Turbo-Jet Engine. J46-WE-2, -4 Relocation of Jacklugs.", **B554V1RG72.3.2NACP**

[39] BuAer Memo, December 19. 1950, "Contract NOa(s) 9670, item 7 – monthly progress report (November 1950) – forwarding of.", **B554V2RG72.3.2NACP**

[40] BuAer Memo, January 16, 1951, "Contract NOa(s) 9670, item 7 – Monthly Progress Report (December 1950) – forwarding of.", **B554V2RG72.3.2NACP**

[41] WAGT Memo, February 11, 1951, "J46-WE-1, -2, -4 Contracts NOa(s) 9670 and 11028; Starting of", **B554V2RG72.3.2NACP**

[42] WAGT Memo, January 30, 1951, "Contract NOa(s) 9670; Model J34-WE-2 engines: Windmilling of.", **B554V2RG72.3.2NACP**

[43] WAGT Memo, February 6, 1951, "Contract NOa(s)-9670 Additional Funds Required, Our Reference WG-60999", **B554V2RG72.3.2NACP**

[44] WAGT Memo, February 20, 1951, "Contract NOa(s)-9670, Additional Funds Required, Our Reference WG-60999", **B554V2RG72.3.2NACP**

[45] BuAer Memo, February 20, 1951, "Contract NOa(s) 9670, item 7, monthly progress report (January 1951); forwarding of.", **B555V3RG72.3.2NACP**

[46] WAGT Memo, February 21, 1951, "Auxiliary Electrical Control for Westinghouse Turbo Jet Engines on Contract NOa(s) 10067, 10114, 10825, and 11028. Material Required for:" **B1095VE1RG72.3.2NACP**

[47] BuAer Memo, March 26, 1951, "Contracts NOa(s) 9670, 10067, and 10114, additional funds required", **B554V13RG72.3.2NACP**

[48] WAGT Memo, March 17, 1951, "Contract NOa(s) 10825; WEC Turbo-Jet Engine Model XJ46WE4; Engine Configuration, Information on.", **B1099VE1RG72.3.2NACP**

[49] BuAer Memo, March 20, 1951, Contract NOa(s) 9670, item 7, Monthly Progress Report (February 1951) – Forwarding of.", **B555V3RG72.3.2NACP**

[50] BuAer Memo, April 11, 1951, "Contracts NOa(s) 9670 and 11028; simplification of XJ46-WE-1/-2 afterburner fuel systems", **B555V3RG72.3.2NACP**

[51] WAGT Memo, January 20, 1951, "Contract NOa(s)-9670 WEC Turbo Jet Engine Models XJ46WE-2 and -4. Incorporation of Auxiliary Electric Control, Survey of. Engine Change Nos. 89 for -2 and 5 for -4 Engines.", **B555V3RG72.3.2NACP**

[52] BuAer Memo, February 12, 1951, "Contract NOa(s) 9670; XJ46-WE-2 and -4 engines; auxiliary electric control; mounting of", **B554V2RG72.3.2NACP**

[53] WAGT Memo, April 14, 1951, "Contract NOa(s) 9670 WEC Turbo Jet Engine Models XJ46WE-2 and -4. Feasibility of Engine Mounting the Auxiliary Electric Control, Survey of Engine Change Nos. 94 for -2 engines and 10 for -4 engines.", **B555V3RG72.3.2NACP**

[54] WAGT Memo, April 25, 1951, "Contract NOa(s) 9670 WEC Turbo-Jet Engine Models XJ46WE-2 and -4. Auxiliary Electric Control, Mounting of.", **B555V3RG72.3.2NACP**

[55] BuAer Memo, April 19, 1951, "Contracts NOa(s) 9670 and 11028; Westinghouse J46-WE-1, -2 and -4 engines; starting of", **B554V13RG72.3.2NACP**

[56] BuAer Memo, April 20, 1951, "Contract NOa(s) 9670; XJ46-WE-2 engine, capacity and tap-off for anti-ice air", **B555V3RG72.3.2NACP**

[57] WAGT Memo, April 17, 1951, "Contract NOa(s) 9670 WEC Turbo Jet Engine Models XJ46WE-1, -2 and -4. Front Mount Redesign, Survey of. Engine Change Nos. 95 and 11 for -4 engines.", **B555V3RG72.3.2NACP**

[58] WAGT Memo, April 19, 1951, "Contract NOa(s) 9670 WEC Turbo Jet Engine Models XJ46WE-2 and -4. Incorporation of a High Energy Ignition System, Survey of. Engine Change Nos. 93 for -2 engines and 9 for -4 engines.", **B555V3RG72.3.2NACP**

[59] BuAer Memo, May 18, 1951, "Contract NOa(s) 9670, Item 7 – Monthly Progress Report (April 1951); forwarding of.", **B555V3RG72.3.2NACP**

[60] BuAer Speedletter, May 7, 1951, "De-rated J46-WE-2 engine performance", **B1099VE1RG72.3.2NACP**

[61] BuAer Memo, May 10, 1951, "Contracts NOa(s) 9791, 11028 and 9670, WEC turbo jet engine models J34WE-32, XJ46WE-1 and -2, removal of afterburner drain boss and afterburner lifting lug and bumper; survey of.", **B555V3RG72.3.2NACP**

[62] WAGT Memo, May 11, 1951, "Contracts NOa(s)-9670 and 10825 Revised Performance of J46-WE-2 and -4 Engines, Our References WG-60999 and WG-64730", **B554V13RG72.3.2NACP**

[63] BuAer Memo, May 21, 1951, "Contract NOa(s) 9670; XJ46-WE-2 engine; exhaust nozzle actuator system", **B555V3RG72.3.2NACP**

[64] WAGT Memo, May 16, 1951, "Contract NOa(s) 9670, 11028 and 10385 WEC Turbo Jet Engine Models XJ46WE-1, -2, -4 and XJ40WE-10. Airframe A.C. Power Requirement for use with Emergency Control Systems, Survey of. Engine Change Nos. 96 for -2 engines, 12 for -4 engines and 67 for -10 engines.", **B554V13RG72.3.2NACP**

[65] BuAer Contract NOa(s) 9670, Amendment 9, **B556VC2RG72.3.2NACP**

[66] BuAer Memo, June 7, 1951, "Contracts NOa(s) 9670 and 11028; simplification of XJ46-WE-1/-2 afterburner fuel systems", **B555V3RG72.3.2NACP**

[67] BuAer Memo, June 15, 1951, "Westinghouse Specification WAGT-24C10-2G revised 13 Mar 1951 covering models XJ46-WE-2, -2A, and -4 engines: approval of", **B554V13RG72.3.2NACP**

[68] WAGT Memo, June 7, 1951, "Contract NOa(s) 9670 and 11028 WEC Turbo Jet Engine Models XJ46WE-1 & 2. Proposed Ejector Configuration, Survey of. Engine Change No. 98 for XJ46WE-2 engines.", **B555V4RG72.3.2NACP**

[69] BuAer Memo, June 28, 1951, "Contracts NOa(s) 9670, 10067, 10114; use of test cell at Aeronautical Engine Laboratory", **B555V4RG72.3.2NACP**

[70] WAGT Memo, June 15, 1951, "Contract NOa(s) 9670 – XJ46-WE-2, clearance for bleed ports.", **B555V4RG72.3.2NACP**

[71] WAGT Memo, June 28, 1951, "Contract NOa(s) 9670, J46-WE-2 engines; Contract NOa(s) 11028, XJ46WE-1 Simplification of Afterburner Fuel System, Survey of. Engine Change No. 101", **B555V4RG72.3.2NACP**

[72] BuAer Memo, July 5, 1951, "Contract NOa(s) 9670 and 11028 WEC Turbo Jet Engine Models XJ46WE-1 and -2. Current Status of the Design of the Simplified Afterburner Fuel System, Information on.", **B555V4RG72.3.2NACP**

[73] BuAer Memo, July 6, 1951, "Contracts NOa(s) 9670, 10385, 10114, and 10067, XJ46-WE-2, XJ40-WE-6, XJ40-WE-8, and XJ40-WE-10 engines; emergency fuel system contractor furnished components", **B554V13RG72.3.2NACP**
[74] BuAer Memo, June 15, 1951, "Contract NOa(s) 9670, item 7; Monthly Progress Report (May 1951); forwarding of.", **B555V4RG72.3.2NACP**
[75] BuAer Memo, July 11, 1951, "Contracts Noa(s) 9670 and 9212, J34-WE-32, J34-WE-38, J34-WE-17 and J40-WE-6 engines, exhaust nozzle actuator", **B555V4RG72.3.2NACP**
[76] WAGT Memo, July 13, 1951, "Contract Noa(s) 9670 Westinghouse Model Specification WAGT-X24C10-2G dated 12 February 1951; Revised 13 March 1951 and 28 April 1951", **B554V13RG72.3.2NACP**
[77] BuAer Speedletter, August 6, 1951, "Simplified Control System for XJ40 and XJ46 engines", **B556VC2RG72.3.2NACP**
[78] WAGT Memo, August 6, 1951, "Contracts NOa(s) 9670, 10385, 10114 and 10067, XJ46-WE-2, XJ40-WE-6, XJ40-WE-8 and XJ40-WE-10 Engines; Emergency Fuel System Contractor Furnished Components", **B555V4RG72.3.2NACP**
[79] BuAer Memo, August 15, 1951, "Contract NOa(s) 9670, monthly progress report (July 1951); forwarding of.", **B555V4RG72.3.2NACP**
[80] WAGT Memo, September 25, 1951, "Contract NOa(s)-9670 Additional Funds Required our Reference WG-60999, "**B555V4RG72.3.2NACP**
[81] BuAer Memo, October 29, 1951, "Contract NOa(s) 9670: grommets or similar devices in support panels", **B555V4RG72.3.2NACP**
[82] BuAer Memo, October 29, 1951, "Contract NOa(s) 9670; XJ46-WE-2 starting at -67°F and power and speed requirements", **B555V4RG72.3.2NACP**
[83] BuAer Memo, October 30, 1951, "Contract NOa(s) 9670; design study for parts interchangeability between XJ46-WE-2 and -4 engines.", **B555V4RG72.3.2NACP**
[84] BuAer Memo, October 16, 1951, "Contract NOa(s) 9670, monthly progress report for month of September 1951; forwarding of.", **B555V4RG72.3.2NACP**
[85] Contract NOa(s) 9670 and 11028 WEC Turbo Jet Engine Models XJ46-WE-1, -2, -3, -4, -8 and -10 Relocation of Oil Cooler.", **B555V4RG72.3.2NACP**

Chapter 4
XJ46-WE-8/A/B and -12/A/B Development

There is more than ample evidence that BuAer sought to ensure similar models of the Westinghouse jet engines were interchangeable in general. They spent time analyzing engines for commonality and installation interchangeability. But for the most part, in spite of their efforts in that direction, the many sub-models of all of their engines were not interchangeable in a given airframe, the engines themselves having gained custom tailored installation features. The many variants of the J34-WE-22, each for a specific airframe, became an issue for the parts support organization who fought to keep the same from happening to the J46.

As the effort shifted from the J46-WE-2 to the de-rated J46-WE-8 (-10) for production, almost immediately changes began to make these "special application" engines tailored for use in the Douglas F3D-2 SkyKnight, Chance Vought F7U-3 Cutlass and then, just a bit later, with yet more unique changes, as the J46-WE-12 for the Convair F2Y Sea Dart. The core sections of these models were identical, and the struggles with the control system applied to both, so they are discussed together in this chapter. The individual installation envelope changes for each will be highlighted.

For the XF7U-3, Vought had asked that a bend ("cant") be incorporated in the A/B extension in their engines. WAGT stated the cant could be incorporated in time for the 150-hour qualification test on the engine. The engines for the XF2Y would retain the straight out A/B design and were designated XJ46-WE-12. Since the engines were identical except for the cant, the qualification of the XJ46-WE-8 would be accepted as the qualification for the -12 as well.[1]

Figure 1: J46-WE-8 Mockup, earliest known picture showing CVA requested cant in exhaust nozzle. WAGT P45882, 12/21/1951. *Courtesy Hagley Museum and Library*

The previously discussed starter system planned by Chance Vought for the F7U-3 required an adapter gearbox to drive the starter-generator and function in reverse to start the engine. In November 1951, BuAer notified WAGT that they were considering making the adapter gearbox an integral part of the engine, using engine oil for lubrication and heat transfer. The part would be made part of the specifications, be qualified with the engine and guaranteed for satisfactory operation under all conditions and limits with the gearbox and starter generator attached. The engine was now considered a special application engine.

A proposal for furnishing this gearbox as an integral part of the engine was requested. Two General Electric starter-generators were to be forwarded from Chance Vought to WAGT to facilitate the preparation of the proposal.[2]

On December 1, 1951, Amendment 11 to the contract was approved changing the General and Administrative Expenses attributable to Headquarters activities to a rate of 2.71%.[3]

In late November, WAGT report A-1274 on the Simple Hydraulic Control was sent over to BuAer. It contains the most complete description of the SHC found and so will be quoted at length here.

The report starts with a brief historical review of the electronic control and its rejection for general service use:

"The electronic control which had been planned for use on the J46 and J40 engines *(and the J34-WE-32/28)* was an advanced system designed to give optimum engine performance under all flight conditions. The control was operated by means of a single lever and was designed to require a minimum of attention from the pilot. In addition, the primary electronic system was provided with full emergency features and the necessary switchover mechanism. Minor manufacturing and flight troubles were of such multitude that it was questionable if they could all be satisfactorily corrected by the time the engine was required for service use. Relatively good performance was obtained, both at this contractor's plant and at the airframe manufacturers, when the quantity of engines was small. However, it was found that when the engine quantity was increased, together with the consequent use of less experienced personnel, the accumulation of these minor troubles contributed to keeping the aircraft on the ground.

"The contributing cause of a multitude of minor troubles was complexity. This complexity was caused by the use of a purely electric control whose acceptance required the use of an emergency hydraulic system with its necessary automatic change-over valves. It is firmly believed that much less trouble would be experienced if this multiplicity of systems was reduced."

With the SHC, the fully modulating automatic control was abandoned. "The exhaust nozzle area is scheduled by throttle position. Thus, the desired turbine temperature is roughly approximated. By the use of an electrical trim on the exhaust nozzle area, this rough temperature approximation can be accurately corrected. Should the electrical trim circuit fail the engine can still be operated with the scheduled exhaust nozzle area. Should the failure cause the exhaust nozzle to close to its minimum area schedule, the pilot can control the maximum temperature by manual manipulation of the throttle. Should it fail to its maximum area schedule the maximum thrust obtainable will be equivalent to that obtainable with a fixed nozzle area which will give *(a)* maximum temperature only at 100°F."

The automatic trim control would not be available on the first engines, so the pilot would be given a manual trim control in the cockpit. This would be replaced with the automatic trim control when it was approved for service use.

Development of the electronic control and alternate hydraulic controls would continue. The alternate hydraulic control was being worked on under Contract NOa(s) 10067. The automatic electrical trim was being developed by Manning, Maxwell, and Moore in cooperation with WAGT.

The first known block diagram of the SHC was published in this report. While the name might lead someone to think that this new control system would actually be "simple", they would be surprised at the remaining complexity of the interconnected circuits needed to control the engine.

At the time of the report, some limited running experience both in the test cells and in the air had been accomplished. The J40-WE-8 had run about 100 hours with 35 hours dedicated to control development. An F3D-1 with an emergency control installed on a ZJ34-WE-38 engine had flown for limited hours. All the major components of the system had been tested for 150 hours or more except for the engine driven fuel booster pump. The overall test results indicated the SHC was basically satisfactory as an engine control.

The booster pump design required more work as indicated by several failures of the vapor element of the J40-WE-8 pump, and an extensive development program was under way to identify a more suitable design. While other design changes in the control overall might be necessary, they were expected to be minor. All drawings were expected to be complete by December 1.

The first production prototype of the automatic temperature control had been received and was in preparation for laboratory testing.

The general operation of the control was explained as follows:

"Any control for the J46 and J40 engines with the present blade paths must hold the exhaust nozzle area open during burst accelerations. This requirement eliminates the conventional fixed area exhaust nozzle in the normal operating range. The 'Simple Hydraulic Control' will, therefore, consist of a Woodward isochronous fuel governor (Speed Relay) controlling military rpm. An electric-hydraulic Exhaust Nozzle Control system arranged to schedule a fully open exhaust nozzle at thrusts from 50% to 100% will be accomplished by progressively closing the Exhaust Nozzle in response to further throttle movement. A fuel flow scheduling Afterburner Control is used in connection with a special Blowout Switch to operate the Exhaust Nozzle. This switch senses a transient pressure rise in the turbine outlet pressure and opens the exhaust nozzle to its fully scheduled position. Conversely, should the afterburner blow out, the switch closes the nozzle. Top military thrust of the engine is secured under conditions of military or maximum thrust by trimming the area of the Exhaust Nozzle from its scheduled position to hold maximum turbine out temperature. As previously stated, early engines will have a manual pilot-operated Temperature Control which will be superseded by an Automatic Temperature Control."

From the pilot's standpoint, their operation of the system entailed:

1. Starting Range (0-21° Throttle Travel) – The fuel flow was manually controlled by the pilot to limit temperature and prevent over-temps.
2. Below Military RPM (21-54°) – Thrust is varied about 50%. The exhaust nozzle was locked in the wide open position. The fuel flow was controlled as a function of compressor pressure rise. At altitude, when the idle rpm increased to the military rpm, the speed relay (topping governor) prevented further increase in rpm, however, thrust could increase through further advancement of the throttle which would progressively close the exhaust nozzle.
3. At Military RPM (45-82°) – Thrust was varied from 50 to 100% at all altitudes by progressively closing the exhaust nozzle to its military position. The pilot used the manual temperature trim to adjust the exhaust nozzle to obtain the allowable turbine out temperature. When available, the automatic

temperature trim accomplished this via an electronic amplifier.

4. Afterburner (82-110°) – The pilot advanced the throttle into this range and the afterburner would light with the full flow being scheduled in response to a compressor discharge pressure measurement. Maximum thrust required the use of the Manual (later, the Automatic) Temperature trim control.

A drawback of the system was that fuel consumption below 75% military was higher than the electronic control, with the maximum variation being about +5.3% at 50% military thrust at sea level and +2% at 35,000 feet altitude. Also, the afterburner could no longer be modulated, now becoming an on/off unit. Under cold day high ram low altitude conditions, the compressor pressure could exceed the safe value for the engine if the throttle was in the military or maximum condition. The pilot had to watch the compressor output pressure gauge in those conditions and retard the throttle as necessary to reduce the pressure.

The emergency provisions of the SHC were listed as:

1. Dual fuel pumps with manual selection and automatic cutover if one failed.
2. Manual override capability of the exhaust nozzle if the AC power from the airframe was lost.
3. If the high pressure hydraulic system failed, the exhaust nozzle locked in the failed position.
4. Upon failure of the electrical temperature control circuit, the exhaust nozzle was still controllable with the throttle.
5. The speed control governor which controlled the fuel flow was a conventional flyball type with proven extremely high reliability.
6. Upon failure of the blowout switch with the A/B in operation and with the exhaust nozzle less than fully open, a bypass switch in the cockpit allowed manual operation of the nozzle.[4]

The November activity report noted a general lack of progress. The redesigned inlet guides had not been tested ("fully exploited") yet, and as a result no performance checks had been made. The turbine out temperature problem was still being experienced with the Nos. 4 and 5 engines being used for development.

The BAR estimated that the J46-WE-8 and -10 qualification tests would be delayed until approximately June, 1952.[5]

In January, BuAer had to request the loan of three starter-generators from the AF for flight testing of the J46-WE-10, scheduled to begin in the "near future".[6]

A full report on the exhaust nozzle control system for the J46 was published January 16, 1952. This concentrated on explaining the design and functions of the exhaust nozzle systems specifically.

The details of the technical design and function are of interest here. In non-A/B operation, the nozzle control position was set from signals from the power scheduler, which was mechanically linked to the throttle. The Auxiliary Electric Control sensed the difference in voltage between the slider of the potentiometer in the power scheduler and the slider of the position feedback potentiometer in the exhaust nozzle control. It then transmitted the appropriate command to the control. The commands were received by a solenoid in the variable displacement pump which, in turn, caused the hydraulic pressure to be applied to the proper side of the control piston to the exhaust nozzle control, thus shifting the nozzle to the new position.

If the speed of the engine was below military, the speed switch overrode the power scheduler until the engine rpm reached military. At that point, the overspeed fuel valve controlled the fuel flow and the speed switch returned to the open position.

For A/B operation, the power scheduler shifted from the exhaust nozzle control to the area position that produced military temperature at 100°F at sea level conditions. Closing the throttle de-energized the various A/B control circuits, closing the fuel valve and returning the exhaust nozzle to its normal schedule.[7]

Figure 2: J46-WE-10 Mockup. Based on -8 compressor/turbine sections, it replaced the -4 engine but the program faded as the resources went to get the -8 series into production. WAGT P46151, 1/31/1952. *Courtesy Hagley Museum and Library*

December's activity report stated that 30 hours of testing were performed that month. Encouraging progress had been made in clearing up the radial temperature distribution at the turbine inlet problem. A J46-WE-10 (equivalent house engine No. 4) had been run at full power, but the engine was damaged in early January by a starter cowl bolt that passed back through the engine. The engine would have to be rebuilt and further checks made to confirm that the configuration could adequately meet the full J46-WE-10 de-rated performance guarantees.

Work continued on the compressor towards achieving J46-WE-2/-4 performance. The best estimate of the BAR was that J46-WE-8/-10 qualification tests could not be completed before June 1952 and that production of those models could not start until July or August 1952.[8]

The incorporation of a thrustmeter was planned, and in response to a BuAer query, WAGT responded that the J46-WE-8 would be the first engine to have them installed.[9]

At this point in time, six engines for the XF2Y-1 were culled out of the -8 orders, the six engines being designated the XJ46-WE-12.[10] These engines were mechanically identical to the -8 engine except for minor changes in attachments. The qualification of the -8 would also be considered the qualification of the -12.

Material shortages began to be felt at this time with numerous follow-ups with vendors and subcontractors needed to ensure that the raw materials and finished goods on orders would be delivered on time. Vendors were under heavy pressure from escalating order volumes for high temperature material resulting from the Korean War. Development programs had to fight hard to get what they needed to keep moving forward. On the J40 program, these follow-ups were done using forms stamped "BRICK BAT". On the J46, these were designated "CUE CAP" for some reason. There does not appear to have been any difference in the motive or function of the programs.[11]

January's report stated testing was up to 460 hours and most of it had been spent trying to match the compressor to the turbine. This effort was holding up other development work.[12]

WAGT had asked that approval of the J46-WE-8 specification be given so it could be sent to Chance Vought. BuAer responded that the specification for the -8, -10 and -12 engines was considered unsatisfactory in many respects and a detailed listing of required changes would be forthcoming shortly. The specification would not be released until it was in a "more satisfactory" state. The performance curves related to the engines could be released when they were received by BuAer and the BAR was asked to expedite the transmittal of the curves on the -8 and -10 engines.[13] BuAer forwarded their requested changes for the specification on February 29. Some of the more significant clarifications were:

1. Mount the power control elements on the engine during qualification testing to test for vibration durability.
2. Provide an alternate fuel test to verify the engines can operate with aviation gasoline.
3. The temperature limits of 165°F continuous and 200°F for five minutes would be more acceptable if the five minute rating was raised to 250°F.
4. Add a requirement to deliver a sample control system with assembly, installation, schematic drawings and operating instructions to be delivered within 30 days after completion of the component qualification tests.
5. The elimination of the electronic control should allow reduced radio interference limits from the -2 specification. Explain the retention of the J46-WE-2 high limits for the -8.
6. The engine should eventually be rated on JP-4 fuel, but JP-3 was acceptable until the fuel program was complete.
7. Advise why the -8 start-up altitude was reduced to 35,000 feet from the 45,000 feet of the J46-WE-2.
8. Specification MIL-E-5008 required that the full throttle operating limits be specified. The engine should incorporate a device to prevent the compressor from exceeding its pressure limits. Explain how the pilot flying the -8 would prevent the engine from exceeding the compressor limits on high ram cold day conditions.
9. Engine acceptance without completion of the high and low temperature tests should state "shall not be required".
10. Explain why the altitude idle conditions are much higher than for the -2.
11. Explain the acceleration and deceleration time intervals required for the engine.
12. Specify the engine starting requirements.
13. Verify the engine length increases have been resolved with Chance Vought.
14. If the fuel pump fails and the fuel system automatically cuts over to the emergency system, provide a light in the cockpit to alert the pilot of the failure.
15. Tighten the idle rpm stability limits and specify the temperature control tolerances for the military and maximum operating conditions.
16. State that rigid fuel lines will have adequate vibration support against failure.
17. The engine power control should be independent of the aircraft electrical system.
18. The manual two position emergency control was unsatisfactory.
19. The high control torque of 35 pound-inches in the A/B range was unsatisfactory.
20. The drive table re-designated the second hydraulic pump drive to a special drive with a speed far too high and torque far too low for the pump.

21. The delivery of performance curves (April 1) was too late and earlier delivery was requested. (It was recommended WAGT increase their drawing staffing.)
22. A meeting on March 11 was requested to work out the changes to the specifications.[14]

BuAer was aware that in part the delay in finishing the design and specifications at that point was the conversion over to the still incompletely designed new SHC.[15]

The manual trim control for the -8 engines with the SHC was surveyed in early February. It was simply stated that it was needed to "prevent (an) over temperature occurring at varied flight speeds and altitudes". It would only be operable when the engine was operating at military power and in A/B.[16]

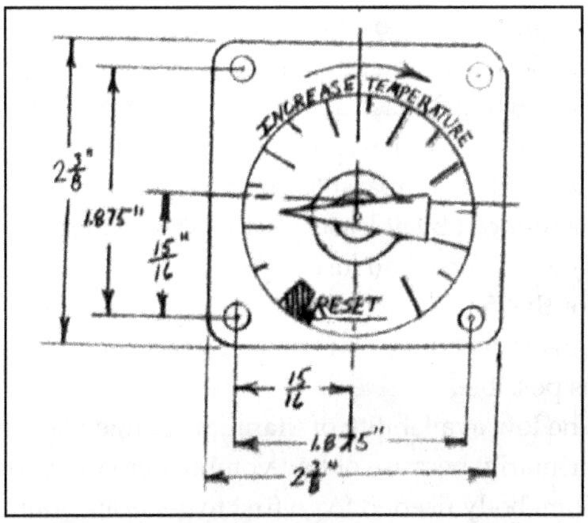

Figure 3: Proposed manual trim switch. From WAGT drawing EDS-M-217344, 2/1/1952.[16]

With the cost to complete the contract increasing over the original committed cost, the director of the Power Plant Division sent a memo over to the Assistant Chief of Research and Development listing a series of programs that could be terminated. Their funds could be reallocated to fund the Westinghouse contracts to completion. Additional Secretary of the Navy emergency funds could also be utilized. He noted that the original engine program costs had been estimated at $9.79 million (J46, J40 programs) and they were now estimating a total of $25.89 million, a 164.5% increase.[17]

In pre-production preparedness, a WAGT plant in Columbus, Missouri had been set up as a second source of jet engine parts for the Westinghouse programs as part of a larger mobilization plan. In February, funding to the tune of $1,395,000 was authorized for set-up and procurement of tools, dies, jigs and fixtures.[18]

The latter part of February, WAGT surveyed the new addition of thermocouples, thermocouple holders and bosses on the A/B diffuser. These were to prepare the engine to use an automatic temperature (trim) control system when it was available. Accommodations for the customer's connection would be added to the junction box.[19]

The starter adapter generator gearboxes for the J46-WE-8 (448 of them) were originally to be ordered directly by Chance Vought from WAGT and be produced to a jointly negotiated layout and design. It was considered this gearbox on the -8 should be considered Government Furnished Equipment (GFE) as part of the engine and BuAer was asked by WAGT to negotiate a contract amendment for WAGT to manufacture and supply the gearboxes with the engines. The weight was 14.5 pounds. The gearbox was mounted on the starter drive pad on the engine centerline at its foremost point and was lubricated with engine oil. The pad output speed to speed ratio was .694:1 giving a pad output speed of 7,000 rpm at an engine speed of 10,100 rpm. It was anticipated that the first three 50-hour flight rated engines would not have the gearbox, but all other -8's would have them. BuAer could supply gearboxes for the first three engines after engine shipment once gearboxes entered production if they so desired.[20]

The tests on a General Electric starter-generator had now been completed. The test results obtained showed that the starter could start a -2, -4, -8 or -10 engine in an average of 62.5 seconds. The report noted that this was longer than considered desirable in the past but if it was acceptable to BuAer, it would be sufficient and satisfactory for the engines. WAGT added that they questioned the durability of the starter, something which had not been tested. They recommended that a search be made for a starter of larger capacity with a goal to obtain the same order of service reliability obtained from the type F2 starter used on the J34 series engines.[21] BuAer forwarded the test report on to General Electric with the warning it could not be used for sales promotion or advertising purposes and could not be reproduced except for internal use within General Electric.[22]

The February progress report states that total running time on the XJ46 was now up to 484 hours. An XJ46-WE-10 had been able to meet guaranteed thrust and specific fuel consumption at military power. At normal power and below, the SFC was slightly above the specification guaranteed volumes. The radial temperature distribution was found to be satisfactory. WAGT reported it might be necessary to use variable inlet guide vanes to obtain stall-free part throttle performance in the -2 engine. The weight estimate for the -8 was now 1,971 pounds vs. the specification weight of 1,863 pounds (a 5.8% increase). A starter cowling bolt had backed out and gone through the #4 engine, cutting testing short.

Figure 4: XJ46 #4 showing compressor damage after starter bolt passed through the compressor. WAGT P45979, 1/8/1952. *Courtesy Hagley Museum and Library*

A test program was being set up at the AEL to determine whether or not the XJ46 compressor had stall characteristics at altitude similar to the XJ40-WE-6. The qualification dates for the XJ46 engines (the -8 and -10) were now set for June and July.

In a last comment, it was stated that the J34-WE-32A engines at Chance Vought were to be modified with J46 and J40 control elements for flight testing in the F7U-1 airframes.[23]

The dual fuel pump for the J46 was approved as qualified to the MIL-E-5009 specification on March 14. The pump was almost identical to the one used on the J34-WE-32.[24]

The Power Plant Division (PPD) acknowledged WAGT's notice to BuAer of the opportunity to incorporate titanium compressor discs in the J46 at a saving of 64 pounds per engine. BuAer had requested PPD recommend and approve the use of titanium in the engines. The PPD reported they had increased pressure for the completion of the necessary development of test data and titanium fabrication ability of WAGT before they could make such a recommendation. The memo noted that if the proposal was accepted, it would be the first direct application of titanium in significant quantities in the production of military equipment. To understand the implications of the use of the metal, the PPD had convened a conference and could report the following information:

1. Each engine would require 750 pounds of sponge titanium to create 7 discs. Production technique improvements might reduce the sponge requirement by 25-35%.
2. 81 pounds of critical chromium per engine would be saved.
3. The additional cost per engine would be about $15,500. This might go down to $14,280 if production efficiencies at WAGT could reduce the need for specimens and testing.
4. The cost of comparable finished titanium parts could be reduced by at least 50% by 1955 or 1956.
5. The Air Force had just requested data on titanium for the J57.
6. BuAer needed to state their requirements as quickly as possible.
7. The low availability of titanium sponge was primarily because of low volume demand to date. Somebody needed to go first to generate volume demand. Was that going to be BuAer?
8. There was not enough (titanium) sponge available to meet the requirements of the J40 program at WAGT and Ford. (Ford was to be the subcontractor for J40 production.) It was not recommended that only part of the J40 run incorporate titanium due to the complexity it would introduce into the logistics.
9. The improvement in performance on an F7U-3 using titanium equipped engines was negligible because the weight savings was only 128 pounds per airplane.[25]

In mid-March, BuAer approved the airframe mounting of the auxiliary electric control for all J46 engines. (For the J40 program BuAer held its ground and required all auxiliary electric controls be engine mounted.)[26]

The March progress report stated that total running time was up to 551 hours. A new compressor configuration (C-50) showed performance characteristics expected to be able to meet the J46-WE-2 performance guarantees. Turbine efficiency was still too low to meet

J46-WE-8 SFC guarantees in the normal range. Changes being made to the -2 were expected to resolve this. The AEL compressor stall investigation test engine had been delivered on March 17. The first production J34-WE-32 was delivered to Chance Vought in late February.[27]

WAGT had sent over the performance curves for the -8, -10 and -12. BuAer notified them that the curves received were identical to the ones for the engines using the electronic control and the ones they wanted were for the engines using the SHC. The earlier curves were not released to the airframe companies.[28]

As WAGT faced the continuing thrust shortfalls in the -8, they came to the conclusion that a larger exhaust nozzle was going to be necessary. The change in design was surveyed in early April. The exhaust nozzle would vary between 180-360 in^2. The provided performance curve shows the new ejector was 23.5 in diameter with an 8 inch overhang.[29]

The Power Plant Division Director sent the BuAer chief a status of the J46 program as of April 3. They summarized the engine positions, stating the latest WAGT production schedule as of February 21 indicated the first -8 would ship during October 1952 and the first -2 during May 1953. The -8 was a 5,720 pound A/B and 3,905 pounds dry engine compared to the -2 at 6,100 pounds A/B 4,080 pounds dry rating with the -2 SFC ratings better than the -8.

"Development progress appears to be on schedule for the -8 engine and it is believed the qualification test date will be met. However, the control system, as a satisfactory service article, is still subject to difficulties which may delay delivery of combat ready engines until December 1952." It was added the -2 appeared to be on schedule and the qualification test date would probably be met with a possible slip of a month due to the redesigned control system.

So in April the chief of BuAer had been told by the PPD Director not to realistically expect the engines according to the current published schedule. All of the engine changes still coming from WAGT must have reinforced the PPD opinion.[30]

Before picking up the development thread again, events related to the WAGT Columbus plant and its funding should be mentioned. The establishment of an expanded parts production capability was part of a larger mobilization plan being put in place for the nation. In our context, the Columbus plant had been authorized to charge their startup costs (in February) to the NOa(s) 10825 contract. But BuAer had cancelled 951 engines from the production contracts at the Kansas City plant only days before that authorization was issued. The original planning included spreading out the Columbus tooling expenses over the first 2,000 engines produced. No instructions had been given to WAGT on how to handle the startup costs now that almost half the planned amortization base was gone. With funding in short supply, a debate within the Navy began regarding the actual costs and rate of development of actual mobilization vs. planning for mobilization itself. Where and how to handle mobilization capability expenses was at the center of the current issue resulting from the engine cancellations. (NOTE: The Ford Motor Company, Lincoln Mercury Division, was to provide a mobilization plant to supply J40 parts and complete engines. Their contract was cancelled when the planned production of the high thrust J40-WE-10 and J40-WE-12 was cancelled.)

A note (dated 4/20) on attached routing sheet states that "WE Co. has been advised by telephone on two separate occasions that since current engine schedules do not warrant use of Columbus plant, 'make-ready' costs incurred there cannot be considered in repricing engines on order. Further, that if and when a sufficient quantity of engines are on order to justify use of Columbus, costs incurred there would be considered in pricing such engines."

A request was sent to move the costs for Columbus to Contract NOa(s) 52-048. This change was not approved. It was recommended that a separate contract for mobilization, tools, jigs, fixtures, etc. be established.[31] Looking ahead a little, the complications of the existing contract obligations regarding Columbus preproduction startup costs reared their head again. With the production schedule slipping, the expenditure plans were now being affected. They triggered a later memo that pointed out the Columbus costs being incurred were for engines not to be delivered until 1953 but they would have to be charged against the 1952 J34 engine shipping contracts. The fact that the preproduction costs were being incurred so far in advance of the planned production was an issue and the schedule should be adjusted accordingly to resolve the expenditure timing issue.[32] Further documentation on the mobilization planning issue was not in the files.

A meeting with Chance Vought and BuAer was held to discuss the desired provisions for emergency control of the exhaust nozzle on the SHC for the -8 if the aircraft electrical system failed. The discussed emergency provisions were only to be needed until the hydro-mechanical control was available.

Four alternatives were discussed and the first was deemed the most desirable. It was stated as: "Provide a manual override so adjusted that the exhaust nozzle may be opened by pilot action. Provide provisions so that the exhaust nozzle will open automatically in case of aircraft electrical power failure, and also, under these circumstances, provide afterburning under wave-off conditions."

WAGT stated they would make every attempt to provide for A/B use during a wave-off but would have to test to ensure consistent A/B ignition with the nozzle wide open during such a wave-off. The design of the system to accomplish manual and automatic emergency control was presented to BuAer in an unnumbered report dated April 9, 1952.[33]

Report A-1317 was sent over to BuAer on April 23, the report supplying the data needed to support the specification requirement for a stated starting torque and speed. It covered the -1, -3, -8 and -10 models. It showed the engine self-sustaining speed was 1,500 rpm with a starting time of 50 seconds to reach that speed.[34]

A major engine design change proposing an increased diameter turbine was surveyed in late April. The redesign was deemed necessary to allow the -2 engine to achieve its performance guarantees. The change would require a new combustion chamber, turbine housing and A/B diffuser combination with larger diameters. The engine mount distance front to rear would remain unchanged.[35]

The change in turbine design of the -2 began to eventually turn this into a different engine. The -2 would emerge from development as the redesigned J46-WE-18 for the A2U-1. The follow-on development of the -2 is covered in full in a later chapter.

For the -8, a survey was done on relocating the ignitors from their present position on the diffuser and placing them on the mount ring. Only two ignitors would be needed at this location. Faster starting and elimination of local hot spots were seen in tests of the new location.[36]

On top of the overrun costs that were already identified, WAGT proposed that the Marquardt Aircraft Company perform tests on the proposed -2 A/B for $150,120.00. The tests would be performed over two, two week test periods.[37]

The April progress report stated the operating time on XJ46 engines was up to 604 hours with 7.68 hours of A/B running. The AEL XJ34-WE-10 compressor test engine was wrecked on April 10 when a starter cowl bracket broke and passed through the engine with the resulting fire destroying the engine. WAGT wrote a report, A-1361, on this failure but a copy was not preserved. The AEL did not publish a report. The AEL cell was repaired by April 21, but another engine was not available until May 14. No significant test results had been obtained from the test engine before the failure occurred. Five J34-WE-32A engines had now been delivered for use in the control system evaluation work. A T-11 larger turbine for the -2 engine was in test at WAGT.[38]

Figure 5: XJ46 #4 wrecked at AEL in April. WAGT P46679, 4/14/1952. *Courtesy Hagley Museum and Library*

With a full court press underway at WAGT, they had had to notify BuAer that all the funds allocated to the two (development and production) contracts had been expended and there was now a projected total shortfall of $6,697,729 for contract completion. WAGT continued development in good faith while the funding issue was being resolved.[39]

In early May, BuAer elected not to jump into titanium material procurement in a big way, telling WAGT they would like to verify the use of titanium discs in five (5) J46 engines before deciding to take action to procure GSA stockpiled titanium. They asked for a proposal from WAGT to include titanium incorporation in the compressors of five (5) J46-WE-8 engines to include a 150-hour verification test of an engine with such a compressor.[40] WAGT had earlier told them that they did not think there was sufficient time to incorporate titanium into J46-WE-8 production but it might be possible to get it into the J46-WE-2. An earlier CID (16371 dated 11/16/51) to build five engines with titanium discs had not been approved.[41] Based on that, it appears that the five -8's to be procured with titanium compressor discs were to be proof of production and design engines, not production prototypes. Skipping ahead a bit, WAGT sent in their proposal in memo form on October 3. The price would be $29,664.00 per engine to replace the steel compressor discs 6-12 with titanium ones, saving 64 pounds per engine. Since these -8 discs were identical in design to the ones being used in the Air Force -3 model (in which discs 3-12 were to be titanium), it was proposed the qualification of the -3 should be accepted as qualification of the -8 using titanium discs as well. Approximately 4,000 pounds of titanium sponge would be needed. It was requested that the contract amendment contain a price redetermination clause since it would be WAGT's first experience with the metal and pricing might not be as accurate as that with production material.[42] BuAer accepted the proposal with the proviso that the test time had be an equivalent number of hours at each of the power settings called for in the -8 specification and have an equivalent number of accelerations, decelerations and starts and that the parts of the test that applied to the verification of the discs be witnessed by a naval inspector.[43]

The latter part of May found the specification battle over the rising weight of the -8 engine continuing. BuAer responded to WAGT's earlier request for a specification weight increase of the -8 from 1,863 to 1,947 pounds. After studying the request, BuAer had allowed the weight to increase to 1,864 pounds.

Now, WAGT had asked BuAer to reconsider their earlier request but raised the requested engine weight to 1,999 pounds and then added another 15 pounds later when the Automatic Trim Control was incorporated. The weight breakdown between the original component weights and the then present actual weight was attached and is of interest. The intent of the chart was to show how much of the weight change was "chargeable" to WAGT, which, in the chart below, is shown in the last column. If WAGT's argument that only 193.7 pounds of weight was due to their underestimating the original weight was accepted, it represented an 11.3% increase. There were no provisions for penalties (or bonuses) in the contract for weight or any other non-fulfillment item, other than withholding final progress payments for each engine accepted.

Component Weight Comparison for J46-WE-8
April 29, 1952

Item	April 7, 1947 Original Wt.	March 31, 1952 Current Wt.	Deviations	BuAer Adj. Deviations
No. 1 Bearing and Inlet Housing	228.0	215.8[1]	-12.2	-52.2
Compressor Rotor Assembly	326.0	315.2	-10.8	-10.8
Compressor Vane Assembly	76.3	110.0	33.7	33.7
Compressor Housing	61.5	68.0	6.5	6.5
Diffuser Assembly	67.7	86.7	19.0	19.0
Engine Mount Cylinder	0.0	31.0	31.0	31.0
Combustion Chamber Housing	76.0	51.8	-24.2	-24.2
Combustion Chamber Liners	55.0	40.5	-14.5	-14.5
Vaporizing Plate	0.0	22.1	22.1	22.1
Turbine Shaft Assembly	50.0	33.5	-16.5	-16.5
Turbine Discs	185.0	151.9	-33.1	-33.1
Turbine Nozzles	63.3	53.6	-9.7	-9.7
Turbine Housing	18.0	25.5	7.5	7.5
Afterburner Assembly (Ext. for F7U-3)	167.0	367.9[2]	200.9	150.9
No. 2 Bearing Assembly	22.0	25.3	3.3	3.3
No. 3 Bearing Assembly	16.0	15.3	-0.7	-0.7
Accessories	241.8	290.3[3]	48.5	97.5
Misc. (seals, clamps, etc.)	0.0	65.3	65.3	65.3
(Margin for Development)	56.4	0	-56.4	-56.4
Total	**1,710.0**	**1,969.7**	**+259.7**	**+218.7**
				-25.0[4]
Total Deviation				**193.7**

BuAer Requests:
[1] Decrease by 40 pounds to allow for requested Generator Drive
[2] Decrease by 50 pounds to allow for requested F7U-3 Afterburner Extension
[3] Increase by 49 pounds to allow for requested change from Electronic to Simple Hydraulic Controls
[4] Decrease to allow for other miscellaneous changes requested by BuAer

WAGT argued that the main reason the weight had gone up was the afterburner and its accessories about which little was known at the time the original proposal was written. The long range weight decrease program was primarily based on using titanium in some of the compressor discs and this would take time to impact the production program. In fact, many of the weight decreases were to apply to later production down the road and the specification weight for production engines should reflect allowances for higher weight at the beginning.[44]

BuAer's response included the following points:

1. The incorporation of titanium discs was no longer a requirement due to high cost and limited availability of material.
2. WAGT was not justified in requesting a weight increase due to the use of steel instead of titanium. *(A close reading of the WAGT memo shows that WAGT never said or implied that. – Auth.)*
3. The two pound weight increase for the Manual Trim Control was not allowed as no 150-hour production engines would have the control.
4. The 15 pound increase for the Automatic Trim Control was allowed.
5. The use of rigid tubing for weight saving was disapproved due to problems in the past with such tubing.
6. BuAer was encouraged by the weight saving program now being carried out.
7. The approved Model Specification weight for the J46-WE-6 was increased from 1,864 to 1,879 pounds.

Perhaps the written comments on the back of the routing sheet for the WAGT memo reflect BuAer's feelings on the situation better than their memo and are worth quoting here.

"This is the same old story with Westinghouse all over again. If we approve 1,999 (pound) spec. wt. we can be sure that additional weight increases will be requested. The inertia of development weight increases usually doesn't subside until the 150-hour endurance test has been approved. March '52 progress reports specified a 43 (pounds) for CVA A/B extension. March '52 progress report specifies also a total expected wt. increase of 63.6 (pounds). I am beginning to believe that future wt. redirection projects are mainly a smokescreen to help justify current wt. increases. Note wt. reduction item which changes fuel lines from flex hose to rigid tubing. Question if we want this in view of early experience with rigid tubing failures on jets. Believe PP211 should take action on this ltr in view of recent policy discussions on action correspondence. Note that this engine now weighs almost as much as J48-P-8 which puts out 7,250 (pounds) thrust dry. Recommend we don't approve wt. increases and let Westinghouse sweat out some wt. reductions."[45]

The weight issue was far from closed, but at least both parties had defined the boundaries of the discussion. It should be noted the potential weight savings from using titanium in the engine was very attractive and the prospect of using it was never lost sight of by both parties.

WAGT surveyed adding 14 turbine temperature traversing bosses to be used during green runs and the placement of cover plates over the bosses after the green runs were completed.[46]

At this point, the May summary progress report could not have been very encouraging to BuAer. Total run time was up to 638.5 hours but no additional A/B test time had accumulated. The XJ46-WE-8 engine designated to start the 150-hour qualification test in late May was being assembled but it was not believed the engine would meet specification maximum thrust or SFC guarantees at normal power settings or below.

Another test engine had been delivered to AEL, with testing to re-start on or around June 4. Most of the development effort was going into turbine development of the -2.[47]

WAGT surveyed the new T-13 turbine and its dimension changes on June 6. This was for the -1 and -2 engines. It added approximately a quarter of an inch additional diameter to the turbine section which had previously been surveyed. The overall length of the engine was still unchanged.[48]

In early June, it was recommended by BuAer that the oil level ("sounding") gauge be removed from the engine as no one was going to use it.[49] WAGT surveyed the removal at a weight savings of 0.8 pounds. The gauge would be replaced with a light metal cover with early engines having the gauge until the covers could be procured.[50]

Contract Amendments 10 and 12 were approved in June, adding additional funds to the contract.[51]

Late in the month, WAGT surveyed changes to the A/B simplified fuel system. A cross-over line was moved to allow withdrawal of both oil tank dipswitches. An engine envelope issue concerning the hex nut for the fitting of a hose was clarified by explanation that the hose was flexible during installation, giving the proper clearance as necessary. (BuAer was to require a mock-up to prove this and this was done satisfactorily.[52]) The planned single check valve for the system was found to be inadequate and was replaced with a double swing-check valve to prevent fuel flow from the A/B line through the crossover line to the high pressure side of the booster pump. Various parts were increased in length and the changed drawings were included for review.[53]

From a contract point of view, all the extra testing and delays in getting engines ready for qualification meant that Westinghouse was incurring a lot of financial expenses and not getting paid any final payments for deliverables. They asked in June that progress payments be changed from 75% to 90% of their fee. BuAer denied the request on the grounds that the increase was not justified.[54]

The June progress report states the total J46 runtime was now 725.6 hours with 8.46 hours of A/B testing. An XJ46-WE-8 engine had been assembled and was undergoing tests prior to the start of a qualification test in July. The A/B durability was questionable and the engine was not expected to meet the specification guarantee for maximum thrust. The AEL -8 test engine had not encountered stall within the engine operating limits, having been tested up to 45,000 feet altitude. A larger turbine was now planned for the -2 because the -8 turbine was known to have an efficiency "substantially below" design value.[55]

In early July, WAGT surveyed the incorporation of a new power control linkage beginning with the 60th production -8. The power lever gearbox was no longer needed with the SHC since the emergency fuel scheduler was no longer used. Its removal would save four pounds of weight and allowed simplified piping arrangements between the control components.

In addition to the new power control linkage, a new lighter, smaller and improved A/B control having a single discharge port would be used.[56]

The BAR forwarded to BuAer a WAGT June 5th request for overtime approval after explaining that he had been approving overtime requests on a piecemeal basis totaling 1,500 hours to date. This request would make a nine hour work day standard practice in the engineering department of WAGT. BuAer approved the request in mid-July.[57]

J46-WE-8 50-Hour Qualifying Test

The flight substantiation test, which had been scheduled to start on July 20, was pushed out to July 31 to allow the engine (X24C10-6) to be removed from the test cell and dismantled. A fix was to be installed to address excessive oil consumption at the No. 2 bearing. Based on the schedule slippage, the completion of the test was expected to be about August 15 and the start of the 150-Hour Qualification test to be October 31. The extra time in the schedule was included to allow for the possibility that the engine was not completely qualified on the first attempt and another test might be required.

An NACA report had been received which documented test results showing that increasing the distance between the A/B spray rings and the front face of the flame holder assembly increased thrust and high altitude performance significantly. The distance between them was referred to as the "coupling" distance. This is discussed in more detail a bit further on.

The short coupled afterburner had been found to be limited to below 38,000 feet in operation and it was determined the long coupled A/B (which had no such restriction) could be in production in time to be installed on 150-hour production J46-WE-8 engines from January 1953 on. Verification testing of the long coupled A/B was to begin December 1. If the test was completed on schedule and was successful, only the first six 50-hour engines would have short coupled A/B's.[58]

Just after the previous memo arrived at BuAer, another arrived from WAGT giving a development status overview of the -8 engine. The BAR's cover memo stated he did not believe a qualification test could start before August. The development status and plan overview memo very clearly illustrated the trial and error approach WAGT was using.

Development status was by area:

Compressor – The C-50 configuration appeared very promising for production. AEL testing showed that a compressor similar to the C-50, which included production compromises for tooling purposes, appeared satisfactory but was marginal for acceleration stall at sea level. A compressor known to be satisfactory for acceleration stall at sea level would be shortly sent to AEL. The test results would help define the J46-WE-8 compressor in its final configuration. Only developments resulting from on-going tip clearance tests were likely to be incorporated in production, although improvements made on the XJ46-WE-2 compressor would be incorporated whenever it was feasible.

Combustor – Acceptable radial temperature distribution at sea level had been achieved on several engines. Altitude checks at AEL indicated similar results to J34 engines equipped with mixers. The acceptable radial temperature distribution was being achieved by redistribution of fuel to the combustors. The cover memo explained WAGT was using "T" connectors to some of the walking stick evaporators and capping off the unused fuel lines. In other words, each engine was having the radial temperature distribution tuned as necessary by limiting the fuel to some evaporators. *(A pencil note next to the paragraph explaining all of this says "NO!")*

A new combustor liner configuration was showing that a predictable hot-spot pattern was achievable vs. the random pattern being exhibited by the current liner. Further tests were underway to reduce the severity of the hot spots that were created. Some evidence was being found that differential expansion between the compressor housing and outlet guide vane outer shroud might be causing air disturbances which could cause poor circumferential distribution. A modified outlet-guide mixer combination was being procured.

Turbine – The current model needed significant improvement in efficiency. Four new configurations were in test or being procured. Changes being investigated were blade angle changes, modification of existing blade sections, and new blade sections using existing patterns. One house engine was now being used exclusively for turbine testing.

Afterburner – The short-coupled A/B would be qualified and long-coupled A/B's procurement was being expedited. A prototype had been tested in June.

Performance – At the time of the memo, the -8 initial performance results expected during the qualification test were:

J46-WE-8 Pre-Qualification
Expected Performance
At 10,100 RPM

Power	Thrust (lb)	SFC	T4 (°F)
A/B	5,600	2.56	1,525
Military	3,905	1.116	1,525
Normal	3,490	1.115	-
Cruise 1	3,140	1.116	-
Cruise 2	2,620	1.189	-
Idle (4,000 rpm)	190	4.00	-

The projected numbers indicated that take-off thrust would be 120 pounds below the -8 specification and the fuel consumption would be 3-12.6% above guarantees across the power range.

Weight – The requested specification weight was now 1,999 pounds vs. the currently approved specification weight of 1,879 pounds. It was felt about 60 pounds reduction would be possible without the use of titanium alloys.

Durability – The XJ46 engines had accumulated 750 hours of running and the AEL test engine had accumulated an additional 75 hours. Upon disassembly, the engines were found to be in very good condition. The AEL engine had been run for 3 hours at 10,500 rpm (vs. Military speed of 10,100 rpm) without damage.

Controls – Covered in other memos.[59]

With the qualification tests in a delayed status, the fact that the engine was still far from being reasonably developed must have caused great concern to BuAer. It still had operational issues (acceleration stalling) and performance shortfalls below the guarantees for an already de-rated version of the J46-WE-2. Some traces of their frustration and concern are visible in notes attached to memos, as we shall see.

A minor but important detail was taken care of in this time frame when BuAer approved the design of the shipping containers for J46 and J34 engines as suitable for service.[60]

Now another unexpected major change, this time in the A/B, was surveyed. As previously mentioned, NACA testing[61] in Cleveland on afterburners (called tail-pipe burning) had shown that high altitude efficiency of A/Bs was greatly improved if the distance between the fuel rings and the flame holder was increased. The additional distance improved the evaporation of the fuel as it traveled the greater distance. The distance between the two elements was known as the coupling distance.

WAGT surveyed increasing the coupling distance of the -8 A/B to 19 7/16 inches and the use of only two (2) A/B fuel manifold rings instead of three (3). The elimination of the third ring reduced pressure losses in the A/B. The increased distance was gained by moving the fuel manifold upstream. Doing this caused a ripple effect on the rest of the A/B design:

1. The flexible joint moved aft and decreased in diameter by 3/64 of an inch.
2. Only two fuel lines ¾ inch diameter were now needed due to one less fuel ring being used.
3. The fuel manifold ports were repositioned to the entrance of the 18.4 inch extension section.
4. The proposed A/B diffuser housing was cylindrical from a point just aft of the quick disconnect clamp to the flexible ball joints. The housing then tapered abruptly outward from the ball joint to the rear of the A/B diffuser flange.
5. Two push-rod guides were incorporated due to the repositioned flexible ball joint, requiring a redesigned push-rod linkage system.
6. The ten engine control thermocouples were relocated forward.
7. The three A/B diffuser struts were to be mounted to the external part of the diffuser housing.
8. The A/B lifting lug was repositioned to 1 5/8 inches aft of the quick-disconnect flange face.

9. The weight of the A/B was increased by 39.1 pounds and the engine center of gravity moved 1.07 inches aft.
10. The A/B still remained inside the installation envelope.[62]

Some of the survey drawings sent out were in error and were later corrected in October initially via a memo alerting BuAer of the correct measurements.[63]

This late change in A/B design was made to improve both high altitude performance and reliability, but greatly complicated the planning of the soon to begin shipments of engines. Combined with a mish-mash of operating systems available during the early production, a series of sub-models of the -8 and -12 had to be created.

The replacement of the SHC exhaust nozzle control with a better designed fail-safe control prior to the introduction of the hydro-mechanical control was explained in WAGT report A-1379. This same change was occurring over in the J40-WE-22A program, but was far better described by reports found in the J46 contract files. The intent was to reduce pilot workload while in emergency operation if the airframe electrical supply failed.[64]

SHC A/B Exhaust Nozzle Control
Fail-Safe Modifications to Accommodate an Airframe Electrical Power Failure
(Interim Exhaust Nozzle Control System)

Issue: The existing SHC exhaust nozzle was dependent on aircraft electrical power to operate and in the event of failure, the emergency A/B fuel controls did not operate.

Existing emergency operation if electrical power failed:
1. The exhaust nozzle could be placed in the wide open position by means of a manual override lever in the cockpit.
2. Using only Military non-A/B power, a wave off could not be accomplished due to the lower thrust of the engine with the exhaust nozzle in the wide open position.
3. The A/B had to be ignited to accomplish a safe wave off. Tests showed it would safely ignite at Military rpm at low level with normal A/B fuel flows.
4. In an emergency caused by loss of the airframe electrical supply, an A/B start could be accomplished by waiting until the engine rpm reached Military and then moving the control into the A/B range.

To operate the A/B fuel control system independently of the aircraft electrical supply, the system had to provide for:	Design solutions to deal with Items 1-4:
1. Positive ignition of the A/B with a wide open exhaust nozzle. 2. Non-electrical actuation of the fuel shut-off valve at the inlet to the afterburner pump. 3. Non-electrical signal to operate the shut-off valve by the pilot in response to the power lever control position. 4. Fail-safe deactivation of the A/B control electrical circuit to: a. Prevent ignition during burst acceleration until military rpm was reached to prevent an engine stall. b. Shut off the A/B fuel flow in the event of an A/B blowout.	1. No electrical changes were necessary. 2. A fuel pressure activated gate valve modified from the prototype valve to be used on the XJ40-WE-10 was used to replace the electrical fuel shut off valve. 3. On early -8 engines using the J34-WE-32 (modified) A/B control, the A/B selector valve was modified to send a signal to the gate valve. On later -8 engines, the signal to the gate valve was sent by a modified cam operated valve based on the position of the control lever position. 4. To prevent early A/B ignition: a. An engine electrical system powered solenoid was provided to prevent A/B ignition below Military rpm engine speeds if the control lever was moved into the A/B range. However, in an emergency, an A/B start could only be made manually by waiting for the engine speed to reach Military and then moving the control into the A/B range. b. The solenoid above was linked to the blowout circuit and the blowout circuit changed to operate from engine electrical power.

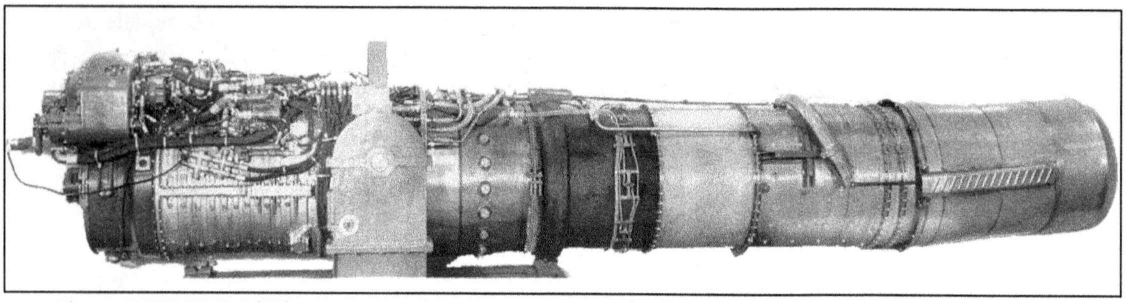

Figure 6: XJ46-WE-8 (#6) type test engine. WAGT P47098, 6/11/1952. ***Courtesy Hagley Museum and Library***

The July progress report more or less covers the issues with excess oil consumption. The BAR did not expect the XJ46-WE-8 to pass the 150-hour qualification test on the first try and agreed it would not meet the specification guarantees for thrust, SFC and weight.

Another step in preparing to move the J46 into service use was made in early August when F7U-3 BuNo. 128468 was assigned to Chance Vought in place of BuNo. 128467 for use in engine equipment demonstrations. The reason for the change in airframes was not mentioned.[65]

The cold cycle component qualification test plan had been approved but BuAer clarified to WAGT in early August that no emphasis should be placed on the time required to make a complete cycle during the test at -67°F. The elapsed time requirement was relaxed due to the very slow control movements of the actuators using MIL-O-6081 oil at that temperature.[66]

Meanwhile the J46-WE-2 design was evolving into another engine in WAGT's attempt to meet the original guaranteed performance. At this point, a survey was sent out that proposed the following changes to the -2:

1. The starter-generator gearbox was to be removed. It would be GFE for applicable airframes using the engine.
2. The compressor bleed bosses had been moved to the compressor diffuser.
3. An installation envelope had been added for the eight ignitor fuel nozzles.
4. The A/B fuel line drawings were shown in the correct location axially for the long coupled A/B.
5. The axial location of the quick disconnect clamp had been fixed in respect to the centerline of the engine mounts.
6. The 3°7′36″ degree bend in the exhaust sandwich section (for the F7U-3 installation) had been eliminated and the equivalent bend incorporated into the exhaust collector section. The A/B combustor section was increased in length from 25 to 28 inches by this change.
7. The ejector was the proposed 200-400 in^2 area nozzle. The ejector was six (6) inches longer making the overhang 14 inches from the trailing edge of the exhaust nozzle in the wide open position.

A note on the distribution sheet states that WAGT had already been told that a longer engine would not be tolerated and it was now surveying something they knew would require a major aircraft redesign.[67]

BuAer gave their approval to start the 150-hour qualification test of the XJ46-WE-8 engine with performance, function and mechanical deviations. The approval contained a number of contingencies:

1. Engines built to the qualification test configuration would be considered only for a 50-hour flight qualification rating. Satisfactory completion of the first 60 hours of the qualification test would be the basis for the 50-hour rating.
2. The following were considered acceptable for the qualification test and six (6) 50-hour rated engines only:
   a. Dry weight 1,990 pounds.
   b. High oil consumption of 4 lb/hr.
   c. A/B altitude limits (38K max/33K re-light feet).
   d. Fuel redistribution to the vaporizing tubes via the use of "T" connectors.
3. A moisture proof test of the ignition system was not required.
4. Transient data recording was waived for this test (only).
5. Prior to delivery of the first J46-WE-8 engine to the Navy, WAGT had to effect at least:
   a. 60 hours of running of the revised compressor diffuser design on any XJ46-WE-8 engine,
   b. The A/B blow-out protection system would be demonstrated by functional and durability standpoints with 20 A/B starts and 20 hours of all-condition operation on any XJ46-WE-8,

c. The planned exhaust nozzle manual-override control would be demonstrated by three (3) cycles of operation on any XJ46-WE-8 engine, and
d. Demonstrate the ability of the XJ46-WE-8 engine to withstand over-temperature operation.[68]

The WAGT memo requesting permission to start the test included a few other configuration and test procedures. The engine would have the manual temperature trim control and testing would start with JP-4 but if problems were encountered, WAGT would shift to using JP-3.

1. The manual over-ride control was not available.
2. The anti-icing valve would be installed but not be operative.
3. The external high pressure relief valve on the open end of the exhaust nozzle actuator stroke might have to be reset from 2,100 psi to 750 psi.
4. The improved diffuser fairing supports would not be on the test engine.
5. Temporary unavailability of minor hardware and detail parts might require substitution(s) with parts that would have no effect on engine durability.
6. At 60-hours, a visual inspection of the engine would be done, using the aft quick disconnect clamp to allow inspection of the inside of the engine and A/B.[69]

The August progress report resumes reporting on the 150-hour testing thread. The qualification test was started on August 26th and at the 26 hour point was interrupted by failure of the diffuser cone which then wrecked the A/B flame holder. A new A/B was being installed. The oil consumption had been reduced through use of baffles around the bearings, but was still above the limits. As expected, the thrust and SFCs did not meet the guarantees. It had been agreed that six (6) 50-hour engines (badly de-rated) and 24 slightly de-rated engines would be built prior to delivery of fully rated J46-WE-8 engines beginning in April 1953.

All was not well in the area of systems testing either. WAGT's request (August 6) for approval of their submitted test plan for the afterburner fuel control component qualification was rejected by BuAer on the grounds that the wording of the descriptions of the components made it impossible to determine exactly which components were being included in each test. To quote BuAer here:

(The referenced examples) "are examples of the apparently loose system by which Westinghouse defines their control systems and components thereof. The only organized information available to this bureau is (a chart) which consists of sketches and a written description of the original so-called "Simple Hydraulic Control", in which descriptive titles such as "ignitor valve" and "relay valve" are given for various subcomponents of the control. Even this information is somewhat ambiguous because of the numerous inconsistencies between the descriptive titles used in the written matter and those used in the sketches. This loose system of identification used by Westinghouse makes it very difficult for this bureau to keep track of the many changes to the many control systems currently under development by the contractor."

A conference had been held earlier on this subject and WAGT had agreed to start using a numerical designation for control systems and subcomponents unique to each version of the control system.

Concluding the rejection memo, BuAer requested "that assembly and schematic drawings be submitted defining the specific components proposed for qualification test, and for any other component qualification test proposed prior to such time when the above program is completed."[70]

In early September, BuAer gave official approval for WAGT to deliver the first six (6) flight rated engines with the deviations specified earlier by WAGT. As agreed, WAGT would later modernize the six engines to the 150-hour Qualification Test specification at no additional cost to BuAer if BuAer desired it should be done.

A written note on the distribution routing sheet states that the first three engines would be used in the No. 3 F7U-3 for early evaluation of the aircraft-engine combination and early flight evaluation. The second batch of three engines would be used in the No. 17 F7U-3 for installation evaluation and conducting ground runs and taxi tests prior to delivery of the production J46-WE-2 in January or February 1953.[71]

On September 4, WAGT notified BuAer they were switching the control cables from the unshielded to shielded type with a metal braided cover. It added strength, durability, moisture resistance, radio noise protection and resistance to abrasion. The change would add four (4) pounds to the engine.[72]

In mid-September, WAGT requested the current AEL testing (compressor, starting and ignition and fuel control for acceleration and stability) be expanded to include windmilling tests, the results of which were urgently needed by the airframe manufacturers. The test would determine the horsepower available for operation of accessory drives actuating airframe control surfaces, landing gear, cabin pressurization, etc. The test was expected to take about a month.[73]

Another specification deviation was requested at this time because the anti-icing valves would not be available for the first six engines, a blanking plate being installed instead to be removed later if valves were to be installed.[74]

The qualification test had been restarted but stopped after completion of the engine examination at the 60-hour point. WAGT's report states that further testing was prevented by two problems:

1. Failure of the fairings in the engine diffuser (reported earlier), which was an obsolete design.
2. Oil consumption increased at the 23 hour point due to a loss of a portion of the No. 2 bearing housing insulation blanket. Fragments from the failed blanket apparently had obstructed the air flow around the No. 3 bearing causing a pressure drop over the housing. As a result, effective scavenging of the rear seal occurred. *(This apparently meant that oil was forced out past the seals as a result.)*

Additional deficiencies were disclosed:

1. The A/B diffuser developed a cracked strut at 26 hours.
2. When going from military thrust (using manual temperature trim) to A/B the iris nozzle assumed a position slightly less than wide open. The manual control was able to compensate.
3. At 39 hours an "O" ring required replacement in the high pressure oil line at the exhaust nozzle actuator.
4. At 10 hours one of the check valves in the A/B fuel lines leaked and was replaced.

The engine had run 67.91 hours prior to test start and had now run 127.91 hours.

Engine changes to be made to a second engine being built up to attempt the test again were:

1. Redesigned engine diffuser
2. Revised baffles for the Nos. 2 and 3 bearings
3. Bearing insulation blanket lashing method to be improved using twisted wires instead of single strand wire.
4. The second A/B used in the first test would be used.
5. Modification of the exhaust nozzle positioning system as necessary based on separate testing.
6. Changed "O" ring arrangement at the actuator to prevent excessive crushing and damage to the gasket. All four ports to be modified with same.
7. The variable displacement oil pump used to control the exhaust nozzle position would include a manual override lever to enable demonstration of the feature.
8. Improved check valves in the A/B fuel lines.
9. While clevis bolts were normally always used where failure could cause engine failure, during the test clevis pins might be used during the test (only) as a result of lack of parts.[75]

BuAer approved the modified engine for testing, reminding WAGT that any provisions of the specifications not met by the test engine configuration would be required prior to delivery of the first flight rated engine.[76]

The BAR had concerns which he expressed in his cover letter when forwarding WAGT's re-test request:

1. The exhaust nozzle actuator guide or crosshead design on the test engine and expected first six engines was not considered by him to be a satisfactory design for service engine use. The design showed a high degree of wear during the first test.
2. The iris design also had durability weaknesses. The cam sections of the cam leaves showed considerable wear and grooving due to bending of the cam surface.
3. The time delay relay (for the Microjet) on the first test was a prototype unit. Five new units were on order but the prototype was to be used on the second attempt. The new units would not be the same as the prototype and were also considered to be prototypes and might not be available for the first production engine. The BAR informed WAGT that production engines without a time delay relay would not be accepted and the final design of the time delay relay would require verification before engines incorporating it were accepted.[77] The delay circuit had been added to eliminate false signals resulting from pressure transients during the afterburner ignitions.[78]

The September progress report states the qualification test on the fuel booster pump would have to be restarted because of a weld failure of a soldered joint that retained a stainless steel ring that rubbed on the carbon ring. The failure occurred while dry running during the checking out period prior to starting the high temperature test. All portions of the test already completed would have to be done over again.

Otherwise, the report added nothing except that the second 150-hour qualification test would start about October 1st and that total XJ46 operating time was up to 1,026 hours.[79] Another delay for reasons unknown caused the test start date to be moved to November 15th with checkout of the test engine starting the 6th of that month.[80]

Figure 7: XJ46-WE-8 qualification engine piping. Note braided cables on actuator side of fire seal. Rubber fuel lines now replace the metal lines. WAGT P47101, 6/11/1952. *Courtesy Hagley Museum and Library*

As was reported earlier, BuAer had requested that WAGT take their weight contingency out of the original bid. Since accepting the proposal at an estimated weight of 1,710 pounds, various changes had increased the weight steadily. With the 150-Hour Qualification Test now approaching, BuAer requested (on August 20) that WAGT informally submit weight estimates for the -8 and -10 engines that reflected the effects of using steel in place of titanium and the switch to the SHC. WAGT responded[81] on September 5 with a detailed weight outlook and asked that a specification weight increase be approved for each engine. (The requested specification weight increase for the -8 was 14.4% higher than the original proposal bid weight for the -2 of 1,701.6 pounds) The data presented were:

Est. Model Weight Changes (lb)		
Change	XJ46-WE-8	XJ46-WE-10
Current Act. Wt.	1,908.0	1,644.0
Use of Steel	64.3	64.3
Use SHC*	-49.0	-49.0
Est. Weight	1,923.3	1,659.3
Cur. Spec. Wt.	1,863.0	1,635.0
Req. New Spec. Wt.*	1,947.0	1,661.0

* Might require temperature trim, weight not included

** The projected weight reductions (below) would not begin to phase in until the 7th production engine, all by the 50th engine.

Weight Reduction Items in Process 7th – 50th Production Engines Phase-in (lb)		
Item	XJ46-WE-8	XJ46-WE-10
Compressor housing wall thickness – better control	5.0	5.0
Inlet housing wall thickness – better control	10.0	10.0
A/B ball joint revision	2.0	----
Use of rigid tubing	8.0	8.0
Turbine disc redesign	5.0	5.0
Total	30.0	28.0

Other weight reduction projects under study at that time were:

Weight Reduction Study Projects (lb)		
Item	XJ46-WE-8	XJ46-WE-10
Titanium compressor discs	64.3	64.3
Titanium hollow compressor discs	20.0	20.0
Smaller exhaust nozzle area	20.0	----
Total	104.3	84.3
Possible future engine weight	1,842.7	1,576.7

In the meantime, other testing was underway. The Phillips Petroleum Company had been investigating the performance of jet fuels in a small scale vaporizer combustor under contract NOa(s) 52-132c. They had reported on the effect of gum and high boiling constituents on vaporizer tube deposits.

BuAer proposed that full scale testing in a J46 using six special fuels be conducted by WAGT in cooperation with Phillips, who would supply the fuels and technical assistance, including additional small scale testing as needed. The fuels would be:

1. A high volatility MIL-F-5624a (JP-4) blended from paraffinic base stocks with minimum gum content. The results would establish the minimum deposition rate limit.

2. A low volatility MIL-F-5624a (JP-4) with low gum content. To be used to measure the effect of average molecular boiling point on the deposition rate of the burner.
3. Fuel 2 with maximum gum content as allowed by the JP-4 specification.
4. Fuel 3 with maximum aromatic content as allowed by the JP-4 specification. To be used to establish the effects of aromatics on deposition rate.
5. Fuel 4 plus 1.15 milliliter of aviation mix tetra-ethyl-lead fluid per gallon.
6. Fuel 4 plus an additive showing promise in Phillips' two inch vaporizing combustor in reducing deposits.

Phillips was to supply 10,000 gallons of fuel No. 4 and approximately 3,000 gallons each of the other fuels. WAGT asked if the arrangement was satisfactory, the quantities of the fuels were correct for the tests planned and whether the No. 4 fuel could be delivered by rail tank car and the balance in drums.[82] WAGT later asked that 20,000 gallons of No. 4 fuel be procured and that Phillips could begin blending in January 1953 with the balance of other fuels arriving in April of that year. Fuel No. 2 was to be shared with participants in the Coordinating Research Council program on coking, supported through the BuAer Ordinance Corps contract DA-30-069-ORD-75.[83]

While the planning for the fuel test was occurring, WAGT extended the Single Port Afterburner Control and New Customer Connection Linkage survey to include the XJ46-WE-12 intended for the Consolidated XF2Y-1 airplane and then decided not to include it. The drawings for the two engines were redrawn accordingly.[84]

A meeting back in August, the purpose of which was a review of the current performance being obtained by the -8 test engine, had been attended by BuAer, the Air Force and Westinghouse. WAGT presented the data along with suggested changes in the specification ratings for production engines.

XJ46-WE-8 Actual Performance Ratings August 12, 1952		
Rating	Thrust (lb)	SFC (lb/hr/lb)
Maximum	5,510	---
Military	4,060	1.094
Normal	3,490	1.095
90% Normal	3,140	1.108
75% Normal	2,620	1.175

WAGT stated that the first production engines would have exhaust nozzle areas of 375 in^2 - 15 in^2 larger than the test engine. The afterburner thrust was expected to be increased by 200 pounds because of the increased exit size and other minor changes.

For the first 21 engines, WAGT asked for a guaranteed performance based on allowances for hot day conditions and a nominal inlet duct loss without adjustment.

XJ46-WE-8 WAGT Recommended Performance Ratings for First 21 Engines August 12, 1952		
Rating	Thrust (lb)	SFC (lb/hr/lb)
Maximum	5,300*	2.53
Military	3,900	1.13
Normal	3,490	1.12
90% Normal	3,140	1.13
75% Normal	2,620	1.13
Idle	241	4.08

* Rising to 5,500 lb thrust after first six (6) engines

The above rating chart would apply to all J46 engines of the -8 type regardless of whether long or short coupled A/B's were used. These ratings eliminated the need to show an 80 pound reduction in thrust for short coupled engines which had the short extension between the A/B diffuser and the flame holder system. The total dry weight of the first 21 engines would not exceed 1,990 pounds.

The first six (6) engines were to be produced in Essington and the balance from the Kansas City plant over the January-March 1953 period.

After those engines were produced, advancements in turbine design were expected to raise the -8 ratings to:

XJ46-WE-8 Projected Ratings Beginning with 22nd Engine August 12, 1952		
Rating	Thrust (lb)	SFC (lb/hr/lb)
Maximum	5,720	2.53
Military	3,960	1.106
Normal	3,540	1.102
90% Normal	3,190	1.111
75% Normal	2,650	1.193

With a slight increase in the A/B rating, BuAer accepted the recommendations on October 9.

Also during the August meeting, the first production J46-WE-2 engines that were to achieve the specification thrust values were discussed. To accomplish this, four main changes were to be incorporated in the design:

1. Use of the larger T-13 turbine,
2. Use of a 400 in^2 exhaust area nozzle,
3. Use of a larger diameter A/B of 28 inches, and
4. Use of a longer engine diffuser.

The engine was the same length and it retained the same mounting point locations, but the installation characteristics were different. The weight would be 2,020 pounds.

The thrusts in the guarantee table were met, but the SFC's would be higher. The development schedule indicated a qualification test could be run in June, 1953 and a second run, if necessary, in September. Production deliveries would be from Essington beginning in September.[85]

A major contractual change occurred in the same time frame when BuAer formalized the modified performance guarantees for both the -2 and -8 engines. (The model XJ46-WE-2A was not affected by the changes.)

XJ46-WE-2 Modified Performance Ratings
October 9, 1952

Rating	Thrust (lb)	SFC (lb/hr/lb)
Maximum	6,100	2.62
Military	4,080	1.074
Normal	3,670	1.035
90% Normal	3,300	1.025
75% Normal	2,755	1.049
Idle	241	4.08

XJ46-WE-8 Modified Performance Ratings
October 9, 1952

Rating	Thrust (lb)	SFC (lb/hr/lb)
Maximum	5,800	2.53
Military	3,960	1.106
Normal	3,540	1.102
90% Normal	3,190	1.111
75% Normal	2,650	1.193
Idle	245	4.000

House Engine Test Results

Engine	Weight (lb)	Blowout (ft)		Relight (ft)	
		Engine	A/B	Engine	A/B
XJ46-WE-8 (Type Test)	1,969.4	60,000	38,000	35,000	33,000
XJ46-WE-8 (5,500 lb Max)	2,025	60,000	48,000	35,000	43,000
XJ46-WE-8 (5,720 lb Max Interim)	2,025	60,000	48,000	35,000	43,000
XJ46-WE-8 (5,720 lb Max Hydro-Mechanical)	2,030	60,000	48,000	35,000	43,000
XJ46-WE-2	2,020	60,000	55,000*	45,000*	50,000*

*Ram pressure ratio 1.25 to 1.35; All others 1.15 to 1.50

In the memo summarizing the details of the August meeting, WAGT included test information (above) that had been obtained at that time for four different house engine builds.

Permission to deliver engines not meeting the specifications had to be requested from the bureau. Each request had to include a plan for qualifying the configuration. The qualification or verification tests of de-rated engine configurations for production purposes were not to be considered applicable to any of the three qualification tests required under the contract.[86]

The ram pressure ratio that would raise the minimum altitude for A/B blowout for the type test engine was 1.4. To raise the relight altitude for the engine to 40,000 feet the ram had to be 1.2 for all -8 engines. To raise the relight altitude of the type test A/B a ram pressure ratio of 1.72 or higher was required.

A production schedule for Essington and Kansas City production of -2, -8 and -12 engines and the production start point for the introduction of the hydro-mechanical was included. These will be discussed in the next chapter on -8/-12 production.

On October 24, WAGT again asked for approval of the simplified A/B fuel system after clarifying a point and stating the benefits of the change. They included a layout chart that showed a 1.25 inch cross-over fuel line, changed to be partly of flexible hose and partly of rigid piping running from the outlet of the engine boost pump to the inlet of the A/B fuel pump. A double swing check valve was installed at the inlet of the A/B pump to prevent reverse flow.

The change would enhance the reliability of the A/B fuel system and eliminate any afterburner interruption as a result of failure in the airframe boost pump. The engine boost pump would function in place of the airframe pump if that pump failed, thus ensuring a fuel supply to the A/B.[87]

In early October, Chance Vought sent in comments on WAGT report A-1373 "Interim Design Report on a Hydro-Mechanical Exhaust Nozzle System dated May 6, 1952. *(A copy of this report was not found in the archive files – Auth.)* The comments were general as many of the components in the proposed design had not been completed.

The points made were:

1. The design replaced the dependence on the airframe electrical system with non-electrical mechanical components.
2. If any of those components failed, the pilot had no corrective action available.
3. If the control system jammed, the engine locked in the power position it was in at the time of failure and the only action the pilot could take would be to shut down via the airframe fuel shut-off valve.
4. The cam box of the exhaust nozzle control enclosed an extensive gearing arrangement which would probably cause considerable backlash between the lever and the control output.
5. The automatic trim would only operate at high power settings and would not be available if there were an electrical power failure.
6. The feedback mechanism would be incorporated in the cam box, adding to the backlash issue, possibly leading to inconsistencies of the throttle setting.
7. The mechanism of the throttle control would require careful design to keep the throttle torque to less than 10 inch-pounds.
8. It was suggested a self-centering cam follower be used to keep the follower from slipping off the cam.
9. Since the control would use engine oil throughout, it was imperative that there be no leaks, as the total usable engine oil supply was only two gallons.[88]

BuAer approved the use of the T-13 turbine in the -2, -1 and -3 engines on October 28 after reviewing the survey comments.[89]

In the background, the use of the J46 in supersonic airframes was being discussed, particularly whether the variable intakes of the airframe should control the power control on the engine or just the opposite. WAGT felt that if the intakes did not make rapid changes, the two should remain unconnected. At that point, the conversation was between WAGT, Douglas and Consolidated.[90]

By way of explanation, both transonic and supersonic aircraft intakes have to prevent shock waves from traveling back to the engine face. They also have to slow the incoming air to subsonic speed and only present as much air to the engine as it can handle without pressures within the engine rising too high. As much as is possible the velocity of the incoming air must be converted to pressure as it is slowed down. This is termed the pressure recovery ratio of the intake, a perfect recovery being 100%. The intakes of transonic and supersonic aircraft became very sophisticated to allow the intake to only take in the proper amount of air under all operating conditions without pilot input. The discussions underway reflected the airframe and engine designers' early appreciation of yet another new aircraft design complexity emerging at the intersection point of the airframe and the engine.

The October Progress Report confirmed the checkout dates of November 6 for the 2nd test engine and the 15th for the test start date. It again repeated the expectation the engine would not meet the guarantees, although which set of them was not clarified.[91]

The A/B fuel pumps and controls component qualification test plan was submitted by WAGT on October 20 and received BuAer approval on November 12, with caveats. The tests would be laboratory tests, not done on a running engine. This memo contains significantly more detail on the components to be tested and the test process itself than any prior such approval request and may have been intended to represent the progress in detailing control systems and components that BuAer had requested. Much of the testing detail reflects the much tougher tests required under the MIL-E specifications.[92]

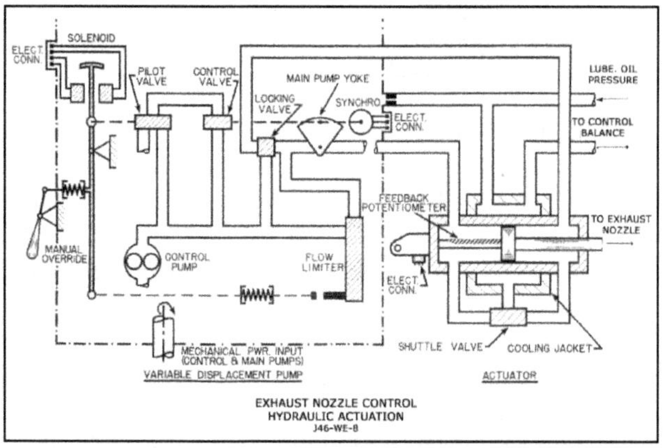

Figure 8: J46-WE-8 hydro-mechanical control schematic for the exhaust nozzle control. WAGT P46048, 1/15/1952. *Courtesy Hagley Museum and Library*

A design change originally included to accommodate the F7U-3 was modified back to the original design when the 3° 7′ 36″ cant of the exhaust nozzle was removed. It was discovered that the stalling speed of the airframe was reduced in the landing configuration if the exhaust was straight out. Due to timing, the cant would be accepted on the first six 50-hour J46-WE-8 engines. All certification testing thereafter would utilize engines with straight out exhausts. And, to add to the confusion here, Consolidated indicated they required the 3° 7′ 36″ on their engines, and these engines were briefly designated the J46-WE-16.[93]

An Aerotec A/B pressure sensitive switch design was surveyed in early November. This switch would be used in the early -8 short coupled A/B's. It had a time delay built in using a ball check valve. It was recommended that the check valve assembly be airframe mounted.[94]

Evidence that WAGT was trying to have the Navy consider their engines for the upcoming Day Fighter Competition (1953) surfaced in the files in the form of a memo containing corrected thrust performance curves for the J34-WE-36 and J46-WE-2 engines. The curves were corrected to account for incomplete expansion in the exhaust nozzle (called "explosion" losses) at pressure ratios above critical.[95] As it happened, neither engine was considered for the competition.

The November and December progress reports state that total XJ46 running time through November was 1,722 hours with 51.56 hours of that in A/B. A second attempt at the 150-Hour Qualification Test had been started on November 13, 1952 but was halted on November 24 when one leaf of the iris nozzle blew out and the diverging exhaust stream warped the ejector and unison ring. (The unison ring had the roller assemblies attached. These ran on ramps and when the ring was pulled axially back and forth the iris opened or closed, with the iris leaves sliding over/or under each other to present a gas tight seal at all positions.) The damaged parts were replaced and the test continued. The oil consumption was within the specification so the problem of excess consumption appeared to have been solved. A strengthened diffuser had been installed for the test after separate durability testing had resulted in observation of significantly improved durability. (This is a good example of a common WAGT practice of last minute modifications of an approved test engine build with "improved" parts without prior approval. The pressure to get the engine qualified was very high and BuAer had no choice but to allow some latitude in this area, but was never happy about it, after the fact requiring re-testing in some cases.)

The second qualification attempt was completed on November 26. The basic gas flow portion of the engine was OK, but the A/B was not. Another test was being scheduled in early February 1953 with a new compressor that had twisted blades to eliminate stall and a new A/B design. The A/B tests to eliminate squeal were successful when the fuel/gas ratio had been increased to .044 (raising the SFC) with the A/B producing a 32% thrust increase over the base engine. (The specification required a 45% increase to meet the guarantee.) The test results were accepted as the 50-hour flight certification of the planned six (6) YJ46-WE-8A 50-hour engines for Vought.

The next test would be for the XJ46-WE-8B engines, production of which would start with the seventh (7) production engine line. The engine would have the following characteristics:

1. Produce 5,500 pounds of thrust in A/B,
2. Use a long-coupled straight afterburner with 375 in^2 exhaust nozzle area,
3. Have a stall free compressor within the model specification limits for the engine, and
4. Use an electrical exhaust nozzle control.

The turbine development was progressing with test efficiencies being about 88% at that point, almost a 3% improvement due to lowering the aerodynamic loading.

The compressor development testing at AEL was now up to compressor configuration C75 which was an attempt at improving the partial load stall margin by closing the first two stages a bit. The improvement in high altitude surge line seen in the C74 using slightly closed seventh through ninth stages was partially lost in the C75 design. Testing was continuing. (This is a perfect example of the trial and error approach that was normal for WAGT.)

The report states that a new 36-hole combustion liner did not improve the hot spot situation and a new standard liner had been reinstalled on the test engine.

WAGT announced they had built another house engine designated X24C10 #1(New) using spare parts and surplus used parts. It was being used for compressor testing. There were now 12 house engines in the test program.

The expected weight of the first production engine was expected to be 2,000.4 pounds, 14.6 pounds over the current specification guarantee.

An interesting manufacturing process detail was contained in a follow-up memo sent to Chance Vought in November regarding fireshield installation information. The memo states "The J46 fireshield is welded to the combustion chamber casing after which the forward combustion chamber flange is welded to the casing and then machined." The fireshield had 65 holes in it.

The December report includes an updated listing of expected weight increase items and weight reduction "projects" to offset the increases.[96]

J46 Projected Weight Increases (lb)		
1	Increased strength flame holder	4.0
2	Allowance for thrust motor adapters	1.0
3	Strengthening of the engine diffuser	6.1
4	Long Coupled 28 in 400 in^2 A/B	20.0
5	Miscellaneous A/B changes	9.3
6	Automatic trim control thermocouple assembly	3.2
7	Turbine outlet straightening vanes	12.5
8	Stiffeners for A/B diffuser struts and inner cone	6.9
9	Added baffles in combustion chamber assembly for improved temperature distribution	3.0
10	Aerotec lines on A/B	2.0
11	Change to Hydro-mechanical controls	5.0
	Total	66.1

J46 Weight Reduction Projects (lb)		
1	Titanium compressor discs	60.0
2	Compressor – rigid control of casting tolerance	5.0
3	Inlet housing – use of metal patterns	10.0
4	Use of 36 rigid tubing fuel distribution lines	8.0
5	Thinner combustion chamber outer liner	1.0
6	Reduced thickness vaporizer plate	2.0
7	Titanium compressor through bolts	3.1
8	Eliminate the A/B extension flanges	6.0
9	Exhaust nozzle actuator yoke change to hollow rectangle	3.5
10	Eliminate 12th stage compressor outlet seal	5.0
11	Use sheet steel hollow compressor blades	22.0
12	Remove fairing at inlet of A/B diffuser	3.5
	Total	129.1

The increased weight of the engine was an ongoing concern of BuAer and was certainly under continuous discussion not only with WAGT but also the airframe companies. The reality was that the actual production weight of the engine would reflect WAGT's ability to design a reliable engine to the performance specifications and at the end of development, with their hand forced by the need for engines, the end weight would have to be accepted. But BuAer felt the need to vent their frustrations to WAGT in a memo (January 9, 1953) on the weight subject. It is worth quoting here so the reader can appreciate their point of view.

After a paragraph describing briefly the WAGT memo back in September and beginning the next paragraph with descriptions of the -8 and -10 engines, they begin to deal with the weight problem. "The original specification weight of the XJ46-WE-2 engine was 1,710 pounds. A weight increase of 153 pounds was approved for this engine ....in early 1950. Of this 153 pounds, only about 65 pounds of this weight increase resulted from changes to the engine requested by this bureau. The remaining increases resulted from miscalculations in original engine weight requirements on the part of Westinghouse. As a result of this increase and large weight increases on other Westinghouse engines the contractor was required to set up a weight control group and affect weight reduction programs for every development engine. However, instead of realizing substantial weight reduction, the trend appears to be that of increasing weight."

The next paragraph, using a "latest weight" of 1,976.2 pounds, adds and subtracts the weight changes for the A/B and various other minor changes. Of these, BuAer stated that 200 pounds of the weight increases were due to Westinghouse miscalculation. (This would be an 11.6% increase if 1,710 pounds is used as the start point.) There is no consideration of any weight contingency in the memo. *(Original proposal or later specifications either-Auth.)*

The memo continues, stating that WAGT is being given a six (6) month temporary waiver to incorporate the 64.3 pound increase for the change from titanium to steel discs. The 49 pound weight decrease from switching from the electronic control was not going to be allowed to offset other WAGT responsible weight increases. "...it is not considered to be in the best interests of the Navy to grant the remaining weight increases requested by WAGT (in their September) memo."

"Approval is granted to increase the dry weight of the XJ46-WE-8 engine from 1,833 to 1,854 pounds. The XJ46-WE-10 engine is to be decreased from 1,635 to 1,586 pounds to account for the 49 pounds saving in weight realized by removal of the electronic control. The temporary waiver granted the subcontractor to increase the weight of the -8 and -10 engines by 64.3 pounds for the change from titanium to steel discs should not be reflected in the model specification."[97]

Weight issues aside, the need for production engines (at any weight) was now becoming urgent and the full court press to get a flight qualified engine to Chance Vought continued, but the documentation of the costs, problems, and discussions now moved primarily to the production contract as, without any earlier discussion, the J46 development contract NOa(s) 9670 was terminated by BuAer in its entirety on December 12, 1952. The final development work, along with final qualification and associated production problems, were henceforth funded and tracked by the production contract NOa(s) 10825 after that date. Residual correspondence discussions begun under the development contract are included in the production discussion in the next chapter.

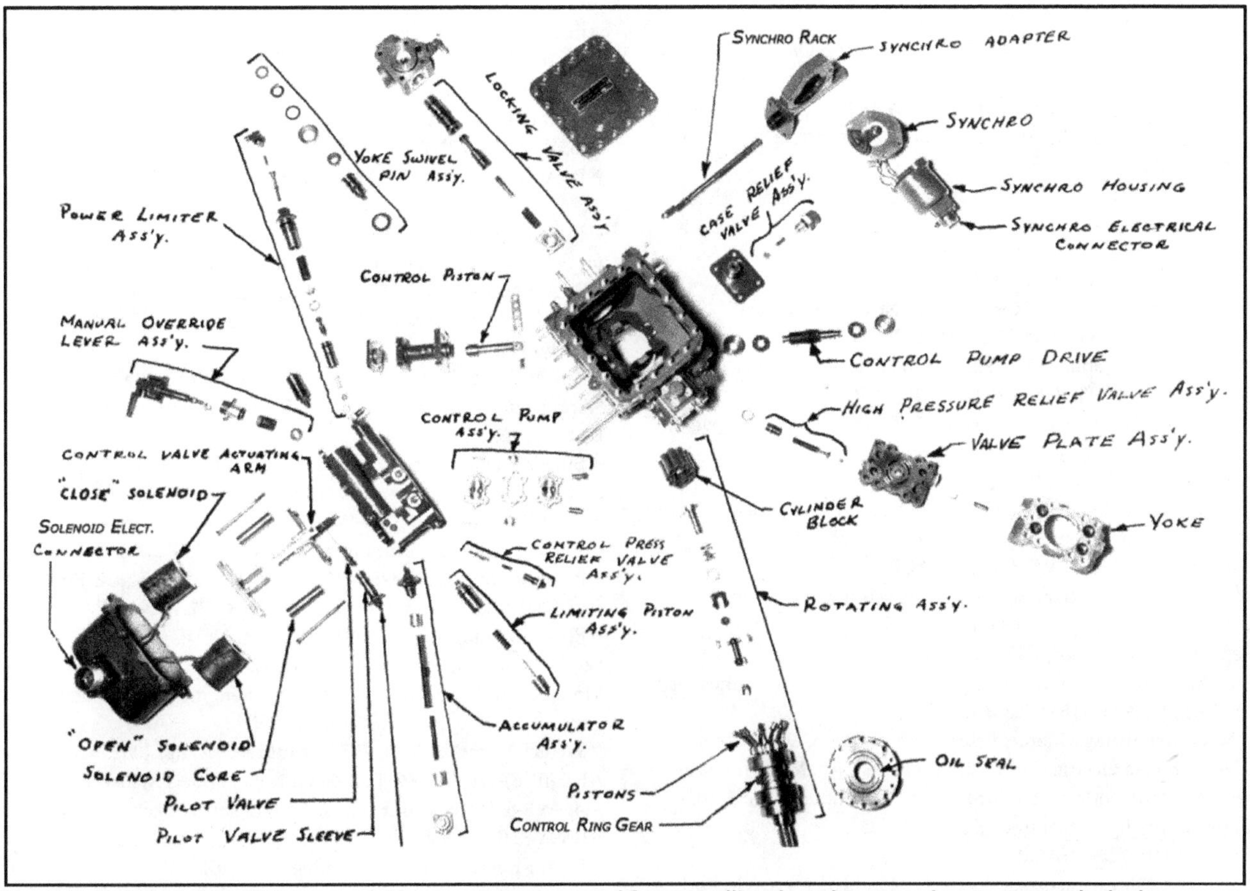

Figure 9: Exploded view of Vickers variable pressure pump, used for controlling the exhaust nozzle actuator on the hydro-mechanical control system. With the intricacy of the parts, it is a wonder that engine oil was used to drive this high precision pump, risking possible contamination. WAGT P48795A, 12/4/1953. *Courtesy Hagley Museum and Library*

Chapter 4 – XJ46-WE-8/A/B and -12/A/B Development

[1] BuAer Memo, November 9, 1951, "Incorporation of Chance Vought engine-afterburner cant for qualification test of J46-WE-8 engine; request for", **B1095VE1RG72.3.2NACP**
[2] BuAer Memo, November 26, 1951, "Contract NOa(s) 9670, Westinghouse Model XJ46-WE-8 engine; starter generator adapter gearbox for", **B555V4RG72.3.2NACP**
[3] Contract NOa(s)9670, Amendment 11, December 1, 1951, "**B555V4RG72.3.2NACP**
[4] WAGT Report A-1274, November 25, 1951, Interim Report on the Simple Hydraulic Controls for the J46 Turbo-Jet Engines and the J40-WE-8 Turbo-Jet Engine", **B557RptA1274RG72.3.2NACP**
[5] BuAer Memo, December 14, 1951, "Contract NOa(s) 9670, Monthly Progress Report for Month of November 1951; Forwarding of", **B555V4RG72.3.2NACP**
[6] BuAer Memo, January 14, 1952, "United States Air Force starter-generator, Loan of", **B555V4RG72.3.2NACP**
[7] WAGT Report A-1320, January 16, 1952, The Exhaust Nozzle Control System for the J46 Turbojet Engine, **B554V13RG72.3.2NACP**
[8] BuAer Memo, January 14, 1952, "Contract NOa(s) 9670 – Monthly Progress Report for Month of December 1951; Forwarding of", **B555V4RG72.3.2NACP**
[9] WAGT Memo, January 3, 1952, "Proposed Incorporation of Thrustmeter Sensing Means on a J46-WE-2 Engines – Our Reference Neg. 93888", **B1099VE1RG72.3.2NACP**
[10] BuAer Funding Notification, January 22, 1952, "Six engines for XF2Y-1", **B1093VA1RG72.3.2NACP**
[11] Navy Department Memo, February 29, 1952, "CUE CAP-66: Westinghouse Electric Corporation, Kansas City, Mo., purchase orders with American Standard Products Company, Hartford, Conn., applying on Contract NOA(S)-10825", **B1095VE1RG72.3.2NACP**
[12] BuAer cover memo excerpt, WAGT Monthly Progress Report for January 1952, date unknown. **B552V1RG72.3.2NACP**
[13] BuAer Memo, January 28, "Installation data, drawings and specifications for the J46-WE-8 and -10 engines", **B1095VE1RG72.3.2NACP**
[14] BuAer Memo, February 29, 1952, "Contract NOa(s) 10825, Westinghouse Model Specification WAGT-X24C10-6 dtd 10 Dec 1951 covering the Models XJ46-WE-8, -10, and -12 engines", **B1099VE1RG72.3.2NACP**
[15] BuAer memo, March 20,1952, "J46-WE-8 engine installation data; availability of", **B1095VE2RG72.3.2NACP**
[16] WAGT Memo, February 1, 1952, "Contract NOa(s) 10825 WEC Turbo Jet Engines J46WE-8 and J46WE-10. Manual Trim Control for the Exhaust Nozzle, Survey of. Engine change Nos. 105 for J46WE-8 and 18 for J46WE-10 engines.", **B1099VE1RG72.3.2NACP**
[17] BuAer Memo, February 21, 1952, "XJ40 and XJ46 engine developments additional cost over-runs, plan for financing", **B554V13RG72.3.2NACP**
[18] WAGT Internal Works Funding transfer notice, February 20, 1952, **B1095VE1RG72.3.2NACP**
[19] WAGT Memo, February 25, 1952, "Contract NOa(s) 10825 WEC Turbo-Jet Engine, J46-WE-8. Addition of Thermocouples, Thermocouple Holders, and Bosses on the Afterburner Diffuser, Survey of. Engine change No. 106 for -8 engine.", **B1099VE1RG72.3.2NACP**
[20] WAGT Memo, March 19, 1952, "NOa(s) 10825, Proposal Amendment to Incorporate a Starter Adapter Gearbox on the J46-WE-8", **B1095VE2RG72.3.2NACP**
[21] WAGT Memo, February 25, 1952, "Contract NOa(s) 9670 – General Electric Starter-Generator (2CM74B1), Starting Tests.", **B555V4RG72.3.2NACP**
[22] BuAer Memo, April 8, 1942, "General Electric Starter-Generator 2CM74B1, test report, forwarding of," **B555V5RG72.3.2NACP**
[23] BuAer cover memo excerpt, WAGT Monthly Progress Report for February 1952, date unknown. **B552V1RG72.3.2NACP**
[24] BuAer Memo, March 17, 1952, "Qualification Test of Dual Fuel Pump for Westinghouse J46-WE-8, -2, -2A Engines", **B555V4RG72.3.2NACP**
[25] BuAer Power Plant Division Memo, March 17, 1952, "Application of titanium to J46-WE-2 engines; proposal for", **B1099VE1RG72.3.2NACP**
[26] BuAer Memo, March 27, 1952, "Contract NOa(s) 10825, 11028 and 10067 Westinghouse Electric Corporation Jet Engine Models XJ46-WE-1, -2, -4 and XJ40-WE-10. Revised Auxiliary Electric Control, Airframe Mounting of", **B1099VE1RG72.3.2NACP**
[27] BuAer cover memo excerpt, WAGT Monthly Progress Report for March 1952, date unknown. **B552V1RG72.3.2NACP**
[28] BuAer Memo, April 1, 1952, "Contract NOa(s) 10825, Westinghouse Specification WAGT-X24C10-6 covering the Models XJ46-We-8, -10 and -12 turbojet engines; performance curves for", **B1099VE1RG72.3.2NACP**
[29] WAGT Memo, April 1, 1952, "Contract NOa(s) 10825 and AF33-(038)-17843 WEC Turbojet Engine Models XJ46-WE-3 and XJ46-WE-8. Increase in Exhaust Nozzle Area, Survey of.", **B1099VE1RG72.3.2NACP**
[30] BuAer Power Plant Division Memo, April 3, 1952, "XJ46 development progress and delivery schedule, status of, "**B1095VE2RG72.3.2NACP**
[31] BuAer Memo, April 21, 1952, "Contract NOa(s) 10825 Westinghouse Electric Corporation; Operations at the Columbus, Ohio Plant, information on", notes on distribution routing sheets, **B1096VE4RG72.3.2NACP**
[32] BuAer Memo, July 27, 1952, "Preproduction Expenses on J-40 and J-46 Engine at Westinghouse Columbus, Ohio Plant.", **B1099VE1RG72.3.2NACP**
[33] WAGT Memo, April 23, 1952, "J46-WE-8 Engines, Provisions for emergency Control of the Exhaust Nozzle Actuator.", **B1099VE1RG72.3.2NACP**
[34] WAGT Report A-1317, April 2, 1952, Calculated Starting Torque and Speed Requirements for J46-WE-1, -3, -8, -10 Turbojet Engines. Contract NOa(s) 10825, **B1099VE1RG72.3.2NACP**
[35] WAGT Memo, April 29, 1952, "Contract NOa(s) 9670 and 11028 WEC Turbo-Jet Engine Models J46WE-1 and -2. Incorporation of Larger Diameter Turbine, Survey of. Engine Change Nos. 3 for -1 engines and 3 for -2 engines.", **B555V5RG72.3.2NACP**
[36] WAGT Memo, April 1, 1952, "Contract NOa(s) 10825 and AF33-(038)-17843 WEC Turbo-Jet engine Models XJ46WE-3 and

XJ46WE-8. Relocation of Spark Plugs and Igniters, Survey of.", **B1099VE1RG72.3.2NACP**
[37] WAGT Memo, May 5, 1952, "Contract NOa(s) 9670 Proposal for Evaluation Testing of J46-WE-2 Afterburner Marquardt Aircraft Company", **B554V13RG72.3.2NACP**
[38] BuAer cover memo excerpt, WAGT Monthly Progress Report for April 1952, date unknown. **B552V1RG72.3.2NACP**
[39] BuAer Memo, May 16, 1952, "Contract NOa(s) and Noa(s) 11028; Expenditures on", **B555V5RG72.3.2NACP**
[40] BuAer Memo, May 6, 1952, "Proposed incorporation of titanium in Westinghouse turbojet engines", **B1099VE1RG72.3.2NACP**
[41] WAGT Memo, February 27, 1952, "Proposed Incorporation of Titanium in Westinghouse Turbojet Engines", **B1099VE1RG72.3.2NACP**
[42] WAGT Memo, October 3, 1952, "Proposed Incorporation of Titanium in Westinghouse J46-WE-8 Turbo Jet Engines", **B1099VE2RG72.3.2NACP**
[43] BuAer Memo, December 15, 1952, "Proposed incorporation of titanium in J46-WE-8 engines", **B1099VE2RG72.3.2NACP**
[44] WAGT Memo, April 30, 1952, "Contract NOa(s) 10825, total Dry Weight of J46-WE-8, Increase of.", **B1099VE1RG72.3.2NACP**
[45] BuAer Memo, May 27, 1952, "Contract NOa(s)10825, J46-WE-8 Model Specification Weight, "**B1099VE1RG72.3.2NACP**
[46] WAGT Memo, May 28, 1952, "Contract NOa(s) 10825, AF33-(038) 17843 WEC Turbo Jet Engine Models J46-WE-1, J46-WE-2, J46-WE-3, J46-We-8 & J46-WE-12 Proposed Turbine Temperature Traversing Bosses, Survey of. Engine Change Nos. 4 for -1 engine, 4 for -2 engine, 4 for -3 engine, 108 for -8 engines and 1 for -12 engine." **B1096VE4RG72.3.2NACP**
[47] BuAer cover memo excerpt, WAGT Monthly Progress Report for May 1952, date unknown. **B552V1RG72.3.2NACP**
[48] WAGT Memo, June 2, 1952, "Contract 9670 and 11048 WEC Turbo Jet Engine Models J46WE-1 and J46WE-2. Dimension changes on WEC Design Layout No 4A3825-A4 Turbine Installation Model T-13. Engine change Nos. 5 J46-WE-1 and J46WE-2.", BuAer cover memo excerpt, WAGT Monthly Progress Report for April 1952, date unknown. **B552V1RG72.3.2NACP**
[49] BuAer Speedletter, June 12, 1952, "Contract NOa(s) 9670; Model J46-WE-2, and -8 Engines; deletion of engine oil sounding gage", **B554V13RG72.3.2NACP**
[50] WAGT Memo, July 3, 1952, "Contract NOa(s) 9670; Model J46-WE-2 and -8 Engines; Deletion of Oil Sounding Gage", **B554V13RG72.3.2NACP**
[51] BuAer Contract NOa(s) 9670, Amendments 10 & 12, **B556VC2RG72.3.2NACP**
[52] WAGT Memo, September 29, 1952, "Contract Noa(s) 9670, 10825; WEC Turbojet Engine Models J46-WE-2, J46-WE-8. Modification of Simplified Afterburner Fuel System, Information On.", **B556V6RG72.3.2NACP**
[53] WAGT Memo, June 25, 1952, "Contract NOa(s) 9670, 10825; WEC Turbojet Engine Models J46WE-2, J46WE-8. Modification of Simplified Afterburner Fuel System, Survey of. Engine Change No. 6 for J46WE-2; Engine Change No. 110 for J46WE-8.", **B555V5RG72.3.2NACP**
[54] BuAer Memo, June 24, 1952, "Contracts: NOas 10385, 10653, 10825, and 52-048", **B1096VE4RG72.3.2NACP**
[55] BuAer cover memo excerpt, WAGT Monthly Progress Report for June 1952, date unknown. **B552V1RG72.3.2NACP**
[56] WAGT Memo, July 1, 1952, "Contract NOa(s) 9670, 10825. WEC Turbo-jet Engine Models J46WE-2; J46WE-8. Single Port Afterburner Control and New Customer Linkage, Survey of. Engine Change No. 7 for J46WE-2; Engine Change No. 111 for J46WE-8.", **B555V5RG72.3.2NACP**
[57] BuAer Memo, July 11, 1952, "Contracts NOa(s) 9670, 10067, and 52-403-c – Request for Overtime Approval", **B555V5RG72.3.2NACP**
[58] WAGT Memo, July 17, 1952, "Contracts Noa(s) 9670 and 10825 J46WE-8 Qualification Testing Our Reference WG-60999 – WG-64730", **B554V13RG72.3.2NACP**
[59] WAGT Memo, July 15, 1952, "Contract NOa(s) 9670 XJ46-WE-8 Engine Development Program", **B554V13RG72.3.2NACP**
[60] BuAer Memo, July 17, 1952, "Reshipment of Shipping Container to Verify Changes", **B1096VE4RG72.3.2NACP**
[61] NACA Report RM E50K22, March 5, 1951, Experimental Investigation of Tail-Pipe-Burner Design Variables, W.A. Fleming, E. William Conrad, and A.W. Young, Lewis Flight Propulsion Laboratory, Cleveland, Ohio
[62] WAGT Memo, July 28, 1952, "Contract NOa(s) 10825, WEC Turbojet Engine Model J46-WE-8. Long Coupled Afterburner, Survey of. Engine Change No. 113", **B1096VE4RG72.3.2NACP**
[63] WAGT Memo, October 8, 1952, "Contract NOa(s) 10825, WEC Turbojet Engine Model J46-WE-8. Long Coupled Afterburner, Information on.", **B1096VE4RG72.3.2NACP**
[64] WAGT Report A-1379, May 13, 1952, Study of Non-Electrical Afterburner Fuel Control System; J46-WE-8 Engines Contract NOa(s) 19825, **B1099VE1RG72.3.2NACP**
[65] BuAer memo, August 2, 1952, "Model F7U-3 Airplane, reassignment of", **B1099VE2RG72.3.2NACP**
[66] BuAer Memo, August 6, 1952, "Contract NOa(s) 10825 and NOa(s) 10385, J46WE-8 and J40WE-22, Engines; Exhaust Nozzle Control, Low Temperature Component Qualification test Scheduling", **B1099VE2RG72.3.2NACP**
[67] WAGT Memo, August 15, 1952, "Contract NOa(s) 10825 Proposed Engine Installation Clearance Envelope J46-WE-2", **B555V5RG72.3.2NACP**
[68] BuAer Memo, August 29, 1952, "Contract NOa(s) 9670; XJ46-WE-8 Qualification Test Engine", **B554V13RG72.3.2NACP**
[69] WAGT Memo, August 7, 1952, "contract NOa(s) 9670, XJ46-WE-8 Engine 50 Hour Flight Rating and 150 Hour Qualification Tests.", **B554V13RG72.3.2NACP**
[70] BuAer Memo, September 23, 1952, "Contract NOa(s) 9670, XJ46-WE-8 engine; afterburner fuel control; component qualification testing of", **B556V6RG72.3.2NACP**
[71] BuAer Memo, September 5, 1952, "Contract NOa(s) 10825, 50 Hour J46-WE-8", **B1099VE2RG72.3.2NACP**
[72] WAGT Memo, September 4, 1952, "Contract Noa(s) 10825. WEC Turbojet Engine Models J46WE-8, J46WE-12. Shielded Control Cable Assembly. Information on.", **B1096VE4RG72.3.2NACP**
[73] WAGT Memo, September 18, 1952, "XJ46 Engine, Simulated Altitude Tests at the Aeronautical Engine Laboratory", **B556V6RG72.3.2NACP**

[74] WAGT Memo, September 22, 1952, "Contract NOa(s) 10825. WEC Turbojet Engine Model J46We-8. Anti-Icing Port Cover, Information on.", **B1096VE4RG72.3.2NACP**
[75] WAGT Memo, September 17, 1952, "Contract NOa(s) 9670, XJ46-WE-8 Engine Flight Rating and Qualification Test", **B556V6RG72.3.2NACP**
[76] BuAer Memo, October 1, 1952, "XJ46-WE-8 Qualification Test engine configuration", **B556V6RG72.3.2NACP**
[77] BuAer Memo, September 1952, "Contract NOa(s) 9670, XJ46WE-8 Engine Flight Rating and Qualification Test", **B556V6RG72.3.2NACP**
[78] WAGT Report A-1565, August 12, 1953, Final Report covering F7U-1 Flight Test Program for Westinghouse J34-WE-32A Engines. On Contracts NOa(s) 9670, NOa(s) 53-556, **B556V7RG72.3.2NACP**
[79] BuAer cover memo excerpt, WAGT Monthly Progress Report for September 1952, date unknown. **B552V1RG72.3.2NACP**
[80] BuAer cover memo excerpt, WAGT Monthly Progress Report for October 1952, date unknown. **B552V1RG72.3.2NACP**
[81] WAGT Memo, September 5, 1951, "Contract NOa(s) 10825, Westinghouse Model Specification WAGT-X24C10-6, covering XJ46-WE-8 and XJ46-WE-10 Turbojet Engines, Total Dry Weight for", **B1099VE1RG72.3.2NACP**
[82] BuAer Memo, October 10, 1952, "Contract NOa(s) 9670, XJ46-WE-8/-2 engine; supply of fuels for full scale combustor deposition investigations", **B556V6RG72.3.2NACP**
[83] WAGT Memo, November 28, 1952, "Contract NOa(s) 9670, XJ46-WE-8/-2 Engine; supply of Fuels and Program for Combustor Deposition Investigation.", **B556V6RG72.3.2NACP**
[84] WAGT Memo, October 8, 1952, "Contract NOa(s) 9670, 10825. WEC Turbo Jet Engine Models J46WE-2, J46-WE-8. Single Port Afterburner Control and New Customer Connection Linkage, Information on.", **B556V6RG72.3.2NACP**
[85] WAGT Memo, September 8, 1952, "Contract NOa(s)-9670 Status of J46 Development Program", **B554V13RG72.3.2NACP**
[86] BuAer Memo, October 9, 1952, "Contract NOa(s) 9670 – XJ46 Engine Model Specifications, changes to", **B554V13RG72.3.2NACP**
[87] WAGT Memo, October 24, 1952, "Contract NOa(s) 9670, 10825; WECO Turbojet Engine Models J46WE-2, -8 and -12. Incorporation of Simplified Afterburner Fuel System", **B554V13RG72.3.2NACP**
[88] Chance Vought Aircraft Memo, October 3, 1952, "Contract NOa(s)-10825, 9670 – WECo Turbojet Engine Models J46-WE-2, -8, -12 – Interim Exhaust Nozzle Control System and Hydro-Mechanical Exhaust Nozzle Control System – Description of", **B1099VE2RG72.3.2NACP**
[89] BuAer Memo, October 28, 1952, "Contract NOa(s) 9670, Incorporation of the T-13 turbine in the XJ46-WE-2 Engine", **B556V6RG72.3.2NACP**
[90] BuAer Memo, October 31, 1952, "Engine control problems at high speed flight", **B554V13RG72.3.2NACP**
[91] BuAer cover memo excerpt, WAGT Monthly Progress Report for September 1952, date unknown. **B552V1RG72.3.2NACP**
[92] BuAer Memo, November 12, 1952, "Contract NOa(s) 9670 engine, afterburner pumps and controls; component qualification tests of", **B556V6RG72.3.2NACP**
[93] BuAer Memo, December 11, 1952, "Contract NOa(s) 9670, XJ46-8/-2 Engine-Afterburner Cant, Removal of", **B556V6RG72.3.2NACP**
[94] WAGT Memo, November 7, 1952, "Contract NOa(s) 10852. WEC Turbojet Engine Model J46-WE-8. Pressure-Sensitive Afterburner Detection Switch, Survey of. Engine Change Number 115.", **B1096VE5RG72.3.2NACP**
[95] WAGT Memo, November 17, 1952, "Thrust Definition Conversion", **B1099VE2RG72.3.2NACP**
[96] BuAer cover memo excerpt, WAGT Monthly Progress Reports for November (date unknown) and December, WAGT Report A-1519, January 12, 1953. **B552V1RG72.3.2NACP**
[97] BuAer Memo, January 9, 1953, "Contract NOa(s) 10825, Westinghouse Specification WAGT-X24C10-6 covering the Models XJ46-WE-8 and -10 engines; total dry weight for", **B1099VE1RG72.3.2NACP**

Chapter 5
J46-WE-8/A/B and -12/A/B Early Production

The original BuAer development contract NOa(s) 9670 only covered development of engines through qualification. A Letter of Intent Contract NOa(s) 10825 was signed on December 29, 1949 to cover the delivery of two J46-WE-2 engines with their shipping containers. This LOI contract was extended and modified by amendment many times to cover the various production models and also to change various order quantities until replaced with a Fixed Price Contract on June 10, 1953. This contract was itself then similarly amended many times until J46 production wound to an end. As was the case with the J40 engine program, significant continuing development costs were charged to the production contract and the correspondence in the related archives overlaps and supplements our knowledge of the development story.

The Air Force versions (-1, -3, -5, -7) of the J46 are dealt with in a separate chapter as they had their own contract.

The earliest production schedule found in the files is dated November 2, 1951. It reflects the contractual commitments at that point in time. A condensed version shows:[1]

J46 Delivery Schedule November 2, 1951								
	Navy South Philadelphia			Navy Kansas City			USAF South Phila.	
Model	-2	-8	-10	-2	-4	-8	-1	-3
First	2/53	4/52	2/52	2/53	12/52	7/52	11/52	4/52
Last	10/53	11/53	12/52	2/53	6/54	6/53	8/53	7/52
Total	94	209	41	721	910	500	5	6

The production schedule began to slip as the development of all the versions lagged. It is of interest to bring the planned production picture up to the point where we left off at the end of the prior development chapter. BuAer had amended the total engine order volume four times from the 2,475 reflected in the above chart and, as we pick up the story in December 1952, BuAer had the following number of engines on order:

J46 Engines on Order As of 12/31/1952[2]	
Model	No.
J46-WE-2	2
J46-WE-4	0
J46-WE-8	707
J46-WE-10	0
J46-WE-12	6
J46-WE-18	815
J46-WE-20	910
Total	2,434

BuAer production planning had obviously moved on to using first the de-rated -8 engines until the fully developed -2 was qualified (with a new designation of -18). The -12 engines with their special tailpipe extension and cant for the XF2Y-2 were almost identical to the -8 engines in all other aspects and BuAer had agreed that certification of the -8 would apply to the -12 as well. That basic approach to the -12 would be applied as well to the further de-rated sub-variants of the -8 as they emerged.

The bulk of the engines were to be produced in the Kansas City plant where the J34-WE-32 and -36 were in production. The start-up of earlier J34 engine production there had been difficult. The bulk of the employees had to be trained and assembly lines set up. The WAGT engine experts were back in the Essington plant and decision-making on engineering issues was impacted by both the large number of change orders and the time and distance between locations. And, specifically related to the first engine produced in Kansas City, the J34-WE-30, there had been over 1,500 change orders that had to be accommodated during the first year of production on that engine alone. As the J46 finally approached production status, a constant stream of change orders related to this new engine would continue as development of the engine continued in Philadelphia.

Engine Shipments Begin

With BuAer's acceptance of the results of the first 60 hours of qualification testing on the XJ40-WE-8A as flight qualifying the engine, the Essington assembly of six engines began. Almost immediately, the Dallas BAR fired off a wire to BuAer's transportation section asking for air shipment of the first J46-WE-8A (Serial WE040001) as soon as it was accepted, stating as a justification that F7U-3 airframes were grounded due to lack of engines.[3]

The target price for J46 engines was now set at $176,287.00.[4]

The complicated production delivery program with various models being produced with variations in A/B's, control systems, turbines and performance led BuAer to publish a table that reflected the differences.[5]

J46-WE-8/-12 Production Program Models
December 24, 1952

Designation	YJ46-WE-8A	J46-WE-8A	J46-WE-8B	J46-WE-8	J46-WE-12A	J46-WE-12B	J46-WE-12
Max Thrust (lb)	5,300	5,500	5,800	5,800	5,500	5,800	5,800
A/B Flameholder	Short Coupled	Long Coupled	Long Coupled	Long Coupled	Short Coupled	Short Coupled	Short Coupled
A/B Axis	Canted	Straight	Straight	Straight	Straight	Straight	Straight
A.B Nozzle Area (in²)	360	375	375	375	375	375	375
A/B Blow-out Limit (ft)	38,000	48,000	48,000	48,000	38,000	38,000	38,000
Exhaust Nozzle Actuator Control	Electrical	Electrical	Electrical	Hydro-Mechanical	Electrical	Electrical	Hydro-Mechanical
Turbine Configuration	Low Perf.	Low Perf.	High Perf.	High Perf.	Low Perf.	High Perf.	High Perf.
35K ft Limit Due to Compressor Stall	Yes	No	No	No	No	No	No
Quantity	6	24 Max	Phase out by -8	Phase out by -2	3	3	16
Est. Avail. Date	Dec 1952	Mar 1953	May 1953	Nov 1953	Apr 1953	Aug 1953	Nov 1953
Time Between O/H (hr)	50	150	150	150	150	150	150

Pre-Production Development

Picking up the development thread, it seems that BuAer had formally objected to the process of "Teeing" the fuel distribution lines to the vaporizing tubes on the grounds of increasing the complexity of the maintenance and overhaul program as well as increasing the costs. They wanted to know: a) the number of engines with "T'ed" lines planned, b) the turbine inlet temperature traverse required during the overhaul of the engines, c) to what extent the hot section components or vaporizing tubes could be replaced in the field with WAGT's requirement for determining the turbine temperature distribution, and, d) what parts would have to be replaced to rebuild the engines to an "un-Teed" configuration. Since only six (6) such engines were to be accepted, early answers to the questions were required.[6]

Another change was surveyed while WAGT prepared their answer to the above request. WAGT proposed moving the axial and circumferential locations of the spark plugs and igniter fuel nozzles on the -2. The experimental results had shown that ignition performance at altitude was considerably improved if those elements were moved to the proposed location. At sea level, ignition performance was essentially the same as the current location.

In addition, the incorporation of the new locations along with the new longer engine diffuser would eliminate the problem of the two upper fuel feed lines to the igniter fuel nozzles passing over the engine mount ring flange (and out of the engine envelope).[7]

The operating temperature limits for the various control systems had to be resolved. WAGT pointed out that the 250°F limit asked for by BuAer was well above the known limit of the solenoid used in the A/B control to operate the fuel shutoff valve. Its limit was 200°F. They asked for an exception to be made for this component, since the component could not be isolated during the test.[8] BuAer did not grant the exception until mid-January, stating "In view of the fact that the subject engine is being developed to the AN-E-30 series specification, which calls for a maximum component qualification test temperature of 165°F, the Bureau of

Aeronautics cannot insist on the 250°F test. In confirmation of (a phone conference between BuAer and WAGT), the high temperature portion of the subject test at 200°F is acceptable to this bureau for the subject engine only."

To supplement their earlier response regarding the need to interconnect the intake scheduling system and engine control system on supersonic aircraft, WAGT submitted a formal review report in memo form at the end of January 1953. They had consulted with the aerodynamics engineers of Consolidated Vultee, Douglas (El Segundo and Santa Monica), McDonnell, North American and Northrup on the issue and after reviewing the opinions of all concerned, WAGT stood by their early response, i.e. that the intake scheduler and engine control should not be interconnected. They added that the expense of doing a detailed study did not appear warranted.[9] Supersonic aircraft being planned were all going to use engines with on/off A/B's, so the challenge of how to control supersonic speed appeared to be one of either regulating the airflow into the engine to modulate thrust or use of the speed brakes on the aircraft. In the future, the development of modulated (variable) A/B's and automatically controlled intakes solved the supersonic flight speed stability problem.

BuAer accepted WAGT's proposal for testing and qualification of the titanium compressor discs in the -8 based on the Air Force's -3 qualification with identical discs. The test plan had to include an equivalent number of hours at each of the power settings called for in specification WAGT-X24C10-6A.[10]

A review of specification WAGT-X24C10-6A brought a new requirement for flight testing of the engines as part of qualification. BuAer justified the "minimal" additional requirements because:

1. WAGT was setting up a flight test facility.
2. The testing requirement was not severe, only intended to demonstrate adequate flight characteristics.
3. Flight test requirements were being integrated into the new MIL-E spec requirements.

The new flight test requirements:

1. The testing should utilize JP-4 fuel of both specific gravities of .81 and .76 without adjustment of the fuel control during the testing.
2. The flight test engine did not necessarily have to be the 150-hour or 50-hour test engine(s).
3. The ram pressure ratio obtainable had to be from an inlet duct whose characteristics had been reviewed and approved by the Procuring Service.
4. At Military Power – Takeoff and climb to the maximum practical altitude (at least 45,000 feet) with the power lever in the military position.
5. At Maximum Power - Takeoff and climb to maximum practical altitude with the power lever in the maximum position.
6. Regulation at Idle Power – Descend from the maximum practical altitude to 5,000 feet with the power lever in the idle position.
7. Normal and Emergency Air Starts – Perform a normal air shutdown and air start at 5,000 foot intervals from 5,000 feet to the maximum guaranteed starting altitude. Simulate a flameout and perform an air start at 5,000 foot intervals to the maximum guaranteed starting altitude.
8. Accelerations and Decelerations – Perform snap accelerations from idle to maximum thrust and snap decelerations from military to idle thrust and from maximum to idle thrust at 5,000 foot intervals from 5,000 feet to the maximum practical altitude.
9. Control Transfers – Perform transfers from primary to emergency operation, check the functional operation of the emergency control features and transfer back to the primary control at 5,000 foot intervals from 5,000 feet to the maximum practical altitude. The transfer from the primary to the emergency control should be at military thrust.
10. Test Data – Make recordings every five (5) seconds while at Military, Maximum and Idle Thrust, every .25 seconds at all other times of the temperatures, pressures and fuel flows listed.

Additionally, WAGT had to advise BuAer in detail why the radio interference limits of specification AN-I-27a could not be closely approached. The then current limits had been carried over from the no longer used electronic control system and were considered unreasonable for the new hydro-mechanical control that would be used on the -8. BuAer's position remained that the prior higher limits were not acceptable.

Many wording changes were required, including the addition of language stating "Snap accelerations shall not cause loss of combustion or objectionable compressor instability" and "Snap decelerations shall not cause loss of combustion."[11]

The flight test program BuAer referred to involved a Douglas F3D-1 SkyKnight modified and instrumented to accept a J46-WE-2 installation envelope compatible engine in the left nacelle and a North American B-45A-5 Tornado bomber modified with a centerline pod that could be raised up tight to the aircraft for ground handling and lowered in flight for testing. The pod was used first for limited J40 testing *(covered in detail in the author's book on the J40)* and then modified for J46 testing. There was also a limited test program using one of the early production F7U-3 airframes and finally a series of thrust increase tests using a F7U-3M. The J46 flight testing program is covered in Chapter 9.

With the shipment of the first (of the six) YJ46-WE-8A engines about to occur, BuAer got around to authorizing the shipment of the starter generator adapter gearbox installed on the engine at the time of shipment. They stated their reasons for installing the "kit" at the factory instead of the field was that it eliminated the necessity of field removal of the front engine oil seal to install the kit. When the oil seal was removed, the spiral gear bearings were exposed to contamination or dirt.[12] WAGT had recommended the kits be factory installed for that reason and intended to ship installed kits as long as they were available at the factory.[13] While the field maintenance manual gave instructions on how to remove the gearbox and replace it, and stated caution needed to be taken to prevent contamination of the accessory gearbox during the process, WAGT never redesigned the kit to be a simple quick change interface that avoided the possibility of contamination.

A later WAGT memo agreed that the risk level of dirt contamination during installation of the starter generator gearbox was about the same as changing the oil filter, but went on to state the real risk issue with installing a starter generator gearbox on the engine for the first time was that the lock nut on the oil seal would not be properly secured in the field. It was recommended that it should be done at a higher level of maintenance than operational unit level.[14]

Looking ahead into March, BuAer decided to have all engines (except the YJ46-WE-8A) that would utilize the starter-generator gearbox shipped without the front oil seal on the starter–generator drive. This included the -8, -8A, -8B, -2 and -2A. The drives would be shipped in separate packaging. (The -2A would have the larger turbine).[15]

More A/B control information was sent out in mid-December when WAGT had to send out supplementary information regarding their pressure sensitive A/B detection switch, specifically about the "time delay" check valve in the system. They explained the check valve was really a damping device to remove the pressure cycling that occurred when the A/B was ignited. The damping minimized the possibility that a false blowout signal might reach the Aerotec switch. They suggested the check valve might be better termed a "pressure-damping" ball check valve in preference to the "time delay" ball check valve. Then, possibly adding to the confusion, WAGT continued to call the orifice in the P1 line to the Aerotec switch the "time delay" orifice. They continued to recommend that the time delay orifice be airframe mounted along with the check valve.[16]

Their continued push to mount some of the additional components needed for the modified control system on the airframe resulted in BuAer responding with a memo that listed all the installation requirements for the Aerotec system. Since the proper operation of the system was the engine manufacturer's responsibility and proper operation required very precise installation, it appeared to BuAer the installation requirements were more easily complied with if the control system elements were engine mounted, also then making the installation of the engine itself simpler.[17]

The automatic trim control mounts were changed to reduce vibration and increase accuracy of the signal and the change was surveyed to alert all airframe companies intending to airframe mount the unit.[18]

The -12 had an issue in mid-January regarding Convair's calculation of a problem with the position of Convair's designed A/B mount stop. They had used J46-WE-8 figures from the thermal expansion tables listed in the Installation Manual and if their calculation was correct, the -12 mount had to be redesigned. The stop on the mount was there to prevent the A/B from departing the airframe if the bellows failed. WAGT did the same calculations and notified Convair via BuAer that their existing mount design was adequate and could remain unchanged.[19]

A wire asking that the second YJ46-WE-8A be shipped was received from the BAR Dallas on January 28, 1953. The serial number was WE040002. Air shipment was again requested due to the urgency of need at Chance Vought.[20] Apparently it shipped almost immediately, as a confirming wire requested in 1955 for some reason states the engine shipped in January of 1953.[21] Shipments continued at a slow pace: WE040003 (likely March), WE040004 (about early April), and WE040005/WE040006 (mid to late April).

The group assembly breakdown parts list for the YJ46-WE-8A was sent over to BuAer on 3 February, representing the build specification for the engine.[22]

WAGT now began to issue a survey log that incorporated every change surveyed for the J46-WE-8x since the mockup daily report was issued. The change numbers started at 74 as a result. The log included the serial number block assignments for the various -8 models:[23]

J46 Engine Serial Number Blocks		
Model	Start	End
YJ46-WE-8A	WE040001*	WE040006
J46-WE-8A	WE040007*	WE040012
	WE404501**	WE404515
J46-WE-8B	WE040013*	WE040080
	WE404516**	WE404684
J46-WE-8	WE404685**	WE405120
* Essington Manufacture		
** Kansas City Manufacture		

WAGT had submitted an Operating Instructions manual for the YJ46-WE-8A to BuAer on January 9, 1953 and BuAer finally approved it for the YJ46-WE-8A engines only.[24] They requested WAGT submit individual manuals for each of the sub-models.[25]

On March 10, WAGT again requested that the Progress Payment cap defined in Amendment 2 of the LOI Contract be lifted from 75% to 90%. The conversion from the LOI contract to the Cost Plus Fixed Fee had not yet been agreed to due to fiscal difficulties within the Bureau. They asked that it be considered:

1. That they were currently delivering engines under the contract,
2. WAGT's expenditures in performing the work under the contract had greatly exceeded the limitation presently under the Letter of Intent, and
3. The definitive Cost Plus Fixed Fee contract might not occur for an indefinite period of time.

The increase would allow WAGT to be reimbursed 90% of their costs to date and 100% of the subcontractors' costs through June 30, 1953.[26]

Early compressor problems with the YJ46 engines brought a response from WAGT. (See Appendix C for examples of the problems and complaints):

1. The engines were satisfactory for flight,
2. They recommended avoiding prolonged flight in the 55-75 percent rpm range,
3. WAGT was willing to twist the C-67 to C-75 compressor configuration (stages 5 through 9 four degrees maximum at tip) at Chance Vought Aircraft using WAGT personnel if BuAer desired, and
4. The C75 configuration eliminated the step in the airflow curve discovered in C67 testing on the #3 house engine. House engine #6 testing of the C75 configuration with strain gauges had no buffeting or break in airflow.[27]

The early performance problems with the YJ46-WE-8A at Vought brought a request to convert the WE040001 and WE040002 engines from the C-67 to the C-75 compressor configuration. BuAer agreed to the modification and asked that the altitude operating recommendations for those two engines be updated.[28] The C-75 compressor was the design to be used in the -8A, -8B and -8 engines.

Open MUB items were still being closed, with BuAer accepting the removal of the oil level indicator. The acceptance memo states that with the termination of all the Air Force series of J46 engines (-1, -3 and -5) there was no longer a requirement. The memo is dated March 23, 1953.[29]

Further requirements from Chance Vought began to be received as they tested the first received YJ46-WE-8A engines. In the latter part of March, they requested that the emergency provisions for the J46-WE-8 Interim Exhaust Nozzle Control be modified. The requests and WAGT responses to BuAer were:

Request: On all engines, provide for opening and closing overrides.

WAGT Response: The manual override was available to open or close the exhaust nozzle. Engines would be provided with opening overrides only or both opening and closing overrides depending on BuAer decisions.

Request: On engines 1-6, provide a customer's connection on the engine for turning afterburner on and off with cockpit switch.

WAGT Response: The first six engines had been provided with a customer's connection to open or close the A/B valve from an AC switch. Full information was forwarded to Chance Vought.

Request: For early flight test provide for preset of exhaust nozzle bias to open or close.

WAGT Response: Information was forwarded to Chance Vought on how to preset the exhaust nozzle bias to either open or closed direction.

Request: Engine #7 and up provide device linked to throttle so that exhaust nozzle will be biased toward being closed below the A/B throttle setting and full open in A/B.

WAGT Response: The decisions on the Interim Emergency Control provisions for engine #7 and up had been agreed. The conclusions reached with BuAer and Chance Vought were:

A. Any change to the existing emergency control system would result in considerable delay in engine delivery.

B. The gain would only be realized in the last six (6) seconds of flight - such a failure being considered remote.
C. The reliability risk increase from the "additional gadgetry" would offset any marginal advantage in the event.[30]

The central district of BuAer requested a summary of the procurement of Kansas City engines by dash model in a wire to BuAer.[31] The data was required for planning.

WAGT forwarded the latest version of the -8 and -12 specification WAGT-X24C10-6A dated March 16 to BuAer on March 31, 1953. It covered the J46-WE-8, -8B, -8A, -12A and YJ46-WE-8A. Some paragraphs (not specified) were still not formally approved.[32]

In early April, the Kansas City BAR notified (reminded?) BuAer they had 627 -8 engines scheduled for production beginning in May at the Kansas City Plant and asked for allocation and shipping instructions.[33]

BuAer asked the Kansas City BAR to "make a thorough evaluation" of the WAGT Kansas City plant's program for the J46-WE-8 engine. The KC BAR sent a memo to the Essington BAR stating that:

"Inasmuch as Westinghouse AGT Division, Kansas City Works, does not have engineering and development cognizance of this engine, your office is requested to furnish the following information:

1. Realistic view of qualification/verification program on the WE-8A to include starting, length of test, and completion date.
2. Your estimate of engineering changes that may delay production once engine has completed qualification.
3. Your comment on the progress to date on the WE-8B/-8 program.
4. Your broad comments on the J46 development program to include schedule of apportionment of engineering and development responsibility and production between Kansas City and South Philadelphia Works.

"In view of the continued slippage of this program, it is requested that any information obtained directly from the Contractor be accompanied by your evaluation of its realism."[34]

The memo is quite revealing, exposing a lack of program status information distribution both from BuAer to the Kansas City BAR and through the Essington BAR to Kansas City as well as a lack of coordination between the KC BAR and the WAGT plant management. It is not clear exactly what everyone involved thought was supposed to be happening, but clearly, with only a few weeks before the planned start of J46 production in Kansas City, the BAR there was not up to date.

The survey sent out in February to add temperature traversing bosses to the engines was rejected by BuAer on the grounds the extra weight on the engine between acceptance and overhaul was not justified. That the engine was already "so much over specification weight" made the additional weight even more objectionable. WAGT had indicated in the survey that all engines would require the bosses and since then WAGT had not responded to BuAer's request with information on the fuel distribution problem that required the bosses for adjustments. The need for the bosses could be eliminated by "proper redesign" of the combustor section. BuAer requested a realistic estimate of the number of engines for which the bosses would be required.[35]

In late March WAGT had proposed raising the maximum thrust rating on the YJ46-WE-8A engines from 5,300 to 5,500 pounds and BuAer agreed to the proposal three weeks later.[36, 37] All other ratings on the engine remained unchanged.

A report (A-1571, 4/14/53)) on limited windmilling characteristics for the J46-WE-8 was sent over to BuAer near the end of April. The report only covered windmilling with no power extraction and no air bleed operating and then with power extraction only. A follow-on report would cover power extraction and air bleed operating. The tests were run at the AEL.[38]

A change order telegram for delivery of engines was received and acknowledged with WAGT taking exception to the -12/16 schedule. They proposed a schedule of one engine a month from September to February.

The earlier Monthly Progress Reports at this time were now replaced by Monthly Production Progress Reports issued from both Essington and Kansas City. The first was sent out in late April for the first time on contract NOa(s) 10825. This and many of the later reports through 1953 were not retained in the BuAer contract correspondence files, most likely being kept in the BuAer Power Plant Division records, which are no longer available.

The Kansas City BAR sent a follow-up memo to BuAer to answer the original memo asking for the Kansas City readiness review. A few fragments from this review stand out to give us an idea of the Kansas City operation at that point:

1. The plant had 470 engineers (including technical writers and supervisors), 89 draftsmen, 196 laboratory technicians, and 160 clerical employees.
2. There was no standard 45 hour work week in effect. All overtime was subject to BAR approval on a departmental "as required" basis. Approvals were

limited to test, development and departments producing "line stopper" items.

3. No engine was in the production line at the time designated as a "back-up" engine to the Kansas City works verification test.
4. In spite of WAGT notifying BuAer that all engineering responsibility for the J40-WE-1, -8 and -22 models had been transferred to Kansas City, the major engineering responsibility remained in Philadelphia.
5. The most recent delivery schedule was based on the qualification test of the J46-WE-8A in Philadelphia by May 1 and completion of the Kansas City qualification test by May 15. The test was not supposed to start until May 8 but it appeared the test could not possibly start before May 10.
6. It appeared that production could start immediately upon successful completion of the tests.

The specification for the starter generator gearbox (WAGT-X24C10-50A) was sent over on May 8.[39]

The qualification test plan for the -12/-12B was forwarded for the BAR's review. As mentioned earlier, the qualification of the -8A and -8B would also be accepted as the qualification of the -12A and /-12B except for the afterburners. The -12A test was modified slightly from the accepted plan for the J34-WE-32. The -12B A/B would not need requalification. The changes for the -12A test are shown in the table to the right.[40]

BuAer accepted the Proposed Qualification Test for the -12A and -12B on May 21.[41]

At the same time that the test plan approval was received, WAGT began to send BuAer tables listing the early engine problems experienced by the YJ46-WE-8A engines and the corrective actions taken or planned.[42] The surviving table examples are reproduced in Appendix C.

Any new engine entering operational service is expected to display minor problems as the technicians operate and install the units. There are a number of events in the list that indicate weaknesses in WAGT's assembly and subcontractor quality control programs. The control system design and components demonstrated a need for a high level of technician training for proper replacement and adjustment. In particular, the use of "O" rings in the design was the cause of many leaks and installation related problems.

The emergency control system for the A/B was clearly not fully developed at that point, with changes in design still being incorporated even as the engines were in pre-flight test at Chance Vought. These ongoing problems caused long delays in getting the F7U-3 with J46 engines prepared for service use.

Development continued in an effort to find a way of meeting the specification performance of the -8 and -2. At meetings on April 20th and May 14th, the -8 model specification performance was questioned due to the failure of a 2% turbine efficiency improvement expected from the T-28 turbine that had not materialized. In addition, BuAer asked for analysis of the feasibility of attaching a 28" diameter A/B to the basic -8 engine.

An analysis was done of the performance of a -8 , -2 and -8 with a 28 inch A/B based on recent test results at AEL using the 74C-75A compressor on -8 house engine #2. The compressor had close tip clearances and non-profiled blading, lacked closed seal clearances (planned for production -8 and -2 engines), and also did not include -2 developed thin inlet guide vanes which might provide a slight increase in air flow.

J46-WE-12A A/B Test Plan*

Five (5) Runs	J34-WE-32 Plan	J46-WE-12A Plan
(1) Idle acceleration to take-off	2 Min	5 Min
(1) Take-off	10 Min	10 Min
(2) Idle Acceleration to Take-off	2 Min	5 Min
(2) Take-off	5 Min	5 Min
(3) Idle Acceleration to Take-off	2 Min	5 Min
(3) Take-off	5 Min	5 Min
(4) Idle Acceleration to Take-off	2 Min	5 Min
(4) Take-off	2 Min	5 Min
(5) Military Acceleration to take-off	2 Min	5 Min
(5) Take-off	5 Min	5 Min
Total Idle Time	2.0 Hours	5.0 Hours
Total Military	.50 Hours	1.25 Hours
Total Take-off	7.50 Hours	7.50 Hours
Total Accelerations	60	60

* Any house J46 engine could run the A/B test.

Engine combustor losses were obtained from -8 house engines #1 and #3. The -8 turbine (T-3) was tested in the #3 house engine earlier that month. Its efficiency was comparable to the previously anticipated T-28 turbine. The -2 turbine (T-19) had been tested in house engine #1 and matched the model efficiency - +½% higher than the T-3 turbine.

The pressure losses in the A/B for the -8 and -2 engines were obtained from tests on the #1 and #3 engines that month. The losses of the -2 28" A/B were the same as estimated but the -8 long coupled A/B losses were higher than expected. The losses for a -8 with a 28" A/B were not obtained from tests and for the study the losses for the 25.5" (standard) -8 A/B were used. It was assumed the expected slightly higher diffuser pressure losses would be largely offset by lower pressures due to lower flameholder velocities.

The A/B combustion efficiency, fuel-gas ratio, exhaust nozzle velocity coefficient, and ejector losses were the same as used in the model specification.

All calculations were made for a turbine inlet temperature of 1,500°F rather than the 1,525°F of the specification. The comparison point was 35,000 feet and 500 knots airspeed at both military non-A/B and A/B.

J46-WE-8, -2 and -8 with 28" A/B Comparison Results	
J46-WE-8	Same as specification with SFC 1.4% higher.
J46-WE-2	Thrust 2.4% higher non-A/B and 1.6% higher in A/B.
J46-WE-8 28"A/B	Non-A/B SFC 1.4% higher than standard -8. A/B thrust 1.5% higher and SFC 1.3% lower.

WAGT recommended the standard -8 specification performance be used for non-AB performance and the -2 specification be used for the A/B performance by reducing thrust by 1.5% and increasing the A/B SFC by 1.3% (6,010 pounds and 2.65 SFC).[43]

A BAR engine acceptance telegram June 5 stated WE040008 was assembled and shipped to Kansas City for its green run and would be accepted by the BAR there.[44]

WAGT notified BuAer (in a response to a March 19 BuAer request) that the A/B on the engines could be cantilevered while attached to the basic engine provided in handling the load did not exceed 1.75 "g". The above limit was to prevent damage to the ball joint on the A/B diffuser. The limit required the A/B always be supported when being transported attached to the basic engine.[45]

Flight limit curves for the new C-75 compressor were sent to BuAer on June 3. These showed smooth lines up to the 60,000 foot limit with no stair-step as experienced by the preceding C-67 design.[46] Shortly after receiving the C-75 compressor charts, BuAer sent over their comments on Amendment 2 to the -8x / -12x specification WAGT-24C10-6B. The included comments reveal that to some extent WAGT was trying to wrap the specification around the different engine sub-model performance and BuAer was pushing back. The main issues BuAer expressed were:[47,48]

1. BuAer accepted WAGT's desire to do more study on MIL-O-8188 corrosion preventative oil before using it in the acceptance test and preservation run.
2. They rejected narrowing the flight re-start limits to a ram pressure range of 1.15 to 1.25 at 35,000 ft. They pointed out the airframe-engine suitability for the fleet was determined based on the guarantees of the bid specification. If the original former limit of 1.5 was not based on substantial data it should have been called to the attention of the Bureau and not inserted into the specification as a firm guarantee. (WAGT had based the 1.5 limit on tests of earlier experimental A/B's on J34 engines.)
3. They requested that the external electrical power requirements be inserted into the specification.
4. The true altitude operating limits of the engines using the C-75 compressor should be shown. The maximum flight starting and flight operating altitudes for the short coupled -12A and -8A afterburner engines differed from the specification and only the guaranteed A/B flight starting and operating limits were acceptable.

And then, having held their ground on specification performance guarantees, BuAer concluded:

5. However, WAGT's proposal to form additional specifications for the subject engines based on separation of the engines by performance differences would be acceptable.
   a. One specification for the J46-WE-8, -8B and -12B engines.
   b. A separate specification for the J46-WE-12B if performance curves for the engine were issued.
   c. One specification for the J46-WE-8A, -12A and YJ46-WE-8A engines.

In a follow-up memo to the one communicating BuAer's disapproval of the addition of inlet temperature bosses, WAGT stated the bosses would be removed beginning with the KC engine WE404030 and Essington engine WE040032. All the -12A engines would retain the bosses but the -12B engines would be produced without them. The weight saved was 4.9 pounds.[49]

Production began on the J46-WE-8A and the Dallas BAR on July 10 requested WE040007, 9 and 10 be shipped to Dallas as soon as they were accepted.[50]

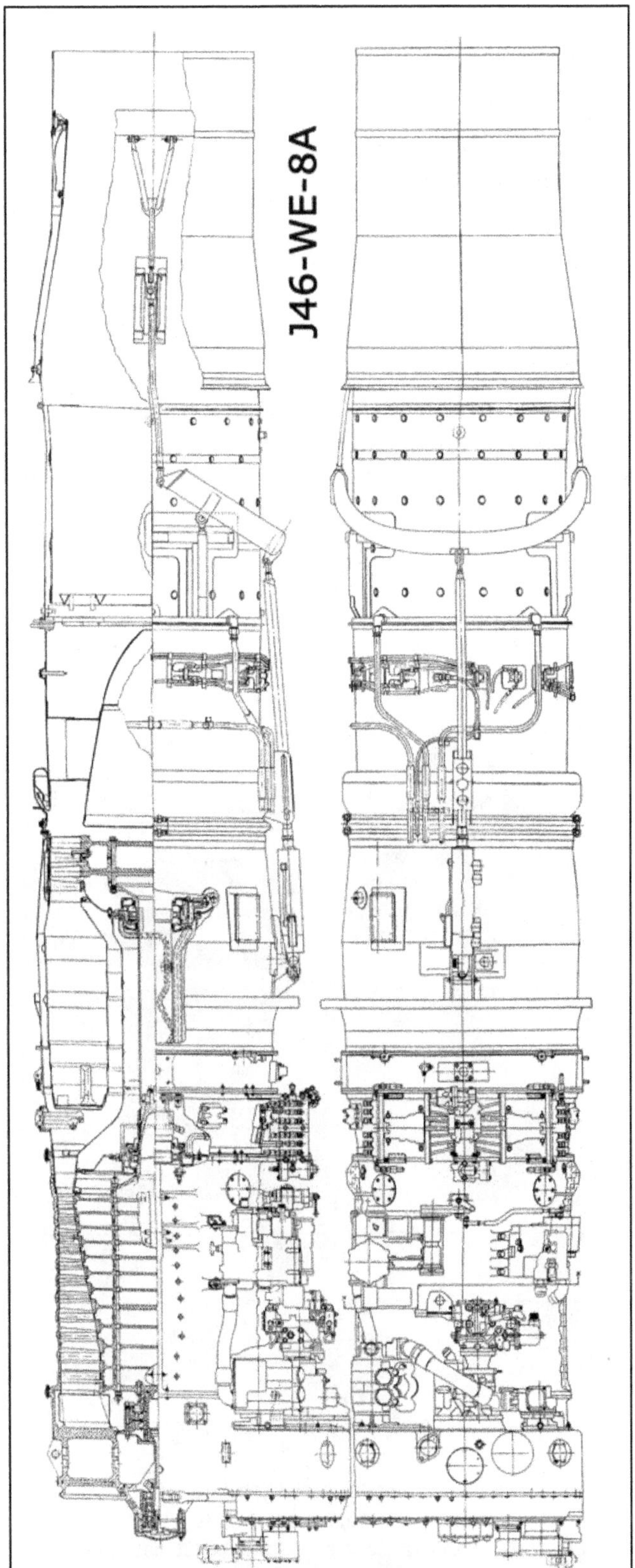

Figure 1: From WAGT J46-WE-8A assembly drawing. ***Courtesy Hagley Museum and Library***

Convair asked that the A/B mounting plate be changed from the current two plate design to a single plate design. WAGT sent BuAer two alternative single plate designs with drawings, materials specifications, bolt dimensions and tightening specifications. These could be constructed at Convair and the two plate mount removed from -12's as received until the single plate design was incorporated in any later production beyond the initial six engines. A written note on the memo circulation sheet states permission was given for Convair to use the design.[51]

A snag in J46-WE-8A engine acceptances occurred when the engines demonstrated excessive oil consumption during acceptance runs. A specification deviation waiver had to be issued to allow the six (6) J46-WE-8A 50-hour engines to be accepted with a maximum oil consumption of two (2) pounds per hour.[52] .

At this point WAGT was now forecasting verification testing would be completed and as a result production engines would be available for shipment during August. The KC BAR sent in an allocation request for 10 engines.[53]

The requirements for flight test support at Chance Vought were covered by a WAGT memo discussing using an F7U-3 airframe for engine flight testing. BuAer had requested six more 50-hour J46-WE-8A engines for flight testing. WAGT had requested F7U-3 BuNo 128453 be bailed to Chance Vought to be used in an engine flight test program (reproduced below) to help with verification. The airframe would be available soon for such testing. It was requested that the six YJ46-WE-8A engines be returned for overhaul or repair and modification necessary to support the proposed engine test program. Due to the press of time, a telegram was requested granting approval to begin immediate overhaul and modification of the three YJ46-WE-8A engines that had been returned to the plant. [54]

EDS-M-226481-A

REV. A By G. PAGE 2/26/53 — Flights Required Under I / " " " III / " " " IV / " " " V / " " " VI

** Tests may be combined
* " " " "

WESTINGHOUSE FLIGHT TEST IN F7U-3

	TITLE OF TEST	FLIGHTS REQUIRED	TEST POINTS	OBJECTIVES	DATA TO BE RECORDED	DATA OBTAINED AS OF 12/29/52	ENG. SER. NO. TO BE TESTED
I	Long Coupled A/B Test on J46-WE-8	20**	ALTITUDE / IAS: 10,000 150-500 Knots; 20,000 150-400 "; 30,000 150-350 "; 40,000 150-270 "	1. Determine A/B Blowout Alt. 2. Detect Any Squeal Tendencies 3. Determine Partial Augmentation Requirements 4. Determine Area Limitations	Time Histories & Plots of Std. Instrumentation	None	Not Determined
II	Single Port A/B Control Test	5*	Same as I	Determine Stability of New Single Port Control	Time Histories & Plots of Std. Instrumentation	None	Same as I
III	28" A/B Test on J46-WE-2	5	Same as I	Determine Engine Stability, T_{t5} Variation, Acceleration into, Deceleration out of A/B, Fuel & Area Trims, Wave Offs	Same as II	None	Not Determined
IV	Hydro-Mechanical Exhaust Nozzle Control	10*	Same as I	Determine Stability of Hydro-Mechanical E/N Control	Same as II	None	Same as I
V	Ejector Cooling Investigation	10**	Same as I	Determine Optimum Ejector Flow & Configuration	Ejector Flow & Temp. Plus Time Histories & Plots of Std. Instrumentation	None	Same as I
VI	Test of Long-Coupled A/B Component Parts in Short Coupled A/B (J46-WE-8 Backup Configuration)	5	Same as I	Same as I	Same as I	None	Same as I
VII	Possible Points for Investigation Based on J34 Test Flight Experience 1. Fuel Manifold Clogging 2. Increasing A/B Maximum Blow-out Altitude 3. Emergency Exhaust Nozzle Actuator Operation	**					

EDS-M-226481-A
G.F. Page 12-29-52

Figure 2: Slightly revised proposed flight test plan for the J46 in the CVA supplied F7U-3. No detailed plans seem to have existed beyond this chart. It was only after testing moved to Olathe, Kansas that detailed test plans came into existence. WAGT EDS-M-226481-A 12/29/1952. Attachment to WAGT memo on Contract NOa(s) 10825 dated 7/16/1953.

Before agreeing to WAGT's proposal, BuAer required that WAGT agree with the following provisions:

1. Replace at no cost the YJ46-WE-8A engines to be used in the flight test program with six new production J46 engines of model designation specified by the bureau, or, overhaul and/or repair and reconfigure at no cost to the Government the six YJ46-WE-8A engines to be provided as necessary to furnish six J46 engines of a model designation specified by this bureau which meet new engine production acceptance test requirements in all respects.
2. Agree to conduct a complete flight suitability test of a J46-WE-8A in the bailment F7U-3 prior to installation of any of the six modified YJ46-WE-8A engines, the schedule and information desired to be subject of separate correspondence upon WAGT accepting the provisions as described. One of the six J46-WE-8A engines would be used in the flight suitability evaluation.
3. Agree to conduct a complete flight suitability test of a J46-WE-8B in the bailment F7U-3 as soon as an engine of that model was provided from production. If a YJ46-WE-8A was installed in the bailment F7U-3 at the time the J46-WE-8B engine evaluation was desired, it would be removed until the J46-WE-8B flight evaluation testing was completed.[55]

While the discussion was underway to determine the bounds of the test plan, the engines to be used, and who was paying for what, WAGT sent over the latest master engine delivery schedule for the J46 (and J40). The plan was dated July 27, 1953 and for the J46 was predicated on the following milestones being achieved:

1. July 1953 qualification of the -8A at Essington.
2. August 1953 verification of the -8A at KC.
3. August 1953 verification of the -12A engine at Essington.
4. August 1953 verification of the -8B engine incorporating a C-75 compressor at Essington.
5. August 1953 verification of the -8B engine incorporating a C-75 compressor at KC
6. September 1953 verification of the -8B engine incorporating a C-89 compressor at Essington.
7. September 1953 verification of the -8B engine incorporating a C-89 compressor at KC.
8. September 1953 verification of the -8 at Essington.
9. September 1953 verification of the -8 at KC.
10. January 1954 qualification of -18 at Essington.
11. March 1954 verification of -18 at KC.

WAGT generously offered to incorporate hydro-mechanical controls on -8B engines if they became available prior to the scheduled incorporation point. If BuAer desired fewer or additional -12 or -20 engines in any particular month, they could be interchanged with -8 or -18 engines on a one for one basis provided notification was received at least five (5) months prior to the desired delivery month.

On August 19, BuAer sent over allocation instructions for August through October covering 29 -8A and 45 -8B engines, all of which were to be shipped to Chance Vought in Dallas.[56] An attached sheet stated seven -8A's had been received as of that date. A written note stated the allocations were all KC production and that Essington would be producing the spare parts.

J46 Master Production Schedule*

July 27, 1953

	1953						1954						
Model	**J**	**A**	**S**	**O**	**N**	**D**	**J**	**F**	**M**	**A**	**M**	**J**	**J**
-8A	2	7	8										
-12A		1	2										
-8B (C-75)			8	17									
-8B (C-89)				19	53	68	41	10	10	7			
-12B (C-89)				2	2	2	2	1	1	2	3		
-8							48	100	100	100	75	50	40
-18									1	2	6	12	23
-20													1

* Complete chart shows planned production out to August of 1955.

The ongoing qualification testing of fuel control components revealed excessive leakage of the fuel regulator at -67°F temperatures. The leakage at each of five external covers on the control was about one (1) drop of fuel per minute. The "O" rings being used could not seal properly when MIL-H-3136 Type I non-aromatic fuel was used as required in the testing.

In March, the control had successfully passed the 400 hour endurance and 50 hour hot test phases of the qualification test, but leakage was then encountered after a 72 hour cold soak and the unit was then cycled. Alternate type "O" rings and sizes had been tried but the problem persisted.

If aromatic fuels were used, the existing "O" rings sealed satisfactorily. As a result, the balance of the 500 hour cold test (50 hours) would be run using Type II aromatic fuel. In all other aspects the control was expected to be fully qualified and WAGT asked that an unspecified number of engines be accepted with a deviation for the -67°F leakage until the problem could be resolved. The BAR recommended that BuAer issue the waiver.[57]

BuAer did issue a waiver for up to 100 -8A/-8B engines, but considered an unlimited quantity to be unrealistic. A written note on the memo routing sheet stated that unofficial sources had stated Bendix had passed the MIL-E-5009A component qualification test. The aromatic content of fuels was expected to rise with large volume production during wartime.[58] *(Perhaps the writer was thinking of avgas?)*

Field problems with the YJ46-WE-8A engines indicated the need to re-examine the limits for rejection to overhaul as a result of the potential of turbine rubs.

Currently, the turbine blade clearance range was .072"-.110"with the lower clearance for the bottom half of the engine. The variation was needed since the first stage turbine nozzle assembly was supported by the No. 3 bearing which itself was cantilevered from the aft flange of the diffuser by the turbine shaft housing, the assembly bending slightly at rest. The turbine rotor was bolted to the turbine shaft which rode in the No. 3 bearing housing with very close clearances. The weight of the turbine rotor pulled the No. 3 bearing housing down. The sagging might be as much as .030". When running hot, the shaft centered itself and the turbine nozzles became tight and the turbine clearances became uniform.

The YJ46-WE-8A engine minimum turbine blade tip clearance was originally set at .063". This was reduced to 0.610" after an engine shipped to Chance Vought had been found to have a tip clearance of 0.0615". It had been shipped with a 0.072" bottom clearance and it was suspected if it had been turned over with the starter the clearance would have been normal, but this had not been done.

As a result of the finding, the shipping limit for bottom blade clearance had been lowered to 0.065" and an engine notice sent out. This allowed an initial cold lower clearance of 0.055" after shipment. An actual turbine rub would be the criterion for rejection during service operation.

Since no engines had actually experienced a rub in service, WAGT asked that BuAer accept the new limits for the -8 and -12 engines.[59]

BuAer's attempt to get the engine contracts modified so that successful completion of flight demonstration tests were included as part of the engine qualification acceptance process met strong rejection on the part of WAGT. In a follow-up memo on August 20 in response to an earlier memo on the 14th, WAGT stated "we do not agree to flight demonstration as a requirement for production acceptance of J46-WE-8A, J46-WE-8B or any other Westinghouse engine model presently under contract." They went on to state that they agreed to the desirability of such a program but did not feel that "Industry" and the "Military Service" collectively were adequately prepared to accept such a requirement at that time. WAGT felt coordination would be needed before consideration could be given to incorporating flight testing requirements in model specification.[60] WAGT had objected in the earlier memo to the request primarily on the grounds that they did not have the facilities and the current contracts did not have the funds to cover the costs of such tests.

The written comments on the routing sheet attached to WAGTs' second memo basically agreed that WAGT had the legal right to not agree to add the requirements and also was correct in suggesting the matter be coordinated as a Service-Industry matter. The Wright Aeronautical division of CWC took the same view point. "We have, in writing, proposed these flight demonstration requirements to the Air Force for both turboprop and turbojet engines, and requested that they be included in MIL-E-5009A by amendment and (also to) the new turboprop qual test spec. However, the Air Force in reply stated the Industry was not ready yet for this step, as there were not sufficient facilities available."

The BuAer comment writer felt that BuAer needed to press the case with the Air Force and be willing and able to appropriate funds for flight or simulated flight test facilities as well as the cost of carrying out the tests themselves.

He continues "we must change the BuAer policy of ordering large quantities of production engines before the experimental engines have been properly tested and evaluated. We usually find ourselves so committed to a production airplane program that we have to buy an engine that can barely struggle thru a 150 hour sea level static endurance test. (Ref J40, J46, J40-WE-22A, J71-A-2 programs)."

(Note: The J71-A-2 was the replacement engine for the failed J40-WE-10. The J71 was delayed two years because of problems with A/B development. This delay placed BuAer in the position of only having the much lower power J40-WE-22A engine available to fit into the initial McDonnell F3H-1N airframes coming off the assembly line. The situation is covered in full in the author's companion book on J40 development.)

The minor issue of early adapter starter-generator gearboxes was reviewed in a telegram in mid-August. The wire stated gearboxes would be installed on Essington engines 011 through 071. Gearboxes had shipped on 001 through 006. Engines 007 and 009 had shipped without the gearboxes. Engine 010 shipped with one gearbox installed and one extra gearbox. The next engine shipped would be sent with one extra gearbox. The BAR requested formal allocation instructions for future gearbox shipments.[61]

This was immediately followed by an allocation wire for the next four engines to be sent to CVA, serials WE040011-WE040014.[62]

The specification covering the YJ46-WE-8A, J46-WE-8A and -12A engines was still not finalized. The latest draft was dated July 10, 1953 and BuAer's review found problems with the clauses on external power requirements, engine weights, and some missing minor performance charts notations. They gave contingent approval of the specification pending the minor changes requested and released it for distribution. A copy of the final draft containing the corrections/alterations was requested to be submitted to the Bureau for final approval.[63]

A rare "internal use only" WAGT memo found in the Hagley Museum and Library recaps the engineering confusion over the J46 test results they were getting. Various tests on seven different engine builds (see table) show variation in compressor pressure ratio results, with some engines showing the correct target point 5.3:1 and others 5.135:1. They did not know if the target value of 5.3:1 was correct, or if the production engines experiencing stall in the field were actually operating at 5.135:1. The question needed to be resolved as the answer had a direct bearing on how much the turbine should be opened as a temporary stall fix and also indicated where the desired stall line for AEL testing should be set.

The memo notes further confusion as the F7U-3 aircraft at CVA were stalling in steady state well below that predicted based on AEL tests. Both production engine #3 and #7 were displaying this behavior.

Model F7U-3 Production Plan
August 25, 1953
J46 Engine Allocation to Airplanes

Engine Set No.	J46 Model	Receipt Date	Avail. For Installation	Airplane Allocation
1	8A	8/4/53	8/14/53	25
2	8A	8/31/53	9/22/53	18
3	8A	9/8/53	9/21/53	21
4	8A	9/15/53	9/29/53	22
5	8A	9/24/53	10/6/53	37**
6	8A	9/30/53	10/13/53	19
7	8A	10/5/53	10/16/53	29*
8	8A	10/7/53	10/20/53	38**
9	8A	10/9/53	10/22/53	23
10	8A	10/13/53	10/26/53	24
11	8A	10/15/53	10/28/53	26
12	8A	10/19/53	10/30/53	20
13	8B	10/21/53	11/3/53	30
14	8B	10/23/53	11/5/53	27
15	8B	10/27/53	11/9/53	28
16	8B	10/29/53	11/11/53	29
17	8B	11/2/53	11/13/53	31
18	8B	11/4/53	11/17/53	32
19	8B	11/6/53	11/19/53	33
20	8B	11/9/43	11/23/53	34
21	8B	11/10/53	11/24/53	35
22	8B	11/11/53	11/25/53	36
23	8B	11/12/53	11/27/53	39
24	8B	11/13/53	11/30/53	40

* To be replaced with -8B engines at a later date.
** To be replaced with -8B engine set 16.

Engineering Department Calculation Sheet
Fuel - JP-4 RPM - 10,100
$T_{t4/t2}$ = 1,985°R (1,525°F)

Engine	C10 #3	C10 #1	C10 #1	C10 # 1	C10 #9*	C10 #11	C10 #4
Test Cell	2	7	7	7	5	8	8
Test Dates	3-25/4-13	4-29/5-12	7-24/7/30	8/14/8-21	8-20	8-21	3-52
Compressor	C-75	C-75	C-89	C-89	C-75	C-75	C-67
Turbine	T-3	T-19	T-3	T-40	T-19	T-3	T-3
Pt3/Pt2	5.3	5.135	5.135	4.92	5.29	5.135	5.31
Pb/Pt3 %	6.6	4.05	4.5	4.9	6.7	6.0	6.21
Compressor Efficiency	82	75.5	81.5	80.7	-	-	81.8
Turbine Efficiency	85	86	85	85.5	-	-	83.8
Thrust - MIL	4,085	4,000	4,120	4,090	3,970	-	4,315
SPC - Mil	1.100	1.102	1.096	1.136	1.150	-	1.019
Thrust - Max	5,840	-	5,975	5,820	6,060	-	
Wf/Fn - Max	2.53	-	2.45	2.45	2.561	-	
Wg t4/t4	29.5	29.5	29.5	31.4	29.5		29.3
Wn t2/t2	74.25	74.0	74.1	74.2	73.3	73.3	73.3

* Data taken 100 running hours after buildup.

Figure 3: Compressor ratio variations test data. Re-typed by the author for clarity. WAGT Memo August 31, 1953, *Courtesy Hagley Museum and Library*

A meeting was held on August 26 with WAGT engineers and a seven step action program was agreed. The steps involved (in part) taking readings from CVA engines to determine exactly where the stall line was, checking an engine in the F3D behind the different intake duct, checking production engines against the blueprints for deviations and checking if compressor seals were leaking.

No follow-up memos on the findings were found, but apparently BuAer was not involved in the loop on all of this, as nothing in the contract files refers to performance inconsistencies of this nature.[64]

The six (6) -12A engines for the YF2Y-1 aircraft at Convair in San Diego were allocated September 2 with shipment of the first engine to be by rail express.[65]

In mid-September, BuAer accepted the results of the 150-hour verification test of the J46-WE-8A built at Kansas City. None of the deficiencies noted during inspection were disqualifying. They were:

1. Combustion chamber inner liner had a burned hole of approximate 1" by 3" size as a result of rotation of a fuel injection tube.
2. The compressor outlet mixer had three vanes broken loose at the outer shroud.
3. The outer liner adapter was moderately burned and cracked.
4. A fuel pump bearing was eroded in the same manner as on the (pump) qualification test part.
5. The aluminum compressor discs were slightly corroded.
6. The engine diffuser had small cracks in the strut welds.
7. The No. 1 thrust bearing was warped (possibly during assembly or disassembly).
8. The junction box had two breaks in the attached electrical cable.
9. During the test one ejector was replaced due to cracking and the final ejector had large cracks. This might have been due to A/B resonance in the test cells.

The BAR was authorized to accept engines using "slave ejectors" during acceptance work until further notice.[66]

In response to a query at a meeting, WAGT sent over a response showing that the maximum turbine inlet temperature rise from power extraction and air bleed that would occur under combat conditions at 35,000, 400 knots on the F7U would be 21°F.[67]

Deterioration of the "O" ring in the A/B fuel manifold had cropped up due to the heat when the units were brazed. WAGT changed the construction process so that in the future, the "O" rings would be removed when brazing was being done.[68]

The Essington BAR reported that six (6) J46-WE-8A engines had been shipped to Dallas during the month of September.[69]

The Kansas City BAR had to report on October 10 that serious production defects had been uncovered while production testing was underway. In one known failure, a notch in the forward end of two horizontal iris leaves allowed a guide clip to disengage, resulting in possible jamming of the unison ring. The pushrod assembly could disengage from the guide channels resulting in loss of tail control. Two instances had been recorded. The two items were considered safety of flight issues and BuAer required all KC manufactured engines be examined prior to flight. No similar problems were known at Essington.[70]

In-flight problems with compressor stalling were being experienced and when an engine allocation and acceptance authorization was sent to WAGT, it was for 42 -8A engines and 58 -8B engines with the proviso that further authorizations would require establishment of an engine configuration with a compressor stall fix and meeting the -8B model specification performance. A schedule for the 100 authorized engines for Essington and Kansas City was requested.[71]

Figure 4: Exploded parts view of a J46 A/B fuel pump (all models). WAGT P49954 8/15/1953. *Courtesy Hagley Museum and Library*

Shortly after the allocation wire above, the Essington BAR wired that -8A engine WE040020 would ship on 10/16/53 and that seven more -8A/B engines would ship by the end of the month (serials (-8A WE040021-23 and WE040026; -8B WE040024-25 and WE040027).[72] A conflicting wire a few days later refers to serials WE040024-27 as -8B's.[73] A second telegram a few days later states 12 more engines in November and completion of the -8A allocation by mid-December if no further design holds occurred.[74]

A memo in the J46 production correspondence from WAGT dated October 14 contains details of a modified J46-WE-8A long-coupled afterburner design at Convair's request. The new design utilized the current 18.4 inch extension of the -8A and added another 18.4 inch straight extension with a 4.5 inch sidewise "pie shaped bend".

The A/B would also have a zero degree 30 minute upward cant taken by the ball joint after installation. The weight increase would be approximately 100 pounds. The memo references BuAer wire #21/1925Z (*not found*). It is not known why Convair asked for the design data in this configuration. No engine designation was assigned.[75]

WAGT sent over the Installation Manual covering the models -8, -18 and -20 on October 22, 1953. The transmittal letter alerted users to look for coming updates and additional information based on flight test results and further design experience.[76] The contractor furnished equipment lists for the -8A and -8B that were sent over. They were reviewed by BuAer who returned one copy marked up to reflect applicable handbook requirements. After the handbook draft was printed it would go through technical review before a reproducible copy was finalized.[77]

As production was beginning at Kansas City, the ability to build acceptable quality accessory gearboxes had to be verified. In September, WAGT forwarded a verification plan to run the test on a test stand (as usual) with the following exceptions:

1. The starter pad had a Starter-Generator Adaptor Gearbox plus a water brake rated at 67.5 hp instead of the required 160 hp. The reason was the test stand drive could not drive the 160 hp plus the full hp load of the other pads without the circuit breakers opening.
2. The tachometer pad would not be loaded since no device was available for the pad and the .46 hp normal loading on the pad could be ignored.
3. One generator pad would be loaded to the maximum combined load of 179 hp for both pads. This approach was selected as there was no room to load each pad individually to 90 hp.

WAGT acknowledged that BuAer also wanted a full 150 hour verification test of the loaded gearbox while mounted on a house engine for a 150-hr run. This would be run as soon as possible, but in the meantime WAGT requested permission to ship KC produced gearboxes if the laboratory test was successful. The laboratory test was to start on September 21 or before.[78]

Before the BAR forwarded WAGT's plan for approval, the laboratory test was run under the conditions described by the plan with the following teardown results:

1. Some nuts were loose.
2. Flakes of steel and/or magnesium were found.
3. Some parts showed signs of rubbing.
4. One bearing did not roll freely.
5. One gear showed burnishing.
6. An "O" ring between an adapter tube and the inlet housing was pinched.
7. All gears were Magnafluxed with no issues.
8. The housings were Zyglo inspected with no issues.
9. The gears in the gear box adapter were in very poor condition, showing fatigue surface defects and possible scuffing. This may have contributed to the flakes of metal found in the gearbox.
10. Two locknuts were found loose but the lock-washers were securely locked.

After the test, the BAR forwarded WAGT's test plan memo and the test results to BuAer HQ stating that the test results were such that no defects were found of such magnitude to warrant Changes of Design. The gearbox adapter gears appeared not to have been sufficiently hardened resulting in their observed post-test condition. He goes on to add that the gearbox was being assembled onto a house -8B engine for a 150-hr verification test.[79]

Now, more than a month after WAGT had submitted their proposal and request for permission to ship KC manufactured accessory gearboxes prior to the 150-hr engine verification test, BuAer responded accepting the test plan but not accepting the request for pre-test shipments of KC units.[80]

From August 12 to October 2, a 150-hour qualification test was run at the Essington facilities using house engine X24C10-13 built up to be a J46-WE-8B. Report A-1697 covering the qualification test results was delayed for over a year. The report itself was dated October 28, 1954. Why it took so long to be forwarded was not mentioned in the report or the forwarding memo. A key thing to note is that the test was to the newer, much tougher MIL-E-5009 standards. The report states that after the test, the T-39 turbine was selected to improve the surge safety margin at altitude and a new design of the outer liner adapter of the combustor was showing excellent results in elimination of burning and cracking. Investigation of improved flameholder materials in the A/B was in process and improved air seals at the turbine inlet location were being investigated in regards to reducing the radial temperature distribution issues related to high thermal shock on the first stage turbine nozzle vanes and blades.

The test engine had many improvements over the -8A test engine, too long to list here. The engine weight was now 2,063 pounds.

The problems encountered during the test or deficiencies noted at post-test teardown were:

1. At 20.0 hours, a hinge failed on one of the flapper valves in the A/B inlet dual check valve causing the

burnout failure of a turbine nozzle and flameholder. The test was stopped and the engine repaired in the shop.

2. The test was restarted from the beginning and at 121 hours a tie wire securing the starter cowl bolt failed and a portion of the wire passed through the engine. The engine was inspected, no serious damage was detected and the test resumed.
3. After the test, ten of the compressor diaphragm halves were found damaged with nicks and cracks along the brazing. The cracks were attributed to "twisting the vanes to (design) drawing" prior to the test.
4. A small crack was found in the weld between the oil cooler bracket and the diffuser housing, attributed to poor welding technique.
5. Several copper gaskets on the fuel adapter were found in a cocked position which occurred during the buildup of the engine. The fuel tube connections were redesigned to eliminate the chance that cocking could occur.
6. High temperature caused one of the silicon "0" rings in the No. 3 bearing to liquefy. Newer type seals were put under test to resist the high temperatures.
7. Cracks were found in the combustion chamber inner liner, attributed to the manual reworking of the liner from an older configuration which left stress risers in the liner.
8. The vaporizer plate assembly revealed cracks from four of the hats, attributed to shop practices since modified.
9. The outer combustion chamber liner had cracks and was distorted at every bolt hole, attributed to excessive heating. Alternative designs were being tested, the best appearing to be one with banana shaped slots with a coop extending 1/6" into the air passage. After 244 hours of testing, an outer liner of this design was found to be in perfect condition.
10. The first stage nozzle showed minor vane erosion and minute cracks on the vanes, attributed to thermal shock.
11. Thirty six (36) first stage turbine blades had cracks in the airfoil section, attributed to thermal shock and creep rupture. Aged blades were now being incorporated in production engines.
12. Cracks were found in the A/B flame holder and the exhaust ejector, attributed to the need for improved materials, thermal shock and poor welding technique.
13. The accessories were in generally good condition with a few minor oil leaks due to excessive depth of "O" ring grooves which prevented the "O" rings from fully sealing the joint.

The results demonstrated that the engine was just able to sustain its performance through the 150 hour test period but that the hot sections of the engine were marginal. It is obvious that supplier quality, assembly quality and design issues were still making the overall quality of the engine suspect.

The test performance against specification WAGT-X24C10-6B was:

J46-WE-8B Qualification Test Results Specification Thrust Ratings				
Rating	Thrust Spec. (lb)	Thrust Actual	SFC Spec.	SFC Actual
Maximum	5,800	5,800	2.53	2.400
Military	3,980	3,980	1.100	1.098
Normal	3,560	3,560	1.078	1.079
90% Normal	3,205	3,205	1.106	1.091
75% Normal	2,665	2,665	1.186	1.176
Idle	245	1139	4.00	4.856

The newer more comprehensive testing required the engine to be adjusted to the maximum turbine intake temperature allowable, in this case 1,525°F, (or any other limiting performance factor) and performance measurements were to be taken both before and after the test to give some indication of performance fall-off over the overhaul life of the engine.

J46-WE-8B Qualification Test Results Maximum Operating Limits				
Rating	Before Test Start	After Test End	SFC Before	SFC End
Maximum	5,990	5,660	2.430	2.430
Military	4,090	4,000	1.082	1.098
Normal	3,780	3,660	1.066	1.080

Starting and Acceleration			
Action	Max.	Ave.	Min.
Start to Light-off (secs)	25.0	15.2	8.0
Start to Idle (secs)	60.0	50.0	34.0
Start Turbine Inlet Temp, (°F)	1,740	1,447	1,300
Idle to Military Thrust, (secs)	10.2	9.0	7.0
Idle to Military Thrust (°F)	1,400	1,313	1,275
Idle to Take-off Thrust (secs)	10.2	9.0	7.0
Idle to Take-off Thrust (°F)	1,370	1,318	1,195

The non-compliance with the model specifications for start times was attributed to the limited power of the

G.E. starter-generator on the Chance Vought adapter gearbox.

WAGT recommended the engine be passed for service use with a 150 hour time between overhauls.[81]

The -8B engines began coming out of Essington in November, serial WE040028-31 being allocated to Dallas that month.[82]

In early November, WAGT sent out the overhaul instructions and parts breakdown for the Starter-Generator Adapter Gearbox Installation Kit. They stated the kit contained 70 parts and the power takeoff kit added 10 more parts including the driveshaft. A power takeoff was used on the -8 engine. WAGT asked permission to expand the overhaul manual to include both kits in one description.[83]

Although a formal amendment had not been processed, BuAer sent a wire to Quonset Point support stating that -18 and -20 engines would be delivered in lieu of the -2 engines. Some -8A and –B engines would be required to undergo overhauls.[84]

WAGT surveyed moving the ignitor plugs and nozzle locations for the -8, -18 and -20 engines in mid-November. The -8 would use two ignitor nozzles in line with the ignitor plugs and the -18, -20 engines would use eight ignitor nozzles located axially forward of the ignitor plugs.[85]

Engine scheduling was still an issue and in a conference with WAGT on November 16, with a total of 36 engines already delivered and the engine shortage deemed critical, the following delivery schedule (noted as being below the actual minimum requirements) was agreed:[86]

J46 Delivery Schedule -8A/8B/8

	N	D	J	F	M	A	M	J	J
KSC	20	26	35	45	60	84	95	95	93
Ess	9	9	9	10	10	1			

The schedule was changed at WAGT's request for KC to deliver 66 engines in March and 85 in April and Essington to deliver 4 in March and 0 in April.[87]

A growing crisis over late delivery of acceptable engines demanded urgent action to resolve. The J46-WE-8A engines were deficient in thrust, had higher SFC's than specification and were subject to compressor stalling. The -8B model intended to replace the -8A's on the production line were late as development testing of compressors and compressor/turbine combinations was pressed to achieve an acceptable solution. Production of F7U-3's was continuing in Dallas and a backlog of airframes without engines was beginning to grow (BuAer called this their "glider pool") on the tarmac at the Chance Vought plant.

On October 5, Chance Vought held a conference with the Dallas BAR to go over the schedules and concerns. The engine delivery schedules received from BuAer had shown that 29 -8A and 45 -8B engines were to be delivered from Kansas City during the period of August through October. Essington delivered their quota early and had built one additional -8A to be delivered in October. Kansas City had only delivered four (4) engines in September, eleven (11) engines short of their quota for the month.

The Kansas City BAR and Dallas BAR met on October 8 to discuss the problem. They were unable to arrive at a firm engine delivery schedule from Kansas City. WAGT had predicted they would be able to ship sixteen (16) -8A engines in October from Kansas City but the BAR, based on observation of available parts, advised Chance Vought to expect nine (9) engines instead. History had shown the BAR's at both WAGT plants were accurate in their delivery forecasts and Chance Vought expressed concern that WAGT management was not fully aware of what the true situations were in their plants.

The BAR Dallas goes on to state: "On Thursday, 8 October, the BuAer Aircraft and Power Plant Divisions agreed in conference that one hundred (100) J-46 engines (-8A with C75-T3 and -8A½ *(SIC)* with C89-T3) would be purchased. The decision to do this was instituted to keep the number of gliders to a minimum. Also, at this same conference, it was decided to start immediate flight testing at Chance Vought of the T-43 turbine (T-43 has improved specific fuel consumption and stall characteristics over T-3 and T-40) with the C-75 compressor. The BuAer Aircraft Division stated that neither the T-3 nor T-40 was an acceptable turbine in the J46 power plant for production. BAR Dallas concluded, from discussions held in BuAer that week that the Cutlass program could be terminated if the T-43 turbine was not an acceptable item for follow-on production after the one hundred (100) engine purchase."[88]

The situation was serious enough that the entire F7U-3 program was now in jeopardy.

On October 21, the BAR repeated the statement that the entire program was in jeopardy due to WAGT failure to deliver acceptable engines. As of that date, there were 28 F7U-3's completed except for engines. The estimates were that by the end of November, there would be a cumulative deficit of 75 engines. The -8A engine with C-75 compressor and T-3 turbine gave good SFC results but had stalling problems. The -8B engine had the C-89 compressor and T-3 turbine (58 engines on order). No further engines would be ordered produced until a compressor with a stall fix was available and meeting the -8B performance specification. A new turbine, the T-43, was being tested with the C-75 compressor. Five engines of that configuration were to be built at Essington and shipped to Dallas in October for testing. A request had been made to allocate two F7U-3 airframes for use in flight testing. Eight C-89/T-3 configuration engines were also in test. The test results would determine the future of the J46 program and the next configuration for future orders.

After reviewing the history of the verification and qualification test problems to date, the BAR continues: "The contractor continues to be plagued with trouble with fuel pumps, VD (variable-displacement) pumps, and other accessories. He is also struggling with the problems of incorporating changes, establishing adequate subcontract sources, engine build-up, and testing incidental to starting a new production contract. *(A reference to the J40-WE-22A program.)* It is assumed that these production problems will be solved in the next few months, but the production from the Kansas City plant will be insufficient to meet the needs of the F7U-3 program for an indefinite time."[89]

BuAer did not respond to the above memo until November 24. They acknowledged the situation and stated WAGT had been "counselled". The flight tests of the various compressor/turbine combinations had been completed and the results presented. A configuration for future orders was selected and more orders were being prepared for -8/-8B engines.[90] The selected configuration is not mentioned in this memo, but the testing results followed in a December 1 memo. The lateness of the memo shows that a final configuration solving the problems still had not been decided upon, even though the F7U-3 program continued.[91]

Development continued at WAGT. In mid-November, WAGT sent over a report A-1591 that covered some preliminary testing of a Woodward X838 all speed (range) governor for air starting. This type of governor was being investigated for the J46 for fuel control and so a J34 with a similar governor was flight tested. The Douglas F3D-2 Bu. No. 124607 was used as the in-flight testbed with a J34-WE-36. It was used for five flights and 110 air-starts were accomplished. The results were satisfactory overall. A condition termed "duct rumble" was experienced at low flight speeds and some failed restarts occurred when such rumble was experienced. The condition was not understood. Some throttle handling had to be done during air-starts to prevent over-temps. Flight tests with a similarly equipped J46 were recommended.[92]

A survey was sent out later in the month regarding the suggested removal the A/B combustion chamber drain plug on the -8, -12, -18 and -20.[93]

On top of all the other problems, shipping containers holding J46-WE-12A engines arrived with loose bolts holding down the engine inside the container. BuAer wanted lock pins to be used in the future but WAGT rejected the suggestion, saying hundreds of engines (J34's and J46's) had been shipped in the previously tested and approved containers without a problem. They wanted the Field Service Representatives to report any future problems noted, particularly with the torqueing of bolts.[94]

WAGT reported on their progress and program to address compressor stalling with a summary:

J46 Stalling Testing Status
Status of Parts: 1. T-39 (T-3 opened up 13%) 2. T-40 (T-3 opened up 16%) now flying. 3. T-43(T-3 opened up 10%) now at Chance Vought. 4. T-45 (T3 opened up 5%) complete and shipped to CVA November 4.
Test Results: 1. A combustion chamber air bleed system which vented air from the combustion chamber and routed it to the A/B exhaust collector eliminated compressor steady state stalling. The SFC's rose by .05 at military with the bleed operating. 2. The C-75/C-89 and T3 combinations produced stall conditions at the same altitude. 3. The C-75/C-89 and T43 showed stall 40 knots lower than the C-75/T-3 combination. 4. C-89/T40 combination reached 47K feet and .66 Mn and was better at altitude, but still not satisfactory.
Stall Test Program moving forward: 1. Continue testing C-89 with T-39, T-43, T-45 and T-40 turbines. 2. Engine W040006 being used to acceleration rate improvement testing was damaged by a foreign object and was returned to Essington for repair.
A/B Dud Light-offs Testing
1. Two fuel lines to inject fuel at the top of the A/B fuel manifold received and being installed. 2. Check valves on the fuel manifolds being tested. 3. Servo valves as backup to the check valves being designed.

The survey on moving the -8 igniter plugs and nozzles was cancelled after it was found the proposed new location did not work well for sea level ignition.[95]

Anti-icing equipment was removed from already delivered and all new -8B engines in early December and stored for possible future use. The notification to remove the anti-icing hose stated that the work on an anti-icing valve for the J46 had been stopped. The weight savings was specified as exactly 9.399(!) pounds.[96] WAGT in Kansas City notified BuAer on January 14, 1954 that -8B engines from WE404572 and on would have the anti-icing hose and associated hardware removed.[97]

In the guise of a memo about proper turbine outlet temperature adjustment differences on Essington vs. Kansas City built engines, WAGT tried to explain why the engines from these two plants were achieving different performances during production acceptance testing. Given that engines built to the same part configuration at the two plants should return consistent performances within a reasonable range of a few percent, WAGT was at a loss to immediately explain why the Kansas City built engines consistently produced 100-200 more pounds of thrust with 3 to 5 percent lower SFC's.

They explained they were investigating every possibility to include: test cell loss differences; fuel meter accuracies and calibration; actual calculated performances based on the accepted formulas; and manufacturing tolerance accuracies.

The majority of the engines produced at Kansas City were demonstrating the inability to mechanically adjust the exhaust nozzle sufficiently far closed (beyond design limits) under Military performance conditions to reach the maximum allowable turbine outlet temperature. It was surmised this might indicate a mechanical difference and possibly improved component performance in the Kansas City engines. In general, they stated that as component performance improved, the exhaust area required was also reduced. If the ambient temperature was reduced, the given turbine outlet temperature also reduced the exhaust area needed. All of the above were being investigated as contributing factors.

Using house engine 24C10#3, a test cell comparison between test cells at the two facilities was half complete.

In spite of the performance differences, the engines at both locations were meeting the specification performance requirements with adequate margin, no mandatory corrective action was necessary at that time. The maximum allowable temperature control adjustments were made during A/B where, in view of the generally open area requirements, sufficient reduction in area permitted maximum temperature operation.

An even more curious situation occurred during the XJ40-WE-6 testing when consistent performance differences on the SAME ENGINE were noted between the AEL test cells and WAGT's at Essington, with AEL's results being consistently lower. A very thorough investigation never uncovered the cause of the differences. The differences in both cases never favored the WAGT test bed results.

No sea level cold day starting problems had been encountered on installed engines at Chance Vought to date. The new T-39 turbine would change the exhaust requirements to a more open area and might automatically relieve the present test area restriction. In the field, use of the -8A was limited by temperature and the -8B was limited by exhaust area at the military throttle position. Adequate service and overhaul information was to be distributed via regular channels.[98]

In the background, the -12A engines were shipping to Convair. One engine, WE041103, was specially treated by spraying the compressor rotating blades with zinc chromate primer and cadmium plating the compressor diaphragm. It was done to determine the reliability of this protection under actual salt spray conditions on just this engine. The current performance specifications for the -12A with the 5° bend were based on the -8A/-8B without a bend. Given that, the take-off thrust of 5,480 pounds determined during acceptance was 20 pounds below the model specification.[99]

An early field engine part failure report back in July was only responded to by WAGT in November. A YJ46-WE-8A engine suffered an iris nozzle segment failure which warped and tore off the engine in flight. Investigation revealed that the iris segments of this and the two prior engines had been removed from the engine for heat treatment. On re-installation, the inner lip was cold worked out into position. The heat from the operation of the engine caused the cold worked section to relax and release the forward end of the segment. Realizing the cold working process on the hot end of the engine was "detrimental", all such procedures were stopped immediately. All three engines were repaired with proper procedures and returned to CVA. No redesign was deemed necessary. The BAR agreed.[100]

This chapter ends with a summary of the acceptance of the verification test run of the Kansas City produced J46-WE-8B (serial WE404101). The 150 hour test was run and completed on November 4 and the engine inspected by representatives of the Bureau on November 23. Based on the parts list used for the Essington built qualification engine for its test, BuAer accepted the Kansas City built J46-WE-8B for production. Some replacement parts specified by WAGT were accepted for production without additional verification.

The following discrepancies were noted upon inspection:

1. Failure of a vaporizer plate retaining bolt.
2. Seized threads on the compressor through-bolts (these held the compressor stack together).

3. Diffuser strut to outer case weld cracks.
4. One vaporizer plate insert was without air holes.
5. Excessive deposit or galling or both on exhaust nozzle linkage clevis bolts and spherical seat joints.
6. Stainless steel afterburner fuel line "B" nut bevel surfaces excessively rough.
7. Fretting corrosion on power take-off gearbox, bevel gear shaft and accessory gearbox bevel drive pinion shaft.
8. Compressor tip rubs in the first three compressor stages.
9. Wear and distortion of unison ring (A/B iris open/close control ring) retaining clips.
10. Compressor housing paint worn.

The following design elements were found to be questionable (and were also problems on J34 engines):

1. Lack of either flexible links or shop assembly inspection limits for internal solid tubing oil lines.
2. Use of springs on quick disconnect clamps.
3. Lack of anti-seize protection for the turbine shaft nut.

The test had been run with a C-75 compressor and T-3 turbine but plans were in place to switch over to a C-89 compressor and T39 turbine once production was underway.

With volume production of the -8B now to begin in earnest, the existing problems of steady state compressor stall, low thrust at altitude due to control problems and the generally lower thrust of the -8B compared to the desired -2 engine were going to have to be lived with as the engines were now being installed in airframes that were to be allocated to service squadrons.

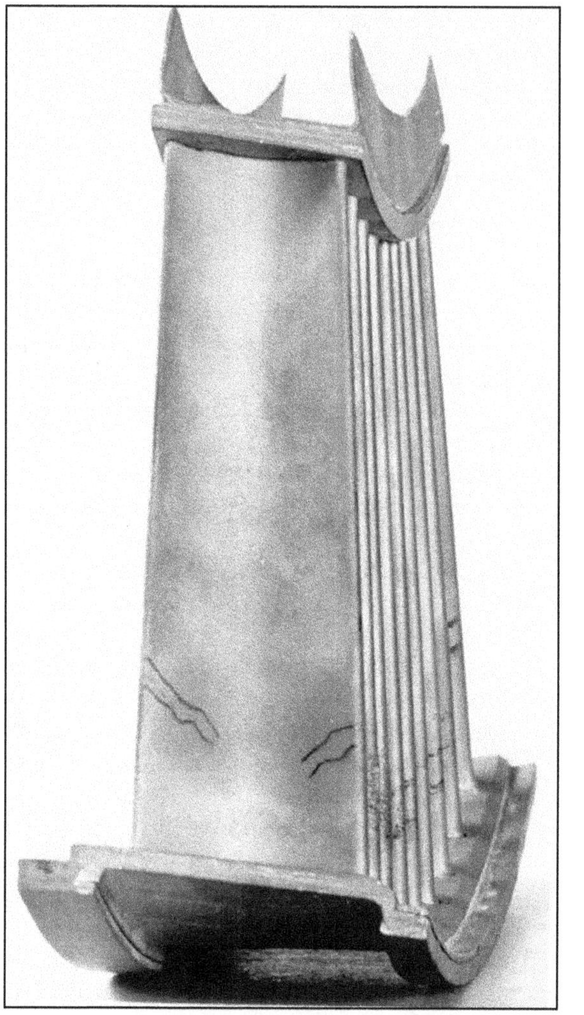

Figure 6: Turbine inlet guide vane segment after testing in engine #4. Note marking around cracks. WAGT P46035, 1/11/1952. *Courtesy Hagley Museum and Library*

Figure 5: Examples of the spring-loaded quick disconnect clamps. New design doubled clamp bolts. Shown on a J34-WE-36 but were the same type as the J46. WAGT 46005, 1/9/1952. *Courtesy Hagley Museum and Library*

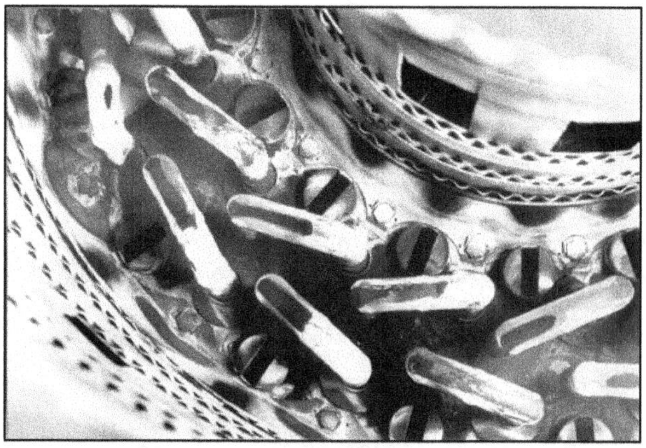

Figure 7: Vaporizer assembly after test running. Coke and carbon deposits were normal. J46 engines ran with very little smoke in the exhaust, rare for the period. WAGT P46122, 1/28/1952. *Courtesy Hagley Museum and Library*

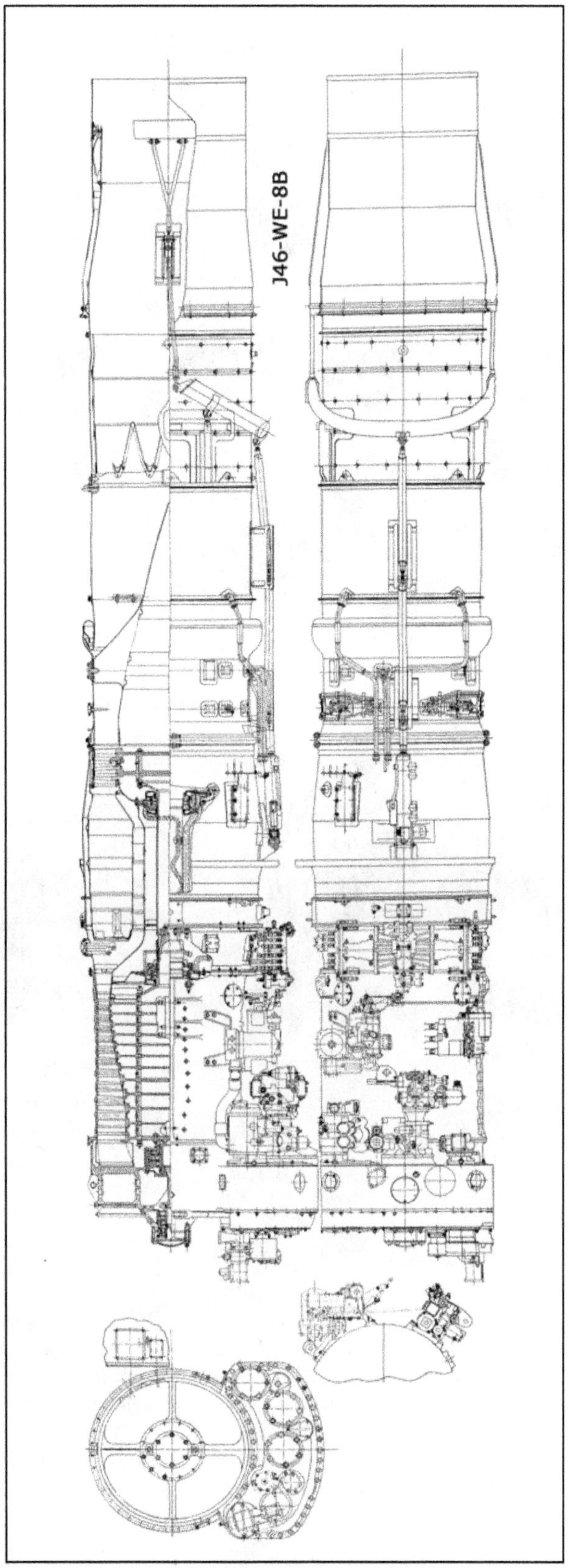

Figure 8: J46-WE-8B drawing from assembly diagram. ***Courtesy Hagley Museum and Library***

Figure 9: J46-WE-8 #5. Metal fuel lines on one side of fuel splitter/oil cooler assembly on #5 house engine. No nicks or scratches allowed or they would fracture at the stress riser point. A potential maintenance nightmare on an installed engine. Rejected by BuAer in a display of excellent judgement. The author's personal experience with broken metal fuel lines on fuel injected military piston helicopter engines fully supports BuAer's call on this WAGT design feature. WAGT P47629, 9/4/1952. *Courtesy Hagley Museum and Library*

Figure 10: WAGT machine shop at Essington, Pennsylvania. At one time this shop was simultaneously supporting the J34, J40 and J46 programs. WAGT P32653-A, 5/29/1945. ***Courtesy Hagley Museum and Library***

Chapter 5 – XJ46-WE-8/A/B and -12/A/B Early Production

[1] WAGT Memo, November 20, 1951, "Contract NOa(s) 10825 Revised Delivery Schedule Our Reference WG-64730", **B1099VE1RG72.3.2NACP**
[2] BuAer Contract NOa(s) 10825 thru Amendment 4, December 31, 1952, **B1093VA1RG72.3.2NACP**
[3] BuAer Dallas BAR Telegram, No subject, December 1, 1952, **B1096VE4RG72.3.2NACP**
[4] BuAer Memo, December 8, 1952, No Subject, **B1096VE4RG72.3.2NACP**
[5] BuAer Memo, December 24, 1952, "Contract NOa(s) 10825, J46-WE-8/-12 engine production program: status of", **B1099VE2RG72.3.2NACP**
[6] BuAer Memo, December 12, 1952, "Fuel systems for J46 engines – request for information on", **B554V13RG72.3.2NACP**
[7] WAGT Memo, December 9, 1952, "Contract NOa(s) 9670; 10825. WEC Turbojet Engine Models XJ46-WE-2 and J46-WE-2. Relocation of Spark Plugs and Igniter Fuel Nozzles, Survey of. Engine Change No. 4.", **B556V6RG72.3.2NACP**
[8] BuAer Memo, December 2, 1952, "Contract NOa(s) 9670 – J46WE-8 Engine, Afterburner Pumps and Controls; Component Qualification Tests of", **B556V6RG72.3.2NACP**
[9] WAGT Memo, January 29, 1953, "XJ46 Engine Control Problems at High Flight Speed", **B556V6RG72.3.2NACP**
[10] BuAer Memo, December 15, 1952, "Proposed incorporation of titanium in J46-WE-8 engines", **B1099VE2RG72.3.2NACP**
[11] BuAer Memo, December 16, 1952, "Westinghouse Specification WAGT-X24C10-6A of 2 Sep 1952 covering the XJ46-WE-8 and XJ46-WE-6A series engines", **B1099VE2RG72.3.2NACP**
[12] BuAer Memo, December 17, 1952, "Starter Adapter Gearbox on J46 engines; incorporation of", **B1099VE2RG72.3.2NACP**
[13] WAGT Memo, December 2, 1952, "Starter Generator Adapter Gearbox Which Mounts on J46 WE-8 in F7U-3 Aircraft, Request for Permission to Place Kit, P/N 58J680-1 on Early Engines Assigned for Shipment to Chance Vought. Reference Contract NOa(s) 10825.", **B1099VE2RG72.3.2NACP**
[14] WAGT Memo, February 2, 1953, "Starter Adapter Gearbox on J46 Engines; Incorporation of", **B1099VE3RG72.3.2NACP**
[15] BuAer Memo, March 10, 1953, "Starter adapter gearbox on J46 engine; incorporation of", **B1096VE5RG72.3.2NACP**
[16] WAGT Memo, December 17, 1952, "Contract NOa(s) 10825. WEC Turbojet Engine Model J46-WE-8. Pressure-Sensitive Afterburner Detection Switch, supplementary Information to Survey of.", **B1096VE4RG72.3.2NACP**
[17] BuAer Memo, January 2, 1953, "Contract NOa(s) 10825, J46-WE-8 engine, mounting of Aerotec system for", **B1096VE5RG72.3.2NACP**
[18] WAGT Memo, December 11, 1952, "Contract Noa(s) 10825; AF-33(038)-17843: 52-403c: 52-048, 10385. WEC Turbojet Engine Models J46-WE-3, -8, -12. Revised Mounts for Automatic Temperature Trim Box, Survey of. Engine Change Nos. 20, 118, 7, 30, 103, 16 for the J46-WE-3, -8, -12 and J40-WE-1, -8, -22 Respectively.", **B1096VE4RG72.3.2NACP**
[19] WAGT Memo, January 14, 1953, Contract NOa(s) 10825 WEC Turbo Jet Engine Model J46-WE-12 Possible Afterburner Mount Redesign, Comments on." **B1099VE3RG72.3.2NACP**
[20] BuAer Telegram, January 28, 1953,No Subject, **B1096VE5RG72.3.2NACP**
[21] BuAer Telegram, February 3, 1953, No Subject, **B1096VE5RG72.3.2NACP**
[22] BAR Memo, February 3, 1953, "Contract NOa(s) 10825, WEC Engine Model YJ46-WE-8A, Group Assembly Breakdown Parts List; forwarding of", **B1096VE5RG72.3.2NACP**
[23] BuAer Memo, February 6, 1953, "Contract NOa(s) 10825. WEC Turbojet Engine Model J46-WE-8. Installation Survey Log Sheet, Transmittal of", **B1096VE5RG72.3.2NACP**
[24] WAGT Memo, January 9, 1953, "Operating Recommendations for YJ46-WE-8A Turbojet Engines; forwarding of:", **B1096VE5RG72.3.2NACP**
[25] BuAer Memo, March 10, 1953, "Operating recommendations for YJ46-WE-8A turbojet engines, approval of", **B1096VE5RG72.3.2NACP**
[26] WAGT Memo, March 10, 1953, "Contract NOa(s)-10825, Request for Increase of Progress Payments", **B1093VA1RG72.3.2NACP**
[27] BuAer BAR Telegram, March 20, 1953, No subject, **B1096VE5RG72.3.2NACP**
[28] BuAer Telegram, March 23, 1953, No Subject, **B1096VE5RG72.3.2NACP**
[29] BuAer Memo, March 23, 1953, "Contract NOa(s) 9670, XJ46-WE-2 and XJ46-WE-8 oil level indicator", **B556V7RG72.3.2NACP**
[30] WAGT Memo, March 25, 1953, , "Contract NOa(s) 10825 WEC Turbojet Engine Models J46-WE-8, Provisions for Emergency control of exhaust nozzle Actuator.", **B1096VE5RG72.3.2NACP**
[31] BuAer Telegram, March 27, 1953, No Subject, **B1096VE5RG72.3.2NACP**
[32] WAGT Memo, March 31, 1953, "Contract Noa(s) 10825, turbojet Engine Models J46-WE-8, -12. Model Specification, Transmittal of.", **B1099VE3RG72.3.2NACP**
[33] BuAer Memo, April 10, 1953, "Contract NOa(s) 10825 – Westinghouse Electric Corporation, Kansas City – J46-WE-8 engines; allocation of, request for", **B1096VE5RG72.3.2NACP**
[34] BuAer Memo, April 10, 1953, "J46-WE-8 Engine Program; request for information on", **B1099VE3RG72.3.2NACP**
[35] BuAer Memo, April 13, 1953, "Contracts NOa(s) 11028 and NOa(s) 53-556; Incorporation of Turbine Inlet Temperature Traversing Bosses on J46 engines; disapproval of", **B1096VE5RG72.3.2NACP**
[36] WAGT Memo, March 27, 1953, "Contract NOa(s)-10825 Change in Performance Guarantees Our Reference WG-64730", **B1099VE3RG72.3.2NACP**
[37] BuAer Memo, April 16, 1953, "Contract NOa(s) -10825, Westinghouse YJ46-WE-8A engines;, increase of afterburner thrust for", **B1099VE3RG72.3.2NACP**
[38] WAGT Report A-1571, April 14, 1953, Preliminary Windmilling Data, J46-WE-8, NOa(s) 10825 SO 4-A-3762, **B1099VE3RG72.3.2NACP**

[39] WAGT Memo, May 8, 1953, "Contract NOa(s) 10825; Amendment-2 to Westinghouse Model Specification WAGT-X24C10-50A which describes the J46 Starter-Generator Adapter Gearbox, transmittal of", **B1096VE5RG72.3.2NACP**
[40] WAGT Memo, May 7, 1953, "Contract NOa(s) 10825, XJ46-WE-12A and -12B Engines, Qualification Test", **B1099VE3RG72.3.2NACP**
[41] BuAer Memo, May 21, 1953, "Contract NOa(s) 10825, XJ46-WE-12A and -12B Engines, Approval of proposed Qualification Test", **B1096VE5RG72.3.2NACP**
[42] WAGT Table, May 20, 1953, "Contract NOa(s) Reports Covering repair and minor rework of J46 Turbo-Jet Engines; forwarding of", **B1096VE5RG72.3.2NACP**
[43] WAGT Memo, May 26, 1953, "Performance Comparison of the J46-WE-8, J46-WE-2, and the J46-WE-8 with a 28" Diameter Afterburner.", **B1099VE3RG72.3.2NACP**
[44] BuAer Telegram, June 2, 1953, No subject, **B1096VE6RG72.3.2NACP**
[45] WAGT Memo, June 3, 1953, "Contract Noa(s) 10825. WEC Turbojet Engine Models J46-WE-2, YJ46-WE-8A, J46-WE-8A, J46-WE-8B, J46-WE-8, J46-WE-12A, J46-WE-12B, J46-WE-12, and J46-WE-16. Engine Handling with Afterburner Cantilevered.", **B1096VE6RG72.3.2NACP**
[46] WAGT Memo, June 3, 1953, "Contract Noa(s) 10825. WEC Turbojet Engine Models XJ46-WE-8A and XJ46-WE-12A. Operating Limits for C-75 Compressor, Information on.", **B1099VE3RG72.3.2NACP**
[47] WAGT Memo, May 19, 1953, "Contract NOa(s) 10825; Amendment-2 (dtd 7 May 1953) to Model Specification WAGT-24C10-6B (dtd 16 March 1953) which describes Turbojet Engine Models J46-WE-8, -8B, -8A, -12A, -12B and YJ46-WE-8A, transmittal of", **B1100VE4RG72.3.2NACP**
[48] BuAer Memo, July 9, 1953, "contract NOa(s) 10825, Amendment -2 of 7 May 1953 to Westinghouse Specification WAGT-25C10-6B covering the models J46-WE-8, -8B, -8A, -12A, -12B and YJ46-WE-8A engines", **B1100VE4RG72.3.2NACP**
[49] WAGT Memo, July 2, 1953, "Contract NOa(s) 10825 and NOa(s) 53-556 Turbine Inlet Temperature bosses on J46-WE-8B, -8 and -12B engines. Deletion of", **B1100VE4RG72.3.2NACP**
[50] BuAer Telegram, July 10, 1953, No subject, **B1096VE6RG72.3.2NACP**
[51] WAGT Memo, July 1, 1953, "Contract Noa(s) 10825. WEC Turbojet Engine Model J46-WE-12. Special Afterburner Mount Fitting.", **B1096VE6RG72.3.2NACP**
[52] BuAer Speedletter, July 30, 1953, "NOa(s) 10825, J46-WE-8A engines; maximum allowable oil consumption for", **B1096VE6RG72.3.2NACP**
[53] BuAer Memo, July 31, 1953, "Contract NOa(s) 10825 – Westinghouse Electric Corporation. Kansas City – J46-WE-8 engines; allocation of, request for", **B1096VE6RG72.3.2NACP**
[54] WAGT Memo, July 16, 1953, "Contract NOa(s)-10825 Request for Allocation of YJ46-WE-8A Engines to Westinghouse for Flight Test Use, Our Reference WG-64730", **B1096VE6RG72.3.2NACP**
[55] BuAer Memo, August 6, 1953, "Contract NOa(s) 10825; allocation of YJ46-WE-8A engines to Westinghouse for flight test use", **B1096VE6RG72.3.2NACP**
[56] BuAer Memo, August 19, 1953, "contract NOa(s) 10825, Lot II, Item 3, J46-WE-8A/8B engines; allocation of", **B1100VE4RG72.3.2NACP**
[57] WAGT Memo, Jun 26, 1953, "Contract NOa(s) 10825 J46-WE-8A Engines, Request for deviation for.", **B1099VE3RG72.3.2NACP**
[58] BuAer Memo, August 4, 1953, "Contract NOa(s) 10825, J46-WE-8A, -8B engines, deviation for", **B1099VE3RG72.3.2NACP**
[59] WAGT Memo, August 4, 1953, "Contract NOa(s) 10825, J46-WE-8 and J46-WE-12 Engines, Field limits for rejection of engines due to Turbine Rubs.", **B1100VE4RG72.3.2NACP**
[60] BuAer Memo, August 20, 1953, "Contract NOa(s)-10825 Allocation of YJ46-WE-8A engines for Flight Test Use Our Reference WG-64730", **B1094VA2RG72.3.2NACP**
[61] BuAer Telegram, August 21, 1953, No Subject, "**B1096VE6RG72.3.2NACP**
[62] BuAer Memo, August 21, 1953, No subject, **B1096VE6RG72.3.2NACP**
[63] WAGT Memo, August 28, 1953, "Contracts Noa(s) 53-56 and 10825; Westinghouse Specification WAGT-24C10-1C of 10 Jul 53 covering the Models YJ46-WE-8A and J46-WE-8A and -12A engines", **B1100VE4RG72.3.2NACP**
[64] WAGT internal Memo, August 31, 1953, (No subject line), Hagley Museum and Library, Accession 1459, Box 18, J46 Folder
[65] BuAer Telegram, September 2, 1953, No subject, **B1096VE6RG72.3.2NACP**
[66] BuAer Memo, September 11, 1953, "Contract Noa(s) 10825; Verification Test of J46-WE-8A", **B1096VE6RG72.3.2NACP**
[67] WAGT Memo, September 9, 1953, "Turbine Inlet Temperature Increase on J46 Engine Model due to Power Extraction and Air Bleed.", **B1100VE4RG72.3.2NACP**
[68] WAGT Memo, September 16, 1953, "J46WE-8A Afterburner Fuel Manifold; Information on", **B1096VE6RG72.3.2NACP**
[69] BuAer Memo, October 2, 1953, No subject, **B1096VE6RG72.3.2NACP**
[70] BuAer Telegram, October 10, 1953, No subject, **B1096VE6RG72.3.2NACP**
[71] BuAer Telegram, October 13, 1953, No Subject, **B1096VE6RG72.3.2NACP**
[72] BuAer Telegram, October 14, 1953, No subject, **B1096VE6RG72.3.2NACP**
[73] BuAer Telegram, October 10/27/1953, No Subject, **B1096VE6RG72.3.2NACP**
[74] BuAer Telegram, October 20, 1953, No Subject, **B1096VE7RG72.3.2NACP**
[75] WAGT Memo, October 14, 1953, "Contract NOa(s) 10825 Proposal for Modification of J46-WE-12 type Afterburner Our Reference WG-64-730", **B1100VE4RG72.3.2NACP**
[76] WAGT Memo, October 22, 1953, "Contract NOa(s) 10825. Turbojet Engine Models J46-WE-8, -18 and -20 Installation Manual, Transmittal of.", **B1100VE4RG72.3.2NACP**

[77] BuAer Memo, October 30, 1953, "Contract NOa(s)-10825, Contractor-Furnished Equipment List for J46-WE-8A and -8B Engines of 1 October 1953", **B1096VE7RG72.3.2NACP**
[78] WAGT Memo, September 18, 1953, "Contract NOa(s) 10825 Kansas City Manufactured J46-WE-8A gearbox; verification of", **B1096VE7RG72.3.2NACP**
[79] BuAer Memo, October 19, 1953, "Verification of Kansas City manufactured J46-WE-8A Gearboxes", **B1096VE7RG72.3.2NACP**
[80] BuAer Memo, October 30, 1953, "Verification of Kansas City manufactured J46-WE-8A Gearboxes", **B1096VE7RG72.3.2NACP**
[81] WAGT Report A-1697, October 28, 1954, Qualification Test of the J46-WE-8B Turbo-Jet Engine on Contract NOa(s) 10825, **B1100VE8RG72.3.2NACP**
[82] BuAer Telegram, November 3, 1953, No subject, **B1096VE7RG72.3.2NACP**
[83] WAGT Memo, November 9, 1953, "Overhaul Instructions and Illustrated Parts Breakdown for Westinghouse Starter-Generator Adapter Gearbox Installation Kit, 58J680", **B1096VE7RG72.3.2NACP**
[84] BuAer Telegram, November 14, 1953, No subject, **B1096VE7RG72.3.2NACP**
[85] WAGT Memo, November 11, 1953, "Contract Noa(s) 10825, 53-556. WEC Turbojet Engine Models XJ46-WE-8, J46-WE-8, XJ46-WE-18, J46-WE-18, J46-WE-20. Location of Igniter Plugs and Igniter Nozzles, Survey of. Engine Change No. 125, 6, 6 for the -8, -18 and -20 Engines Respectively.", **B1096VE7RG72.3.2NACP**
[86] BuAer Telegram, November 17, 1953, No Subject, **B1096VE7RG72.3.2NACP**
[8787] WAGT Memo, November 23, 1953, "Contract NOa(s)-10825 J46-WE-8A/8B/8 Delivery Schedule Our Reference WG-64-730", **B1096VE8RG72.3.2NACP**
[88] BuAer Memo, October 10, 1953, "J-46 Engine Program", **B1100VE4RG72.3.2NACP**
[89] BuAer Memo, October 21, 1953, "Summary of the J46 engine situation and effect of shortage of J46's on the F7U-3 program", **B1100VE4RG72.3.2NACP**
[90] BuAer Memo, November 24, 1953, "J46 Engine situation; status of", **B1100VE4RG72.3.2NACP**
[91] BuAer Memo, December 1, 1953, "CVA Flight Test Report", (The report itself is not dated, from content likely written in last November.) **B1100VE4RG72.3.2NACP**
[92] WAGT report A-1591, October 26, 1953, Air Starting Tests of the Woodward X838 All Speed Governor on Project 4A3762-1609 Contract NOa(s) 10825, **B1096VE7RG72.3.2NACP**
[93] WAGT Memo, November 9, 1953, "Contract NOa(s) 10825, 53-556. WEC Turbojet Engine Models XJ46-WE-18 and J46-WE-8A, -8B, -8, -12A, -12B, -18 J46-WE-20. Removal of Afterburner Combustion Chamber Drain Plug, Survey of. Engine Change Nos. 124, 14, 5, 5 for the -8, -12, -18, -20 Engine Series Respectively.", **B1096VE7RG72.3.2NACP**
[94] WAGT Memo, November 23, 1953, "J46-WE-12a Engine Container, Loose Bolts", **B1096VE8RG72.3.2NACP**
[95] WAGT Memo, November 25, 1953, "Contract NOa(s), 53-556. WEC Turbojet Engine Models XJ46-WE-8 and J46-WE-8. Location of Igniter Plugs and Igniter Nozzles, Cancellation of Survey for.", **B1096VE8RG72.3.2NACP**
[96] BuAer Memo, December 3, 1953, No subject, **B1096VE8RG72.3.2NACP**
[97] WAGT Memo, January 14, 1954, "Contract NOa(s) 10825, Westinghouse Turbo-Jet Engine Model J46-WE-8B anti-Icing Provisions – Deletion of", **B1097VE8aRG72.3.2NACP**
[98] WAGT Memo, November 30, 1953, "Engine Performance; J46-WE-8A, -8B, -8 Engines, on Contract NOa(s) 10825 – Information on Turbine Outlet Temperature Adjustment.", **B1100VE4RG72.3.2NACP**
[99] BuAer Naval Speedletter, December 4, 1953, "WEC Engine J46WE-12A, Ser. #WE041103, Deviations from Model Specification Performance and Acceptance of", **B1096VE8RG72.3.2NACP**
[100] WAGT Memo, November 25, 1953, "Contract NOa(s) 10825 Westinghouse Turbo-Jet Engine Model YJ46WE8A Serial No. WE040003, Iris Nozzle Segment Failure.", **B1096VE8RG72.3.2NACP**

Chapter 6
J46-WE-8/A/B and -12/A/B Production and Service

Kansas City began production of the J46-WE-8B in early 1954. The model of the engine was still de-rated (-8A had been 9.8% lower and -8B 4.9%) compared to the contracted for J46-WE-2 and both had an interim control system that was not totally satisfactory. A new hydro-mechanical system was under development that would emerge over the coming months as the also 4.9% de-rated J46-WE-8.

The early exposure to the J46 in test flying at Chance Vought had revealed significant problems. The steady state compressor stalling was not totally cured by the chosen C-89 (with blade tip and seal reduced clearances) and T-39 compressor/turbine combination. These would begin to be used after -8B production was underway. The chosen combination also increased the SFC's at every power setting. Early -8B (and -8A) engines had to have power management restrictions placed on the engines to avoid compressor stalling at altitude. In addition, a micro (Aerotec) pressure switch was added in a circuit to slow acceleration of the engine at any altitude above 20,000 feet. This one change significantly reduced stalling at those higher altitudes but meant slower engine response times.

In the compressor, "O" rings were now being used instead of Bakelite seals on the through bolts going through the 1st through 11th discs. They prevented air leaks back through the compressor from the high pressure stages into lower pressure areas. On the J40 engine, compressor disc failures had occurred due to thermal stress fractures caused by too much leakage overheating and weakening the aluminum discs. Bakelite was used successfully to eliminate such leakage on both engines, but was later replaced on the J46 with "O" ring seals to ease manufacturing and overhaul difficulties.

Even after all the development, the engine was still overweight, the -8B verification engine at Kansas City proving to weigh 2,074 pounds.

Enough of the engines were now flying on a regular basis to begin to expose other issues with the engines as the service mechanics and crew chiefs began to learn how to care for them. The new complex airframe was also demanding a new level of technical knowledge and skill.

BuAer was trying to deal with the open service issues and concerns regarding engines that had passed qualification tests but were being pressed into service with significant development still occurring. It was not a satisfactory situation for anyone involved.

WAGT's recommendations for setting field limits for rejection of engines due to turbine rubs were rejected. BuAer stated that they agreed that any engine that actually experienced turbine blade rubs should be returned for overhaul, but then stated (confusingly, after rejecting WAGT's limits for this) that some minimum clearance should be established for rejection to overhaul. They asked that further information be provided relative to safe limits for turbine disc and blade stretch.[1]

Already mentioned earlier, in early January 1954, BuAer approved Kansas City's request to modify the fuel regulator acceleration valve on a no cost basis. The modification allowed for the integration of the Aerotec signal to restrict acceleration at altitudes above 20,000 feet on all -8A/B and -12A/B engines. This helped reduce surging and stalling at altitude.[2]

Changes in design or assembly procedure being incorporated in the Kansas City -8B engines that differed from the verification engine were:

1. Two additional eyebolts were added to the compressor housing flange.
2. Eliminated the welding of the A/B diffuser inner cone retainer pin.
3. Use of silicone "O" rings to seal the compressor thru bolts.
4. A changed torqueing procedure on the A/B quick disconnect clamp.
5. Incorporated a new fuel check valve.
6. Increased the size of the pin in the oil pump drive.
7. Incorporated a welded type oil cooler.
8. Incorporated a new front bearing seal and baffle.
9. Changed the spring in the accessory gearbox relief valve.
10. Increased the diameter of the rubber cord (a seal) in the accessory gearbox.
11. Altered the knife-edge filter in the #2 and #3 bearing oil supply.
12. Added a new plug and snap ring to the shaft gear at axis "F" in accessory gearbox.
13. Changed the bolt lengths in the lube pump.
14. Changed the material of the clamp used for the inverted flight scavenge tube.
15. Changed the A/B fuel shutoff valve.

16. Added spacers to the A/B unison ring bearings.
17. Reduced the cracking (opening) pressure of dual relief valve.
18. Changed the dimension of the rigid tube between the A/B pump and A/B fuel control.
19. Changed the turbine housing, rear turbine liner, and 2nd stage turbine nozzle by chamfering or widening notches.
20. Incorporated a new tubing assembly.
21. Changed the tubing assembly configuration.
22. Added washers to the shutoff valve spring in the A/B fuel control and changed the manufactured bracket.
23. Changed the oil cooler mounting brackets.
24. Would incorporate the new -10 fuel control in place of the earlier -4 version beginning with engine WE404691.

The list may give the reader some indication of the detailed development still going on to both improve engine reliability and/or ease production problems. None of these changes improved the performance of the engine.[3]

While individual allocation of engines became less important as the production line output increased, the early allocation of -8B engines is of interest if simply to give a full picture of how the various supply lines were being filled to support the field deployment of the F7U-3 aircraft. A January 6, 1953 allocation of early -8B's was:[4]

J46-WE-8B Allocations Jan/Feb 1954	
Qty	**Allocation**
1	BAR Kansas City for factory training
41	BAR Dallas for NOa(s) 51-643 (F7U-3)
3	RecO NAS Patuxant River marked "For Spares"
7	RecO NAS San Diego marked "For ComAirPac Spares"
1	S.O. NAS Memphis Tenn. Marked "For NATechTraCen, Memphis, Tenn."

The engine handling instructions for the pilot of the F7U-3 had to cover the emergency procedures in the event of various engine component or electrical failures. WAGT submitted report A-1748 (stamped "Preliminary") in early January to detail those procedures for such electrical failures.

The report assumed the aircraft would have two engines and only one would suffer an electrical AC/DC related failure. It also assumed that the combined military rated power on both engines was 100%.

In general, the stated conclusion was that if such a failure occurred on one engine, the pilot could open the exhaust nozzle with the manual override lever and then go into afterburning on the engine having the electrical failure. He would then modulate the power on the other engine. Exactly how this would be done if the aircraft was being landed on a carrier was not mentioned, but presumably the affected engine would be shut down and a single engine approach made. Events would prove single engine recovery of an F7U-3 to the carrier was so risky that if an engine failure occurred at night and an alternate land runway was not reachable, airframe abandonment at the pilot's judgement was authorized.

Failures in the main would be noticed by the pilot by T_5 temperature variations, auto-trim failures, A/B failure to light, and/or exhaust nozzle hunting or abnormal behavior. Appendix A of the report lists 75 individual possible failures, the indication to the pilot of such a failure and the proper pilot action in response. The list does not include any description of how to handle general AC or DC failures which would affect both engines.[5] The report was accepted by BuAer on April 26.[6]

Following up a BuAer request dating all the way back to November 1951, WAGT reviewed for BuAer the history of attempts to provide for a GFE thrust meter on the J46 engine. They noted that they had been making an effort to provide a mounting on the -8 and -18 engines, but that drawings of the thrust meter were not available. The BuAer thrust meter supplier had apparently shifted from being Schaevitz Engineering Company to Aviation Engineering over the past two years. WAGT had not been able to obtain copies of the installation drawing of the thrust meter.

They stated that since the hydro-mechanical control had taken precedent over the thrust meter, various planned mounting provisions on the J46 had been removed to make room for the hydro-mechanical control components. The only items of the thrust meter mountings remaining were a bracket on the exhaust nozzle actuator and a turbine outlet pressure rake. The weight of the thrust meter mounting attachments had been estimated at three (3) pounds which was still included in the J46 model specification weight.

WAGT requested that BuAer let them know if the thrust meter provisions were still required, and if so, on receipt of drawings, they would make every effort to modify the engine to allow the installation.[7]

In early January WAGT sent over a chart on the calculated performance of a J46-WE-8B using the new C-89/T-39 compressor/turbine combination vs. the older C-89/T-3 combination. The chart stated in a footnote that if the air bleed was less than the maximum 6.9%, the C-89/T-39 performance approached the earlier C-89/T-3 performance. The chart itself is not dated, but must have been produced in late 1953.

J46-WE-8B Performance Comparison Chart Compressor/Turbine Combinations C-89/T-3 vs. C-89/T-39								
			Thrust			SFC		
Altitude (ft.)	Flight Speed (kt.)	Engine Condition	T-3	T-3 w/Bleed 6.9%	T-39	T-3	T-3 w/Bleed 6.9%	T-39
S.L.	120	Max.	5,624	5,624	5,432	2.654	2.654	2.75
S.L	120	Max. 85°F Day	5,132	5,132	4,900	2.733	2.733	2.86
S.L	170	Fn-825#	825	825	825	1.667	1.667	1.76
S.L	325	Mil.	3,528	3,528	3,470	1.390	1.390	1.432
S.L	600	Max.	7,219	7,219	7,072	2.837	2.837	2.90
15,000	375	Mil.	2,514	2,514	2,420	1.349	1.349	1.41
35,000	440	Mil.	1,468	1,395	1,400	1.297	1.27	1.35
35,000	500	Mil.	1,556	1,556	1,480	1.315	1.315	1.38
35,000	540	Max.	2,742	2,742	2,585	2.485	2.485	2.62
35,000	480	Normal	1,111	1,028	1,000	1.334	1.347	1.46
35,000	510	Mil.	1,358	1,358	1,295	1.332	1.322	1.39
48,000	550	Max.	1,411	1,411	1,329	2.628	2.628	2.80
S. L.	0	Max.	5,800	5,800	5,725	2.53	2.53	2.56
S. L.	0	Mil.	3,980	3,980	3,940	1.100	1.100	1.14
S. L.	0	Normal	3,560	3,560	3,520	1.097	1.097	1.137
S. L.	0	90%	3,205	3,205	3,170	1.106	1.106	1.162
S. L.	0	75%	2,665	2,665	2,640	1.186	1.186	1.285

WAGT noted that although the C-89 compressors would begin to be installed in production engines earlier in the build blocks, the T-39 would begin to be installed at Essington on WE040043 and at Kansas City on WE404567. On the J46-WE-12B, the new turbine would appear first on Essington built WE041108. The performance chart data above was to be incorporated into the -8B / -12B engine specification.[8]

With the finalized compressor/turbine configuration for the -8B and -12B finally chosen, attention was then turned to trying to help the compressor stalling on the -8A engines. A contributing factor to the stalling problem on some of the -8A's was that some of them could not be trimmed to produce the correct temperature of 1,525°F at the turbine inlet, the actual maximum temperature being as much as 60°F lower. These engines were being called "cold" engines. Of the 91 engines accepted to date, as many as 25 were deemed cold engines. The reason this was happening was that on some of the engines, the exhaust nozzle would not close far enough to raise the temperature to the maximum 1,525°F, even though the engines met the thrust and SFC guarantees. At altitude, these colder engines were more likely to have stalling problems. The inconsistent compressor ratio problem might have also been contributing to the issue but is not mentioned as part of the problem description.

The problem could be fixed either by changing the A/B unison ring, installing a new actuator with a longer actuator rod range or opening up the turbine by installing the T-39 model. WAGT recommended the latter as the simplest and more immediately available solution.

The recommended solution of the field installing the T-39 turbine also required that the fuel control had to be adjusted to maintain the proper air-fuel ratio with the new turbine air flow. The adjustment had to be done differently depending on which control version was installed on the engine. Alternatively, the fuel control could have a new constant flow orifice installed and the unit recalibrated by the supplier.

If the field adjustment solution was used, the engine would then have to be flight tested to determine the A/B blowout altitude of the engine after the adjustment was made. If it was determined to be too low, the control would be returned for the orifice installation and calibration.

It was calculated that at 540 knots at 35,000 feet, a 60°F (low temp) cold engine with the T-39 turbine installed and A/B fuel control adjusted should produce about 2% increased thrust.

A written note on the memo states the Power Plant (PP) section at BuAer recommended the field adjustment approach be used. It was also noted that PP had recommended all J46's be retro-fitted with the C-89/T-39 components.[9]

That any given engine required significant adjustments or could even exceed the ability of the engine to be properly adjusted for turbine inlet temperature was a strong indicator of the lack of consistent manufacturing tolerances and/or assembly accuracy on the engine.

Figure 1: J46 Fuel Regulator. The internal complexity required extreme care in manufacture and assembly. WAGT P46193, 2/1/1952. *Courtesy Hagley Museum and Library*

WAGT sent out some minor modifications to the 1952 version of the operating recommendations covering the YJ46-WE-8A and -8A/-8B models in mid-January. The main changes were to include an inspection requirement for the compressor if any possible damage was visible or suspected after looking through the intake vane assembly. The blade minimum tip clearances were set as 0.031 inches for the 1st stage and 0.032 inches for the 12th stage. For stages 1 thru 3 and 6 thru 11 the minimum clearance was 0.025 inches. For stages 4 and 5 the minimum was 0.032 inches.[10]

In early February, Kansas City requested more blank engine log books be sent for inclusion with each engine shipped and BuAer responded that the log book was out of stock and copies would be sent as soon as they were again available.[11]

Two days later, a telegraph memo was received by both the Philly and Kansas City BARs expressing concern over the continuing delay in delivery of service instructions, overhaul manuals and parts catalogs covering the J46-WE-8A, -8B and J40-WE-22 and -22A engines. The memo asked that they advise the appropriate departments of the cause of the delay and expedite the delivery of the publications.[12] No BAR responses were found but it is likely the constant flow of changes to the engines that involved updates to the manuals made it hard to freeze the volumes for reproduction.

Verification testing reports covering the subcomponents of the Kansas City J46-WE-8B were still being submitted to comply with the contract. WAGT report A-1701 covering the verification test of the -8B inlet housing assembly was sent in on January 20. The inlet housing was tested at full load and military speed for 150 hours on a test stand. All parts were found to be in good condition. A considerable quantity of fine steel particles was found in the bottom of the gearbox. These were found to have come from excessive wear of gear teeth in a slave adapter gearbox used for loading the starter pad. *(They would have shared the oil supply, ending up in the gearbox.- Auth.)* After the test, the gearbox was built back up and used on the Kansas City verification test engine for the 150-hour test. On teardown, the gearbox parts were found to be in excellent condition.[13]

A survey was issued in mid-February covering an increase of the front engine mount clearance by +/- 1/64 inches on all J46 series engines. The stated reason for the change was given as not being able to achieve the previous tighter tolerance in quantity production without a high rejection rate.[14]

A field servicing issue was discovered when a FOD-ed engine was returned to WAGT and on teardown the gearbox was found to have corroded elements. It was found that preservation oil had been properly added to the engine according to the servicing instructions, but since the engine could not be operated (turned) the oil had not covered all the elements in the gearbox. The memo described an alternate lubrication method to be used in such cases. The oil discharge line should be removed from the lube pump and connected to the source of the preserving oil. The oil pumped in via this method would fill the reservoir, gearbox, bearings, bevel gear assembly, lube pump, actuator, and most of the remaining parts of the system. The oil should be pumped in until most of the old oil was flushed out and the gearbox and reservoirs were filled.[15]

Less than a week later, another oil lubrication servicing change had to be made. To reduce the chance of overfilling the reservoir, the oil level on refilling the reservoir was changed from 13 to the 11 quart level. After the change to the 11 quart level method, the oil level in the reservoir should be checked as soon as possible but not later than four (4) hours after engine shutdown.[16]

An engine allocation wire was sent in late February to the BARs allocating -8B engines serials WE040057 thru WE040064 to Dallas.[17]

In compliance with MUB Item 43, the results of cold starting of the J46 were forwarded in Report A-1630. The -67°F cold start was accomplished 10 times to 85% of military speed without ignition. The oil pressures were high at first but did not damage anything. However, the oil flow to the exhaust nozzle actuator was extremely low and it took a long time for it to begin to move the exhaust nozzle. It was found that increasing the size of the 1/8" control pressure line to 3/16" greatly reduced this delay and the change was recommended for all engines at first overhaul. Teardown of the engine after the test was complete found no wear or other problems.[18]

An investigation using Zyglo tests was done to determine the limits of the forward bearing mount of the engine. The test found that the limit for rejection could be raised from 1.5 to 2.5 inches. Thirty three Essington manufactured front engine inlet housings were tested against the new limit and all but seven were found usable in flight engines. Those seven housings were set aside to be used in house test engines and to provide further data on the performance of housing bearing mount defects after being used in testing.[19] It is likely the entire issue was over raising the production acceptance yield of inlet housings to speed production, but the memo is silent on the reason for the need to change the acceptance limit.

WAGT issued report A-1424 (a complete copy was not located) that contained an analysis of the possible failures of the two pressure switches in the A/B and how the pilot would recognize them and what actions could be taken should a switch fail. WAGT concluded the switches should be considered fail-safe since nothing of a dangerous nature could happen to the aircraft or pilot and manual procedures to manage the engine in the case of such a failure allowed the mission to continue. The report was forwarded to the Power Plant Division for review and approval.[20]

Report A-1741-Z was submitted covering the successful qualification tests on the Microjet pressure switch.[21] BuAer accepted the report as qualification of the switch on April 28.[22]

Another report (A-1700) arrived at BuAer describing the effect of removing (or capping off) the power limiters in the variable delivery oil pump on the A/B. Other than a temperature rise of 75°F at full pump flow, no other change was noted in pump behavior and the engine functioned normally. The conclusion was that elimination of the pump power limiters would not pose any restriction on the pump, the engine or the oil temperature limits. The suggested removal of the power limiters would eliminate the necessity of tailoring the actuator line relief valve setting during engine acceptance testing and simplified the pump itself.[23]

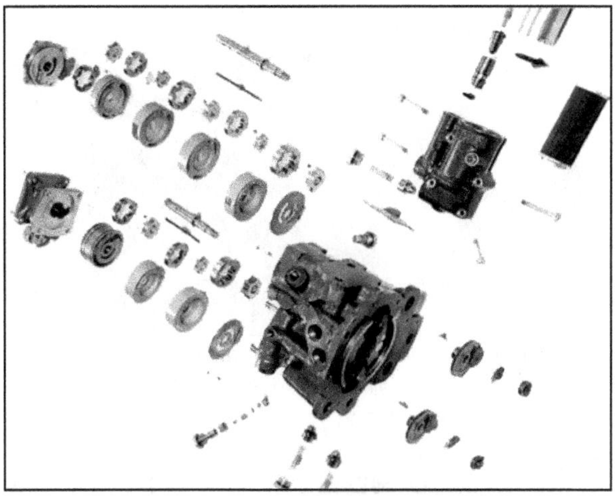

Figure 2: Dual fuel pump with internal emergency cut-over. Either side could supply 100% non-A/B fuel needs at all altitudes. WAGT P46376, 2/29/1952. *Courtesy Hagley Museum and Library*

The next report (A-1701) covering the Kansas City manufactured inlet housing assembly verification test results arrived the same day and it stated that the housing teardown showed it to be in good condition after the 150-hour verification test. The gearbox parts were all in excellent condition after a total of 357 hours of operation in this and other tests. The successful completion of this test was needed so that Kansas City production of inlet housings could begin.[24]

In mid-March WAGT sent over the adapter gearbox overhaul instructions and an illustrated parts breakdown listing of the materials needed for reproduction and distribution.[25]

Alternate suppliers of parts were being sought and a few are mentioned in memos. On March 3, WAGT asked for approval for the following suppliers to be based on various component verification tests:

1. Indiana Gear Works – gearbox shaft gears
2. Bendix Aviation, Eclipse Machine Division – vaporizer tubes, ignitor nozzles
3. Universal Metal Products, A/B fuel pump case
4. ITE Circuit Breaker – A/B exhaust collector[26]
5. American Bosch Corporation – exhaust nozzle actuator[27] (these were formally made by the Westinghouse Special Products Development Division)

The BAR in Dallas sent a wire in early April requesting that 54 adapter gearboxes (along with 4 sets of spares) be sent ASAP. The reason for the request was not mentioned in the wire.[28]

Figure 3: Vickers variable pressure hydraulic pump used for exhaust nozzle control. WAGT P46198, 2/4/1952. *Courtesy Hagley Museum and Library*

The requested bailment of an F7U-3 to Westinghouse for flight testing of the J46 ran into a problem due to the fact that Westinghouse had no F7U related flight test facilities. The aircraft would have to be bailed to Chance Vought instead. That process was held up due to the problems experienced with getting the North American B45 flight testing program resolved. By this time, that program was resolved and BuAer asked that WAGT restate their need for the necessity of using an F7U-3 for engine testing.[29]

Early engines began to be returned for repairs for various problems. YJ46-WE-8A WE040009 with 38.99 hours was returned for repair because of overtemp related damage which had occurred at 400 feet altitude. Later, the engine demonstrated a high altitude steady state stall outside of the stall boundaries.[30]

Engine WE404511 with 46.06 hours was returned after finding steel chips in the fuel system. Engine WE404527 was returned at 10.5 hours after steel and magnesium chips were found in the gear box.[31]

Engine WE040028 with 27.0 hours was returned after evidence was found of an oil seal failure of the No. 1 bearing.[32]

Days later, engine WE404579 with 8.71 hours was returned for an oil leak at the pressure plug on the forward side of the gear box housing.[33]

In mid-April, the Kansas City plant notified BuAer that in response to BuAer's query the plant could reuse the plywood crates in which the A/B's had been shipped to Dallas instead of building new ones as long as the used crates were not damaged. They could use as many as 100 crates ASAP. At that point, Dallas had 32 used crates available.[34]

On April 22, 1954, BuAer terminated their order for the J46-WE-2 engine under the contract NOa(s) 10825.[35] This engine would be replaced by the J46-WE-18 developed (in part) under contract NOa(s) 53-556. See the Chapter on the -18 engine.

As evidenced by WE040009's return, engines were experiencing transient overtemps in service causing turbine blade clearances to fall below the minimum. This resulted in various engines being rejected to overhaul well before their planned 150-hour rejection to overhaul target life was reached. With these occurrences in the background and with WAGT seeking to reduce production costs and complexity, WAGT proposed a 150-hour endurance test of a -8B that would utilize a Westinghouse East Pittsburgh produced T-39 first stage turbine nozzle and first stage turbine rotating blades of both aged and unaged ("as cast") Kansas City manufacture. The blades were to be distributed equally around the disc. To compensate for the air bleed and gearbox accessory loading experienced in actual service (and not normally included in a qualification test) and both of which raised the turbine inlet temperature, a third of the test would be run at a turbine inlet temperature of 1,540°F instead of the normal maximum of 1,525°F. The engine would be manually adjusted to achieve the higher temperature. Teardowns would be done at 60, 100 and 150 hours to determine engine condition. House engine J46-WE-8B #10 was going to be used for the test. If the unaged turbine blades performed satisfactorily, they could cut down on a significant production cost and speed blade production overall.[36]

A written note on the back of the memo routing sheet was related to a current weakness in the engine testing at that time. "Under the current scheme of things, turbojet engines are endurance tested with no air bleed or shaft power extraction for airframe accessories. We have no turbojets with turbine-in-temp controls, so the desired limit turbine-in-temp can be exceeded under conditions of air bleed and shaft power extraction. In the past, this has not been considered too serious, because the power take-off quantities have been modest, and the durability effects attributed thereto have been presumed to be masked by the engine-to-engine life history differences. However, this problem is pointed up by BLC (Boundary Layer Control) air bleed requirements."

It should be noted here that none of the F7U series of aircraft were ever considered for use of BLC. The comment must have been referring in general to the other aircraft under development or being submitted in proposals at that time, many of which planned to use BLC in some form.

The subject of pressure limiters on the engines was a growing concern. Higher speeds in dives, including transonic performance and high speeds in very cold low level conditions meant the intakes could provide air pressures at the face of the engine so high the compressor might exceed its design strength. Most airframe intakes of late 1940's design had no provisions either for shock wave management or pressure recovery limiting. For the -8 (and -18) program, provisions for a pressure limiter were under investigation. In this case, the engine alone was considered in determining the feasibility of putting a limiter on it. After considering six points, it was determined that on the -18 engine, a pressure limiter should be included in the qualification test. Chance Vought was asked to continue to provide space for a limiter on the F7U-3 and A2U-1 airframes. WAGT was requested to develop a limiter to be available as an "off-the-shelf" for future use.[37]

In September, BuAer expanded the possible requirement for pressure limiters to the -8B and -8 and WAGT notified BuAer that the engine installation drawings were being modified to show the location and dimensions of a "phantom" pressure limiter device and they would be forwarded as soon as they were available.[38]

The qualification test on the latest version of the starter-generator gearbox was accepted on May 18, no parts being found excessively worn or damaged after 150 starts.[39]

On May 20, WAGT acknowledged BuAer's instructions to re-designate eight -8 engines into -12 engines. Why this was done is not clear.[40]

The qualification test of the Kansas City manufactured J46-WE-8 engine using serial number J46-WE-8-102 with the hydro-mechanical control system was run at Kansas City from February 19 to March 5th. The test took a total of 201 hours and was run against specification WAGT-24C10-6C (July 1, 1953) test requirements. The tests of control transients, compressor air bleed, and alternate fuels were conducted separately and reported later in report A-1778. As tested, the engine was in good condition and production approval was recommended.

The tested engine used many earlier version parts and some parts previously used in production testing. They were used in this case to allow the qualification test to be conducted as soon as possible and were not considered to differ in function or expected reliability from the production build parts actually used once production was approved. The dry weight of the engine as tested was 2,074 pounds. Per an exchange between WAGT and BuAer, on December 11, 1953 BuAer had approved a maximum specification weight of 2,125 lbs.

Various problems were encountered during the test.

1. The tubes in the electric control assembly were replaced at 236.1 hours.
2. The oil filters had to be replaced due to contamination caused by disintegration of the "O" rings in the hot oil system pumps. (The hot oil test was a new addition and the pumps were separate from the usual engine pumps.)
3. Three (3) A/B ejectors had to be used. The failures were attributed to the Kansas City test cell configuration.
4. Due to the limited power of the Bendix E-2 starter, the cranking speeds were not met.

At the end of the test the engine was re-calibrated and found to meet the deterioration limits of MIL-E-5009 as follows:

J46-WE-8-102 Post Test Performance Calibration vs. Specification		
	Thrust Percent	**SFC Percent**
Maximum (A/B)	98.6	101.7
Military	97.2	103.5
Normal	98.4	101.9
90% Normal	99.8	101.2
75% Normal	100.0	102.3
MIL-E-5009 Spec. Limit	90.0	107.5
Engine starts: • Avg. to Light-off – 14 seconds (range 6-23 seconds) • Idle to A/B avg. – 6.3 seconds (range 5–10 seconds)		
Throttle Torque (inch- lb): • Cutoff to Idle (Spec 23) – 19.2 • Idle to Military (Spec 15) – 14.1 • Military to A/B (Spec 35) – 19.4		
Oil consumption was 0.215 lb/hr		

Post-test teardown revealed the following discrepancies:

1. Cracks were found using Zyglo tests in 10 of the 66 first stage turbine blades.
2. Further examination using a 30X binocular found 40 other blades with cracks from 1/16 to ½ inch long. All were the result of creep rupture. A study was planned on the creep rupture phenomenon.
3. The two failed tubes were likely caused by the test cell's 110 Volt, 400 cycle power source which had been found to indicate a supplied voltage of 130-135 volts. The continuous operating limit of the power control was 125 volts. Voltage regulators were in-

stalled in all of the Kansas City test cells to prevent these failures in the future.[41]

BuAer disapproved the qualification of the engine due to the failures in the first stage turbine blades. Because of the urgent need for engines, the Bureau stated they were willing to accept fourteen (14) engines built to the test engines' build standard plus any approved changes to the parts list. The temporary acceptance was made on the understanding that WAGT would make every effort to correct the first stage turbine issues. Any consideration of further engine production approval would be given only if the needs of the Navy indicated such an approval was required and that WAGT could show concrete evidence of progress toward eliminating the previously noted deficiency.[42]

Engine WE404519 was returned on May 27 due to being FODed by an unknown object which caused turbine blade failures.[43]

A survey was sent out on May 27, 1954 concerning the gradual increase in length of the A/B ball joint up to ¾ of an inch due to thermal expansion of the component parts of the joint. Four effects could be noted:

1. Maximum increase in engine length as a result of the ¾ inch permanent set.
2. A/B mount moved aft by a maximum ¾ inch.
3. Change in the engine exhaust nozzle settings.
4. Damage to the A/B combustion chamber housing caused by the yoke striking the shell.

No actual damage to the ball joint occurred and the A/B diffusers could be used regardless of the growth. Because of the fixed position of the exhaust nozzle actuator relative to the yoke, the iris nozzle tended to open resulting in a lower turbine inlet temperature (and lower thrust). The linkage could be adjusted and such adjustment would not affect the performance, although such adjustments could not extend beyond the linkage inspection holes.

To prevent damage to the yoke, newly manufactured engines would have their A/B housing dimpled prior to shipment. Engines in the field would be retroactively dimpled.[44]

The May memo was followed up in August with another specifically referencing the ball joint expansion on the -12A and -12B. WAGT explained that at military power in A/B, the actuator rod in the guide slot directly aft of the ball joint could conceivably bottom on the forward end of the slot. This could happen on a hot day with maximum ball joint expansion. If this occurred, the exhaust nozzle might not open to the fullest extent. The effect on the engine performance of that occurring was still being investigated, but no examples of it actually happening had been reported. If the engine user found that full military power could not be achieved, the actuator rod linkage should be adjusted to compensate for the ball joint expansion amount discussed in the prior memo. The possibility of eliminating the ball joint completely was still being investigated.[45]

The first standardized Monthly Production Progress Report (MPPR) forms appear in the files just after May. This first set reports the cumulative WAGT engine deliveries against current contract orders through the end of May.

MPPR May 1954 Total Production to Date		
Essington	**On Order**	**Produced**
YJ46-WE-8A	12	12
J46-WE-8A	45	45
J46-WE-8B	23	23
J46-WE-8	0	0
J46-WE-12A	0	0
J46-WE-12B	4	4
Kansas City	**On Order**	**Produced**
J46-WE-8A	27	27
J46-WE-8B	400	266
J46-WE-8	204	5
J46-WE-18	472	0

Of even more interest, the monthly reports contain comments relative to current problems being experienced and their impact on production deliveries for the reporting month and the next month. The May report for Kansas City reports the following -8B problems that caused delays:

1. Leaks at the gearbox pads and accessories were still prevalent, but fewer because of redesign of the gearbox adapter pad.
2. Turbine blade tip clearances below the low limit were occasionally still encountered but in decreasing numbers.
3. All engines were still being temperature traversed during the green run, in some cases requiring retraversing causing delays. Hot spot problems were still being encountered.
4. Some lube pump nuts were tightened to the incorrect torque while shipping engines. The problem was resolved.
5. Cracks in an A/B exhaust collector were found causing a delay.

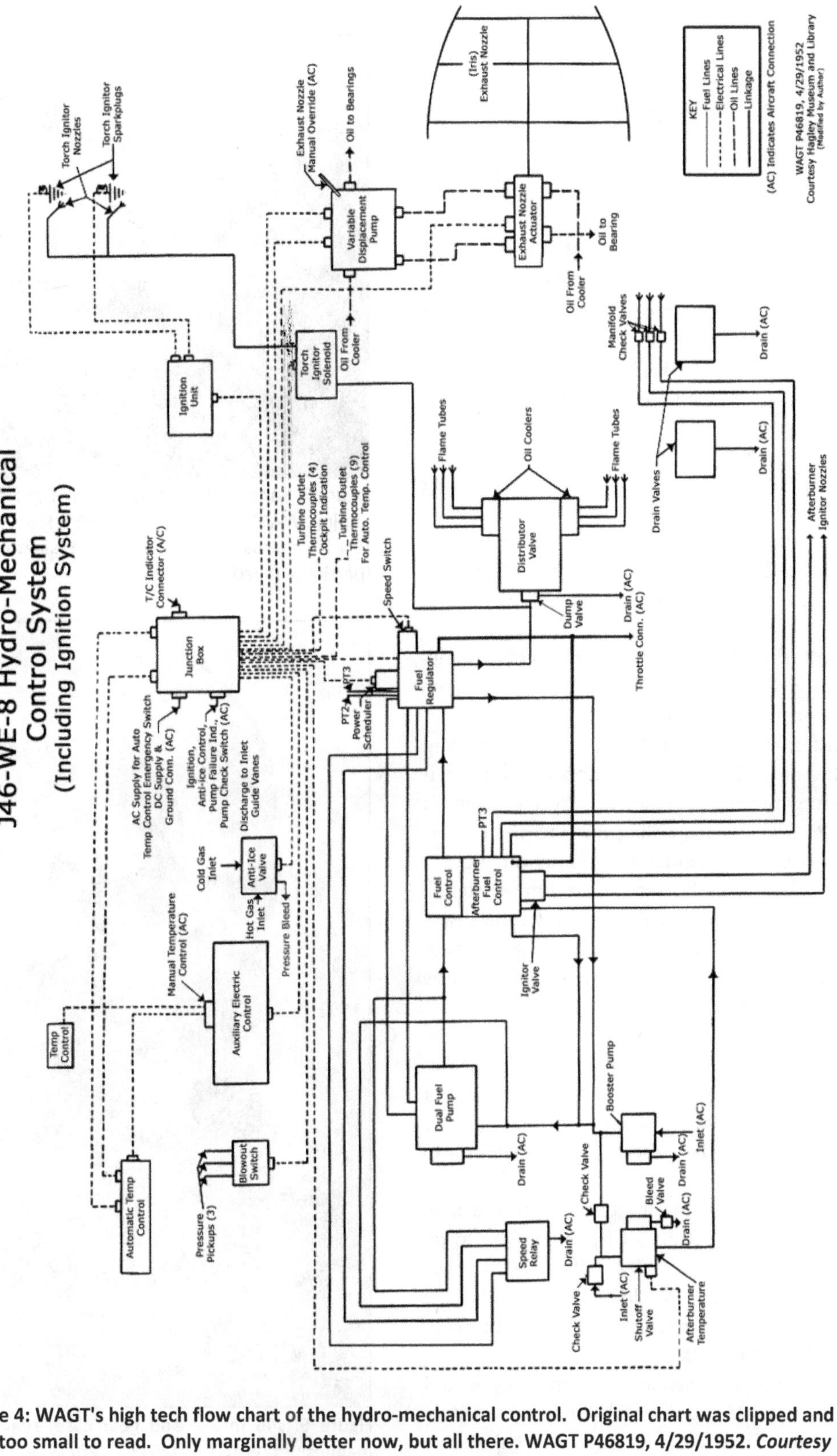

Figure 4: WAGT's high tech flow chart of the hydro-mechanical control. Original chart was clipped and text too small to read. Only marginally better now, but all there. WAGT P46819, 4/29/1952. *Courtesy Hagley Museum and Library*

For the -8, the report stated there was a lack of availability of hydro-mechanical controls due to a subcontractor failing to meet commitments. June's delivery schedule for -8 engines would not be met.

The other anticipated delay was going to be caused by the lack of trained personnel in the assembly area. This had caused a delay in May and would continue through June in spite of intensive training going on and an all-out recruiting program already underway. The delivery of 39 engines was anticipated in June.

Additional personnel had been hired at Essington to facilitate production of publications. Subcontractors had been employed to assist in the writing of Service Instructions and overhaul manuals.[46]

Engine WE040013 (a YJ46-WE-8A) had been sent back to WAGT in February after being FODed at Pax River. The engine had been submitted to a large number of catapult shots and arrested landings and was included in the engines studied as part of a program investigating front bearing support cracks. In mid-June it had been rebuilt to repair damage and was ready for shipping, an allocation request then being made by the BAR.[47] The engine was then shipped to Quonset Point, Rhode Island on August 25.[48]

As a result of verification testing, in June BuAer accepted the American Bosch Corporation as an alternate provider of exhaust nozzle actuators.[49]

Notwithstanding the extra effort WAGT was putting into getting the needed manuals and documentation out to the service organizations, BuAer was forced to send a wire on June 25 urgently requesting that the preliminary drafts of the overhaul manual and parts catalog for the -8A and -8B engines be forwarded immediately.[50] Three days later WAGT hand delivered three copies each of the preliminary manuals to the Naval Maintenance section. Some drawings were missing and were to be forwarded in Early July.[51]

BuAer had allocated 241 -8B engines as of June 29. In an allocation memo covering the planned July and August production, they allocated another 106 -8B engines of which 26 were designated as spares.[52] In a separate memo on the same day, they allocated 16 -8 engines to Dallas for the same time period.[53]

Also on the same day as the prior memos, a telegram was sent to the BAR Kansas City telling him to redesignate 39 -8's on order as -8B's. The total number of -8B's on order at that point would be 180 and 159 of the -8 model. A contract modification would have to be negotiated with WAGT after the fact.[54]

WAGT responded immediately, stating they would order 39 sets of -8B parts and discuss the revised schedule with BuAer in July. They understood from verbal discussions with BuAer that 39 -18 engines would be changed to be -8 engines in the schedule so that the -8's that were to be changed to -8B's could be reinstated in the schedule.[55]

Figure 5: Kansas City 150-hour intake/gearbox qualification testing. WAGT KC14034, 1/20/1954. From Report A-1710. Few KC photos still exist, only the Essington photo library was (partially) saved.

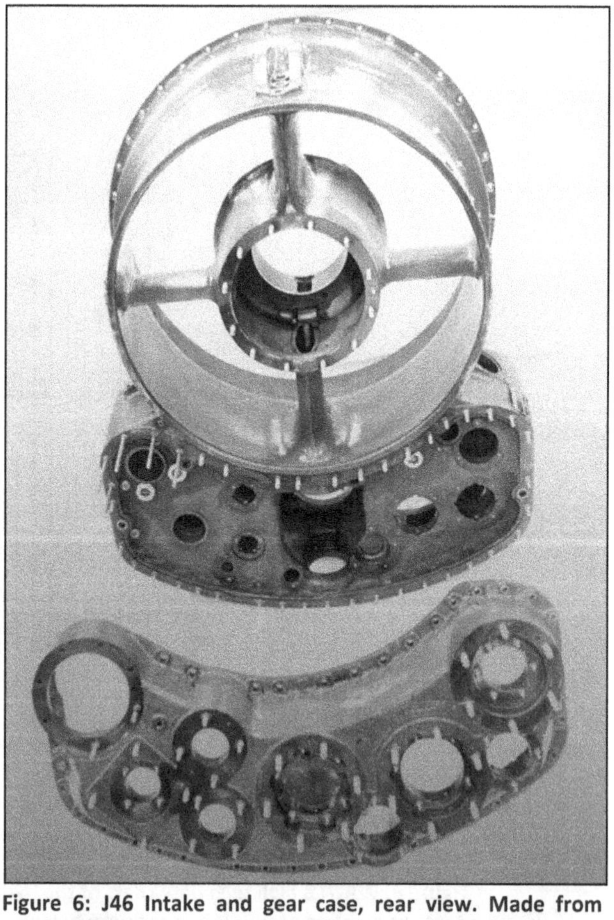

Figure 6: J46 Intake and gear case, rear view. Made from magnesium. WAGT KC14035, 1/20/1954. Report A-1701.

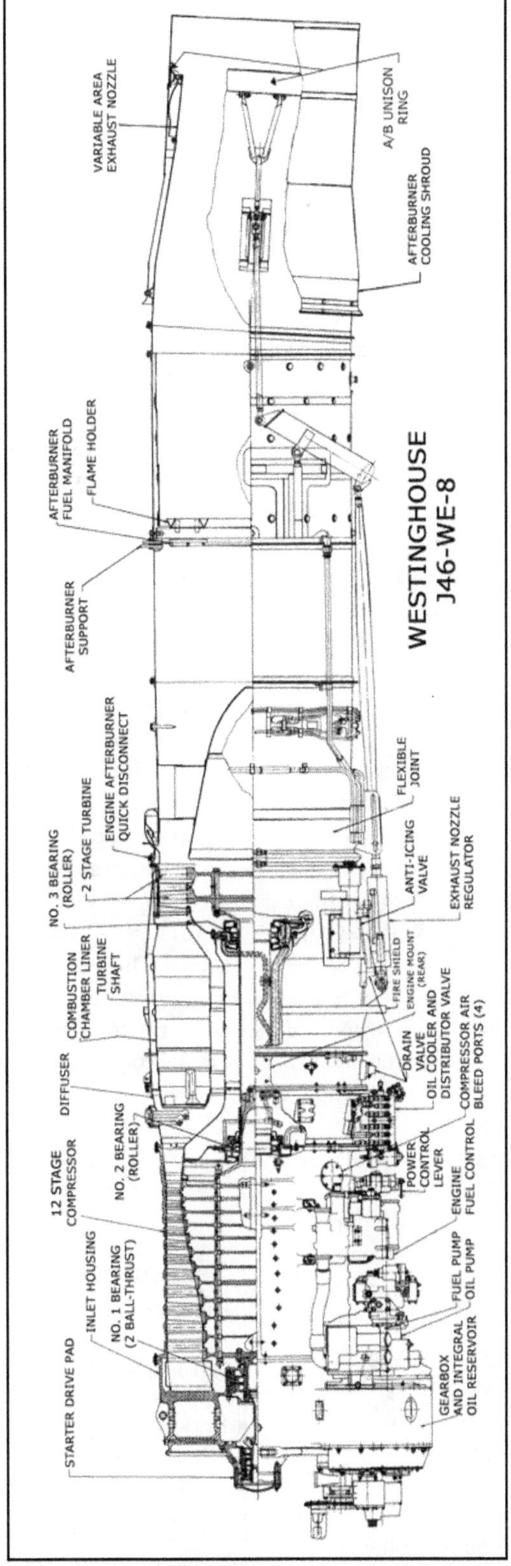

Figure 7: The final de-rated J46. With modernization and the final operational adjustments it approached the original contracted specifications. Even as a line drawing, the external complexity compared to the original J46-WE-2 mockup is readily evident. Note location of the later removed anti-icing valve. WAGT P46977, 5/20/1952, Drawing 58J-790. *Courtesy Hagley Museum and Library*

It is obvious that with a limited supply of -8 hydromechanical controls being available, to avoid another delay in engine deliveries, BuAer was tactically making changes to keep airplanes flowing into service use.

On June 29, given the delay in completing the qualification test of the -8, now scheduled on August 15, BuAer authorized accepting another 21 pre-qualification -8 engines. They would be accepted with the understanding that they would be retrofitted by WAGT with any necessary parts needed to comply with the official parts list after the qualification test was successfully completed.[56]

The June MPPR states that Kansas City failed to produce any -8 engines for that month due to "Engineering Design" problems. These problems would jeopardize the July -8 delivery schedule.[57]

In early July, BuAer modified the contract to eliminate the packing crates being supplied by WAGT and substituting steel containers to be supplied by BuAer. This change affected 96 -8 engines and 469 -18's.[58]

Before the end of June, WAGT had sent out a survey on making some wiring changes to the Automatic Temperature Control. It had been found that when the automatic temperature control was first turned on, the control would move the exhaust nozzle to the full closed position, overtemping the engine. The problem could be avoided if the throttle was reduced below the military position before turning the trim on.

The proposed solution was to add a wire going to the trim circuit to keep the tubes in the control continuously warm during flight. Tests had shown if the tube filaments were kept warm, the erroneous signals to the control were eliminated. A circuit breaker in the new circuit would allow the current to be turned off when the engine was not running. The pilot would turn the breaker on when starting the engine and turn it off at the end of the flight.[59]

BuAer countered with a simpler solution of changes inside the engine junction box. They suggested moving one single connection inside the box and changing the power supply to the pin from 110 volt AC to 28 volt AC. The change would allow the tubes to begin to warm up as soon as the engine start cycle was begun and turned off automatically when the electric supply to the engine was turned off during normal engine shut down. No new wires or circuit breaker circuits would be needed and the pilot task of pushing a breaker in and pulling it out would be eliminated.[60]

In a follow-up memo a few weeks later, WAGT clarified the situation, stating that overtemping of the engine was not occurring during start up because the automatic temperature control was on during starting and took control once the engine reached sufficient speed. The length of time involved during starting was sufficient to make sure the tubes were warmed up and spurious signals were not generated. The problem only occurred if the auto-trim was turned off for any length of time and after the tubes cooled it was turned back on.[61]

In mid-July, BuAer authorized the scrapping of the J46 engine mock-ups that had been returned from Chance Vought and Convair. They asked that they be verified as worthless before scrapping. Some confusion over the -18 mock-up existed, with BuAer admitting it should be accepted as the replacement for the prior -2 mockup under contract NOa(s) 10247 (for the XA2U-1), the -2 mock-up having been returned to WAGT previously.[62]

On July 19, BuAer continued modifying the quantities of various engine models on order. They now redesignated 123 -18 engines as –8 engines. They asked WAGT to advise them of the change in unit price and the recommended delivery schedule.[63] A written note on the routing sheet states the change was made due to the switch of 50 A2U-1 aircraft to be F7U-3M aircraft. It brought the total -8 planned procurement to 280 engines.

A written note dated August 2nd on the back of WAGT's acknowledgement memo states that because the -8 had not been qualified as yet, the -8 engines accepted prior to the qualification would have be accepted as derated -8B equivalent engines. The BAR added a note on August 20 that it was intended that only 14 -8 engines be accepted prior to qualification.[64]

The problem of leaking "O" rings at low temperatures was studied by WAGT and resulted in report A-1811 being submitted on the testing methods and results. The report (undated) has WAGT's cover memo dated July 22. Parker and Goshen "O" rings were included in the fuel regulator test using fuels with different aromatic content down to temperatures of -65°F.

Report A-1811 Parker and Goshen O-Rings Test Specification MIL-E-9007A

Ring Type	Fuel Aromatic Content %	Temp °F	Seal Type	Result
Parker 443-12	0	-60	Static Rotating	No Leak Leaked*
Parker 443-12	2	-65	Static Rotating	No Leak No Leak
Goshen 1120	0	-60	Static Rotating	No Leak Leaked**
Goshen 1120	2	-65	Static Rotating	No Leak No Leak

* One drop of fuel per every two (2) minutes
** One drop of fuel per every 29 minutes

The resulting recommendation was that an alternative type of seal needed to be investigated for the rotating throttle shaft on the fuel regulator.[65]

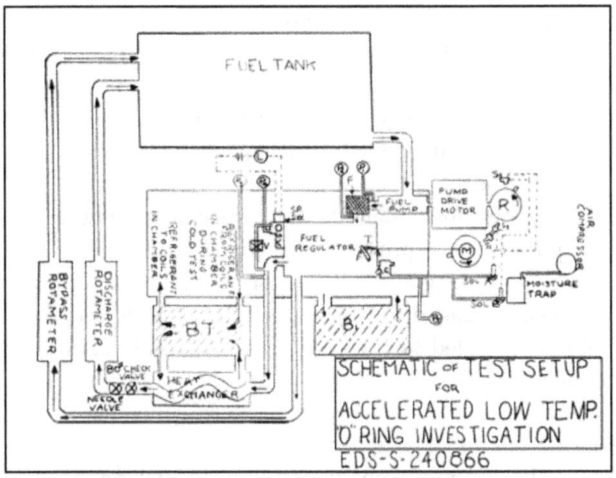

Figure 8: "O"-rings test schematic. From WAGT report A-1811.

The J46-WE-12B engines serials WE041114-17 were ready for allocation in August of 1954.[66]

The maintenance of an adequate supply of A/B shipping boxes at WAGT was to become a persistent supply problem, with repeated memos over the next 18 months from the Essington BAR to the Dallas BAR all urgently requesting that boxes be returned to WAGT ASAP.[67]

The July MPPR shows that Kansas City was now shipping -8 engines in quantity with the -8B version now scheduled to ship in almost identical quantities for the next 6-8 months. The problems which had been the cause of engine shipment delays in July had been: some under tip clearances but it had been resolved; some engines showed low A/B thrust which was being investigated; and some gearbox covers showed excessive porosity and required additional sealing impregnation.[68]

The verification test of unaged vs. aged turbine blades was run in July but had to be stopped when, during an acceleration at the 125 hour mark, the engine badly overtemped (over 2,000°F) causing major damage to the second stage turbine blades. Inspections up to that point had found no evidence of creep rupture in any of the aged blades but significant numbers of unaged blades were presenting the condition.

Inspection of the engine discovered the primary fuel control calibration was outside the specification limits and the servo-valve was scratched. This would cause it to stick during acceleration and could have caused the over temperature.

The plan was for the test to be rescheduled and rerun.[69]

WAGT had run an A/B verification test on the -12B A/B back in October, 1953. They submitted their report to BuAer in June of 1954, the reason for the delay not being mentioned. BuAer accepted the report and the test results, releasing the engines for 150 hours of service use.

The -12B A/B on the back of a -8B engine differed from the -8B A/B in having an additional 18 inch straight extension and one (1) four (4) inch extension canted on a 5 degree angle. These extensions were placed between the combustion chamber assembly outlet and the standard -8B extension leading into the A/B diffuser. The design allowed the five degree cant to be assembled either to the left or right. The A/B weight was 476 pounds (43 heavier than the -8B afterburner).

The test was 7.5 hours of A/B operation over a 20 hour engine run. On tear-down some cracking was noticed and was either repaired by welding successfully or by minor changes in design. Interestingly, about half of the cracks were determined to have been the result of manufacturing problems and they existed in the A/B before the test had begun.

The engine's performance against Specification WAGT-24C10-6B is listed here to make possible comparison to the standard -8B results:[70]

J46-WE-12B A/B Verification Test Results				
Rating	Thrust Spec.	Thrust Actual	SFC Spec.	SFC Actual
Maximum	5,720	5,730	2.53	2.356
Military	3,960	3,998	1.106	1.107
Normal	3,540	3,710	1.102	1.079
90% Normal	3,190	---	1.111	1.083
75% Normal	2,650	---	1.193	1.090
Idle	245	185	4.00	4.032

BuAer approved the test results on November 10.[71]

The August MPPR reported continuing Kansas City shipment delays that were caused by turbine tip clearance issues and leaking gearbox covers. They added that the impellers for the gearbox were defective as received from the supplier and had to be replaced on the assembled engines. Receiving good engine controls was a problem and the production forecast for September was lowered from 65 to 51 engines.[72]

A wire to the Essington BAR in the record and dated September 3 shows that BuAer was considering cancelling 78 -8 engines and 96 -18 engines. The Kansas City BAR was not to be alerted prior to further notice being sent. The BAR was to try to estimate the cost of such a termination and recommend the most economical quantities of engines to be terminated.[73] The BAR responded that for the -8B, the cost would be 60% of the total planned and for the -18, the cost would be 25%.[74]

The September MPPR was much like August's in the causes for engine shipping delays. In addition to the other on-going issues, Solenoid Valve problems, A/B Fuel Controls, Lube Pumps and Exhaust Nozzle Actuator's requiring replacement resulted in a large number of penalty runs in acceptance tests. Only 30 -8B engines shipped against the forecast of 38.

For the -8, only 5 engines shipped vs. a forecast of 13, the shortfall caused by the same problems affecting the -8B delivery schedule.[75]

Lubrication pumps were found to be inadequate on quality testing at Kansas City but checks at the supplier and Kansas City found that all of the pumps met the Eclipse-Pioneer pump specification at the time they were shipped to WAGT. An investigation showed that only in September had the rejection rate for pumps become significant. A review of the actual needs at Kansas City found that they needed of a pump of greater performance over and above the existing requirements to which Eclipse-Pioneer was manufacturing. The rejections at Kansas City were due to the mismatch between the existing specification and the actual needs. Pumps were quality tested at Kansas City to the new requirements and being failed accordingly.

It was apparent that the specifications needed to be changed and possibly new design work would be needed to meet the new requirements. Both parties were working to resolve the situation.[76]

Back in August, WAGT sent their comments to BuAer regarding problems being reported with exhaust nozzle hunting. It was found the hunting was being caused by over-heating of the germanium rectifiers in the Auxiliary Electric Control. This was happening because of hot air leakage from the cockpit pressurization line which was routed past the Auxiliary Electric Control. WAGT noted that the latest design Auxiliary Controls would use selenium rectifiers in place of the germanium ones; the selenium rectifiers functioned properly under the higher temperature conditions if such a leak were present. The information was not forwarded to BuAer HQ until October.[77]

In September, WAGT surveyed replacing the A/B surge check valve with a new one that contained a more positive seal. In addition, the valve would be relocated to make it more accessible at the customer connection. New engines would contain the new valves at the new location as follows: -8B serial WE404901 and up, -8 serial WE405561 and up. Engines to be retroactively fitted were: -8A serials WE040007-020 and WE404503-529, -8B serials WE040021 and up, WE404530-690 and WE404691-900.[78]

In a memo on October 6 in response to a WAGT request for drawings describing the thrust meter to be used with the J46 engines, BuAer stated that no drawings describing the meter or its mounting provisions existed. WAGT need take no further action on the matter. However, they asked that items related to the thrust meter be retained on the engines. BuAer stated they would let WAGT know if further requirements related to thrust meters arose.[79]

Shortly thereafter, WAGT issued a clarification on the purpose of a project being conducted by Soundrive Engine Company at the behest of WAGT. The company had been engaged to investigate alleviation of acoustical vibration related to compressor surging and heat driven oscillations in A/B's, both abnormal conditions, not the general work of suppression of noise generated during normal engine operation.[80]

The contract went through its annual revision for 1954 prices and Amendment 22 was issued to adjust the contract. The amendment shows there were still 209 -8 engines still on order (and 346 of the -18's). The amendment reduced the total contract price by $20,015,075.00.[81]

J34-WE-8A engine WE404522 had been returned in July after suffering a severe overtemp (1,832°F) for a number of seconds in flight, resulting in turbine damage. (Confusingly, the early wires refer to this engine as WE040552, which was a -8B. In October WAGT corrected the error in their damage investigation results memo.)

The pilot had been in A/B and retarded the throttles out of A/B and then felt a vibration. Just after the A/B run at 20,000 ft, the dive brakes had been opened.

WAGT concluded the overtemp was caused by a compressor stall due to the left hand engine exhaust nozzle moving rapidly to a closed position immediately following shutdown of the A/B. The stall was aggravated by a reduction in airspeed following the opening of the dive brakes. The condition permitting the exhaust nozzle to go fully closed on one engine could be caused by the throttles not being evenly retarded or the throttles being unevenly adjusted so that the left engine remained in the military flat area while the right engine throttle was retracted below the military flat spot.

At the time of the failure, it was not known that engine expansion and airframe distortion under flight conditions caused slightly opposite motions of the engine control linkage at any fixed throttle setting. To account for this, airframes now had the linkage adjusted to favor the low end of the idle and military flat spots on the port engine and the high end of the same flat spots on the starboard engine. The service instructions had been updated to give the appropriate rigging adjustments necessary.

WAGT added that an EC in the works would eliminate the necessity for the pilot to retract the throttles well below the military range when coming out of A/B at altitude.[82]

Contaminated fuel filters were found in July and WAGT reported on their findings and planned actions in October. BuAer asked if a micronic fuel filter needed to be added to the inlet side of the engine boost pump. Chance Vought stated there was no room for such a filter to be added. WAGT concluded the dirt, which was found to have non-magnetic ferrous, magnesium and plain dirt components, likely came from the airframe fuel system. This conclusion was reached since a new fuel filter cartridge was always installed in the inlet of the boost pump after a final factory test run, precluding the dirt source from the engine itself. The airframe boost pump manufacturer was asked to improve the manufacturing quality of their pumps and investigate changing the bellows seal material from brass to stainless steel.[83]

Engine WE404579 had been found to be consistently running 31°C overtemp at military power settings. Investigation showed that the relay igniter surge check valve was stuck in the open position, allowing fuel to flow into the hot combustor air flowing to the turbine, raising the exhaust gas temperature. Damage was found to be an "H" shaped crack on an iris leaf of the A/B.

The explanation given for why the engine control did not detect the overtemp condition and reduce fuel flow accordingly is very confusing. On one hand, it seems to state that the engine control, receiving its signal from the four-thermocouple harness, was seeing only 689°C temperature instead of the 720°C being seen in the cockpit, which also received its signal from the same harness. For some reason, the cockpit reading would be higher if fuel was flowing (and burning) from the relay ignitor nozzle. This only makes sense (to the author) if the cockpit reading was from a single thermocouple located directly behind the relay ignitor nozzle position, and not from the entire set of harness readings.

The memo states that steps were being taken to replace the four-thermocouple harness with a nine-thermocouple harness "to eliminate T5 discrepancies". It is not explained how the specific failure of the control system to detect the higher temperature would be prevented. To the author's mind, only ensuring that the engine control input and the cockpit indicator received identical inputs from the thermocouple harness would accomplish this.

No comment was made on the reason for the stuck ignitor valve itself.[84]

Improved control components were being evaluated and WAGT took steps to accelerate their evaluation by including real life testing of these components at an F7U-3 squadron on the east coast, VF-81. The components, an additional experimental Automatic Temperature Control and new Auxiliary Electrical Control were installed on a J46-WE-8B and flown until October 29. WAGT reported the preliminary results were very satisfactory but included no details.[85]

A November 18 wire to the BAR at Kansas City requested a change in the existing engine shipment priorities. "Change nec to pvnt further increase of glider pool".[86] Presumably, the recipient understood the abbreviations.

In response to urgent requests, WAGT forwarded five copies of the tool drawings prints for the temperature traversing tools for the -8A and -8B. The urgent need was for the overhaul shops to be able to do the temperature traversing and speed up overhauls of engines.[87]

Three separate suspected compressor stall incidents resulting in engine damage were investigated by WAGT who summarized their findings in a memo in November.

1. The engines (-8B WE040037 and WE404534) experienced violent compressor stall in F7U-3 BuNo. 129548 when being flown in formation on another F7U-3. The engines overtemped (+1,000°C) as a result. The maneuver was a two to three "G" pullup from an A/B dive at 400 knots. The aircraft had experienced a lack of lateral control for three to five seconds. An engine check showed everything to be adjusted properly. WAGT concluded the stalls were the result of either partial aircraft stall or disturbed inlet airflow from the wake of the lead aircraft.
2. The right engine of F7U-3 BuNo. 128474 (-8A Serial WE404513) experienced an overtemp while in steady state operation at an outlet temperature of 800°C. It was not known if the automatic temperature control was on or off. The temperature control was replaced and the subsequent engine operation was normal.
3. Aircraft BuNo. 128475 (-8A Serial WE040020 starboard and -8A Serial WE404523) experienced compressor stalls during a wave off. The records showed that the Auxiliary Electrical Control's shock mounts had failed and the AEC was replaced. Compressor stall was again experienced 4 days later. A hot day adjustment on the Auxiliary Electric control lowered the T5 setting, the minimum flow adjustment was lowered, the acceleration time was increased, and the resistance setting on the Microjet was increased. Subsequent engine operation was satisfactory.

In every case, WAGT stated the related damage to the engine(s) was found to be consistent with that expected from the overtemp the engines had experienced.

An engine allocation memo on December 1 covered the final allocation of -8B engines. It shows that 487 engines had been contracted for, 451 previously allocated

(42 from Essington and 417 from Kansas City production) and the memo allocated the final 46 -8B engines from Kansas City.[88]

On the same day the final allocation of -8B engines was sent out, BuAer sent out a partial contract termination confirmation letter (for a previous telegram sent on November 18) reducing the contract by 275 -18 engines. This cancellation reflected the termination of the -18 program in its entirety.[89]

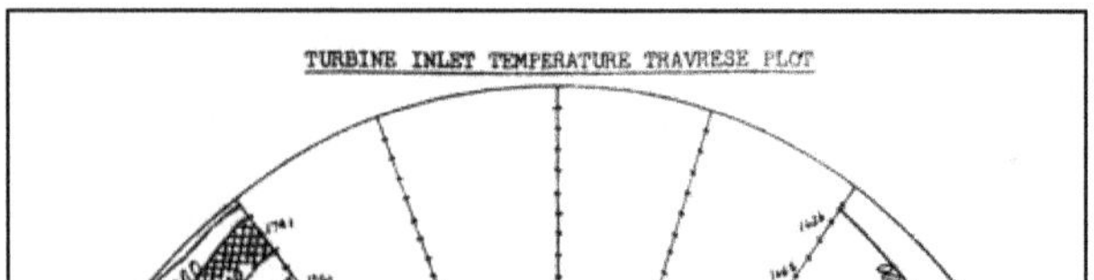

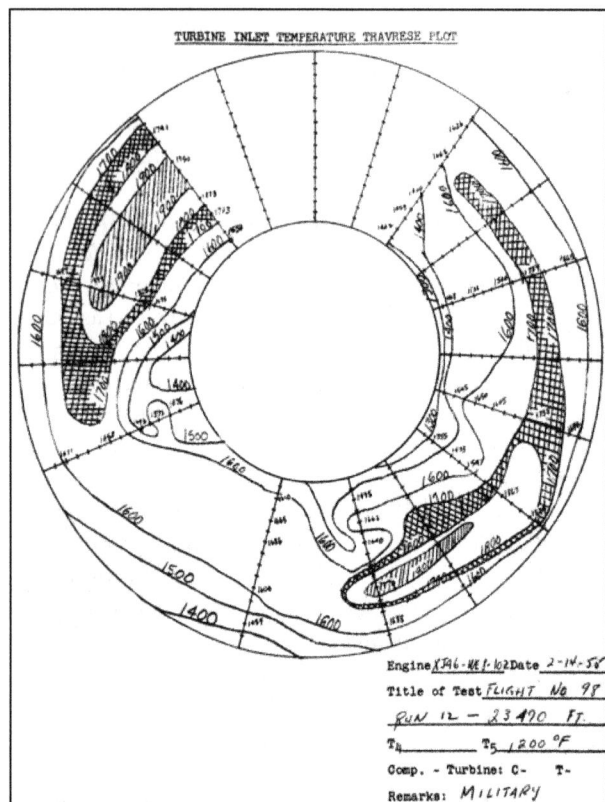

Figure 9: Sample of a temperature traverse taken during altitude testing.

In spite of being able to ship 32 -8B engines against a forecast of 25 and a contract schedule of 16 in November, prior Kansas City production problems continued with turbine tip rubs experienced, malfunctioning primary fuel controls, A/B fuel controls and VD pumps. Supplier problems were at the root of the controls and pump problems. WAGT was working with the suppliers on the defects being seen, but these were expected to continue for several more months.

Kansas City shipped 24 -8 engines against a forecast of 20 and a contract schedule of 27. Porous gearbox covers and fitting leaks, lube system contamination, malfunction of the controls during testing, malfunctioning variable displacement pumps and parts shortages were the main reason for delays.

Forty-seven (47) percent of the lube pumps received during the month were found to be defective. The A/B fuel control had a design change made during the month delaying receipt of new design compliant controls. Controls that were received had to be checked out on the green runs because they failed to pass quality acceptance tests.[90]

On December 2, BuAer added eight (8) -8 engines and fourteen (14) -8B engines to the contract on Amendment 25. The price for the -8's and -8B's was to be $157,015.00 and $146.105.00 respectively. Delivery of the -8's was to be in July of 1955 and the -8B's in August and September. WAGT was to use as much of the parts inventory from the terminated -18's as possible in building these engines.[91] This order was to extend the production lines before WAGT shut them down.

A bit later WAGT sent a wire to BuAer stating that engine WE040009 was ready for shipment. The reason for it being repaired was not given in the wire.[92]

Back in October, BuAer asked WAGT to propose a location and provide space and hose connections for a fuel flow transmitter on -8 engines. A preliminary location had been identified and on December 6, WAGT sent out a survey showing the recommended space was on the lower half of the compressor housing on the left side directly above the afterburner fuel control.[93]

At the end of November, the Essington BAR recommended to BuAer that they approve the transfer of J46 design cognizance to the BAR Kansas City. WAGT had pointed out to the Essington BAR that WAGT's HQ had been moved to Kansas City and other engine models had already had their engineering cognizance moved to Kansas City. In a telegram on December 9, BuAer's Contracting Office authorized the transfer.[94]

In an echo of past activity, a memo was placed in the correspondence files explaining the reason proposed October 3, 1952 contract Amendment 6 was cancelled. Since the amendment had been executed to include titanium discs in compressor discs six (6) through twelve (12) on thirteen (13) J46-WE-8 engines, when the decision was taken not to include titanium discs, the amendment was cancelled as of the date of the memo.[95]

A turbine blade had failed on WE404593 in early December and WAGT (via the BAR) forwarded pictures of the break on the blade without explanation of the suspected cause. Rupture cracks on turbine blades had been occurring and were an ongoing concern, but this memo does not relate any information about the specific cause of this particular broken blade.[96]

The field was reporting that Chance Vought was shifting Auxiliary Electric Controls and Automatic Temperature Controls from one engine to another to salvage functional controls.[97] (The effect of this would mean that engines being returned for repair would have defective controls installed and this may have masked the true cause for which the returned engine was originally rejected, delaying root cause analysis.) The field complained of having to reject numerous Automatic Temperature Controls and Auxiliary Electric Controls for quality issues and BuAer asked WAGT to respond on how many were being returned and their findings upon inspection. WAGT responded on December 21 for the period of September 8 to November 11 and they reported:

Automatic Temperature Controls – 44 received and 24 tested by the time of the memo. Twelve (12) were found to be satisfactory after adjustments and returned to service.

Auxiliary Electric Control – 30 received and 15 tested to date. Ten (10) units were found to be satisfactory.

The field rejection rates of satisfactory units were most likely due to a lack of proper test and checking facilities, information and trained personnel in the field.

The memo does not give percentages of assembly rejections vs. the number of assemblies available on engines or in the parts pipelines. Nor does it address what was wrong with the units that had to be returned to the supplier. It was expected that modifications to both assemblies that were in process would "substantially reduce the field rejection of these components."[98]

It was reported on December 22 that engine WE040032 was ready for shipment from Essington. The reason for return and repair was not given.[99]

The December MPPR shows that 21 -8B engines had shipped that month against a forecast of 28 and a contract schedule of 10. Processing delays were encountered due to oil and fuel leaks, malfunctioning controls – particularly A/B fuel regulators and fuel test stand problems. Replacement controls were hard to obtain due to a backup of controls at the Receiving Inspection department caused by personnel not being at work due to contractual issues. Lube pumps (Eclipse – Pioneer), Boost Pumps (Nash Engineering), Automatic Temperature Controls and Auxiliary Electric Controls (Manning, Maxwell & Moore) and A/B Fuel Controls (Excello) were in short supply.

During December, twenty eight (28) -8 engines had shipped against a forecast of 32 and a contractual schedule of 28. The same problems and shortages affected this engine as had the -8B with the additional problem of a shortage of Exhaust Nozzle Actuators (American Bosch).[100]

The stream of engines coming from production had been steadily introducing new improved components but the result was a field engine inventory with widely varying parts assemblies. This greatly added to field servicing and maintenance efforts and complexity.

At a conference held between BuAer and WAGT on January 18, 1955, WAGT proposed a Phase I and Phase II "Modernization and Improvement Program" for the J46-WE-8B. Under this program, BuAer generated Amendment 29 to the contract to cover the additional cost of $528,000. The balance of the -8B engines yet to be delivered was to incorporate the changes and be delivered out through September.[101]

Amendment 29 covered the Phase I and II modernization of a total of 194 -8B engines. Phase I covered the inspection of the turbines and replacement of turbine blades which were cracked as a result of excessive temperature, plus other changes in the settings procedures which would more closely control the turbine temperature level and reduce the tendency of the engine to overtemp. This phase of the program would raise the safe turbine life to 120 hours. (At this point, engines were being rejected to overhaul mostly for turbine damage at an average of 60 hours.)

Phase II would make changes to reduce the turbine temperature level, plus other changes improving engine durability. This phase was designed to bring the average turbine service life to 240 hours.

The average price per engine for Phase II was to be $11,927.00. Phase I was estimated at about $100.00 per engine but would have to be re-determined once examination and repair of engines began.[102]

Only a week after approval of the modernization program, the BAR in Kansas City had to report that some engines (WE404945 and WE404964) would be accepted and shipped minus several of the program upgrades and would be upgraded in the field as soon as the parts became available. For a period, engines would continue to be accepted minus some upgrade level parts and these would be upgraded in the field as well.[103]

The following is a listing of the parts upgrades necessary to bring a -8B engine build to the Phase II level which is designed to achieve the 240 TBO durability goal. While the drawing numbers and the actual parts themselves are of little interest, the dash number after the drawing number indicates how many design changes had been made to that part or assembly up to the point the list was created (June 30, 1955). The last 14 engines in July were still awaiting parts to complete the modifications.[104]

J46-WE-8B Phase II Modernization Program Parts/Assemblies Build Level					
No.	Drawing Number	Description	No.	Drawing Number	Description
1	64J313-3	Variable Displacement Pump	18	64G376-1	"U" Fitting
2	58J554-21	Afterburner	19	64G639-17	Lead*
3	46F573-1	Pressure Switch	20	58J650-12	Auxiliary Electric Control
4	64J318-14	Primary Fuel Regulator	21	61E635-9	Automatic Temperature Control
5	63F645-8	Fuel Valve	22	67G113-1	Lead*
6	61E550-5	Power Scheduler	23	58J105-5	Fuel Pump*
7	58J292-13	Lube Pump*	24	63F673-7	Oil Relief Valve*
8	63F746-6	Manifold	25	67G990-2	Tube*
9	41E435-2	Cable	26	62F727-3	Cover*
10	AN832-12D	Elbow*	27	63E65-2	Vane Assembly 5th Compressor Stage*
11	60E788-11	Exhaust Nozzle Actuator	28	61E973-2	Vane Assembly 3rd Compressor Stage*
12	65G760-2	Restrictor*	29	61E980-2	Vane Assembly 10th Compressor Stage*
13	63E625-2	Valve (relief)*	30	62F252-2	Overspeed Relay*
14	60E734-12	Fuel Distributor	31	58J893-35	Gearbox*
15	63F918-1	Cross*	32	64J240-4	Combustion Chamber Housing
16	22E694-16	Boost Pump*	33	22E628-2	Vane Assembly 8th Compressor Stage*
17	58J650-12	A/B Fuel Control			
* Part common with the J46-WE-8					

No record remains in the contract correspondence files as to the success of this effort to increase the TBO on the turbine section to the 240 hour target.

While engines were being cycled through the program, back in early February, BuAer ordered that all -8B engines be shipped to the Dallas BAR with the exception of the first four (4) Phase II engines which were to go to ComAirPac for evaluation.[105] Two other critically needed -8B's were requested to be released for fleet use by shipping two -8's to WAGT to replace two -8B's being used for training.[106] As it happened, only one -8 (WE405566) was allocated for training since Essington no longer produced engines and -8B engines WE404548 and WE414712 were shipped to Norfolk.[107]

Critical shortage of -8B's or not, the modernization program had an immediate impact in reducing the number of engines shipped in January. The program caused 19 engines in the late stages of processing at the plant to be returned for rebuilding and retesting. New rules were in place at the plant that now required complete re-runs on penalty engines. Also, the continuing problem of parts shortages of control units and fuel regulators, leaks and labor issues had their usual effect on slowing output in January.

The modernization work on the -8B engines pulled workers from -8 production output and only 20 -8 engines shipped that month against a forecast of thirty-four. The reduced -8 output forced the Navy to ship needed engines from Dallas to the Norfolk, Virginia naval base for fleet outfitting instead of supplying them directly from Kansas City.

The Kansas City BAR noted on the production report that the WAGT production schedules always assumed 100% labor availability and doubted the forecasts in the future would prove to be attainable.[108]

The Essington BAR submitted a final telegram stating that no engines had been accepted by the BAR in January and all production in Essington ended with the partial termination of the contract. The wire was the final Essington production status report.[109]

With the J46's in the field proving to have short service life between overhauls, any other perceived defects on the engine, regardless of cause, seemed to heighten increased demands for redesign. A good example was -8B WE404644 which experienced an oil pressure loss in flight. Inspection showed the oil filter cover was cracked. The repair depot recommended redesign of the cover. When WAGT looked into the issue, they found the cover had been over-torqued on installation. They pointed out that the cover was sealed by an "O" ring and did not need a lot of torque when tightened and recommended a tightening torque value of 100 inch-pounds on cover installation and updated the field service manuals accordingly. Since the report of a cracked filter cover was an isolated occurrence, WAGT felt no redesign was needed and planned no further action.[110]

Proof that some -8A engines were converted in the field repair depots to -8B configuration was contained in a telegram shipping J46-WE-8A WE404527 to Kansas City for investigation and repair. When the repairs were complete, it was shipped to San Diego for conversion to a -8B.[111]

WAGT requested that the Essington conducted verification test on the Starter-Generator Adapter Gearbox be accepted as verifying the component. BuAer responded that until the Kansas City BAR forwarded his comments, approval would be withheld.[112] The Kansas City BAR sent his comments on February 23 recommending acceptance as fully verified.[113] BuAer finally accepted the verification test results and accepted the gearbox for use on J46 engines on March 4.[114]

Figure 10: J46-WE-8B trigger anticipator circuit solenoid location (arrow lower left). Note the complexity of the hosing. WAGT KC15800, 3/21/1955. Memo attachment.[115]

A proposed survey got BuAer's prompt attention when WAGT suggested adding a trigger anticipator circuit to the A/B ignition system. It included a solenoid in the primary A/B ignitor line. In conjunction with the solenoid, a pressure differential switch would be incorporated to energize the solenoid at a pre-determined pressure differential between the A/B fuel manifold and the turbine outlet total pressure. This would ensure an adequate fuel delivery for A/B ignition.[115] BuAer felt the change was of primary importance and wanted the survey conducted on a top priority basis. Since there were reports the circuit might cause interference with the current flow meter installation plan, such possible interference was to be examined as well.[116]

Engine Bulletin No. 107 was issued in February covering the inspection and rejection limits for J46 turbines. The EC lists no less than 10 different turbine configurations on the various -8A, -8B, -8, -12A and -12B engines. Inspections were necessary if the engine had the following overtemp conditions observed:

J46 Turbine (T4) Over Temperature Limits J46-WE-8A, -8B, -8, -12A, -12B		
Temp. Range (°C/°F)	Time (sec)	No. Occurrences After Which Inspection is Mandatory
950 and up (1,742)	5	1
900-950 (1,652-1,742)	5	3
850-900 (1,562-1,752)	5	4
816-850 (1,500-1,562)	5	6

Up to 15 first stage turbine blades and 10 second stage turbine blades could be changed on a rotor before the rotor assembly needed to be replaced. Inspections could be by either Zyglo or dye-check methods, whichever was available locally.[117]

Other durability issues than the turbine section remained. Two engines had been found (-8B's WE040064 and WE404718) with cracked outer and inner combustion chamber liners. The outer liner cracking problem was reported upon in report A-1747 and a new design had been tested for over 244 hours with no cracking evident. The inner liner damage was determined to have occurred as a result of the aft end of the liner not being properly supported by the support bracket at operating condition. Only two instances of the latter problem had been reported so they planned no corrective action.[118, 119]

Work stoppages at the plant again restricted the flow of engines. In February only two (2) -8B's were accepted against a forecast of 20 and eighteen (18) -8's against a forecast of twenty-eight. Some parts shortages for the -8's were still continuing, including lube pumps and fuel regulators. Leaks at the adapter pad had been reduced to seepage in most cases, with development continuing to eliminate the problem completely. High idle speeds were expected to be resolved through the use of the latest fuel regulators.[120]

Reporting on engine damage to -8B WE404584 that occurred in November, it had been found that a nut had passed through the turbine. This was due to the locking tabwasher having straightened, allowing the nut to loosen. Inspection showed that there was an interference existing with the combustion chamber outer liner. As a result, the engine assembly procedure was modified to check for such interference and Engine Bulletin No. 92 was sent out.[121]

A request for the final allocation of -8 engines was met on March 9 when the final 91 engines under contract were allocated with all but 16 going to Dallas. The 16 otherwise allocated engines were sent to be spares to COMAIRPAC and COMAIRANT.[122]

Engines -8B WE404612 and WE040049 had been rejected to overhaul back in June 1954 for cracks in the first stage turbine blades. WAGT did not get around to commenting on the damage until March 1, 1955 and basically referred BuAer to the modernization program underway for -8B's to reduce blade cracking due to overtemp excursions of the engines.[123]

Throttle flat spots were adjusted at WAGT request with BuAer approving revised military flat spots on the performance curves: 80° – 85° for the -8 and 82° – 87° for the -8B, -12B.

The next paragraph in this memo seems to criticize both WAGT and BuAer design and specification personnel, stating "Other situations have occurred in the past in the Westinghouse J46 engine program where it was necessary to effect changes to the model specifications because the engines were designed with features or component items which differed from the requirements in the specifications. It appears that closer liaison between engineering design and engine specification personnel is necessary."[124]

In early March, WAGT formally asked BuAer for help in establishing a steady flow of shipping containers back to Kansas City to enable WAGT to ship accepted engines. At that time, even with the labor problems they were experiencing that were limiting delivery of engines, a shortage of containers at the plant existed. To complete the contract, WAGT had asked for 200 GFE shipping containers to be supplied at the rate of 10 per week but they were not arriving. Alternate containers that could be modified were also not being supplied. Shipments of J46 engines would be halted from March 14 on until containers were received in Kansas City.

WAGT requested BuAer permission to accept engines less the shipping container under the contract so that the engines accepted could be invoiced and WAGT paid. The engines would then be shipped as soon as containers arrived. The BAR endorsed the plan.[125]

The reader will be left to imagine the frustration everyone must have felt at this point, after the struggle to get engines of any sort through qualification, verification, a new plant into production, the airframes ready to deploy and the engines into full scale fleet service, only to now find that the lack of simple low-tech shipping containers was threatening to bring things to a halt once again.

Responses to the survey regarding the addition of a trigger anticipator circuit to the J46 began to arrive, the first from Chance Vought. They pointed out that they currently used the left hand turbine out total pressure probe for the Microjet pressure switch on the left hand engine and for the thrustmeter pickup on right hand engines. The new circuit was to use the same total pressure out probe and they felt it was undesirable to attach the Microjet and Trigger Anticipator switch to the same probe. They proposed a solution involving moving the thrustmeter pickup lines and moving the Microjet pickup to the right side of the left engine.[126]

Looking ahead a bit, when WAGT responded to Chance Vought's concerns, they suggested using a new switch that referenced ambient pressure instead of PT5. The PT5 piping was eliminated and a new location for the switch proposed. Chance Vought accepted the new design and asked for early implementation.[127]

Back in October reports came stating that inspection of 12 A/B's in the field had found that five (5) had iris leaves and unison ring failures at the 3:30 o'clock position. Early testing by WAGT showed that they were encountering iris leaf cracking at the 3:00 and 9:00 o'clock positions due to the blanketing of the cooling air by the unison ring push rods. Later they discovered that the currently specified A/B adjustment limits could leave gaps between iris leaves. After review, Engine Notice 32 was updated with new A/B adjustment limits that eliminated the possibility of gaps occurring.[128] It was not clear if the gaps were causing the leaf cracking or the overheating - an example of a typically vague answer.

A request to review all the engine bulletins regarding the Primary Fuel Regulator and the A/B Fuel Regulator caused WAGT to remind BuAer that at the time of the request in December, the design of the Primary Fuel Regulator was being changed and WAGT no longer felt a conference was necessary. The two new alternate Primary Fuel Regulators were physically and functionally identical and could be used on all -8 and -8B engines and also -8B engines if they had been modified by Engine Bulletin 13. New Engine Bulletins were prepared as follows:

EB 96 – provided for the use of the new Power Scheduler.

EB 97 – provided for the overhaul modification of all presently used fuel regulators to the new configuration.

EB 98 – provided for overhaul modification of the Fuel Valve presently used on all Fuel Regulator assemblies.

EB 108 – was an information bulletin with information on field replacement of individual fuel regular assembly components.

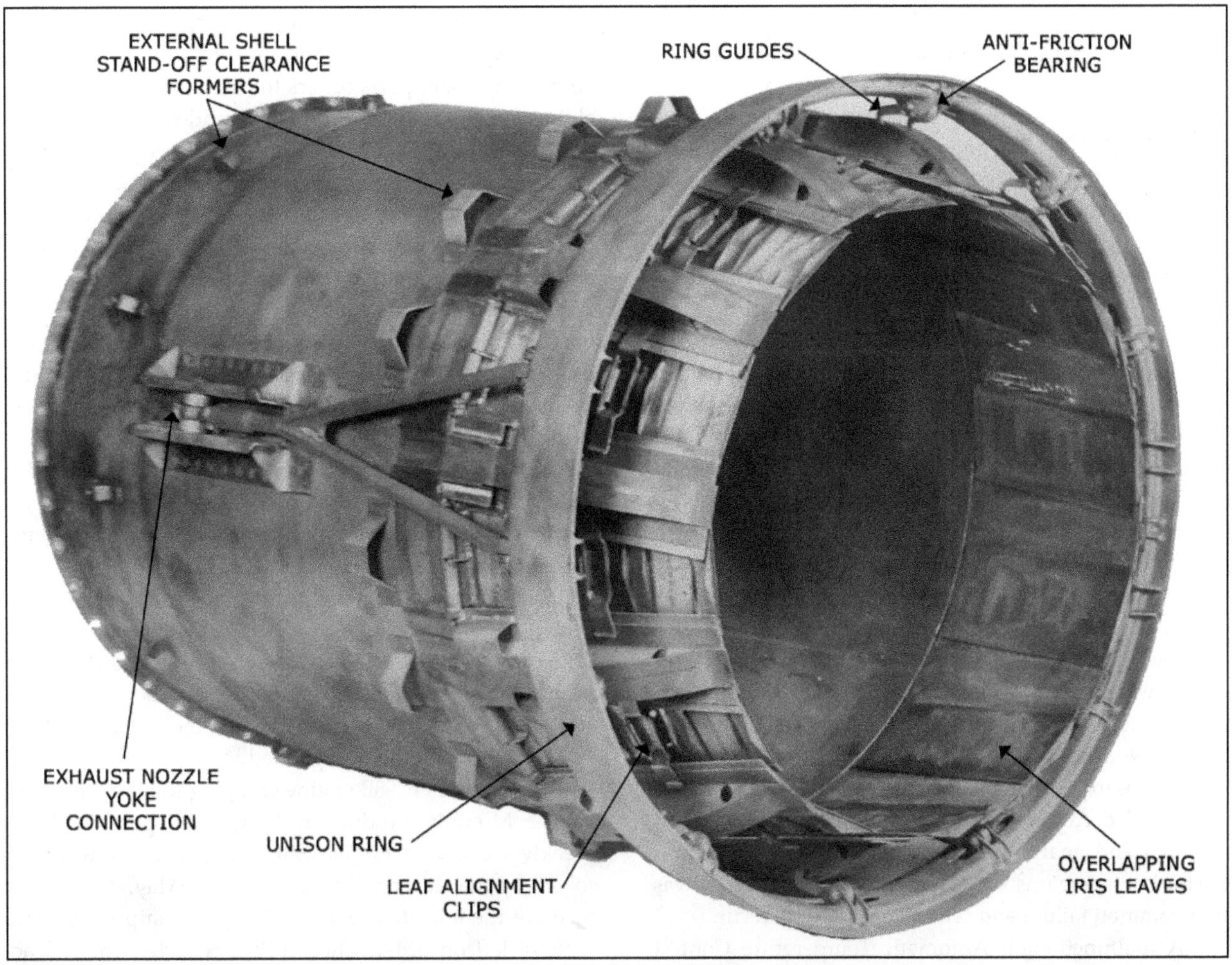

Figure 11: J46-WE-8A iris type afterburner nozzle. Rarely seen exposed without its cooling shell, here it is in the fully closed position since no gas is passing through it. Exhaust gas pressure kept the iris leaves pushed open as far as the unison ring would allow for any given power setting. The -8B, -8 and -18 A/B differed in minor changes to the spray rings and flame holders, not shown here. WAGT P49513, (not dated). ***Courtesy Hagley Museum and Library***

Additionally, the existing EB 65 provided instructions for modifying all presently used A/B fuel controls to the latest configuration. However, if the Bureau thought a conference was still necessary, WAGT would be happy to comply with the request.[129]

The March MPPR showed 26 – 8 engines shipping against a forecast of 19, the increase due to reduced labor problems. Control testing restrictions and lack of exhaust nozzle actuators were still throttling deliveries. Twenty one -8B's shipped during the month against the forecast of five.[130]

In mid-April, -8A WE404511 was repaired and shipped back to Dallas. The reason a repair was needed was not mentioned.[131]

A sticking pilot valve in the Variable Delivery Pump was suspected of causing the pump from -8B Serial WE040049 to be rejected, but verification of the suspected cause was not possible as the serial number of the pump was never identified and WAGT had no record of receiving the pump back for examination.[132]

In one of the stranger formal survey requests, WAGT requested an after-the-fact formal survey approval on their recommendation to replace the first stage aluminum compressor disc with a steel one for durability issues. The change would add 8.98 pounds to each engine. The CID's had already been processed to accomplish the change.

The CID's would be effective on engines:

1. J46-WE-8 WE405666 and up;
2. J46-WE-8B WE404969 and up;
3. J46-WE-8 retro-active to first overhaul engines WE405501-665, -8B engines WE404530-968, -8B engines WE040021-068, -8A engines WE040007-020, WE404503-529; and
4. J46-WE12A, -12B retroactive when aluminum discs were exhausted on engines WE041101, 041103 and up.[133]

The survey resulted from WAGT's analysis of first stage compressor disc failures that had found that failure was due to air flow conditions over the first stage blade foils during operation at certain RPM ranges below military. The vibrations exceeded the strength of the aluminum resulting in fatigue cracks. While the first investigation concerned -8 engines, they found that the -8A and -8B engines had identical strength problems and later expanded the recommendation to all J46 type engines.[134]

The April MPPR stated four -8B and twenty three -8 engines were shipped. Fuel pumps and cam box shortages continued to constrain production. Procurement of Exhaust Nozzle Actuators was still a problem and purchasing was expediting an increase in future supplies.[135]

A failed Automatic Electrical Control was returned in March and analysis showed that a lead wire from a resistor had come unsoldered at the terminal post. It was an uncommon failure and seen as not likely to recur.

A malfunctioning Automatic Temperature Control was analyzed and it was suspected that a defective aneroid switch had operated intermittently at altitude. In support of this conclusion was Chance Vought's report that the temperature trim was fluctuating at altitude on the engine. Due to lack of proper test facilities at WAGT, the control was sent to the manufacturer Manning, Maxwell, and Moore for analysis. No other failure of an aneroid switch was known, so this too was interpreted to be an isolated case.[136]

The forced landing of F7U-3 Bu. No. 129561 in Linkhorn Bay (just offshore from Norfolk, VA) on April 15 caused BuAer to request the engine controls on both engines be examined to determine the cause of both engines failing to produce power as reported by the pilot.

After examining and testing all of the components, WAGT reported "although they do not meet specification requirements in all respects, they were flight acceptable". Detailed analysis followed and other than slight changes noted due to either field adjustment or salt water corrosion, the units functioned satisfactorily in testing after corrosion attributed to salt water immersion was removed.

The final conclusion was that the engines were likely starved of fuel rather than any control malfunction.[137]

In April, WAGT notified BuAer that after surveying all existing spare parts orders to determine general availability for retro-fitting -8B engines under the modernization program, they had determined that very few parts had been ordered to provide for both support and retrofit.

The Aviation Supply Office had indicated they would order retrofit parts separately from those for general support, but no orders had been received as yet. While funding approvals were in process, WAGT requested immediate dispatch approval to order 200 sets of parts specified by the Phase II CID's. The BAR would designate parts to the applicable CID and handle shipping to the appropriate destinations.[138]

The WAGT memo was forwarded by the BAR and BuAer HQ, who then forwarded it to the Supply Office commanding officer, stating that of the 25 different items covered by Phase II, engine bulletins 85, 86, 101, 102, 111 and 115 warranted higher priority action than the other items. In addition, all boost pumps below part number 22E694-13 would be immediately modified to the latest dash number. All such pumps part numbers 22E694-13 and above would be modified when the engine went to overhaul or when the pump failed.[139]

Differences in -8B engine shipping totals as reported by the Monthly Production Progress Reports and the Weekly Report of Engine Shipments Reports were resolved by the Kansas City BAR in early May. Of interest here are the adjusted totals, 487 engines shipped to date with 14 left on order. The MPPR's reported engines accepted and the weekly engine shipment report reflected actual shipments, meaning they occasionally differed.[140]

WAGT submitted report A-1863 concerning the results of a study to determine if an altitude temperature inversion occurred on -8B or -8 engines when used in the F7U-3 aircraft. The report, dated April 29, 1955, concluded no such inversion occurred and that engine use at altitude imposed no restrictions on turbine durability.[141] A -8 engine would be similarly tested as soon as possible.

The report states the tests were conducted to determine the possible cause of excessive quantities of creep rupture type cracks in turbine blades of -8A and -8B engines. Wind tunnel altitude testing did not indicate that a temperature inversion condition existed, but conceivably the inlet ducting could alter the temperature profile by causing a radial shift. Ground running with the engine installed in an F7U-3 airframe revealed a mild radial shift of the temperature profile towards the engine hub. Flight testing occurred from February 9th to the 17th in Bu. No. 128472. It was determined the temperature profile stayed constant during the varying flight conditions although there was a change in magnitude of the temperatures. In no case were spot temperatures in excess of 2,100°F recorded.

The Navy investigated the possibility of using a mix of jet fuel and aviation fuel in existing jet engines should the condition arise in war-time of a shortage of jet fuel on an aircraft carrier. It was recognized that the two fuels had different effects on engine durability and operability. A fuel mix of 3/1 JP5 (MIL-F-5624B) jet fuel to 115/145 performance number (MIL-F-5572) aviation fuel was termed "JFC" fuel and was tested by WAGT in engine XJ46-WE-8 #7 for 50 hours. The summary report stated that JFC use was acceptable for service operation of the engine. Minor control adjustments might be necessary if JFC fuel were used instead of JP4. No changes in radial or circumferential temperature distribution were noted in any degree when JFC fuel was in use. Fuel consumption was 2-4% lower using JFC fuel, depending on throttle setting.[142]

Turbine inlet temperature effects on the turbine blade cracking problem were the subject of several memos in May, first with WAGT confirming that a minor error in their calculations for compensating for the effect of fuel temperature on measured fuel flow had resulted in the calculated turbine inlet temperature being 10°F or lower than previously reported. Subsequent testing of a turbine rotor run for fourteen hours at a true 1,525°F inlet temperature had resulted in no discrepancies being found, the turbine being in excellent condition.

They also reported that inspection of turbines from Phase I and Phase II engines to date were showing good durability, with no cracks found on five rotors with 60 hours run time on four and 120 hours on a fifth. Based on the reported observations, WAGT recommended BuAer increase the turbine temperature to a true 1,525°F on production engines and on overhaul engines.[143]

The BAR endorsed the recommendation and added additional information. Two test runs of 28 hours each had been run at turbine inlet temperatures of 1,525°F and 1,550°F with satisfactory blade condition. He reported that a change to 1,525°F turbine inlet temperature would result in a thrust increase of approximately 45 pounds at military and A/B throttle settings. Continuous testing indicated the primary cause for turbine difficulties was the result of greatly overtemp operation above 1,525°F.

A written note on the routing sheet states that BuAer agreed in conference on June 6, 1955 to accept engines at a true 1,525°F, confirmed by a telegram on June 24.[144], [145]

In trying to close out a matter dating back to 1953, WAGT wrote that they had discovered it was not possible to modify early engine ignition units to the new configuration as previously committed. They had estimated that 70 such early ship units would need to be modified at no expense to the government. As only 30 early units had actually shipped, WAGT suggested that they simply replace 25 such units at no cost, the slight numerical difference being the estimated units that would have been lost due to engine loss, fleet operation or other damage.[146] BuAer accepted the proposal on May 26, 1955.[147]

Alternate parts suppliers were still being qualified. A new supplier (I.T.E. Circuit Breaker Company) for the A/B Exhaust Collector submitted a sample part (64J300-5) for test and it failed. WAGT modified the part to the -15 configuration and on re-test it passed. Since I.T.E Circuit Breaker Company had already shipped 22 -5 model collectors to WAGT, WAGT offered to modify them all to the -15 configuration and qualify them for use on production engines. Confusingly, the memo subject line refers to the units as verified at the -17 part revision level, but the text states they were verified at the -15 level.[148] BuAer accepted the modified units as verified at the -17 level for production engines.[149] The acceptance memo did not address qualifying the I.T.E. Circuit Breaker Company as an alternate supplier, only qualifying the WAGT modified parts.

A Microjet problem on a -8B was reported when the engine's exhaust nozzle did not close when successful A/B light-offs did not occur at altitude. It was interpreted that the Microjet was not signaling blowout of the A/B. The Microjet was set at 56 ohms. The replacement of the unit with one of the same part number resulted in ground operation of the A/B with ignition and then immediate blow-out. A controls engineer went out and inspected the unit and found the engine was operating with the latest (-4) design of Microjet, making it likely the engine had been operating earlier with the first (-1) design. The installed unit was set to 42 ohms instead of the 60 ohms setting everyone had previously assumed was correct. It was reset to 65 ohms with satisfactory operation on the ground but did not sense blow-out at altitude. When reset to 56.5 ohms, it operated properly on the ground and at altitude thereafter.

WAGT added that the Phase II program, which replaced the control systems, added the trigger anticipation circuit and replaced the -1 Microjet with the -4 version. This would materially reduce the type of A/B related exhaust nozzle malfunctions.[150]

The MPPR for May shows no -8B engines were shipped but engines scheduled for August were beginning to be assembled in May. Twenty-five (25) -8 engines were shipped against a forecast of twenty-nine (29) engines. Problems were static leaks in lube pumps, short supply of exhaust nozzle actuators, and leaks and malfunctioning accessories requiring penalty runs and delays in final packaging. Installation of quad ring seals and extra care in preventing scratched sealing surfaces were reducing the leakage problems.[151]

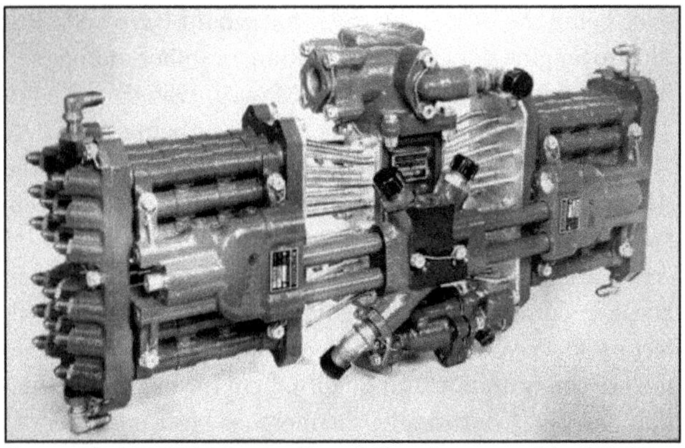

Figure 12: J46-WE-8 Fuel splitter/oil cooler assembly. WAGT P49288, (not dated). *Courtesy Hagley Museum and Library*

Another problem report came in on a dual fuel pump that indicated insufficient pump output. The pump had been installed on -8B WE404711 but was installed on engine WE040039 when it was returned to WAGT.

Investigation of the pump showed no discrepancies in materials, assembly or function. It was installed on -8B engine WE040023, and the engine ran in a test cell satisfactorily and the pump returned to active service.

WAGT suspected the erratic operation was caused by a sticking valve in the primary fuel regulator. A foreign substance could cause the valve to stick in the open position, allowing the engine fuel to by-pass to the pump inlet - thus creating the indication of a malfunctioning pump. Upon removal of the pump, the back wash could free the sticking valve of the foreign substance in the control.[152]

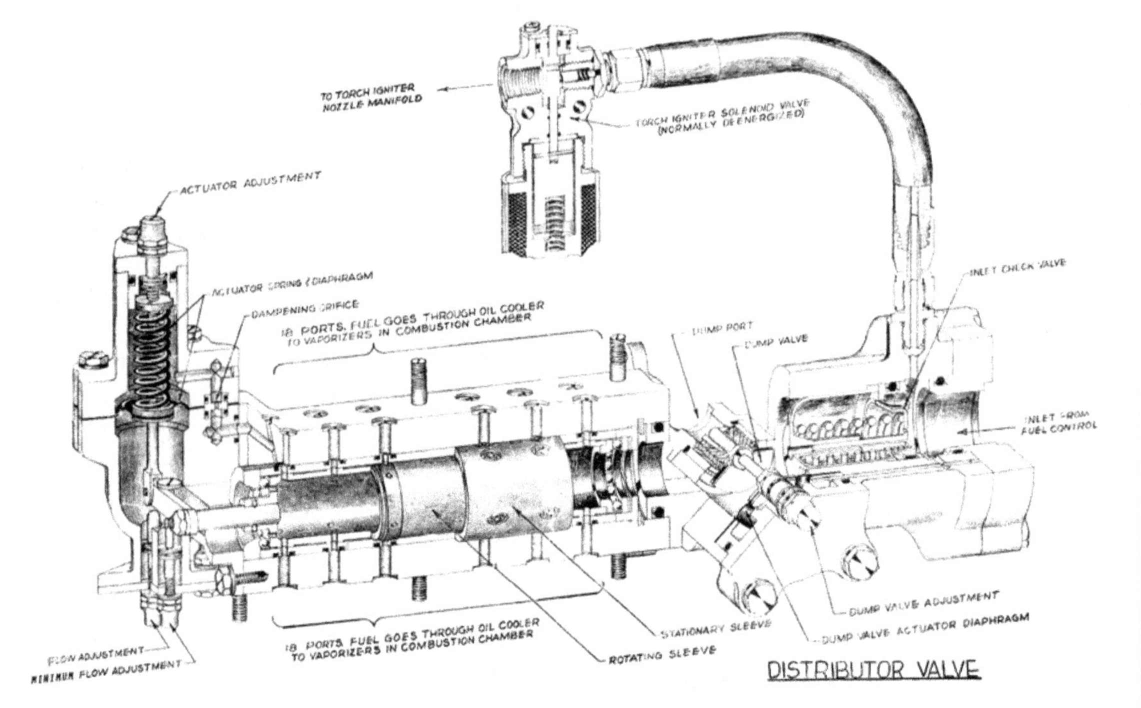

Figure 13: The fuel distributor valve that went between the two oil coolers in the prior figure. Just getting the right volume of fuel to each evaporator walking stick required high mechanical precision. WAGT P49317, 5/4/1953 from EDS-M-211704. *Courtesy Hagley Museum and Library*

A financial report dated June 30, 1955 showed the value of the NOa(s) 10825 contract as $127,576,827.02.[153]

The MPPR for June reported only nine (9) engines were shipped in June due to a work stoppage that occurred beginning on June 12th and that was still in effect.[154]

In early June WAGT surveyed a change to the current method of securing the vaporizer plate retaining bolts by tack-welding them to the tab washer and replacing that process with one using a modified tab washer with adequate clearance to ensure a tight fit to the bolt. The existing process made it very difficult to replace securing bolts in the field.[155]

As of July 6, 1955, only fourteen (14) -8B's remained to be modified and were awaiting parts in order to be completed.[156]

Back in May WAGT had suggested changing the military rpm speed range in the specifications from 10,100 +/- 100 rpm to 10,150 +/- 50 rpm and the off-speed performance correction factors be waived if the engine speed was maintained at 10,150 +/- 25 rpm during acceptance testing. BuAer accepted the specification change on July 8.[157, 158]

Isolated problems continued to demand WAGT analysis. A -8 engine, WE405556, suffered three consecutive fuel distributor diaphragm failures. The failures resulted in the engine attaining a maximum obtainable RPM of 90% at the Military Power Setting. The valve was replaced and the defective one analyzed in Essington. Metal particles of unknown source were found on the rotary valve and in the valve actuator chamber. When the valve was flushed out and reassembled, it operated normally. Such particles could have caused the rotary valve to stick and create a sufficient pressure differential across the diaphragms to cause their failure. WAGT recommended the fuel system of the engine be examined for contamination.[159]

Next in line was an examination of failed high pressure oil tube assemblies on -8 engines. Examination of these lines on operating field engines found the same condition as the failed lines: a tooling scratch in the root of the bellows convolution, which, when followed by the ballooning of the section, could cause a stress concentration in the root and eventual failure of the bellows at that point. Further analysis showed that the current 3,000 psi oil pressure was exceeding the yield strength of the bellows but that the pressure could be lowered to 2,000 psi with no ill effect on engine operation. A parts kit and Engine Bulletin 133 were sent out to allow the field to lower the oil pressure to 2,000 psi on the engine without removal of the engine from the airframe. A written note on the routing sheet states that WAGT was asked to investigate replacing the rigid tubing with synthetic oil resistant hose as requested by the Maintenance Division.[160]

Auxiliary Electric Control failures were attributed to a defective A/B lockout solenoid, where the failure to pull in within the normal time would cause it to draw over 8 amperes of current, causing various control circuits to burn out. The cause of the sticking solenoid was attributed to minute particles in the fuel.

A new one (1) ampere solenoid had been designed and satisfactorily tested which eliminated the high current characteristics of the previous type. It was recommended the new solenoid be retroactively fitted to all J46 engines.[161]

The MPPR for August states that no -8B engines were shipped that month due to the continuing work stoppage while eight -8 engines were accepted but none shipped due to the same labor work stoppage.[162]

Reports of engine fires after engine shutdown had been coming in from the field and WAGT sent out a survey for a proposed fix to eliminate the cause. It had been found that from one to twelve minutes after engine shutdown a fire could start, the result of fuel boiling out of the A/B fuel system after shutdown and then being ignited in the hot environment. Such fires were most likely to occur when the engine was shut down directly after A/B operation. A surge check valve was to be incorporated in the A/B fuel system to prevent hot fuel from being boiled out of the system into the engine after shutdown. The change would drain an additional 600 cubic centimeters of fuel into the dump tank in addition to the normal 50 to 100cc normally pumped by the A/B pump at engine shutdown. The new system had been tested in a test cell as well as in an F7U-3 and no fires had occurred under conditions known to have always consistently caused a fire in the past.[163]

Chance Vought objected to the additional use of the dump tank to take the purged fuel, as the tank was sized to take approximately 1,500cc of liquid for two false starts, one shutdown and 200cc for two A/B starts. Calculations showed that two preflight A/B shutdowns would dump 1,200 cc of A/B fuel instead of 200cc giving a total in the circumstances of 2,900cc. The current F7U-3 dump tank could only hold 2,570cc of liquid, so an overflow could occur.[164]

BuAer determined that the dump tank would be sufficient for a normal pre-flight run. If field activities determined that in normal usage the tank was inadequate, the situation would be re-reviewed. Until then, the appropriate maintenance manuals were to be updated to require draining of the dump tank with the manual drain valve prior to duplicating pre-flight or other engine run-up tests.[165]

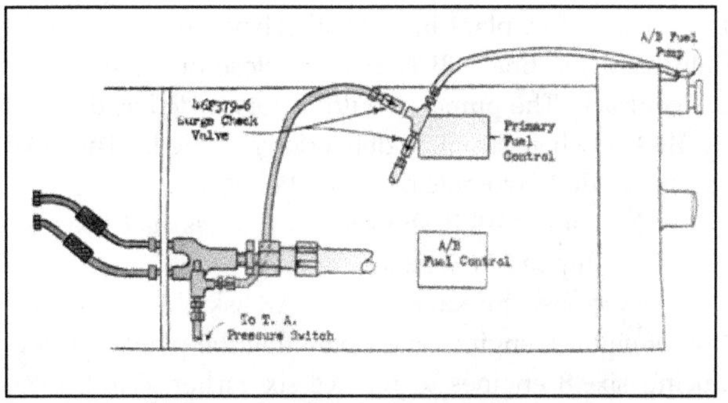

Figure 14: A/B fire prevention fix check valve locations. WAGT EDS-K250640, 8/2/1955.

A series of reported engine control failures were also investigated:

1. Failure of a Microsen unit was determined to be due to an open circuit in the feedback coil lead wire due to a mechanical failure of the wire.
2. A faulty electric control assembly for which the cause could not be determined but was felt to be an isolated case.
3. A dual fuel pump that malfunctioned had been found to have small steel particles inside. These were found to be shot blast material that entered the pump housing during the supplier's manufacturing process. Once cleaned, the pump operated properly.

WAGT determined the above failed items were within the contract guarantee and repaired or replaced them at no cost.[166]

A more serious turbine failure occurred to the starboard engine of Bu.No. 129658 on June 13, 1955. The engine (-8B WE404937) was examined at China Lake and was found to have massive damage to both the turbine guide vanes and blades. The A/B was coated with molten metal. At the time of the incident, the engine had 26.3 operating hours of which 4.25 had been in A/B. The engine had not had the Phase-I examination yet ("pre-Phase I") but had had Engine Bulletin 100 applied.

The aircraft had been climbing while in A/B in a turn to the left at 28,000 ft. The pilot described a rumble "like the beginning of a compressor stall" as the turn was executed.

After examining each of the damaged components, WAGT was able to determine that the likely cause was a compressor stall while in A/B resulting in a severe overtemp of the engine. The initiating cause was the likely blocking of the intake (from an out of trim condition) while entering the turn.[167]

In the first part of September, WAGT had requested authority to deliver the final 14 -8B engines minus their variable displacement pumps due to labor stoppages at the Vickers, Inc. plant in Detroit. They were desirous of delivering the final -8B engines to close out that part of the contract. The pumps would be forwarded as directed by BuAer when supplier deliveries resumed. The BAR recommended favorable consideration.[168]

The latest MPPR showed no engines had been accepted or shipped in September.[169]

In October the Kansas City BAR asked for an allocation of eight -8 engines and noted that the plant was producing six -8 engines to replace six earlier YJ46-WE-8A engines that were loaned back to WAGT.[170] All 14 engines were sent as spares to ComAirPac and ComAirLant.[171] There were 290 -8 engines produced in total.

The October Monthly Production Progress Report showed two -8B engines shipped with 14 remaining to be produced to complete production. Twenty -8 engines were produced with six engines left to complete production. Delays due to strikes at the Excello Corporation (A/B Fuel Controls, Igniter Valves and Exhaust Nozzle Controls) would affect deliveries in November at a minimum due to the need for these parts for green runs.[172]

WAGT issued a survey regarding the removal of the fuel boost pump from the engine. Bearings on the pump had been continually failing during actual service and WAGT determined the cause was due to the washing of the grease out of the bearing due to fuel in the seal drain cavity. It was suspected that the airframe vent system did not vent the cavity completely under all flight conditions. As well as studying how to improve venting and sealing of the pump, complete removal of the pump itself was also studied. The survey proposed removal of the pump and rerouting fuel flows to provide proper cooling flows to the A/B fuel pump. Engine weight was to be reduced by 3.1 pounds. It was noted removal of the pump would eliminate a high fuel pressure problem that occasionally prevented the speed switch from closing. With the airframe boost pumps operating normally, flight operation of the engines would be normal.

Flight tests showed that with the pump removed and the airframe pumps inoperative:

1. Operation below 10,000 ft would be normal except for reduced acceleration rates and a slightly reduced idle rpm. Acceleration from idle to full power (A/B) would take 1.5 seconds longer at sea level due to the lower idle rpm. Military rpm would drop to 99.2% at 6,000 feet.
2. Above 10,000 feet the maximum military rpm would be 95% of normal. Rough A/B would occur from about 15,000 feet and up with flameout occurring at a minimum altitude of 33,000 feet. Engine flameout would occur at 35,000 feet.
3. If aviation fuel was used (an emergency alternative fuel), the dual fuel pump would cavitate at sea level with fuel temperatures at or slightly above those normally encountered.[173]

Chance Vought agreed to the pump removal as long as BuAer agreed that the performance of the normal operation of the engines would be as represented by WAGT and that specification MIL-E-5007A was met. This required "performance with no assistance from the Airplane Boost Pump" be satisfied. They noted an engineering change to the airframe would be needed for removal of existing boost pump vent and drain systems and plugging, capping or changing of T-fittings to elbows to close openings in the remaining systems.[174]

Looking ahead, BuAer apparently agreed to the pump removal as follow-on memos discuss further developments. It was discovered that without the boost pump in the circuit, a smooth flow adapter was necessary to eliminate low pressure issues, but none was immediately available. As a result, WAGT had the internal workings of the boost pump removed and used it temporarily as a flow-smoothing device, ultimately planning on replacing the pump with a purpose designed flow-smoother adapter and then rerouting the various fuel lines.[175], [176]

Months later, BuAer complained that the changes to the engine specifications submitted for the -8B and -8 did not include the engine weight adjustment for the removal of the boost pump. They asked that the specifications be adjusted or an explanation of why no weight reduction was appropriate be submitted.[177] WAGT responded a few weeks later and stated the weight reductions on the engines were .08 pounds for the -8 and .45 pounds for the -8B. The reduction was considered too small to warrant a modification of the specification.[178]

The November MPPR shows twelve (12) -8 as well as eleven (11) -8B engines shipped in November.[179]

December's MPPR shows five -8 engines shipped against a forecast of eight, leaving only three engines left to complete the contract.[180]

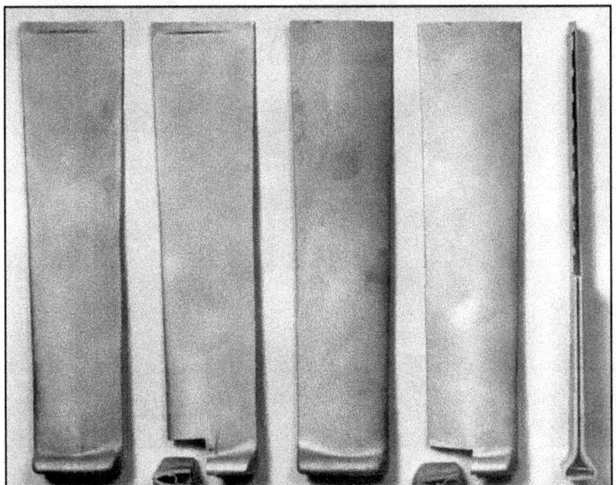

Figure 15: Experimental hollow aluminum-silicon alloy sheet metal compressor blades. WAGT also tried similar blades on the J40 (also unsuccessfully). WAGT P46998, 5/27/1952. ***Courtesy Hagley Museum and Library***

It was found that certain missions of the F7U-3 airplanes made it necessary to operate the -8 beyond its design limits. As a result, WAGT was asked to conduct tests on the -8 engine beyond design its limits to determine the durability of the hot section parts. Deemed "Hot Shot" tests, they were performed on a F7U-3M at Olathe by WAGT, the tests costing BuAer $183,234.00.[181] While test result documentation was not identified, the author believes the results of these tests led to the earlier increase in the standard turbine inlet temperature limit to 1,525°F as the tests to support that increase were run at up to 1,550°F without damage to the turbine elements.

On the same subject as hot section durability, Chance Vought wrote a memo to BuAer stating that in November and December of 1955, BuAer had sent out dispatches placing limits on the tactical operation of the F7U-3M aircraft through restriction on military and maximum power duration. They pointed out that the -8 and -8B engines had been cleared for unrestricted operation in A/B since February 1955 but Chance Vought had inadvertently failed to make that clear in the applicable Flight Handbook.

They added that BuAer, WAGT and themselves had been aware for several months that a high percentage of the operating time on F7U-3 and -3M aircraft was at military and maximum power. WAGT tests had shown this was not particularly detrimental to the engine at the current turbine inlet limits because the bulk of hot section deterioration was the result of thermal shock rather than continuous high temperature operation. The 120-hour inspections of engines flown in service without restrictions on continuous operation had shown no increase in hot section deterioration. They added that -8 and -8B service engines operating at turbine inlet temperature limits between 1,490-1,525°F had demonstrated exceptional durability.

They asked that BuAer issue a dispatch to all affected activities that no time limit was required on military or maximum power at 1,525°F or lower turbine inlet temperature other than that imposed by airframe fuel availability.[182]

January 1956's MPPR shows all J46 new engine production was now complete with no annotations.[183]

In late December, WAGT proposed updating the J46 performance specifications based on all the performance increases realized to date and also the incorporation of a new "combination flameholder" in the A/B. A written note on the routing sheet states "will back-fit or not prior to revising spec. Recommend we push for these new ratings at earliest practicable date. This will further weaken CVA's arguments about the thrust deficiency of F7U-3 power plants."

Another note reads: "These ratings are based upon test cell data taken on 35 consecutive production engines operated without modified afterburners. The 'Combination Flameholder' comprises a redesigned fuel manifold plus reworking part of the flame holder bars and removing others. A/B's with some bars removed have been evaluated by flight tests. The new (oval) type fuel manifold will be verified by tests in the very near future. This modification provides +50# thrust at Military and +90# in A/B."[184] With the A/B modifications installed, the -8B and -8 would produce 97.9% of the original XJ46-WE-2 proposed A/B thrust.

J46-WE-8, -8B, -12B Proposed Changes in Performance Ratings (plus additional A/B redesign modifications)

Rating	Reference RPM	Recommended New Thrust Specification	Current Thrust Specification	Projected After A/B Modification	New SFC	Current SFC Specification
Maximum	10,150	5,880 (lb)	5,725 (lb)	5,970 (lb)	2.54	2.56
Military	10,150	4,140	4,020	4,190	1.09	1.12
Normal	10,150	3,730	36,20	---	1.083	1.115
90% Normal	10,150	3,355	3,260	---	1.095	1.135
75% Normal	10,150	2,795	2,715	---	1.182	1.260
Idle	4,000	245	245	---	980 lb/hr/lb	980 lb/hr/lb

Clean-up of old qualification test discrepancy items continued into 1956. The cold tests of the fuel control system had shown that while Goshen "O" rings had solved most of the leak problems on the controls, unless additional changes were made in design to the throttle shaft seal on both the main control and A/B control, those items could only meet the specification if set to -40°F with 2% aromatic fuel. WAGT requested the specifications be changed to reflect that limit until the design changes could be made and successfully tested. Also, the dual fuel pump needed to be retested to the -65°F limit.[185]

Figure 16: Early first stage turbine blade design with ground leading edge from mean diameter to tip section. J46 model unknown. WAGT P47021, 5/29/1952. *Courtesy Hagley Museum and Library*

WAGT issued report A-2140-1 that covered the investigation of a -8 Phase II second stage turbine disc failure that occurred in the field. Materials and stress testing of the disc did not reveal a cause for the crack failure in the disc. Test cell testing of an identical disc at 140 hrs of a planned 240 hour test had not turned up any discrepancies on the disc. All house engine second stage turbine discs had been examined and none showed any problems. While testing was still continuing, WAGT felt the problem was an isolated one and not a concern as to safety for the discs in use.[186]

The low-pressure boost pump flow-smoother adapter design was completed April 24, but a note on the back of the memo states it did not appear it would be available in time to be of much value.[187]

An incident occurred on a -8B where two hoses with same size connectors were cross-connected to the fuel and oil systems, resulting in an engine fire. At BuAer's request, WAGT responded with a table of possible methods to reduce the chance of such interconnection errors.

Methods for Reducing Interconnection Errors

Method	Fuel	Lube Oil	Hydraulic	Air
I – Sizes	-4, -12, -16, -20	-8 (Supply -10 (Scavenge	-6	-3
II – Threads	Rt. Hand Fine Thread	Right Hand Coarse Thread	Left Hand Coarse Thread	Left Hand Fine Thread
III - Special Fittings	Radial sealing spiggoted "O" ring	AN or WECO flange fittings	AN 37° cone seat fittings	Eremeto flareless fittings
IV – Color* coding tape	Red	Yellow	Yellow w/blue stripe	Orange

* Matched to the color dyed anodized on the mating fittings.

WAGT recommended going back to Method IV which had previously been used (but discontinued for some unexplained reason by WECO CID 46007). The method was easiest to implement with lower cost, considering the current status of production.

They warned that any solution to the problem of inadvertent errors during interconnection of hoses had to be complete, as a partial solution would not preclude such errors on other parts of the engine aside from the two interconnections that triggered the study. Everyone involved with maintenance and/or overhaul had to be made aware of the importance of making the corrections accurately and the likely results of any given misconnection.[188]

Late engine improvements continued in June with WAGT surveying the replacement of flanged and radial sealing fittings in the A/B fuel system. The change was being made to reduce the chance of fuel leakage in the system. The change would use flanged and special radial sealing fittings, both using "O" rings at the connections of the A/B fuel lines.[189] BuAer rejected the change on the basis of cost and the delays accompanying redesign and procurement unless the change was absolutely necessary.[190]

With the improved A/B ignition sequence using a trigger anticipator circuit, a study was made to determine the need for the A/B ignitor nozzle. The study (report A-2186) showed that in 552 A/B ignition attempts with the fuel ignitor nozzle inactivated over 108 flights, 521 attempts (94.3%) were successful, with and without the trigger anticipator. It was recommended the ignitor nozzle be removed from production engines for system simplification, weight conservation and elimination of a possible fire hazard.[191] WAGT recommended the ignitor relay nozzle be removed at first overhaul and BuAer accepted the recommendation.[192]

At the same time as the study of the need for the ignitor relay nozzle, it was brought to WAGT's notice that flight experience was showing that if the engine was operated in A/B, then the throttle pulled back quickly to below 50% and then a slam acceleration back to A/B was attempted, the A/B would not ignite. It was determined that since the ignitor charging system only activated between 50% and military throttle settings, a throttle slam from below 50% after the previous action would not give the A/B ignitor time to recharge. Pilots were using various techniques to work around the deficiency, but they were not considered good practice. WAGT suggested adding a fuel line directly from the dual fuel pump outlet to the ignitor to keep it charged at all times (except when actually in A/B) instead of routing the fuel through the A/B control. The change had been tested with good results.[193] WAGT was told the change did not require a survey but then discovered it did and sent one out in the latter part of June.

In September WAGT surveyed replacement of the metal flex lines on the hydraulic actuator and actuator cooling lines with "Resistoflex" Teflon hose lines. The new lines could withstand pulsating pressures better than the current lines.[194]

An engineering change was processed in August to replace the existing A/B fuel manifold with a newer design that increased the life of the manifold by up to five times over the current design.[195]

Another CID in December recommended removing the new low pressure smoothers in the five -8 engines that received them. It was found they introduced hot spots in the engines that the older design did not. Since the older design worked well, BuAer accepted the change.

Of note in the removal acceptance memo, the BAR stated that all J46 engines during overhaul now incorporated forty ¼" diameter cooling holes in the turbine nozzle seal instead of twenty 1/8" diameter holes. The increase in cooling air would reduce the turbine disc temperature approximately 50°F.[196]

On February 1, 1957, BuAer accepted all the changes to the fuel controls for low temperature acceptance including the change to Goshen "O" rings. They stated no further testing was necessary.[197]

In March WAGT submitted a CID 44742 to replace the Teflon seal in the A/B shutoff valve with a seal split "T" Teflon seal of improved design. In a test of 15,000 cycles duration the new seal did not fail even once and in a limited service squadron test no failures occurred. BuAer approved the new design on April 26.[198]

With production ended, BuAer discovered that the contract had no clause in it for the delivery of production jigs and tooling to BuAer at the end of the contract. A price for the tooling and jigs was requested. If BuAer had no use for the tooling or jigs, WAGT was to be authorized to dispose of the material in their usual manner.[199]

So, with all production ended, even engineering work to improve the engines ground to a halt. Such work had materially improved performance and reliability but the engine was still marginally below the performance of the XJ46-WE-2 proposal and fuel performance was also marginally worse.

It is ironic that in the J46-WE-18 (next chapter) they had a much better, (almost) fully qualified engine that met the original performance guarantees (although to AN specifications for the most part), but it had matured too late and was not interchangeable with the installation envelope of the -8B or -8 without airframe changes. The only airframe that was targeted to take it, the Chance Vought A2U-1, was cancelled before first flight after one airframe was complete, in final checkout and the second about to be completed. All the time the J46 program played out, aviation advances eclipsed the airframes designed for its thrust output and economy. The military had moved on to require far greater power for supersonic flight and significantly improved SFC's for range extension.

Figure 17 A/B Inner heat shield showing A/B attachment points. *Courtesy Al Casby / Project Cutlass. Used with permission.*

Figure 18 A/B heat shield outer view. *Courtesy Al Casby / Project Cutlass. Used with permission.*

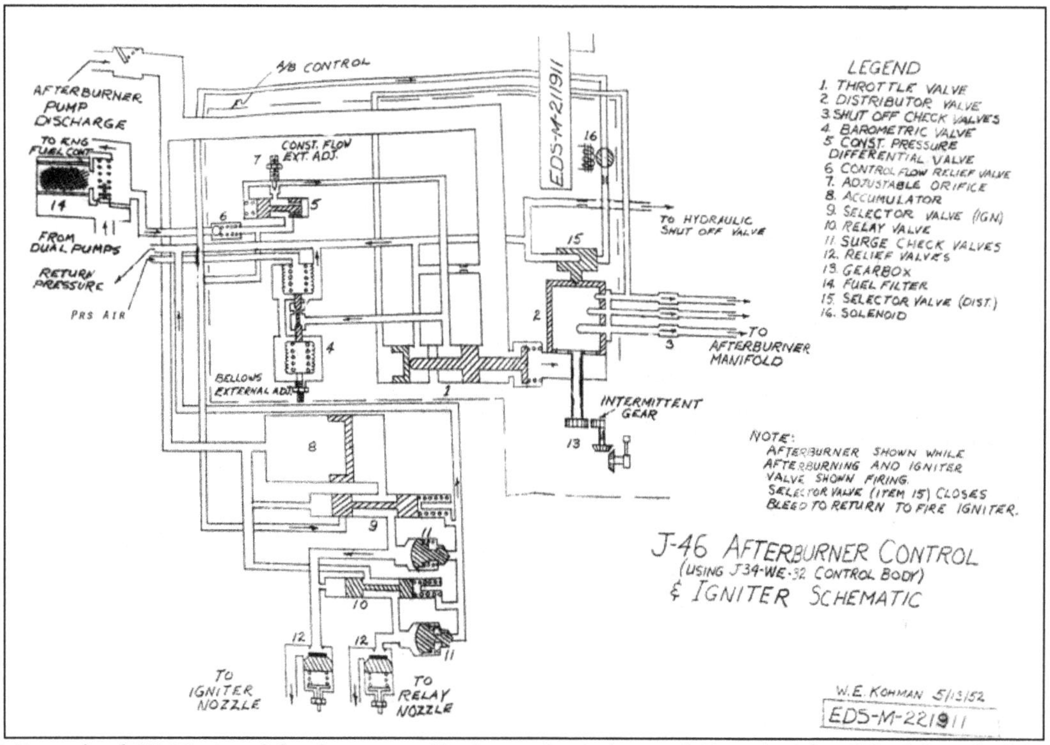

Figure 19: Example of WAGT artwork for the proposed hydro-mechanical control. Note the A/B still had three fuel rings. The bellows would later prove to fail repeatedly until extensively developed. No two charts used to describe the versions of the control used the same symbols for various components, adding to BuAer frustrations.
WAGT EDS-M-221911, 5/13/1952. Report A-1379

Figure 20 Intake extension installed on J46-WE-8 for F7U-3. Courtesy Al Casby / Project Cutlass. Used with permission.

Chapter 6 – J46-WE-8/A/B and -12/A/B Production and Service

[1] BuAer Memo, December 30, 1953, "Contract NOa(s) 10825, J46-WE-8 and J46-WE-12 engines, field limits for rejection of engines due to turbine rubs", **B1096VE8RG72.3.2NACP**
[2] BuAer Telegram, January 5, 1954, No subject, **B1097VE8aRG72.3.2NACP**
[3] WAGT Memo, December 21, 1953, "Bill of Materials, Transmittal of", **B1097VE8aRG72.3.2NACP**
[4] BuAer Memo, January 6, 1954, "Contract NOa(s) 10825 Lot III Items 3,8,9a and 9b, J46-WE-8B engines; partial allocation of", **B1097VE8aRG72.3.2NACP**
[5] WAGT Report A-1748, January 12, 1954, Failure Analysis of Electrical controls on J45WE8B Engines, **B1100VE6RG72.3.2NACP**
[6] BuAer Memo, April 26, 1954, "J46-WE-8B engine; Failure analysis of electrical controls", **B1100VE6RG72.3.2NACP**
[7] WAGT Memo, January 14, 1954, "Thrust Meter Interference", **B1100VE8RG72.3.2NACP**
[8] WAGT Memo, January 1, 1953, "Contract Noa(s) 10825. WEC Turbo Jet Engine Models J46-WE-8B and J46-WE-12B Operating Limits for C-89, T-39 Compressor - Turbine Combination, Information Concerning.", **B1100VE5RG72.3.2NACP**
[9] BuAer Memo, January 14, 1954, "J46WE-8A and J46WE-12B Engines Utilizing T-39, Recommendation for", **B1100VE5RG72.3.2NACP**
[10] WAGT Manual, Operating Recommendations for Models YJ46-WE-8A, J46-WE-8A and J46-WE-8B, January 15, 1954 revision. **B1097VE8aRG72.3.2NACP**
[11] BuAer Memo, February 10, 1954, "Turbo-Jet Engine Log Book, NAVEAR 2033, information concerning", **B1097VE8aRG72.3.2NACP**
[12] BuAer Naval Speed Letter, February 12, 1954, "Publications for Westinghouse Gas Turbine Engines - Contracts NOa(s) 10385, Noa(s) 52-048 and NOa(s) 10825", **B1097VE8aRG72.3.2NACP**
[13] WAGT report A-1701, January 20, 1954, Verification Test of Kansas City Manufactured J46-WE-8B Inlet Housing Assembly, **B1097VE8aRG72.3.2NACP**
[14] WAGT Memo, February 16, 1954, "Contract Noa(s) 10825, 53-556. WEC Turbojet Engine Model J46-WE-8A, J46-WE-8B, J46-WE-8, J46-WE-12A, J46-WE-12B and J46-WE-18. Front Mount Tolerance, Survey of. Engine Change Nos. 126, 15, 9 for the -8, -12, -18 Series Respectively.", **B1097VE8aRG72.3.2NACP**
[15] WAGT Memo, February 23, 1954, "Contract Noa(s) 10825, 53-556. WEC Turbojet Engine Models YJ46-WE-8A, J46-WE-8A, J46-WE-8B, J46-WE-8, J46-WE-12A, J46-WE-12B and J46-WE-18. Lubrication System Preservation of. Information On.", **B1097VE8aRG72.3.2NACP**
[16] WAGT Memo, February 19, 1954, "Contract Noa(s) 10825, 53-556. WEC Turbojet Engine Models YJ46-WE-8A, J46-WE-8A, J46-WE-8B, J46-WE-8, J46-WE-12A, J46-WE-12B and J46-WE-18. Oil Reservoir Capacity, Information On.", **B1097VE8aRG72.3.2NACP**
[17] BuAer Telegram, February 25, 1954, No Subject, **B1097VE8aRG72.3.2NACP**
[18] WAGT Memo, February 11, 1954, "Transmittal of Report of Cold Starting Performance of the J46-WE-8 Lubrication System", **B1100VE5RG72.3.2NACP**
[19] BuAer Memo, March 8, 1954, "J46WE-8A/-8B engines; Acceptance limits for fluorescent penetrant indicators on center hubs of front bearing support", **B1097VE9RG72.3.2NACP**
[20] WAGT Memo, March 8, 1954, "Contract NOa(s) 10825 Failure Analysis of Afterburner Pressure Switches for Westinghouse J46 Engines", **B1100VE6RG72.3.2NACP**
[21] WAGT Report A-1741-Z, March 15, 1954, 500 Hour Laboratory Qualification Test of the Microjet Pressure Switch. Contract NOa(s) 53-556 and NOa(s) 10825; J46-WE-8B, 8 and 18 Engines Contract NOa(s) 10385; J40-WE-22 Engine., **B1096VE4RG72.3.2NACP**
[22] BuAer Memo, April 28, 1954, "Contracts NOa(s) 53-556, 10825, and 10835, J45-WE-8B, -8, and -18 and J40-WE-22 engines; Microjet Pressure Switch Qualification Test", **B1097VE10RG72.3.2NACP**
[23] WAGT Report A-1700, January 22, 1954, A Study of the Removal of the Power Limiters from the J46-WE-8A, 8B, 12A, 12B Variable Delivery Oil Pump, **B1097VE9RG72.3.2NACP**
[24] WAGT Report A-1701, January 20, 1954, Verification Test of Kansas city Manufactured J46-WE-8B Inlet Housing Assembly, **B1097VE8aRG72.3.2NACP**
[25] BuAer Reproducible Manuscript Transmittal, March 18, 1954, Adapter Gearbox Overhaul Instructions and Illustrated Parts Breakdown, **B1098VE13RG72.3.2NACP**
[26] WAGT Memo, March 3, 1954, "Contract NOa(s) 10825. Westinghouse Turbo-Jet Engine Model J46-WE-8B, -8 and -18. Supplier Verification of Various Engine Components.", **B1097VE9RG72.3.2NACP**
[27] WAGT Memo, March 30, 1954, "Verification Test of Alternate Supplier for 60E788-8 Exhaust Nozzle Actuator for use on J46-WE-6A, -8B Engines on Contract NOa(s) 10825.", **B1097VE9RG72.3.2NACP**
[28] BuAer Telegram, April 2, 1954, No Subject, **B1097VE10RG72.3.2NACP**
[29] BuAer Memo, April 12, 1954, "Expediting of Westinghouse J46 Flight Test in F7U-3 Airframes at Chance Vought Aircraft, Incorporated", **B1097VE10RG72.3.2NACP**
[30] BuAer Shipment Order, April 13, 1954, BASO-35512/54, **B1097VE10RG72.3.2NACP**
[31] BuAer Shipment Order, April 13, 1954, BASO-35514/54, **B1097VE10RG72.3.2NACP**
[32] BuAer Shipment Order, April 14, 1954, BASO-35513/54, **B1097VE10RG72.3.2NACP**
[33] BuAer Shipment Order, April 14, 1954, BASO-35521/54, **B1097VE10RG72.3.2NACP**
[34] BuAer Memo, April 19, 1954, "Reuse of J46-WE-8 Afterburner Plywood Boxes, **B1097VE10RG72.3.2NACP**
[35] BuAer Memo, April 29, 1954, "TN-5", **B1094VA2RG72.3.2NACP**
[36] WAGT Memo, April 29, 1954, "Program to verify the use of a T-39 turbine "As Cast" first stage turbine blades on J46-WE-8B engine under Contract NOa(s)-10825.", **B1100VE6RG72.3.2NACP**

[37] BuAer Memo, May 4, 1954, "Contracts NOa(s) 10825 and 53-556 – Engine Pressure Limiters for J46-WE-8 series and J46-WE-18 engines; desired action concerning", **B1100VE6RG72.3.2NACP**
[38] WAGT Memo, September 9, 1954, "Contract Noa(s) 10825. Westinghouse Turbojet Engine Model J46-We-8B and -8. Retention of space for the Pressure Limiter, Information concerning.", **B1097VE12RG72.3.2NACP**
[39] BuAer Memo, May 18, 1954, "Contract NOa(s) 10825, qualification tests on the Westinghouse 61E370 starter-generator adapter gearbox", **B1097VE10RG72.3.2NACP**
[40] WAGT Memo, May 20, 1954, "NOa(s) 10825 Change of Eight J46-WE-8 Engines to J46-WE-12 Engines Our Reference WG-64730", **B1100VE6RG72.3.2NACP**
[41] WAGT Report A-1708, March 22, 1954, Qualification Test of J46-WE-8 Turbo-Jet Engine., **B1096VE8RG72.3.2NACP**
[42] BuAer memo, May 21, 1954, Contract NOa(s) 10825, Model J46-WE-8 Engines; limited acceptance of", **B1100VE6RG72.3.2NACP**
[43] WAGT Shipping Notice, May 27, 1954, BASO-37449/54, **B1099VE15RG72.3.2NACP**
[44] WAGT Memo, May 27, 1954, "Contract Noa(s) 10825. WEC Turbo-Jet Engine Models J46-WE-8A, -8B, -8, -12A, -12B. Thermal Expansion of Afterburner Ball Joint, Survey of. Engine change Nos. 131 and 19 for the -8 and -12 Respectively.", **B1097VE10RG72.3.2NACP**
[45] WAGT Memo, August 16, 1954, Contract NOa(s) 10825, WEC Turbojet Engine Models J46-We-12A, -12B, Afterburner Ball Joint Thermal Expansion.", **B1100VE7RG72.3.2NACP**
[46] Monthly Production Progress Report, May 1954, **B1100VE6RG72.3.2NACP**
[47] BuAer Memo, June 16, 1954, "NOas 10825. Engine Serial WE040013. Request for disposition of", **B1097VE11RG72.3.2NACP**
[48] BuAer Telegram, August 25, 1954, No Subject, **B1097VE11RG72.3.2NACP**
[49] BuAer Memo, June 21, 1954, "Contract NOas 10825; Verification test of alternate supplier for 60E788-8 Exhaust Nozzle Actuator for use on J46-WE-8A, -8B engines", **B1097VE11RG72.3.2NACP**
[50] BuAer Telegram, June 25, 1954, No Subject, **B1097VE11RG72.3.2NACP**
[51] BuAer Telegram, June 28, 1954, No Subject, **B1097VE11RG72.3.2NACP**
[52] BuAer Memo, June 29, 1954, Contract NOas 10825, Lot III, Items 8a and 9c J46-WE-8B Engines; partial allocation of", **B1097VE11RG72.3.2NACP**
[53] BuAer Memo, June 29, 1954, "Contract NOa(s) 10825, Lot III, Item 9C J46-WE-8 engines; partial allocation of", **B1097VE11RG72.3.2NACP**
[54] BuAer Telegram, June 29, 1954, No Subject, **B1097VE11RG72.3.2NACP**
[55] BuAer Memo, July 1, 1954, "Contract NOa(s) 10825 – J46-WE-8 engines; redesignation of", **B1097VE11RG72.3.2NACP**
[56] BuAer Memo, June 29, 1954, "Contract NOa(s)-10825, Model J46-WE-8 engines, continued acceptance of", **B1097VE11RG72.3.2NACP**
[57] Monthly Production Progress Report, June 1954, **B1100VE6RG72.3.2NACP**
[58] BuAer Telegram, July 7, 1954, No Subject, **B1097VE11RG72.3.2NACP**
[59] WAGT Memo, June 22, 1954, "Contract Noa(s) 10825, 53-556. WEC turbojet Engine Model J46-WE-8A, -8B, -12A and -12B. Revision of the Automatic Temperature Control Wiring. Survey of. Engine Change Nos. 134 and 22 for the -8 and -12 respectively.", **B1097VE11RG72.3.2NACP**
[60] BuAer Memo, July 8, 1954, "Contract NOa(s) 10825, 53-556, WEC Turbojet Engine Model J46-WE-8A, -8B, -12A and -12B. Revision of the Automatic Temperature Control wiring, Survey of, Engine Change Nos. 134 and 22 for -8 and -12 respectively.", **B1097VE11RG72.3.2NACP**
[61] WAGT Memo, July 27, 1954, "Contract Noa(s) 10825, 53-556. WEC Turbojet Engine Models J46-WE-8A, -8B, -12A, and -12B. Revision of the Automatic Temperature Control Wiring, Information on.", **B1097VE11RG72.3.2NACP**
[62] BuAer Speed Letter, July 12, 1954, "J46 engine mock-ups", **B1097VE11RG72.3.2NACP**
[63] BuAer Telegram, July 19, 1954, No Subject, **B1097VE11RG72.3.2NACP**
[64] WAGT Memo, July 27, 1954, "Contract NOa(s) 10825 change in -8 and -18 order quantities", **B1097VE11RG72.3.2NACP**
[65] WAGT Report A-1811, No Date (cover memo dated July 22, 1954), Accelerated Low Temperature Test of Parker 443-12 and Goshen 1120 O-rings in a J46 fuel regulator Contract NOa(s) 10825 Project 1739, **B1097VE11RG72.3.2NACP**
[66] BuAer Telegram, July 28, 1954, No subject, **B1097VE11RG72.3.2NACP**
[67] BuAer Telegram, July 29, 1954, No Subject, **B1097VE11RG72.3.2NACP**
[68] WAGT Monthly Production Progress Report, July 1954, **B1100VE7RG72.3.2NACP**
[69] WAGT Memo, July 23, 1954, "Verification of unaged first stage turbine blades in combination with T39 turbine on J46-WE-8B engine under Contract NOa(s)-10825", **B1100VE7RG72.3.2NACP**
[70] WAGT Report A-1696, June 10, 1954, Verification Test of J46-WE-12A and -12B Afterburner on Contract NOa(s) 10825, **B1100VE7RG72.3.2NACP**
[71] BuAer Memo, November 10, 1954, "Contract NOa(s) 10825 – Verification Test of J46-We-12A and -12B Afterburner", **B1097VE12RG72.3.2NACP**
[72] WAGT Monthly Production Progress Report, August, 1954, **B1100VE7RG72.3.2NACP**
[73] BuAer Telegram, September 3, 1954, No Subject, **B1097VE12RG72.3.2NACP**
[74] BuAer Telegram, September 10, 1954, No Subject, **B1097VE12RG72.3.2NACP**
[75] WAGT Monthly Production Progress Report, September 1954, **B1100VE7RG72.3.2NACP**
[76] BuAer Memo, October 1, 1954, "Contract NOa(s) 10825, Westinghouse Electric Corp., Kansas City Purchase Orders 55-YKC-9519-R, 62072, 5909R, 80696RM, 80592RM, 80668R, 66381RM, and 89162R on Eclipse-Pioneer Division, Bendix Aviation Corporation, Teterboro, N.J.", **B1097VE12RG72.3.2NACP**

[77] WAGT Memo, August 20, 1954, "Contract NOa(s) 10825, Westinghouse J46-WE-8A and WE-8B Turbo-Jet Engines; Exhaust Nozzle Hunting – Contractor's Comments On.", **B1097VE12RG72.3.2NACP**
[78] WAGT Memo, September 9, 1954, "Contract Noa(s) 10825 Westinghouse Turbojet Engine Models J46-WE-8A, -8B, -8, and -18. Incorporation of New Afterburner Fuel Pump Surge Check Valve and relocation of Surge Check Valve Customer Connection, survey of.", **B1097VE12RG72.3.2NACP**
[79] BuAer Memo, October 6, 1954, "Model J46 engines; thrust meter provisions", **B1100VE8RG72.3.2NACP**
[80] WAGT Memo, October 12, 1954, "Further Description of Project at Soundrive Engine Co.", **B1097VE12RG72.3.2NACP**
[81] BuAer Contract NOa(s) 10825, Amendment 22, October 20, 1954, **B1095VA6RG72.3.2NACP**
[82] WAGT Memo, October 26, 1954, "Contract NOa(s) 10825, Westinghouse J46-WE-8A and WE-8B Turbo-Jet Engines, Compressor Stall Resultant Overtemperature – Contractor's Comments On.", **B1097VE12RG72.3.2NACP**
[83] WAGT Memo, October 20, 1954, "Contract NOa(s) 10825, Westinghouse J46-WE-8A and WE-8B Turbo-Jet Engines; Fuel Boost Pump Part No. 22E694-13, Contractor's Comments On." **B1097VE12RG72.3.2NACP**
[84] WAGT Memo, October 26, 1954, "Contract NOa(s), Westinghouse J46-WE-8A and WE-8B Turbo-Jet Engines; Overtemperature operation of J46 engines – Contractor's comments on.", **B1097VE12RG72.3.2NACP**
[85] WAGT Memo, November 8, 1954, "Contract NOa(s) 10825, Westinghouse J46-WE-8A and WE-8B Turbo-Jet Engines, J46-WE-8B Engines; Evaluation of Improved Control Components.", **B1097VE12RG72.3.2NACP**
[86] BuAer Telegram, November 18, 1954, No Subject, **B1097VE12RG72.3.2NACP**
[87] WAGT Memo, November 18, 1954, "Contract NOa(s) 10825, Westinghouse J46-WE-8A and WE-8B Turbo-Jet Engines; Tools required for temperature traversing.", **B1097VE12RG72.3.2NACP**
[88] BuAer Memo, December 1, 1954, "NOa(s) 10825, Lot III, Item 9c, J46-WE-8B Engines; final allocation of", **B1097VE12RG72.3.2NACP**
[89] BuAer Memo, December 1, 1954, "TN-7", **B1094VA2RG72.3.2NACP**
[90] WAGT Monthly Production Status Report, November 1954, **B1100VE8RG72.3.2NACP**
[91] BuAer Contract NOa(s) 10825, Amendment 25, December 2, 1954, **B1095VA6RG72.3.2NACP**
[92] BuAer Telegram, December 3, 1954, No Subject, **B1095VA6RG72.3.2NACP**
[93] WAGT Memo, November 29, 1954, "Contract NOa(s) 10825, WECO Turbojet Engine Model J46-WE-8 Flowmeter Installation, Survey of", **B1097VE12RG72.3.2NACP**
[94] BuAer Telegram, December 9, 1954, No Subject, **B1097VE12RG72.3.2NACP**
[95] BuAer Memo, December 13, 1954, "Proposed Amendment No. 6 to Contract NOas 10825 (P.D. EN11-2675-52 – Production Division)", **B1094VA3RG72.3.2NACP**
[96] BuAer Memo, December 17, 1954, "Contract NOa(s) 10825 – Westinghouse J46-WE-8A and -8B Turbo-Jet Engines Turbine Blade Failure on J46-WE-8A Serial No. 404593", **B1097VE12RG72.3.2NACP**
[97] WAGT Memo, December 8, 1954, "Contract NOa(s) 10825, Westinghouse J46-WE-8A and WE-8B turbo-Jet Engines; Rejection of Engine Fuel Controls, Afterburner Fuel Controls, and Exhaust Nozzle Controls – Contractor's Comments On.", **B1097VE12RG72.3.2NACP**
[98] WAGT Memo, December 21, 1954, "Contract NOa(s) 10825, Westinghouse J46-WE-8A and WE-8B turbo-Jet Engines; Rejection of New Control Components, Contractor's Comments On.", **B1097VE12RG72.3.2NACP**
[99] BuAer Telegram, December 22, 1954, No Subject, **B1097VE12RG72.3.2NACP**
[100] WAGT Monthly Production Progress Report, December 1954, **B1100VE8RG72.3.2NACP**
[101] BuAer Telegram, January 25, 1955, No Subject, **B1098VE13RG72.3.2NACP**
[102] BuAer Contract NOa(s) 10825, Amendment 29, **B1095VA7RG72.3.2NACP**
[103] BuAer Telegram, January 26, 1955, No Subject, **B1098VE13RG72.3.2NACP**
[104] BuAer Memo, July 9, 1955, "Contract NOa(s) 10825, Phase II J46-WE-8B engines; status of", **B1099VE15RG72.3.2NACP**
[105] BuAer Telegram, February 1, 1955, No Subject, **B1098VE13RG72.3.2NACP**
[106] BuAer Memo, February 1, 1955, "Allocation of J46 engines for factory training program", **B1098VE13RG72.3.2NACP**
[107] BuAer Naval Speed Letter, February 4, 1955, No subject, **B1098VE13RG72.3.2NACP**
[108] WAGT Monthly Production Progress Report, January 1955, **B1098VE13RG72.3.2NACP**
[109] BuAer Telegram, February 1955, No Subject, **B1098VE13RG72.3.2NACP**
[110] WAGT Memo, November 8, 1954, "Contract NOa(s) 10825, Westinghouse J46-WE-8A, WE-8B, and WE-8 Turbojet Engines; Oil Filter Cover, Contractor's Comments On.", **B1098VE13RG72.3.2NACP**
[111] BuAer Naval Speed Letter, February 1955, No Subject, **B1098VE13RG72.3.2NACP**
[112] BuAer Memo, February 15, 1955, "61E370-8 Starter-Generator Adaptor Gearbox; Verification Test Report", **B1098VE13RG72.3.2NACP**
[113] BuAer Memo, February 23, 1955, "61E370-8 Starter Generator Adaptor Gearbox, Verification Test Report", **B1098VE13RG72.3.2NACP**
[114] BuAer Memo, March 4, 1955, "61E370-8 Starter Generator Adaptor Gearbox; verification test", **B1098VE13RG72.3.2NACP**
[115] WAGT Memo, December 30, 1954, "Contract NOa(s) 10825, WEC Turbojet Engine Models J46-WE-8B, -8. Trigger anticipator Afterburner Ignition System, Survey of.", **B1101VE9RG72.3.2NACP**
[116] BuAer Memo, February 18, 1955, "contract NOa(s) 10825 – WECO Turbojet Engines Models J46-WE-8B, -8 Trigger Anticipator Afterburner Ignition System; survey of", **B1098VE13RG72.3.2NACP**

[117] WAGT Engine Bulletin No. 107 (Preliminary Copy), February 16, 1955, **B1098VE13RG72.3.2NACP**
[118] WAGT Memo, February 11, 1955, "Contract NOa(s) 10825, Westinghouse J46-WE-8A and WE-8B Turbo-Jet Engines; Combustion Chamber Liners – Contractor Comments On.", **B1098VE13RG72.3.2NACP**
[119] WAGT Report A-1727, November 16, 1954, Endurance Test of a J46-WE-8A Turbo Jet Engine in accordance with MIL-E-5009A Specification on Contract NOa(s) 53-556, **B3123V9RG72.3.2NACP**
[120] WAGT Monthly Production Progress Report, February 1955, **B1098VE13RG72.3.2NACP**
[121] WAGT Memo, February 21, 1955, "Contract NOa(s) 10825, Westinghouse J46-WE-8A and WE-8B Turbo-Jet Engines; Turbine Damage due to relaxing of 68H38-1 tabwasher – Contractor's Comments On.", **B1098VE13RG72.3.2NACP**
[122] BuAer Memo, March 9, 1955, "NOas 10825, Lot III, Item 13a, J46-WE-8 Engines; partial allocation of.
[123] WAGT Memo, March 1, 1955, "Contract NOa(s) 10825, Westinghouse J46-WE-8A and WE-8B Turbojet Engines; J46-WE-8B Engines First Stage Turbine Blade Defects during field operations; Contractor's comments on.", **B1098VE13RG72.3.2NACP**
[124] BuAer Memo, March 11, 1955, "Contract NOa(s) 10825 – change in Military Flat Positions Specified in J46-WE-8, -8B, and -12B Model Specifications, request for approval of", **B1098VE13RG72.3.2NACP**
[125] WAGT Memo, March 10, 1955, "Contract NOa(s) 10825, Government Furnished Engine Shipping Containers", **B1098VE13RG72.3.2NACP**
[126] Chance Vought Aircraft Memo, March 1, 1955, "Contract NOa(s) 10825 – WEC turbojet Engine Models J46-WE-8B, -8, Trigger Anticipator Afterburner Ignition System, Survey of", **B1098VE14RG72.3.2NACP**
[127] Chance Vought Aircraft, April 20, 1955, "Contract NOa(s) 10825, Model WECO turbo-Jet Engine Models J46-WE-8B, -8 Trigger Anticipator Afterburner Ignition System, Survey of", **B1098VE14RG72.3.2NACP**
[128] WAGT Memo, March 1, 1955, "Contract NOa(s), Westinghouse J46-WE-8A and WE-8B Turbojet Engines, Afterburner Failures, Comments On.", **B1098VE13RG72.3.2NACP**
[129] WAGT Memo, March 17, 1955, "Contract NOa(s) 10825 – Westinghouse Turbo-Jet Engine Model J46 – Fuel Control Components, Engine Bulletins For.", **B1098VE14RG72.3.2NACP**
[130] WAGT Monthly Production Progress Report, March 1955, **B1098VE13RG72.3.2NACP**
[131] BuAer Naval Speed Letter, April 11, 1955, No Subject, **B1098VE14RG72.3.2NACP**
[132] WAGT Memo, April 1, 1955, "Contract NOa(s) 10825, Westinghouse J46-WE-8A and WE-8B Turbo-Jet Engines; Exhaust Nozzle and Oil Pressure Fluctuation, - Contractor's Comments On.", **B1098VE14RG72.3.2NACP**
[133] WAGT Memo, April 18, 1955, "Contract NOa(s) 10825 Westinghouse Turbojet Engine Models J46-WE-8, -8A, -8B, -12A, -12B Incorporation of Steel 1st Stg. Compressor Disc, Survey of.", **B1098VE14RG72.3.2NACP**
[134] WAGT Memo, July 11, 1955, Contract NOa(s) 10825, Westinghouse J46-WE-8A and J46-WE-8B Turbo-Jet Engines; Unsatisfactory Reports, Contractor's Comments On.", **B1099VE17RG72.3.2NACP**
[135] WAGT Monthly Production Progress Report, April 1955, **B1098VE14RG72.3.2NACP**
[136] WAGT Memo, April 14, 1955, "Contract NOa(s) 10825, Westinghouse J46-WE-8A and WE-8B Turbo-Jet Engines, ATC and AEC Failure, Contractor's Comments On.", **B1098VE14RG72.3.2NACP**
[137] WAGT Memo, April 29, 1955, "Contract NOa(s) 10825, Westinghouse J46-WE-8A and WE-8B Turbo-Jet Engines, Controls Investigation.", **B1098VE14RG72.3.2NACP**
[138] WAGT Memo, April 20, 1955, "Modernization of J46-WE-8B Engines", **B1098VE14RG72.3.2NACP**
[139] BuAer Memo, May 4, 1955, "Modernization of J46-WE-8B engines", **B1098VE14RG72.3.2NACP**
[140] BuAer Memo, May 9, 1955, "Contract NOa(s) 10825, Weekly Report of Engine Shipments, information on", **B1098VE14RG72.3.2NACP**
[141] WAGT Report A-1863, April 29, 1955, Altitude Temperature Inversion Testing of a J46-WE-8B Engine, **B1101VE9RG72.3.2NACP**
[142] WAGT Report A-1904, April 14, 1955, J46-WE-8 50-Hour Alternate Fuel Test Using JFC Fuel, Contract NOa(s) 10825, **B1101VE9RG72.3.2NACP**
[143] WAGT Memo, May 18, 1955, "Contract NOa(s) 10825, J46-WE-8 and WE-8B Engines, Turbine Temperature Revision", **B1101VE9RG72.3.2NACP**
[144] BuAer Memo, May 25, 1955, "Contract NOa(s) 10825 – J46-WE-8 and -8B Engines Turbine Temperature Revision", **B1101VE9RG72.3.2NACP**
[145] BuAer Telegram, June 24, 1955, "Contract NOa(s) 10825 – J46-WE-8 and -8B Engines Turbine Temperature Revision", **B1099VE15RG72.3.2NACP**
[146] WAGT Memo, March 1955, "Contract NOa(s) 10825, Westinghouse Turbo-Jet Engine Model J46-We-8A, Modification of (SNT) 10-59750-2 Ignition Units.", **B1099VE15RG72.3.2NACP**
[147] BuAer Naval Speed Letter, May 26, 1955, "Contract NOa(s) 10825 – Westinghouse Turbo-Jet Engine Model J46WE8A, modification of (SNT) 10-59750-2 ignition units", **B1099VE15RG72.3.2NACP**
[148] BuAer Memo, May 11, 1955, "Contract NOa(s) 10825 – Westinghouse turbojet Engine Model J46-WE-8 Alternate Supplier of Afterburner Exhaust Collector Weldment 64J300-17, Verification of", **B1099VE15RG72.3.2NACP**
[149] BuAer Memo, May 27, 1955, "Contract NOa(s) 10825 – Westinghouse Turbojet Engine Model J46-WE-8 Alternate Supplier of Afterburner Exhaust Collector Weldment 64J300-17, Verification of", **B1099VE15RG72.3.2NACP**
[150] WAGT Memo, April 19, 1955, "Contract NOa(s) 10825, Westinghouse J46-WE-8A and WE-8B Turbo-Jet Engines; J46-WE-8B Engine Microjet Pressure Switch Discrepancy; Contractor's Comments On.", **B1099VE15RG72.3.2NACP**
[151] WAGT Monthly Production Progress Report, May 1955, **B1099VE15RG72.3.2NACP**

[152] WAGT Memo, April 19, 1955, "Contract NOa(s) 10825, Westinghouse J46-WE-8A and J46-WE-8B Turbo-Jet Engines; Unsatisfactory Report, Contractor's Comments On.", **B1099VE15RG72.3.2NACP**
[153] BuAer Financial Status Report, June 30, 1955, **B1095VC1RG72.3.2NACP**
[154] WAGT Monthly Production Progress Report, June 1955, **B1099VE16RG72.3.2NACP**
[155] WAGT Memo, June 9, 1955, "Contract NOa(s) 10825, Westinghouse J46-WE-8A and J46-WE-8B Turbo Jet Engines; Insecurity of Vaporizer Plate Retaining Bolts, Contractor's comments on.", **B1099VE16RG72.3.2NACP**
[156] BuAer Memo, July 6, 1955, "Contract NOa(s) 10825, Phase II J46-WE-8B engines; status of", **B1099VE15RG72.3.2NACP**
[157] WAGT Memo, May 25, 1955, "Contract NOa(s) 10825 Westinghouse Turbo-Jet Engine Models J46-WE-8 and -8B. Engine Performance – Testing of", **B1099VE16RG72.3.2NACP**
[158] BuAer Memo, July 8, 1955, "Contract NOa(s) 10825, Westinghouse Models J46-WE-8 and -8B engines; performance during acceptance testing", **B1099VE16RG72.3.2NACP**
[159] WAGT Memo, July 7, 1955, "Contract NOa(s) 10825, Westinghouse J46-WE-8A and J46-WE-8B, Turbo-Jet Engines; Engine Fuel Distributor Valve Investigation, Contractor's Report On.", **B1099VE16RG72.3.2NACP**
[160] WAGT Memo, July 12, 1955, "Contract NOa(s) 10825, Westinghouse J46-WE-8A and J46-WE-8B Turbo-Jet Engines; Unsatisfactory Report, Contractor's Comments On.", **B1099VE16RG72.3.2NACP**
[161] WAGT Memo, July 12, 1955, "Contract NOa(s) 10825, Westinghouse J46-WE-8A and J46-WE-8B Turbo-Jet Engines; Auxiliary Electric Controls Failures; Contractor's Comments On.", **B1099VE17RG72.3.2NACP**
[162] WAGT Monthly Production Progress Report, August 1955, **B1099VE17RG72.3.2NACP**
[163] WAGT Memo, August 2, 1955, "Contract NOa(s) 10825 WECO Turbojet Engine Models J46-WE-8B, -8 Afterburner Shut down Fire Fix", **B1099VE17RG72.3.2NACP**
[164] Chance Vought Aircraft Memo, September 1, 1955, "Contract NOa(s) 10825 – Westinghouse Turbojet Engine Models J46-WE-8B and -8 Afterburner Shut-down Fire Fix, Survey of", **B1099VE17RG72.3.2NACP**
[165] BuAer Memo, September 29, 1955, "Contract NOa(s) 10825, Model F7U-3-3M Aircraft, Westinghouse Turbojet Engine Models J46-WE-8B and -8 Afterburner Shut-Down Fire Fix, comments on", **B1099VE17RG72.3.2NACP**
[166] WAGT Memo, August 23, 1955, "Contract Noa(s) 10825, Westinghouse J46-WE-8A, J46-WE-8B, and J46-WE-8 Turbo-Jet Engines; Investigation of Defective Engine Controls, Contractor's Comments On.", **B1099VE17RG72.3.2NACP**
[167] WAGT Memo, August 10, 1955, "Contract NOa(s), Westinghouse J46-WE-8A, WE-8B, and WE-8 Turbo-Jet Engines; Turbine Failure of J46-WE-8B Engine Serial WE404937, Contractor's Comments On.", **B1101VE9RG72.3.2NACP**
[168] WAGT Memo, September 1, 1955, "Contract NOa(s) 10825; Acceptance of J46-We-8B Engines", **B1099VE17RG72.3.2NACP**
[169] WAGT Monthly Production Progress Report, September 1955, **B1099VE17RG72.3.2NACP**
[170] BuAer Memo, October 3, 1955, "Contract NOa(s) 10825, Allocation J46-WE-8 engines; request for", **B1099VE17RG72.3.2NACP**
[171] BuAer Memo, October 17, 1955, "NOas, Lot VII, Item 29, J46-WE-8 engines; final allocation of", **B1099VE17RG72.3.2NACP**
[172] WAGT Monthly Production Progress Report, October 1955, **B1099VE17RG72.3.2NACP**
[173] WAGT Memo, November 4, 1955, "Contract NOa(s) 10825, WECO Turbo-Jet Engine Models J46-WE-8A, -8B and -8 Engine Boost Pump Removal, Survey of", **B1101VE9RG72.3.2NACP**
[174] Chance Vought Aircraft Memo, December 6, 1955, "Contract NOa(s) – 10825 – WECO Turbo-jet Engine Models J46-WE-8A, -8B and -8 Engine Boost Pump Removal, Survey of", **B1099VE17RG72.3.2NACP**
[175] WAGT Memo, June 21, 1956, "Contract Noa(s) 10825. Westinghouse Turbojet Engine Models J46-WE-8B, J46-WE-8. Incorporation of Smooth Flow Adapter, Information Concerning", **B1101VE10RG72.3.2NACP**
[176] WAGT Memo, June 21, 1956, "Contract NOa(s) 10825. Westinghouse Turbojet Engine Models J46-WE-8B, J46-WE-8. Incorporation of Smooth Flow Adapter, Information Concerning", **B1101VE10RG72.3.2NACP**
[177] BuAer Memo, November 27, 1956, "Dry Weight of the J46-WE-8 and -8B Engines", **B1101VE10RG72.3.2NACP**
[178] WAGT Memo, January 3, 1957, "Dry Weight of the J46-WE-8 and -8B Engines", **B1101VE10RG72.3.2NACP**
[179] WAGT Monthly Production Progress Report, November 1955, **B1099VE17RG72.3.2NACP**
[180] WAGT Monthly Production Progress Report, December 1955, **B1099VE17RG72.3.2NACP**
[181] BuAer Memo, January 4, 1956, "High Temperature Test of J46-WE-8 Engines", **B1099VE17RG72.3.2NACP**
[182] Chance Vought Memo, January 17, 1956, "Contract NOa(s)-10825, Model WECO turbo-Jet Engine – Models J46WE-8 and -8B – Operating Restrictions", **B1101VE10RG72.3.2NACP**
[183] WAGT Monthly Production Progress Report, January 1956, **B1101VE10RG72.3.2NACP**
[184] WAGT Memo, December 22, 1955, "Contract NOa(s) 10825; Westinghouse J46-WE-8 Model Specification WAGT-24C10-6D and J46-WE-8B, -12B Model Specification WAGT-24C10-14, proposed changes in performance ratings for", **B1101VE10RG72.3.2NACP**
[185] WAGT Memo, March 6, 1956, "Contract NOa(s) 10825, Amendment 2 to Westinghouse Specifications WAGT-24C10-6D and WAGT-24C10-14 covering the Models J46-WE-8, J46-WE-8B and J46-We-12B Engines respectively", **B1101VE10RG72.3.2NACP**
[186] WAGT Report A-2140-1, February 29, 1956, J46-WE-8 Phase II Second Stage Turbine Disc Failure, **B1101VE10RG72.3.2NACP**
[187] WAGT Memo, April 30, 1956, "Contract NOa(s) 10825, WECO Turbojet Engine Models J46-WE-8B and J46-We-8, Low Pressure Drop Boost Pump Adapter, Status of", **B1101VE10RG72.3.2NACP**
[188] WAGT Memo, June 8, 1956, "Contract NOa(s) 10825, WECO Turbojet Engine Models J46-WE-8B, J46-WE-8, Interchangeability of Fluid Line Connections.", **B1101VE10RG72.3.2NACP**
[189] WAGT Memo, June 19, 1956, "Contract NOa(s) 10825. Westinghouse Turbojet Engine Models J46-WE-8, J46-WE-8B. Incorpo-

ration of Flanged and Radial Fittings in A/B fuel system, Survey of", **B1101VE10RG72.3.2NACP**
[190] BuAer Memo, January 7, 1957, "Contract NOa(s) 10825; Westinghouse Turbojet Engine Models J46-WE-8 and J46-WE-8B. Survey of Incorporation of Flanges and Radial Sealing Fittings in Afterburner Fuel System; additional information concerning", **B1101VE10RG72.3.2NACP**
[191] WAGT Report A-2186, April 16, 1956, Flight Test of the Effects of J46 Relay Hot Streak Removal, **B1101VE10RG72.3.2NACP**
[192] BuAer Memo, June 28, 1956, "Westinghouse Engineering Change Proposal Number 44286, Removal of Hot Streak Relay, J46 Engines; approval of", **B1101VE10RG72.3.2NACP**
[193] WAGT Memo, May 2, 1956, "Contract NOa(s) 10825, WECO Turbojet Engine Models J46-WE-8B, J46-WE-8 Afterburning Ignitor Valve Accumulator Recharge Improvement, Survey of", **B1101VE10RG72.3.2NACP**
[194] WAGT Memo, September 11, 1956, "Contract NOa(s) 10825, Westinghouse Turbojet Engine, Models J46-WE-8B. Installation survey of Incorporation of "Resistoflex" Teflon Hose Lines between the Firewall and the Exhaust Nozzle Actuator.", **B1101VE10RG72.3.2NACP**
[195] WAGT CID 3343, August 23, 1956, **B1095VB1RG72.3.2NACP**
[196] BuAer Memo, January 16, 1957, "Westinghouse Change Proposal; submittal of", **B1101VE10RG72.3.2NACP**
[197] BuAer Memo, February 1, 1957, "Qualification of J46 Fuel Controls Incorporating Goshen "O" rings; approval of", **B1101VE10RG72.3.2NACP**
[198] BuAer Memo, April 26, 1957, "Westinghouse Engineering Change Proposal No. 44742, J46 Engines, Modification of Afterburner Fuel Valves; approval of", **B1101VE10RG72.3.2NACP**
[199] BuAer Memo, June 28, 1957, "J46 Tools, Dies, Jigs and Fixtures", **B1101VE10RG72.3.2NACP**

Chapter 7
J46-WE-18 Development

As previously mentioned, the ongoing effort to extract the guaranteed performance from the -2 design had uncovered the fact that a larger turbine design was needed. It was known at that point that the -2 compressor stall boundary was very close to the operating line and while the -2 engine could operate at or near sea level without stalling, above 20,000 feet it suffered from stalling under acceleration and was very sensitive to inlet flow disruption when the aircraft was maneuvering or flew through disturbed air. The enlarged turbine would improve turbine efficiency which improved the stall boundary margin.

When BuAer elected to accept a de-rated -2 as the -8, work continued on the -2 to meet the original performance goals, but the engine (and its successor the -18) languished as the -8 moved center stage in the effort to get the delayed F7U-3 into the air and into service use.

It turned out that additional design changes were going to be necessary on the -2 to achieve a reliable performance. The move to a larger turbine was matched to a 13 stage compressor with a compression ratio of 5.30:1. Air flow per second was slightly increased. The combination finally lowered the workload per compressor stage. It moved the surge line far enough to greatly reduce altitude compressor stalling while meeting the -2 performance guarantees with a comfortable margin. The new configuration was eventually to be designated the J46-WE-18. The extra effort to bring the new -18 model to qualification status delayed it from being ready for any but the Chance Vought A2U-1, the ground attack version of the F7U-3. When the A2U-1 was cancelled the engine was left an orphan with no other application and it was also cancelled. Advanced versions of the -18 were schemed by WAGT in an attempt to maintain and broaden the appeal of the engine. These advanced engines incorporated features carried over from the Rolls Royce engines. Today the -18 is a forgotten program and WAGT's schemed successors all but unknown.

Development of the -18 was under its own contract, NOa(s) 53-556, although some items remained chargeable to NOa(s) 10825. The specification started out as WAGT-X24C10-8 covering the -2 and -16. The J46-WE-16 replaced the J46-WE-12 and was later replaced itself by the J46-WE-20 as the planned production engine for the Convair XF2Y-1 had the Sea Dart gone into production.

Basically, the -18 was an aerodynamic redesign of the earlier -2. During the early -2 development, it appeared that the basic -8 could be made to perform at close to the -2 guarantees. On June 25, 1953, BuAer wired WAGT (through the BAR) that they could accept modified -2 performance ratings of:

(J46-WE-8 w/ J46-WE-2 A/B)		
Performance	**Rating**	**SFC**
Maximum	5,950	2.70
Military	3,980	1.10
Normal	3,560	1.097
90% Normal	3,205	1.106
75% Normal	2,660	1.186
Idle	245	980 lb/hr

To achieve those new objectives, the engine was now to be the basic -8 with the hydro-mechanical control system and a new larger A/B design. The wire asked the BAR for an estimated cost for such an engine and indicated a separate engine model designation might be established if it was deemed warranted.[1]

Additional details on the design emerged in the WAGT survey of EC 21 at the end of July 1953. The engine (now referred to as the "J46-WE-2 (modified) configuration" would be the -8A/-8B to the aft part of the diffuser with a new engine combustion chamber design and the new, larger -2 turbine. The A/B would be the new -2 design (this had a larger nozzle). Length would now be 202.1 inches.[2]

The same day the above survey was sent out, BuAer informed WAGT that they intended to cancel the -2 (and -4) from Contract NOa(s) 10825 and the development of the J46-WE-2 (modified) would be continued under NOa(s) 53-556 under the likely designations of J46-WE-18 and J46-WE-20. BuAer believed the existing -8/-12 specification would in large part apply to the new engines.[3]

The updated specification for the -18 and -20 was the WAGT-124D and the first version was dated August 21, 1953. The WAGT transmittal gives a bit more information on the engines, stating the turbine would be the T-19 and they would use the improved altitude performance -2 A/B. The -20 would be a -18 with a three (3) degree cant to the A/B and was described in Appendix A of the specification.

On October 14, WAGT surveyed changing the -8, -18 and -20 to a thirteen stage compressor to address marginal altitude stall characteristics of the twelve stage compressor and "to provide additional engine growth characteristics not possible with the twelve stage compressor". The change added 1.7 inches to the engine length and 49.0 pounds of weight. The Microjet switch pressure pickup would be cast integral with the compressor casing.

The 13 stage compressor would not be on the first J46-WE-8's but would be incorporated in some future production block. (As events turned out, all -8 versions retained a 12 stage compressor throughout their production lives.)

For the -18 and -20, it would be incorporated in the first engines (for the -18, WE400001 and up, WE040501 and up; and for the -20 WE040301 and up).[4]

(NOTE: The -20 would have been the production version of the -18 with the features needed for the production Convair F2Y-1 Sea Dart.)

WAGT requested when the specification would be reviewed and BuAer responded that until the altitude performance, specific fuel consumption and operating limits curves had been received and reviewed they would not evaluate the specification.[5]

In the meantime, WAGT surveyed changes to the size of the Electric Control Assembly in order to conform to the actual size of the assembly. This was EC 4 for the -18 and -20.[6]

Since the -2 had had many EC's sent out and -8 engines were still having EC's issued, WAGT published lists of EC's for the -2 and -8 that were applicable to the -18 as well. They provided lists of EC's beginning from EC 74. Only those marked as applicable to the -18 are included in the table.[7]

On December 10, BuAer accepted the qualification test results (WAGT Report A-1634-1) for the engine fuel boost pump for all J46 engine models. This test was run to specification MIL-E-5009. In other running, the pump design had been through five other 150-hour qualification tests on various engines and exhibited excellent performance.[8]

The November progress report states the #9 engine was testing turbine blades with various radii on the trailing edge to find a solution to the cracking problem. Failures in the first stage turbine were still not understood. The blades came from Westinghouse, East Pittsburgh and the manufacturing techniques were under review. As a work around, blades were being seasoned in brief engine runs, then Zyglo inspected to weed out the defects. The proofed blades went into test engines to allow testing to proceed. Engine #10 was used for cyclic testing to model

J46-WE-2/-8 EC's Applicable to the -18 Engine

From -2	From -8	EC No.	Description
*	*	80	Change to Vaporizing Combustion Chamber
*	*	81	Change to Fireshield Contour
*	*	82	Incorporation of Iris Nozzle
*	*	85	Relocation of E/N Control & Aft Combustion Chamber Drain Valve
*	*	90	Change in Design of Engine Front Mount
*	*	91	Change Oil Filter Removal Clearance
*	*	92	Incorporate Spring-Loaded Quick Disconnect Clamp
*	*	94	Change in Front Mount Design
*	*	95	Incorporation of a High Energy Ignition System
*	*	105	Addition of Thermocouples, Thermocouple Circuit & Bosses
*		107	Dimension Changes On Larger Diameter Turbine
*		109	Single Port A/B Control & New Customer Contact Linkage
	*	110	Single Port A/B Control & New Customer Connection Lines
*	*	114	Rotation of Overspeed Relay
*		115	Forward Engine Mount Tolerance
*	*	116	Relocation of Flexible Joint
*		117	Removal of Jack Lugs & Forward Combustion Chamber Drain Valve
	*	118	Engine Jacklug Information & Removal of Forward Combustion Chamber Drain Valve
*	*	119	Turbine Temperature Traversing Bosses
	*	121	Accessory Gearbox Shaft Seals Vent Pressure Limits
*		121	New Engine Configuration J46-WE-2 Modified (J46-WE-18)
	*	122	Change Basic Engine Configuration to a 13 Stage Compressor
	*	123	Revision to Electric Control Assembly
	*	124	Removal of A/B Drain Plug
	*	125	Location of Ignitor Plugs and Nozzles

specification requirements but testing had to stop after 35 hours when (unspecified) compressor damage occurred.[9]

In January of 1954, BuAer received a proposal from the Marquardt Aircraft Company for seven (7) weeks of A/B test time in their facilities based on a proposal made to Westinghouse. BuAer, in a memo to WAGT, said before they could consider the proposal, WAGT needed to inform them if the testing would be for planned improvements to the current specified J46 A/B and whether or not the results of such a program could be realized in time to "significantly" benefit J46 engines. Action on the proposal would be deferred until WAGT's response was received and evaluated.[10]

WAGT sent a memo in response stating that they had asked Marquardt for an estimate since Marquardt's A/B test cells were in demand and scheduled far in advance. Since that request, WAGT had investigated the cost of upgrading a test cell at a WAGT laboratory ($345,000.00) which could be complete in time to benefit advanced J46 engine development. They concurred that until such time as the J46 advanced engine program was more firmly established and a decision taken on upgrading the WAGT test cell for A/B testing, BuAer should take no action to provide Marquardt test time.[11]

Amendment 1 to specification WAGT-124D changed the engine weights for the -18 from 2,130 to 2,163 pounds and for the -20 from 2,135 to 2,168 pounds. The compressor configuration was changed from 12 to 13 stages and the initial test run was changed to be 2.5 hours long. The oil and fuel specification was changed to use MIL-C-8188 Grade A corrosion preventive oil (was MIL-C-7853) and MIL-F-5624A JP-4 fuel designated for all acceptance testing. The engine A/B cooling shroud airflow requirement was changed from 4 to 6.5%.

In the section describing the Power Control, the afterburner ignitor pressure switch was added and the following paragraph added to describe the A/B control operation: "Upon selecting afterburning operation, provided the engine has reached military rpm, fuel is admitted into the afterburner through the afterburner fuel manifold rings. Simultaneously fuel is injected by the ignition nozzles. The afterburner ignitor pressure switch detects ignitor fuel pressure and transmits a signal to the exhaust nozzle actuator control which positions the exhaust nozzle to a predetermined open position just prior to ignition. After afterburner ignition, the automatic temperature control modulates the exhaust nozzle to maintain maximum turbine outlet temperature at maximum afterburning conditions as described above. Afterburner blowout is detected by the afterburner blowout switch and the exhaust nozzle is automatically rescheduled to the military 100° day position. If the ignitor pressure switch fails, the blowout switch will reschedule the exhaust nozzle upon detection of increased gas pressure in the afterburner after ignition."[12]

BuAer accepted the new weights for both engines on January 21.[13]

The cover letter on the progress report for January relayed that XJ48-WE-18 #14 (likely a house engine numbering carried over from the cancelled -2 program) was fully assembled and test run from January 13th to the 25th. Immediately the engine displayed poor turbine temperature distribution and the issue went away when the fuel lines were T'ed but this was only a temporary solution. The first calibration showed the projected performance could be attained. After 23 hours of running, inspection showed the second stage blading had failed across the trailing edge, a problem seen earlier in testing the T-19 turbine. A 3/8 inch bevel on the trailing edge design proved in test the best way to eliminate the problem.

Figure 1: J46-WE-18 #14 prior to testing. Thirteen stage compressor can be identified by vane stud count (+1) on the compressor case. WAGT P50856, 1/11/1954. *Courtesy Hagley Museum and Library*

Further examination of the first stage turbine revealed the nozzles had some cracks which were attributed to turbine temperature distribution unevenness.

The XJ46-WE-18 #15 engine had the new beveled edge blades and was in build, scheduled for preliminary running as of February 5th in preparation for a qualification test attempt. The BAR felt the engine was not ready for a type test. The engine could not reach maximum turbine temperatures at temperatures below 25°F. It needed a smaller nozzle area and it was planned to do this by reducing the existing nozzle from 400/200 in² to 388/180 in² by using longer iris leaves. Afterburner squeal on the engine had been eliminated by supplying more fuel to the inner fuel ring than the outer.

Figure 2: J46-WE-18 compressor disc balancing rig. Blades were individually weighed and closely similar ones matched on each disc. WAGT P50895, 1/13/1954. *Courtesy Hagley Museum and Library*

The #15 engine exhibited acceleration stall for an unknown reason. Engine #14 had not exhibited the problem. A control system inspection had not revealed a cause.

The #1 -18 test engine had exhibited some high vibratory stress in the first three stages that was below any danger area but the condition was being watched. If it developed into a problem then the 13 stage C94-1 compressor would have to have its aerodynamic characteristics revisited.

It was not likely that the start of production in May would occur since the qualification test possibilities in March and April were marginal.[14]

The use of the hydro-mechanical control system was found to require a redesign of the junction box on the engine. This was surveyed in mid-March.[15]

At a conference with BuAer on March 19, WAGT discussed the latest performance curves for the -18 and they raised questions as they showed lower performance than the preliminary charts sent earlier. BuAer asked for a detailed explanation of the difference and WAGT responded two weeks later.

"As discussed in the conference, the performance data submitted by letter on March 2 was based on a performance calibration of a prototype J46-WE-18 built in South Philadelphia. The thrust was obtained by calculation from the internal performance data obtained in a closed cell, i.e. with muffler in place and the front doors closed. (*The thrust*) did not agree with the thrust measured on the same engine when tested in an open cell. A correction was, therefore, made to the internal performance data so that the altitude data would not be penalized by this cell discrepancy.

"It has now been found, by comparison with other relevant engine data that this correction has resulted in optimistic thrust values at altitude. The magnitude of this error under normal flight conditions is approximately six percent in net thrust. The specific fuel consumption data is still felt to be realistic for the configuration tested."

They planned to test a -18 with improved turbine inlet nozzle shroud seals which had been tested and would be qualified. The estimated performance of the engine with the seals would be four percent less thrust and two percent better SFC's than the preliminary estimates.[16]

February's MPR shows the BAR still saying that he felt the -18 was not ready for a qualification test. In addition to the problems reported in the prior month's report, the problem of ball joints failing in test had to be added. The hinge section of the exhaust nozzle was being burned. A design change to the hinge was currently being tested on engine #15. Depending on resolution of the ball joint problem, the qualification test of the -18 would be delayed beyond May.

The qualification engine itself would be either the first production engine or XJ46-WE-18 #17.

Based on the severity of the problems encountered, the BAR estimated the full qualification of the engine would not be accomplished before September.[17]

On April 14, BuAer accepted the current version of the specification as amended by Amendment 2 with the exception of the 6.5% shroud cooling requirement and until the updated performance curves were submitted.[18]

WAGT surveyed a new cylindrically-tapered A/B diffuser without a flexible ball joint in place of the current design. Analysis had indicated such a design change could be incorporated without detriment to satisfactory engine flexibility. Removal simplified manufacture and eliminated the necessity for exhaust nozzle linkage readjustment. The weight reduction would be approximately 30 pounds.[19]

The 500 hour laboratory qualification test results for the Microjet pressure switch were submitted in report A-1741Z and BuAer accepted the results and agreed the part was fully qualified.[20, 21]

Removal of the lubrication oil temperature resistance bulb was surveyed as EC No. 13 for the reason that none of the airframe companies were using the bulb to measure oil temperature. The resistance bulb receptacle provisions would be retained.[22]

Amendment 3 to the specification primarily covered the new ejector thrust correction curves. Some additional alterations to other curves were due to be delivered by June 1. The lubrication specification changed again to MIL-L-7808B and the fuel to MIL-F-5624B-1. Any reference to Block III advanced performance ratings was no longer included. [23]

The performance ratings at standard sea level static conditions were to be:

J46-WE-18 Performance Rating Specifications

Rating	Thrust	RPM	SFC	Fuel Flow lb/hr
Maximum	6,100	10,100	2.60	-
Military	4,100	10,100	1.080	-
Normal	3,690	10,100	1.065	-
90% Normal	3,320	10,100	1.060	-
75% Normal	2,765	10,100	1.090	-
Idle	245	4,000	-	980

As expected, the March MPR announced the qualification test of the -18 would be delayed beyond May 1954. The problems encountered while testing the #14 and #15 engines had been:

1. Burning of the A/B flameholder,
2. Ball joint "yielding",
3. Clogging of the engine igniter nozzles,
4. Drift in the hydro-mechanical control, and
5. Second stage turbine blading and disc failure in #14 in the root section of the blades and one blade disc serration.

Design changes to be tested were unaged blades along with a combustor shroud added for better temperature distribution by allowing more cooling air to flow to the root of the blades. Four possible solutions to the ball joint problem were: 1) redesign the ring with a slightly heavier section to take more thermal loading, 2) change the current material of the ring from 25 Cr 20 Ni to Inconel "W" (15 Cr 73 Ni), 3) change the material to Hastelloy "C" (5W 17 Cr 54 Ni 17 Mo), or 4) eliminate the ball joint (which was out for survey).

Figure 3: J46-WE-18 13 Stage compressor disc stack. WAGT P50896, 1/13/1954. *Courtesy Hagley Museum and Library*

Figure 4: J46-WE-18 compressor stack balancing. Hosting tool in the middle would be removed prior to spinning up. Minor balance vibration at this stage could be adjusted by using different weight through-bolt nuts. WAGT P50897, 1/13/1954. *Courtesy Hagley Museum and Library*

The BAR's concerns continued regarding WAGT's likelihood of being able to conduct an official 150-hour Qualification Test on the -18 anytime soon. He noted that no XJ46-WE-18 had undergone any long endurance test due to interruptions by design failures. After engines #14 and #15 encountered problems eliminating them from running an early 150-hour unofficial qualification test in March, WAGT had built up #15 again for such a test run in May. It suffered a turbine disc failure similar to the #14 engine. WAGT was then rebuilding engine #14 with unaged second stage turbine blades and would use it as the official qualification test engine if a preliminary 150-hour run using #15 was completed with mishap.[24]

Figure 5: Flame holder assembly damage from March 1954 testing. WAGT P51444, 3/31/1954. *Courtesy Hagley Museum and Library*

The qualification test results of the hydro-mechanical control system components were accepted by BuAer on May 18, 1954 as qualifying the system for both the -8 and -18.[25]

A day later WAGT surveyed the relocation of the ignitor plugs and nozzles as EC 14. This was done for improved accessibility. The ignitor plugs were to be positioned one on either side of the engine and located 20° 40′ and 19° 20′ respectively below the engine horizontal centerline. The ignitor plug leads could be positioned so as to run forward along the engine to provide airframe clearance.[26]

In response to a question asked during a conference between BuAer and WAGT which discussed the J46 improvement program (discussed in Chapter 10), WAGT stated that it was anticipated the first (of four) (*improvement program*) test engines with all of the improvements would be running in July with one more each month thereafter. Test flying would begin in November 1954.[27] This is the last mention in the records of the J46 improvement plan that involved long range planning for future versions with more power and better durability. It is likely the long range thinking of BuAer for airframe development and procurement at that point in time was uncovering the fact that the J46 engine, improved or not, had no likely future home in BuAer or the Air Force.

There had to be questions raised as well as to where WAGT was finding the engineering resources to do the work on the future program while the improvement progress of the initial J46 engines lagged. Unfortunately, the records are silent on this.

Figure 6: A/B spray rings from aft end. Both the 4 and 9 thermocouple groups are partially visible. WAGT P51449, 3/31/1954. *Courtesy Hagley Museum and Library*

The May MPR indicated the qualification test had been delayed to July. The BAR believed full qualification could not be completed until September. On May 1, engine #15 had completed 150 hours of running to the endurance schedule. The second stage turbine wheel was replaced twice. Blades were replaced due to casting flaws. Three A/B iris nozzles were burned out during the test. The electric control assembly had to be readjusted twice back to the 1,525°F setting. Otherwise the engine was in good condition. Another 150-hour run was underway on engine #14. Inspection at 50 hours had found one iris leaf burned out (the iris nozzle was replaced) and some turbine deterioration.

Engines were being assembled with selected frequency checked blades. Failed blades on a failed 2nd stage disc on engine #14 in April were found to have frequencies in the 1,020-1,030 range. Blades in test now had frequencies of 1,070 and above.

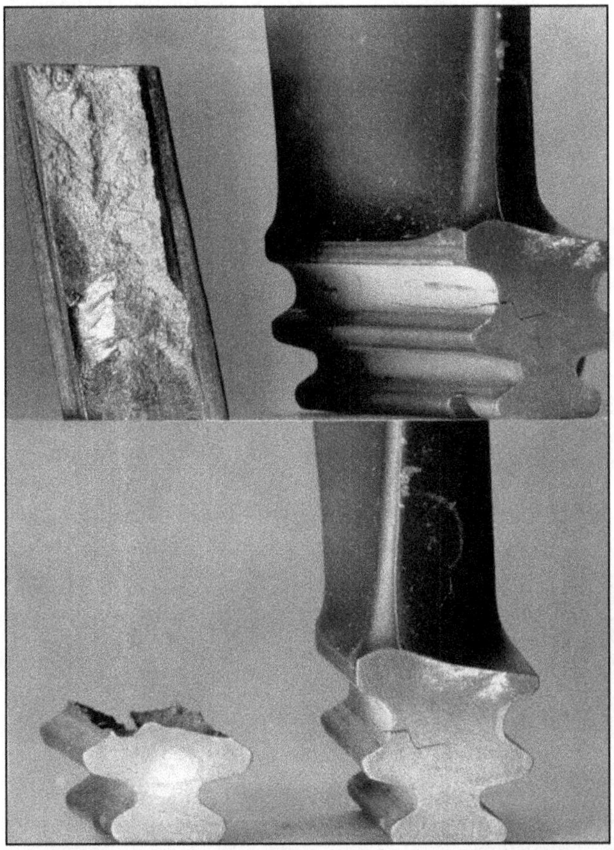

Figure 7: Cracked and broken 2nd stage turbine blades. WAGT P51712, 5/3/1954. *Courtesy Hagley Museum and Library*

The programs currently in place to resolve the turbine problems were:

1. Use blades with a frequency of 1,070 and above.
2. Apply more cooling air at the root.
3. Perform rigid inspection of root and groove lands for parallelism, flatness and spacing.
4. Grind blades to raise their frequency.
5. Test low frequency blades to reduce failure.

Long range programs were:

1. Test three land root blades in lieu of the two land root design.
2. Study the use of shrouded blades.
3. Test the Rolls-Royce T66 turbine design.
4. Find the source of the exciting force which was apparently associated with the blade failures (+6x engine speed frequency).

The hydro-mechanical control temperature drift still needed to be resolved. The use of internal shock mounts on some internal components within the control had not reduced the problem.[28]

The test of the XJ46-WE-18 was moved up to begin on June 24 using house engine #17. The configuration was to include the T-46 turbine (T-19 with 3% closed 1st stage nozzle); a five segment turbine-out straightening vane design in lieu of the six segment design, a low pressure drop liner and a pre-stretched ball-joint. The use of the T-46 turbine and low pressure drop liner were used to regain performance lost due to the use of beveled second stage turbine blades. Receipt of an official answer from Chance Vought on the removal of the ball joint survey in process, was a contingency before testing could be started.[29]

On June 30, 1954 the handbooks for the operating recommendations for the J46-WE-8 and J46-WE-18 were forwarded to BuAer.[30]

WAGT had surveyed an additional change to the high pressure filter element normal inspection period to every 30 hours and possibly to every 60 hours. Chance Vought requested information from WAGT on any engine failures that might be missed (*not detected*) if the high pressure filter was inspected less frequently. One known possibility was that failure of the variable displacement pump would not be detected in the main oil filter only. Verification was asked for both the 30 hour and 60 hour inspection periods.[31]

A BuAer evaluation showed that a compressor pressure limiter would not be needed on the -18 because the airframes planned to use the engines could not go fast enough, even in a dive, to generate ram pressures high enough to endanger the engine. When notified of deletion of the requirement, WAGT supplied the space needed to be reserved in the airframe in case a pressure limiter was found to be needed in the future. The weight saved by the deletion would be 13.0 pounds.[32]

A qualification test of the -18 using the #17 engine was started and had run 100 hours as of July 12. Engine performance and control were good. The turbine inlet temperature was set to 1,525°F and the test was expected to be completed on July 14.[33]

Five deviations from the 1st production engine at South Philadelphia were requested:

1. Of the 24 roller bearings in the iris exhaust nozzle, 12 would be the standard bearing and 12 would be

new stainless steel bearings. Both had been verified but the stainless steel type would eliminate the slight corrosion tendency of the standard bearings. A CID was in process to use the stainless steel type in early production engines.

2. The second stage turbine nozzle segments would be nicro-brazed in addition to being welded to the shroud segments. Welded-only segments tended to be adequate but the nicro-brazing helped resist the small manufacturing cracks at the joint of the shroud and vane.
3. A thin .032 inch outer combustion chamber liner (light-weight) was to be used on the engine while the first production engines would have the .050 inch shell.
4. A titanium 7th stage compressor disc was used in #17 and in production engines beyond the 10th engine. Steel discs would be used in the first ten engines for the 7th stage.
5. A spacer between the thrust bearing and the housing would be used beginning on the 10th production engine.

For the other final qualification tests, WAGT requested that the air bleed, hot oil and control transient tests be run on an engine other than #17 and run in parallel on the other engine. The parallel tests would include BuAer's request for the cell muffler thrust correction check.

The fuel control would be switched to the new Goshen 1120 "O" rings as soon as possible, but they were not on the test engine. The switch to Goshen rings would have to be qualified and it was not possible to complete that until October 1954.

While the test engine had the normal 2nd stage turbine tip clearances, experience with engine #15 showed the engine ran with no damage (*did they really mean no performance change?-Auth.*) after a tip rub removed .025 inches from the blade tips. All -18 engines other than #17 had their clearances opened up by .025 as a result and tip rubs were no longer experienced.

The surveys for moving the spark plugs and removal of the ball joint were still in process and were not on the test engine. The spark plug change was estimated to lower the engine restart altitude no more than 3,000 feet.[34]

In addition to the above data, attached to the BAR's preliminary qualification test status report was a WAGT memo explaining the design and configuration changes made to the qualification test engine before the test began:

1. Spacers were inserted in the A/B combustion chamber liner to ensure concentricity with the housing to give more even distribution of cooling air and reduction of shell hot spots.
2. Metal pins were added to the A/B nozzle leaf joints to prevent deterioration of the hinge.
3. A defective tube in the electric control assembly was removed. It had interfered with the hot day setting.
4. The 2nd stage turbine was populated with higher frequency 1,100 and above cycles per second to resolve blade and disc cracking.
5. Additional cooling air (.5% of engine flow) was supplied to the blade and disc roots by allowing compressor air to pass through an added hole in the center of the 1st stage turbine disc and through opened-up curvic clutch teeth.
6. The inner shroud of the turbine outlet guide vanes were pinned together making it a solid shroud in the axial direction.
7. New Goshen 1120 "O" rings were NOT installed as waiting for them would result in a three month delay.
8. The ball joint was retained because the issue of removal was not yet resolved. All testing showed sufficient flexibility in the engine remained with the ball joint removed even under loading. An interference at the aft aircraft seal at the ejector could occur if both the engine and aircraft nacelle were at the extreme limits of their manufacturing tolerances.[35]

The specification was reviewed by BuAer and the most importance changes requested were:

1. The ignition system high temperature rating for operation at 168°F could not be accepted unless Chance Vought assured BuAer that the maximum operating temperature would not be exceeded for the ignition system.
2. They objected to changing the fuel pressure transmitter thread specification from the standard 7/17- 20 NF-3 threads, which had been used from the J34-WE-22 on. They did not understand the reason for needing to change this thread and requested that WAGT change the specification back to the normal thread.

BuAer accepted the updated performance curves and asked that an amended copy of the specification be sent for approval.[36]

A full day after the first update on the qualification test of the #17 engine arrived, BuAer's authorization of the test arrived. The authorization contained some qualifications:

1. WAGT had to be satisfied that previous issues were fixed to a point where recurrence of the same difficulties was improbable.
2. The engine model test deviations previously noted by WAGT were acceptable.
3. The tested engine, if it successfully completed the test, would establish the type's parts list. Any deviations from that list for production engines must be subject to any appropriate CID's then being processed for incorporation in production engines.
4. The specification, while not yet approved, was sufficiently firm to be a basis for the start of the qualification test.[37]

Two days after the above letter was sent, the qualification test results were updated by the BAR. The test of #17 had been completed on July 14. He confirmed that the test had been started based on the conditions stated in the after-the-fact memo, even though it arrived after most of the test was run. (BuAer must have verbally passed along their qualifications prior to test start.)

No control adjustments had been made during the test. At 142 hours, one iris leaf cocked and jammed with burning of the hinge being evident. Blistering of the skin just ahead of the hinge had been observed before the test but did not progress during the 150 hours. Two other nozzles of the same design had run 10 hours in A/B without a problem. At 140 hours, the dual fuel pump seal began to leak through the drain port and the leakage was excessive. It was not felt this was a major concern since the same fuel pump had accumulated 19,000 hours of operation on other J46 engines without problems.[38]

Chance Vought finally submitted a memo stating that they understood BuAer's request as to whether the A2U-1 could be revised to use a straight section (no ball joint) A/B in all or a portion of the airplanes. They responded that they were conducting a study to show if the A2U-1 tail cone and A/B mount could be modified to accommodate either type of A/B interchangeably.

They added that before a "stiff" A/B could be put into service, considerable testing should be done which would take a great amount of time and might preclude the design from being used in all but a few airframes.

Chance Vought had no service experience with the redundant mounting principle and would not accept responsibility during testing or during service until sufficient operating experience had been accumulated to permit an accurate evaluation.[39]

Upon receipt of the qualification test results and tear-down inspection findings, BuAer rejected the test as qualifying the engine. Their reasons were:

1. Unsatisfactory turbine design. In order to achieve minimum acceptable turbine durability, the following discrepancies had to be corrected:
   a. Second stage disc cracking.
   b. Second stage blade cracking.
   c. First stage nozzle cracking and bowing.
   d. First stage blade cracking and tip curling.
   e. Second stage nozzle cracking. (Seal bending, cracking and breaking.)
2. Unsatisfactory exhaust nozzle design. The burning, warping and cracking of the several components of the exhaust collector needed to be corrected.
3. Failure of the A/B fuel manifold.

In addition, the engine was considered deficient at least with regard to the following and WAGT should correct as many of the deficiencies as was practicable in the subsequent test engine:

1. Damaged seals on the compressor vanes.
2. Galled and damaged threads on the compressor through-bolts.
3. Possible damage to the compressor discs due to scribe markings.
4. Cracking of, and apparent weld flaw in, the vaporizing tubes.
5. Blistering and cracking of the outer liner adapter.
6. Cracking of the inner liner.
7. Cracking of exhaust nozzle actuator casing.
8. Cracking and breaking of the exhaust cone inner stiffeners.
9. Cracking of, and bold seizing at, the squeal baffle.
10. Cracking of the flame-holder assembly.
11. Erosion and leaking of dual fuel pump.
12. Cracking of the ejector.
13. Oil leakage at an unused gearbox pad.
14. Apparent disengagement of the metering pin in the A/B fuel control bellows.
15. A faulty air seal around the spark plug.
16. Rubbing of exhaust nozzle actuator with tubing.
17. Clogging of the ignitor nozzles.

The subsequent endurance test engine(s) should incorporate spark plugs re-located in accordance with CID 40634 (previously surveyed) and should incorporate the same basic afterburner diffuser design (that is, without ball joint) unless advised by BuAer to the contrary.[40]

It must have been a shock to BuAer that an engine that could run through a 150 hour qualification test with only two noted physical problems during the test could have suffered so much physical damage upon inspection. Clearly, after all the years of research and development, the basic J46 engine was still subject to significant cracking and burning in both the stationary and rotating parts.

Preparation for -18 production went on, with WAGT asking permission to verify products from Indiana Gear Works (starter-generator gearbox), Westinghouse Nuttal Works in Pittsburg (alternate supplier of gearbox), and Pattern Products Manufacturing Company (shaft, splined).[41] BuAer approved the verification procedure on July 22.[42]

WAGT notified BuAer that the dry weight of the engine would increase by six (6) pounds due to the net weight change (+8 pounds) from using a steel 7th stage compressor disc and decreasing the outer compressor liner to 0.032 in (-2 pounds).[43]

A survey of a new fuel booster pump sent out in June was responded to by Chance Vought who noted that the new pump required a different connection hose since the old pump connected toward the rear of the engine and the new one connected to the right of the engine looking from the rear. Hoses would have to be supplied with the new pump kit so that if any given engine changed a pump and a different hose was necessary, it was provided.[44]

A WAGT request that the Holley Carburetor company be considered a qualified producer of A/B controls on the -18 based on their verification of the A/B control for the -8 was rejected by BuAer, who noted that the A/B control for the -18 had not yet been qualified. If the equipment to do the test was not available at the Kansas City site, the control could be delivered to Essington and tested there.[45]

With the prototype A2U-1 airframe almost finished at the Dallas plant and a second machine following close behind, on August 6 BuAer authorized the production of eight (8) YJ46-WE-18 engines to be built to the same parts list as the #17 qualification engine just tested. Parts changes had to be approved by BuAer prior to use. The delivery schedule was to be three (3) each in August and five (5) in September. They would be 100 hour flight rated and would be brought up to the final configuration of the approved qualifying engine by WAGT. The cost of modification and any change in contract price would be negotiated.[46]

On September 10, WAGT submitted a deficiency correction plan for the -18, which stated:

1. Second Stage Blade and Disc Cracking – use high frequency blades having increased thickness in the lower airfoil section. Strain gauges on test blades had shown that if the blade was of sufficiently high frequency it would not resonate with high stress under any operating condition.
2. First Stage Blade Cracking and Tip Curling – The leading edge of the tip of the first stage rotor blades would incorporate a small 45 degree bevel. Tests had shown this eliminated curling. Trailing edge cracking had been demonstrated to be a creep-rupture problem and better temperature distribution would alleviate it. Sealing of combustion liner leakage would be used to lower temperatures at the blade root.
3. Second Stage Nozzle Cracking and Seal Bending – Extra tabs to strengthen the seal strips would be used. The cracking had been in the weld at the nozzle shroud and was known to be a manufacturing problem. Brazing after welding was expected to eliminate the problem.
4. Exhaust Nozzle Design – Wide cam leaves would be incorporated as well as minor changes to reduce cooling air blockage on the collector shell. Testing with the changes had revealed no tendency of the modified leaves to cock or jam.
5. Afterburner Fuel Manifold – The failure was attributed to the final forming operation to flatten the supporting struts after welding. These would now be flattened prior to welding. The inlet legs would be welded to both manifold rings to decrease bending stresses in the rings.
6. Damaged Seals on Compressor Vanes – Was traced to the passage of a foreign object through the engine. No changes were planned.
7. Galling and Damaged Threads on Compressor Through Bolts – One was caused by mishandling during assembly and three appeared to have the thread cuts misaligned with the bolt axis.
8. Possible Damage to Compressor Discs Due to Scribe Markings – Manufacturing instructions issued to avoid using scribe markings on the discs.
9. Cracking and Apparent Weld Flaw in Vaporizing Tubes – Had not previously occurred and was felt to be an isolated case. The weld was found to be incomplete on one turbulence pin.
10. Blistering and Cracking of Outer Liner Adapter – Cooling slots to be incorporated in the next test engine.
11. Cracking of Inner Liner – Small weld cracks not seen previously occurred and were not thought to be a serious problem. Forward tabs on the liner to open the mounting holes into slots to prevent binding of tab washers and possible cracks would be used.
12. Cracking of Exhaust Nozzle Actuator Casing – A faulty braze on the actuator was found to be the cause. The brazing technique was changed.
13. Cracking and Breaking of Exhaust Cone Inner Stiffeners – Small additional stiffeners to be added.

14. Cracking of and Bold Seizing at Squeal Baffle – A chronic problem still undergoing study.
15. Cracking of the Flameholder Assembly – Not previously seen and felt to be an isolated case.
16. Erosion and Leaking of Dual Fuel Pump - Erosion had been apparent for some time and #17 had used a better material and suffered reduced erosion. Seal leakage was due to defective part from the supplier.
17. Cracking of Ejector –Vibration caused by test cell and muffler not seen in field service of engines.
18. Oil Leakage at Unused Gearbox Pad – The carbon seal and rub-ring were found scored by a particle of foreign matter. Better handling to be used.
19. Apparent Disengagement of Metering Pin in the Afterburner Fuel Control Bellows – Caused by complete failure of the A/B fuel manifold inlet leg and not the control.
20. Faulty Air Seal Around Spark Plug – Tack-welds added to the air seal to restrict its motion.
21. Rubbing of Exhaust Nozzle Actuator with Tubing – More careful checking of tubing dimensions and clearances at assembly.
22. Clogging of Ignitor Nozzles – Caused by location in a hotter area than previously. Swirl passage area changes were under investigation.

The second qualification test attempt would be made in the week of September 27.[47]

The BAR's forwarding memo stated the second stage turbine disc and blades remained the primary weakness and that the planned solution would not solve the problem. Also, dual fuel pump bearing seal erosion had been seen on every test but service use had not revealed it to be a problem. The exhaust nozzle design was doubtful to him and needed a complete endurance test to verify it. He expressed the opinion that if the turbine exhaust nozzle fixes were adequate, the #19 engine qualification test would be satisfactory.[48]

The August monthly progress report was summarized by the BAR and gives some background not in WAGT's memo or the BAR's covering memo above. WAGT had tested the J40-WE-24 second stage (*design*) turbine blades on the #19 engine and had gotten the same result as a test on the #15 engine; a two (2) to three (3) percent loss in performance but the turbine exhibited good mechanical durability. The blade design was temporarily abandoned for the -18. They were currently testing redesigned blades on the T-19 second stage turbine on #14 and were being put through a 50 hour qualification equivalent test in regards to military and A/B time. While it was being inspected after the test, the #14 engine would receive the turbine being built up for the qualification test and "seasoned" on the engine. If the turbine test engine showed good results, the turbine on #14 would be moved to the #19 engine and the engine would be ready for test start on September 20.

The #19 engine would incorporate a perforated squeal baffle. The design reduced temperatures around the exhaust nozzle by 200°F. The new baffle design reduced thrust by 100 pounds at the military power settings. This loss is considered critical because the -18's tested to date had only exceeded the maximum model specification thrust performance by about one (1) percent.[49]

While resolution of the #17 qualification test issues was being worked on, WAGT had to prepare to build eight (8) YJ46-WE-18's. They asked acceptance of the following deviations from the #17 test engine be granted for the eight engines.

1. The use of the earlier tool steel material bearings on the A/B unison ring. The stainless steel bearings would be incorporated via CID 40572 and would be used on some future block point.
2. To use either the nozzle leaves that were nicrobrazed at the junction of the vane and shroud at the leading and trailing edge of the vane or an alternate design used successfully on #14 with full weld penetration but no nicrobrazing, depending on availability.
3. Use the 0.050 inch combustion chamber design. All future engines would have the 0.032 inch "lightweight" design.
4. Use steel for the number seven (7) compressor disc. Titanium would be used in all subsequent engines.
5. Eliminate the secondary A/B ignitor nozzle and its related piping.
6. Based on Chance Vought's answer to the ball joint removal survey, provide four (4) engines with ball joints and four (4) with the "stiff" A/B.
7. Use the old design of thrust bearing spacer on the eight engines and the new design on all subsequent engines.
8. Reroute two lines found to be within the engine envelope but causing airframe interference on the A2U-1 when tested using a metal shell mock-up of the -18. All subsequent engines would retain the revised routing.[50]

BuAer approved the requested deviations and added that WAGT should submit any necessary CID's for any not already in the pipeline.[51]

On September 2, WAGT submitted operating recommendations that were special to and supplemented the normal J46-WE-8 operating recommendations for the YJ46-WE-18:

1. "Do not operate the engines under steady state conditions between the engine speeds of 5,200 and 8,100 rpm. This also applied to J46 engines in general but was repeated here to be sure it was complied with on the -18 as well.
2. "After thirty (30) hours of total running time or two and one half hours at top temperature operation, the second stage turbine should be removed and replaced. The replacement disc should not have more than 5.8 hours of total temperature operation before being inspected by WAGT and found suitable for continued service. WAGT would supply replacement wheels through the normal WAGT (not Navy) spares channels. A total running time record must be kept with each turbine wheel. To facilitate replacing the second stage turbine wheel without rebalancing the complete rotating turbine unit, the discs would have the high-point marked with an "H" etched on the aft side of the rim of the disc. The high-point of a disc should coincide with the high-point of the disc that is being replaced.
3. "The thirty (30) hours running and 2.5 high temperature times are to start at the time the engine is delivered. The 100 hours to total running time and 8.3 hours top temperature operation, however, would include the total time accumulated since last build-up, which includes the time accumulated during final testing at WAGT.
4. "Inspection of the A/B iris nozzle was recommended at every post-flight inspection. Special attention should be given to the minimum 1/8 inch overlap of adjacent leaves and any indication of detrimental burning at the floating leaf hinge points. At the same time the second stage turbine wheel was replaced, the A/B fuel manifold be removed, cleaned, and visually inspected for cracks, particularly at the welds on the fuel inlet and support tubes.
5. "The main engine ignitor nozzles were to be removed at the end of every thirty hour period, inspected for any possible clogging, and cleaned or replaced if necessary."[52]

WAGT got around to responding to BuAer's concern with the specification relating to the use of a non-standard thread on the pressure tap at the fuel distributor valve. They explained that the non-standard thread was used for the ground check only tap because there was no room for the standard tap. The cockpit pressure tap used the standard thread tap. The standard cockpit pressure tap was added to the list of customer connections.

They also asked if BuAer's statement that "the performance curves were reviewed and found to be in conformance with the rated performance and operating limits" constituted approval of the curves.[53]

On September 22, WAGT sent out a survey covering the addition of fuel check valves on both sides of the A/B Fuel Shut-off Valve actuator piston to prevent inlet pressure surges from opening the Shut-off Valve and passing fuel into the overboard tank. The valves would be included on WE040501-29 and WE-400001-346.[54]

The decision on the ball joint continued to be a focal point. WAGT sent a memo to BuAer telling them they planned to equip all the -18's after the first eight with ball joints and a strengthened afterburner mount until flight testing proved the ball joint could be dispensed with. The first three flight engines (WE400001, WE400003, and WE400005) would have the ball joint and WAGT urged that airframe #120 be allocated before it was modified and instrumented for other flight test activities.

They added that they had tested an engine without a ball joint in a test cell and applied a 10-g load on the A/B mount with no effect on turbine tip clearance while the engine was operating at full power.

They noted that the engine was not designed to sustain aircraft loads but the loadings and deflections previously agreed upon would be acceptable.

A redesigned ball joint with Iconel X rub rings was being procured for test in case the other programs were delayed, but the elimination of the ball joint from a weight and cost standpoint was the ultimate goal.[55]

By way of an information only letter, WAGT notified BuAer that beginning with engine WE400061 a bleed orifice and related piping would be incorporated to provide a means of limiting engine fuel flow during the starting cycle. The limiting of fuel flow during the starting period would eliminate excessive turbine out temperatures and increase the field suitability of the engine.[56]

On October 5, BuAer approved the replacement of the hydraulic locks on the hydro-mechanical exhaust nozzle control system with check valves. The approval was given based on two reports that showed the thrust loss would not be as much as previously anticipated.[57]

Qualification – Second Attempt

The second J46-WE-18 150-hour Qualification Test using engine #19 was stopped after 62 hours when the A/B fuel ring at the fuel inlet boss failed. On examination, slight overtemp indications were noted at two points on the iris exhaust nozzle leaves. The trigger anticipator, which had been part of the initial build of the engine, had been removed earlier in the test following its malfunction.

The fuel ring failure was caused by thermal stresses induced by binding of the fitting in the housing. In addition, one of the three remaining fittings expanded beyond the housing which caused the three remaining fittings to carry the load. Metallurgical analysis confirmed the failure was a high temperature tensile failure.

The new fuel manifold had increased clearances between the fittings and housing boss, extended housing bosses to prevent disengagement, increased radii to the manifold bosses to relieve possible galling and graphite was added to the bosses to improve the sliding actions. The spray distribution in the manifold was altered so as to give less fuel to the outer ring and more to the inner ring which would lower the shell temperature by about 50°F.

Engine #19 was fitted with a new turbine, a new A/B and the new A/B fuel ring design. The trigger anticipator was removed pending the completion of development and flight testing had demonstrated that it was ready for use on production engines.

A new start date for a third qualification test run was set to on or before October 15. Production engine WE040509 was considered the test backup engine.[58]

BuAer sent a wire approving the rebuild of engine #19 as described in WAGT's memo.[59]

On October 13, BuAer issued partial termination TN-6 for the contract, removing 94 -18 engines from the production contract NOa(s) 10825.[60]

On October 15, BuAer forwarded a memo accepting specification WAGT-124D-A for the J46-WE-18. They had finally accepted the maximum continuous operating temperature for the ignition system as 165°F.[61]

Qualification – Third Attempt

The third attempt was started on October 12 and as of October 15th the engine had run 55 hours. Performance to that point had been satisfactory with no problems of any kind occurring. The test was expected to be completed by October 25 and teardown and inspection during the week of November 1.[62]

While the test was occurring, WAGT sent a follow-up memo to BuAer asking when they could expect a reply to their earlier memo regarding the thrust corrections for test cell losses in the Essington works test cells.[63]

BuAer responded with acceptance of the adjusted curves and SFC's but only for engines run in the test cells at Essington.[64]

Also while the qualification test was underway, WAGT sent over their plan to run a 150-hour verification test on a Kansas City manufactured J46-WE-18 in the near future. They asked that certain specification requirements be waived as they would be tested during the on-going qualification test in Essington.[65]

BuAer did not respond to that memo until November 12, 1954. The qualification test was complete and a preliminary inspection of the test engine showed that the performance of the first stage compressor disc, the first stage turbine nozzle and the A/B was unsatisfactory.

In view of the situation (A2U-1 program cancellation and the pending cancellation of the -18 contract), another test would have to be run to correct those and any other errors found during the formal teardown inspection then scheduled for November 15.[66]

After the formal teardown inspection WAGT submitted report A-1890 dated November 15 to BuAer. The summary report of the test results stated that four major discrepancies were found:

1. Root cracks in the 1st stage aluminum compressor disc.
2. Trailing edge cracks in the 1st stage turbine nozzles.
3. Weld cracks in the afterburner fuel manifold.
4. Burnout at the trailing edge of the afterburner perforated squeal baffle.

WAGT stated "it is concluded that the basic engine, less the afterburner, satisfactorily meets the durability requirements for the J46-WE-18 engine."

The discrepancies noted required the retest of the afterburner to correct the deficiencies of the fuel manifold and squeal baffle. While the first stage turbine nozzle cracks appeared somewhat severe in appearance they were (by WAGT) considered acceptable for service use. The statement was justified by the results of five endurance tests, each one including all the military afterburner time for a full 150 hour qualification test, which had been completed on two sets of first stage turbine nozzle assembly parts. The nozzle did show some cracking but the cracks had progressed only slightly from one endurance run to another with a maximum crack length of 3/8 inch after the third endurance test.

They recommended acceptance of the XJ46-WE-18 model with the following limitations:

1. The first stage aluminum compressor disc should be replaced with the steel disc on the first production engines and not at a later block point.
2. Accept the 1st stage turbine nozzles for 150 hours as is. After 150 hours, WAGT would establish inspection limits that reflected experience in service.
3. Pursue a modified A/B design with modified support bosses and a perforated squeal baffle to be introduced on production engine WE040509 and up.[67]

On November 17, 1954, BuAer notified WAGT by telegram that the J46-WE-18 engine program contract NOa(s) 10825 was terminated. BuAer was authorized to run one more qualification test on the -18 after the engine had incorporated all of the improvements listed in their problems resolution plan for the last #19 engine test attempt as long as it could be started no later than December 15.[68]

A detailed test plan had been sent over on November 9 covering the qualification test of the Override Selector assembly for the -18.[69] The Kansas City BAR noted that the same selector was to be used on the J40-WE-22 and possibly on late J46-WE-8 engines. Since the -18 program was (at that point) cancelled, he recommended the verification test of the Override Selector be run on a J40-WE-22.[70]

The same day, WAGT forwarded an update on the status of the J46-WE-18 improvement program being run under Contract NOa(s) 10825. It is clear that the memo was a bit out of date and the writer unaware of the recent partial cancellations of production -18 engines and the development contract termination in its entirety. In April and June of that year the Chief of BuAer had recommended that WAGT proceed with improvement developments that would provide design performance improvements according to curves submitted by WAGT (*not found*). The improvement would also have to be capable of incorporation into a substantial portion of the production engines under contract and be readily back-fitted at overhaul into previously delivered J46-WE-18 production engines with no installation changes.

The possibilities had been narrowed down to a new 13 stage compressor design (the C-115) that operated properly when equipped with an air bleed device. The changes needed to the control system required further study to determine their impact on installation interchangeability. The timing and likelihood of installation issues now made this approach unattractive.

The configuration would be a C-115-4 compressor with a T-47 turbine (10% closed). The performance improvement on the -18 it would have offered was:

C-115-4/T-47	Thrust	SFC
Maximum (A/B)	+1.01 %)	-1.15 %
Military	+0.73 %	-1.85 %

The recommended improvement path included adding a 14th stage to the back of the 13 stage compressor. If the go-ahead was given by December 1st, 150 of the engines under contract could receive the new compressors. The main engine mounts would have to be moved back 1.7 inches. No performance improvement was noted on the project schedule attached.

The program would be in two stages, the first being to grant approval to build up a -18 with a 14 stage compressor for testing to be ready for test start on December 31. The second stage would be a production proposal.[71]

Based on their understanding that a final fix for the deficiencies in the first stage nozzle and the turbine had not yet been put in place, BuAer sent a wire on December 9, 1954 asking for the status of the substantiation test and final fix established correcting the first stage nozzle and turbine deficiency.[72] The memo that triggered the telegram reminded WAGT that a qualification test had to start before midnight on December 15th.[73]

While WAGT raced to put a memo together to answer their test readiness status, the BAR sent a speed letter covering the status:

- All available engines at Essington and two engines in Kansas City had been running tests on fixes, with over 250 hours of running accomplished on twelve engines.
- The first stage nozzle trailing edge cracks would be eliminated by drilling cooling holes in the turbine housing and perforating the concave side of the vanes, and the material changed from Hastelloy "C" to Refractalloy 80. Tests were run in conjunction with a one percent cooling air seal at the turbine inlet instead of the 7.5% seal previously used. The change had improved the combustor turbine inlet distribution and lowered the maximum hot spots, as well as eliminating first stage turbine blade trailing edge cracks.
- Another 150-hour Qualification Test was started on December 13 using the modified engine #27.[74]

A WAGT memo followed, dated December 15, and to the above it added another development improvement:

- An improved design of the A/B fuel manifold made in AMS 5572 (25-20) instead of the previous AMS 5570 (18-8) was to be used. Shims at each supporting strut to allow torsional movement at assembly and during operation, beefed up support tubes and tees for added strength and reinforced scarf welds at the critical points had been incorporated.[75]

When the BAR forwarded WAGT's memo, he stated "This activity is of the opinion that the time element that remained between 15 November and 15 December was not sufficient to fully evaluate and resolve all the major problems. However, the contractor has accomplished a tremendous task within a short period of time........Although this activity is of the opinion that this all-out effort should have been applied earlier, it was considered impossible since production commitments and other development commitments had to receive their share of equipment to meet scheduled dates."[76]

With the contract terminated, the BAR asked how the continuing development costs for the -18 should be charged. BuAer answered that the Contract NOa(s) 10825 included funds for development and that the qualification test report, parts list, final design drawings and material, process and test specification would still be required to be delivered.[77]

As of December, the qualification test had run 62 hours with WAGT using two shifts to run the engine. No control problems had been encountered and the engine was meeting the model specification guarantees with some reserve.[78]

Even as the last WAGT qualification test was beginning, the final disposition of the small number of -18 engines was underway. One -18 had been shipped to the NACA Lewis Flight Propulsion Laboratory for performance testing under project TED-NACA-PP-2083. BuAer, in light of heavy testing requirements at Lewis, then decided to reship the engine to the Naval Air Turbine Test Station, Trenton, New Jersey for the tests. The engine was shipped back to WAGT first. Two -18 engines were to be shipped to Trenton with spares to accomplish 100 hours of testing. The engines were to be to the latest configuration possible and through their final acceptance runs to be selected for this purpose.[79]

WAGT sent an inquiry to BuAer asking if the eight (8) YJ46-WE-18 engines had to be reworked to the final configuration of J46-WE-18 as specified in the contract. They carried serial numbers WE040501 to WE040508. The comments on the distribution sheet were in agreement that no one had any use for the engines and they should be sent to O & R North Island for cannibalization to recover as many usable -8B compatible parts as possible.[80]

Upon inspection of all the available -18 engines in various completion states, it was found that only the first production engine, WE040509, met the requirements for testing in Trenton. WAGT was instructed to ship that engine to Trenton and hold WE040510 as an uncompleted spare should it be needed. It lacked a number of components that could be supplied at no cost from WAGT spares, but the labor to complete the build and acceptance would be approximately $35,000.00.

The spares to be shipped to Trenton included one spare turbine shaft balanced with a high frequency bladed disc, one spare set of 1st and 2nd stage turbine nozzle vanes assembled in the housing including the turbine liners, and one complete A/B assembly. Controls and accessories would be delivered as needed from termination inventory or on a loan basis.

The differences between the qualification test #27 engine and WE040505 were:[81]

Test Engine #27	WE040509
Squeal Baffle 6% open area.	40% open area. Should be changed at Trenton to 6% to extend life of the burner.
A/B Fuel Manifold "25-20" scarf welded design.	Has original "D" Block A/B fuel manifold.
One piece wide leaf exhaust nozzle collector.	Welded wide leaf exhaust nozzle collector.
1st stage turbine nozzle vanes perforated for air cooling, made from Refractalloy "80".	Material is Hastelloy "C"

With the program cancelled and the residual engines being shipped off, WAGT did not publish its official 150-hour Qualification Test report (A-1902) until January 21, 1955.

The report's summary states:

- The qualification test was completed on December 28, 1954.
- The engine completed the 150 hour test without trouble of any kind.
- No control or engine adjustments were required during the test.
- The general condition of the engine at the end of the test was excellent with no serious defects.
- Minor defects found after the test were judged to be non-detrimental to further engine operation.
- WAGT "estimated" the engine, with only minor replacements, could be submitted to the full MIL-E-5009 qualification test.
- Engine performance was above the guarantees by a large (5-8%) margin.
- Production engines run through "green" run had also demonstrated higher than guarantee performance.
- Acceptable durability for all component parts had been demonstrated.
- Cracking of the mixer vanes in the compressor guide vane assembly was attributed to twisting them at their outer radial length to closer tolerances. (A new mixer vane would not have this done.)
- Inner liner cracking was due to an adverse tolerance buildup between the inner liner air seal and 1st state turbine nozzle inner shroud. They butted together but the design called for them to overlap. The CID to prevent the problem was never processed due to program cancellation.[82]

The BAR recommended the engine be accepted as qualified.[83]

BuAer sent their response on March 1. They stated they were pleased with the performance and endurance exhibited by the engine during the test. The defects noted during the inspection on 13 January were considered minor in nature and amenable to correction by changes in various fabrication methods and processes.

The completion of successful J46-WE-8 control component qualifications would satisfy the J46-WE-18 requirements. However, completion of the -8 requirements for the alternate fuel test, hot oil test, air bleed test and transient data were not acceptable alternatives for similar requirements applicable to the J46-WE-18 model. As a result, the J46-WE-18 was not considered fully qualified and, due to termination of the work, additional testing to complete the qualification of this engine was not contemplated.

They noted that while specification WAGT-124D-A referred to specification MIL-E-5009 as applicable, it specified deviations to the extent that the actual qualification test closely resembled that specified in the superseded specification AN-E-32. The older specification required 12.5 hours of top temperature operation whereas MIL-E-5009 required a much more rigorous test including more than 31 hours of top temperature operation. While AN-E-32 was in effect when the J46 program began, MIL-E-5009 and 5009A became effective and other manufacturer's engines were now required to fully conform to the later specifications. For that reason, the endurance capabilities of the J46-WE-18 appeared to be uncertain when compared with other engines under the new specification.

As a result, Westinghouse report A-1902 was accepted as partial fulfillment of the requirements. They requested that WAGT forward the remaining data, namely: the engine parts list, final design drawings, and material, process and test specifications as soon as possible so the engineering records could be brought up to date.[84]

And with that, the -18 faded from view. It would appear that BuAer had moved the goalposts on WAGT in their critique of the engine. Given that position, one wonders what the purpose of all the push to do qualification testing was for if the specification no longer met BuAer's needs or requirements.

WAGT forwarded the last items over the next two months to complete the contract deliverable items.

Upon review, approximately 60 percent of the -18 parts forward of the combustion chamber were found to be usable as -8B and -8 parts and a list was sent over by WAGT on March 21 to BuAer. The same memo listed the run-times on the eight YJ46-WE-18 engines to be cannibalized: [85]

Operating Time of YJ46-WE-18 Engines

	Since New			Since Last Build-up		
Serial No.	**Total**	**MIL**	**A/B**	**Total**	**MIL**	**A/B**
WE040501	13.22	3.38	0.67	8.60	2.25	0.67
WE040502	21.18	3.38	2.14	15.83	2.08	1.97
WE040503	22.14	---	---	17.77	---	---
WE040504	14.48	3.18	1.42	7.20	1.18	0.87
WE040505	13.62	3.23	1.48	10.35	1.93	1.48
WE040506	14.77	2.92	0.66	10.40	1.25	0.58
WE040507	13.51	2.21	0.65	10.08	1.43	0.65
WE040508	12.52	2.73	1.25	7.12	1.53	1.05

At least two partially assembled J46-WE-18 engines survived the end of the program for some period of time. A J46-WE-18 (50% complete) in its container was shipped to the Indiana State Agency for Surplus Property on November 5, 1957. The shipping authorization states it was for use at Purdue University.[86] Another -18 (85% complete) in its engine container was shipped on December 28, 1957 to the Wisconsin State Agency for Surplus Property for use at the University of Wisconsin.[87] The final disposition of these engines is unknown.

No J46-WE-18 is known to exist in any form today.

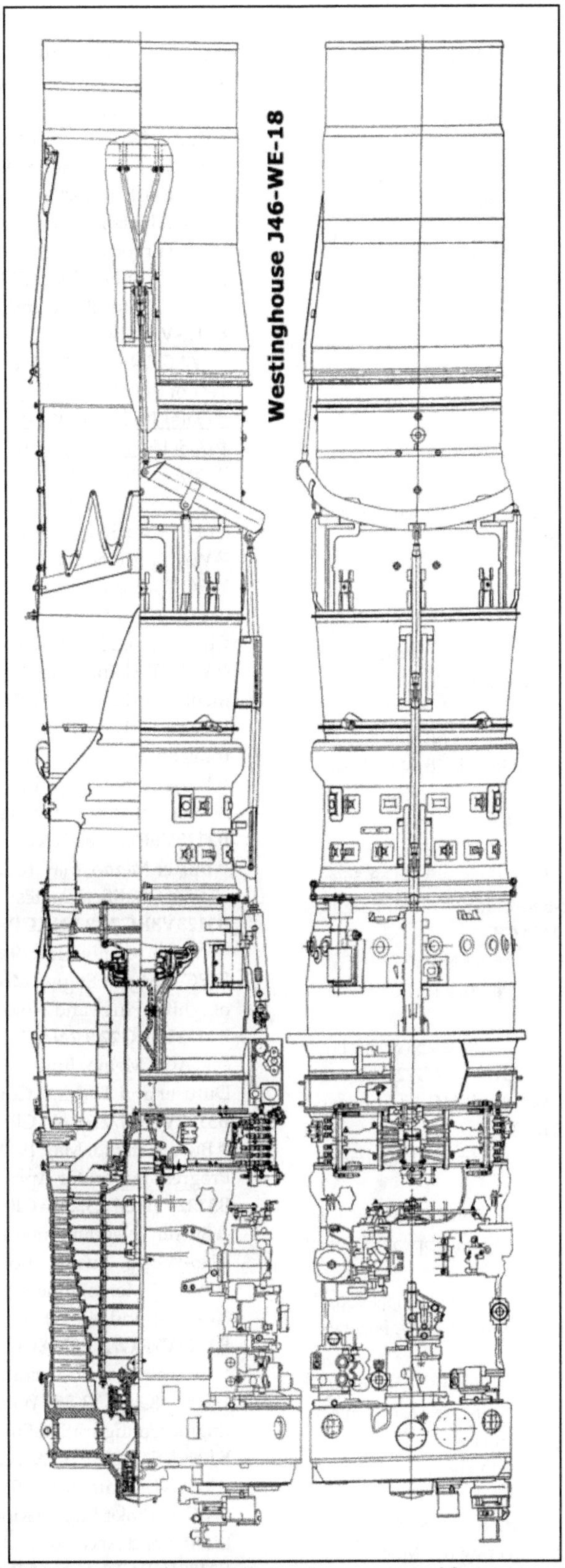

Figure 8: J46-WE-18 sectioned assembly drawing, recovered from unnumbered WAGT drawing dated 4/23/1954. ***Courtesy Hagley Museum and Library***

Chapter 7 – J46-WE-18 Development

[1] BuAer Wire, June 25, 1953, No Subject, **B3123V1RG72.3.2NACP**
[2] WAGT Memo, August 6, 1953, "Contract NOa(s) 53-556, 10825. WEC Turbojet Engine Model J46-WE-2 Modified. New Engine Configuration, Survey of Engine Change No. 21.", **B3122V2RG72.3.2NACP**
[3] BuAer Memo, August 7, 1953, "Contract NOa(s) 53-556, Westinghouse Specification WAGT-24C10-8 covering the Models J46-WE-2 and -16 engines", **B3122V2RG72.3.2NACP**
[4] WAGT Memo, October 14, 1953, "Contract Noa(s) 53-556, 10825. WEC Turbojet Engine Models J46-WE-8, XJ46-WE-18, J46-WE-18, J46-WE-20. Proposed Changes to Basic Engine Configuration, Survey of. Engine Change No. 122, 3 , 3 Respectively.", **B3123V1RG72.3.2NACP**
[5] BuAer Memo, November 7, 1953, "Contract NOa(s) 53-556, Model Specification WAGT-124D of 21 Aug 53 which describes the J46-We-18 and -20 turbojet engine models", **B3123V1RG72.3.2NACP**
[6] WAGT Memo, November 16, 1953, "Contract NOa(s) 53-556, 10825. WEC Turbojet Engine Models XJ46-WE-18, -20. Revision to Electric Control Assembly, Survey of. Engine Change Nos. 123, 4, 4 for the -8, -18, -20 Engines Respectively", **B3123V2RG72.3.2NACP**
[7] WAGT Memo, November 23, 1953, "Contract NOa(s) 53-556, 10825. WEC Turbojet Engine Models XJ46-WE-18, J46-WE-18. Applicability of J46-WE-8 Engine Series Surveys to J46-WE-18 Engines, Installation survey Summary of.", **B3123V2RG72.3.2NACP**
[8] BuAer Memo, December 10, 1953, "Contract NOa(s) 53-556; engine fuel booster pump P/N 22E-694-12 for J46 series engines, qualification of", **B3123V2RG72.3.2NACP**
[9] BuAer Memo, December 18, 1953, "Contract NOa(s) 53-556, Item 7, Monthly Progress Report for November 1953.", **B3123V2RG72.3.2NACP**
[10] BuAer Memo, January 4, 1954, "Contract NOas 53-556; Marquardt J46 afterburner development", **B3123V2RG72.3.2NACP**
[11] WAGT Memo, March 15, 1954, "Marquardt J46 Afterburner Development.", **B3123V2RG72.3.2NACP**
[12] WAGT Specification WAGT-124D, Amendment 1, January 8, 1954, **B3122V5RG72.3.2NACP**
[13] BuAer Memo, January 21, 1954, "Contract NOa(s) 53-566 and 10825, Westinghouse Model J46-WE-18 engine; approved specification weight for", **B3122V4RG72.3.2NACP**
[14] BuAer Memo, February 18, 1954, "Contract NOa(s) 53-556, Monthly Progress Report for Month of January, 1954, Forwarding of", **B3122V5RG72.3.2NACP**
[15] WAGT Memo, March 12, 1954, "Contract Noa(s) 53-556, 10825. Engine Models XJ46-WE-8, -18 and J46-WE-8, -18. Junction Box for Hydro-Mechanical Control Engines, Survey of. Engine Change Nos. 127, 10 for -8 and -18 Respectively.", **B3123V2RG72.3.2NACP**
[16] WAGT Memo, April 2, 1954, "Contract NOa(s) 53-556; Westinghouse Specification WAGT-124D Covering the Models J46-WE-18 and -20 Engines.", **B3122V6RG72.3.2NACP**
[17] BuAer Memo, March 15, 1954, "Contract NOa(s) 53-556, Monthly Progress Report for Month of February, 1954; Forwarding of", **B3122V6RG72.3.2NACP**
[18] BuAer Memo, April 14, 1954, "Contract NOa(s) 53-556; Westinghouse Specification WAGT-124D covering the models J46-WE-018 and -20 engines", **B3122V6RG72.3.2NACP**
[19] WAGT Memo, April 23, 1954, "Contract Noa(s) 53-556, 10825. WEC Turbojet Engine Models XJ46-WE-8, -18 and J46-WE-8, -18. Removal of Flexible Ball Joint, Survey of. Engine Change Nos. 130 and 12 for the -8 and -18 Respectively.", **B3123V3RG72.3.2NACP**
[20] WAGT Report A-1741Z, <u>500 Hour Laboratory Qualification Test of the Microjet Pressure Switch. Contract NOa(s) 53-556 and NOa(s) 10825; J46-WE-8B, 8 and 18 Engines Contact NOa(s) 10385; J40-WE-22 Engine.</u>, **B3123V3RG72.3.2NACP**
[21] BuAer Memo, April 28, 1954, "Contracts NOa(s) 53-556, 10825, and 10385, J46-WE-8B, -8, and -18 and J40-WE-22 engines; Microjet Pressure Switch Qualification Test", **B3123V3RG72.3.2NACP**
[22] WAGT Memo, May 13, 1954, "Contract NOa(s) 53-556, 10825. WEC Turbojet Engine Models XJ46-WE-18 , J46-WE-18. Removal of Lube Oil Temperature Resistance Bulb, Survey of. Engine Change No. 13.", **B3123V3RG72.3.2NACP**
[23] WAGT Memo, May 4, 1954, "Contract NOa(s) 53-556; amendment-3 (dtd 27 April 1954) to Westinghouse Model Specification WAGT-124D covering turbojet engine models J46-WE-18 and -20, transmittal of", **B3122V6RG72.3.2NACP**
[24] BuAer Memo, April 15, 1954, "Contract NOa(s) 53-556, Monthly Progress Report for March 1954; forwarding of", **B3122V6RG72.3.2NACP**
[25] BuAer Memo, May 18, 1954, "Contracts NOa(s) 52-048 and 53-556; Qualification test of hydromechanical control system", **B3123V3RG72.3.2NACP**
[26] WAGT Memo, May 19, 1954, "Contract Noa(s) 53-56, 10825. WEC Turbojet Engine Models XJ46-We-18, J46-WE-18. Relocation of Igniter Plugs and Nozzles, Survey of. Engine Change No. 14", **B3123V3RG72.3.2NACP**
[27] WAGT Memo, June 9, 1954, "J46-WE-18 Improvement Engines, Diffuser and Turbine, Component Description of.", **B3122V7RG72.3.2NACP**
[28] BuAer Memo, May 19, 1954, "Contract NOa(s) 53-556, Monthly Progress Report for April 1954; forwarding of", **B3122V7RG72.3.2NACP**
[29] BuAer Telegram, June 10, 1954, "Contract NOas 53-556 – XJ46-WE-18, Qualification Test on", **B3122V7RG72.3.2NACP**
[30] WAGT Memo, June 30, 1954, "Operating Recommendations for J46-WE-8 and J46-WE-18; forwarding of", **B3122V7RG72.3.2NACP**
[31] Chance Vought Aircraft Memo, July 2, 1954, "Contract NOa(s) 10825, NOa(s) 53-556 WEC Turbojet Engine Model J46-WE-18; Engine Configuration Survey of Engine Change No. 8", **B1099VE15RG72.3.2NACP**
[32] WAGT Memo, June 15, 1954, Contract Noa(s) 10825, 53-556. WEC Turbojet Engine Model J46-WE-18. Removal of Pressure Limiter and Accessories, Information Concerning.", **B3123V3RG72.3.2NACP**
[33] BuAer Memo, July 12, 1954, "Qualification of J46-WE-18 Engine, Contract NOa(s) 53-556c", **B3122V7RG72.3.2NACP**

[34] WAGT Memo, May 28, 1954, "Qualification of J46-WE-18 Engine Contract 10825", **B1100VE6RG72.3.2NACP**
[35] WAGT Memo, July 8, 1954, "Qualification of J46-WE-18 Engine, Contract NOa(s) 53-556c", **B3122V7RG72.3.2NACP**
[36] BuAer Memo, July 13, 1954, "Contract NOa(s) 53-556; Westinghouse Specification WAFT-124D-A of 27 May 1954 covering the J46-WE-18 engine", **B3122V7RG72.3.2NACP**
[37] BuAer Naval Speed Letter, July 15, 1954, "Contract NOas 53-556, XJ46-WE-18 Engine Serial No. 17; qualification test on.", **B3122V7RG72.3.2NACP**
[38] BuAer Naval Speed Letter, July 15, 1954, "Contract NOas 53-556, XJ46-WE-18 Serial No. 17; qualification test on.", **B3122V7RG72.3.2NACP**
[39] Chance Vought Memo, July 26, 1954, "Contract NOa(s) 53-556, 10825 J46-WE-18 Engine Afterburner Diffuser Configuration", **B3123V3RG72.3.2NACP**
[40] BuAer Memo, July 27, 1954, "Contract NOas 53-556-c; XJ46-WE-18 engine endurance test completed 14 July 1954", **B3123V3RG72.3.2NACP**
[41] WAGT Memo, June 24, 1954, "Contract NOa(s) 10825. Westinghouse Turbo-Jet Engine Model J-46-WE-18. Supplier Verification of Starter Generator Gearbox and Shaft, Spline.", **B1097VE11RG72.3.2NACP**
[42] BuAer Memo, July 22, 1954, "Contract NOa(s) 10825, Westinghouse turbo-Jet Engine Model J-46-WE-18, Supplier Verification of Starter Generator Gearbox and Shaft, Spline.", **B1097VE11RG72.3.2NACP**
[43] BuAer Memo, July 15, 1954, "Contract NOa(s) 10825, 53-556. WEC Turbojet Engine Model J46-WE-18. Establishment of the Nominal Dry Weight, Information Concerning.", **B1100VE7RG72.3.2NACP**
[44] Chance Vought Aircraft Memo, July 23, 1954, "Contract NOa(s) 10825 WEC Turbojet Engine Model J46-WE-18 Incorporation of New Fuel Booster Pump. Survey of Engine Change 17.", **B1100VE7RG72.3.2NACP**
[45] BuAer Memo, August 6, 1954, "Contract NOa(s) 10825 – J46-WE-18 Engine – Supplier Verification of Afterburner Fuel Control", **B1097VE11RG72.3.2NACP**
[46] BuAer Telegram, August 6, 1954, No Subject, **B1097VE11RG72.3.2NACP**
[47] WAGT Memo, September 10, 1954, "Qualification of J46-WE-18 Engine, Contract NOa(s) 53-556", **B3122V7RG72.3.2NACP**
[48] BuAer Memo, September 14, 1954, "Qualification of J46-WE-18 Engine, Contract NOa(s) 53-556", **B3122V7RG72.3.2NACP**
[49] BuAer Memo, September 17, 1954, "Contract NOa(s) 53-556c, Item 7, Monthly Progress Report of August 1954; forwarding of", **B3123V8RG72.3.2NACP**
[50] WAGT Memo, August 13, 1954, "Westinghouse turbo-jet Engine Model J46-WE-18 Contract NOa(s) 10825. Configuration of eight (8) YJ46-WE-18 Engines – Request for Approval of", **B1100VE6RG72.3.2NACP**
[51] BuAer Memo, August 30, 1954, Westinghouse turbo-jet Engine Model J46-WE-18 Contract NOa(s) 10825. Configuration of eight (8) YJ46-WE-18 Engines – Request for Approval of", **B1100VE7RG72.3.2NACP**
[52] WAGT Memo, September 2, 1954, "Contract NOa(s)-10825 Westinghouse Turbo-Jet Engine Model YJ46-WE-18, Operating Recommendations for", **B1100VE7RG72.3.2NACP**
[53] WAGT Memo, September 16, 1954, "Contract NOa(s) 53-556; Amendment-1 dated 24 August 1954 to J46-WE-18 Confidential Model Specification WAGT-124D-A dated 27 May 1954, transmittal of.", **B3122V7RG72.3.2NACP**
[54] WAGT Memo, September 27, 1954, "Contract NOa(s) 10825, Westinghouse turbo jet engine model J46-WE-18, incorporation of check valves to afterburner fuel shut-off valve lines and minor piping revisions to allow this incorporation, survey of.", **B1097VE12RG72.3.2NACP**
[55] WAGT Memo, September 22, 1954, "Removal of Flexible Ball Joint on J46-WE-18, Contract NOa(s) 53-556, 10825.", **B3122V7RG72.3.2NACP**
[56] WAGT Memo, October 1954, "Contract Noa(s) 10825. Westinghouse Turbojet Engine, Model J46-WE-18. Incorporation of a Bleed orifice and Related Piping Information Concerning.", **B1100VE8RG72.3.2NACP**
[57] BuAer Memo, October 5, 1954, "Contracts NOas 10385, 10825; Hydraulic Lock (Exhaust Nozzle Control System), Removal from J40-WE-22, J46-WE-8, and -18 engines", **B1097VE12RG72.3.2NACP**
[58] WAGT Memo, October 7, 1954, "Contract NOa(s) 53-556 J46-WE-18 Qualification Test Our Reference WG-70126", **B3123V8RG72.3.2NACP**
[59] WAGT Telegram, No Subject, October 12, 1954, **B3123V3RG72.3.2NACP**
[60] BuAer Contract Termination Notice TN-6, October 13, 1954, **B1094VA2RG72.3.2NACP**
[61] BuAer Memo, October 15, 1954, "Contract NOa(s) 53-556; Westinghouse Specification WAGT-124D-A of 27 May 1954 and Amendment-1 thereto covering the Model J46-WE-18 engine; approval of.", **B3123V3RG72.3.2NACP**
[62] BuAer Memo, October 18, 1954, "Contract NOa(s) 53-556c, Item 7, Monthly Progress Report of September 1954; forwarding of", **B3123V8RG72.3.2NACP**
[63] WAGT Memo, October 25, 1954, "Contract NOa(s) 10825, J46-WE-18 Engines, Thrust Corrections for Cell Losses in South Philadelphia Works Test cells", **B1097VE12RG72.3.2NACP**
[64] BuAer Memo, October 28, 1954, "Contract NOa(s) 10825, J46-WE-18 Engines, Thrust Corrections for Cell Losses in South Philadelphia Works Test Cells", **B1100VE8RG72.3.2NACP**
[65] WAGT Memo, September 28, 1954, "Contract NOa(s) 10825, Verification Test of Model J46-WE-18 Turbo-Jet Engine, Information On", **B1100VE8RG72.3.2NACP**
[66] BuAer Memo, November 12, 1954, "Contract NOa(s) 53-556, Turbojet Engine Model J46-WE-18, Qualification Test", **B3123V8RG72.3.2NACP**
[67] WAGT Memo, November 15, 1954, Qualification Test of the XJ46-WE-18 Turbo-Jet Engine on Contract NOa(s)-53-556, **B3123V8RG72.3.2NACP**
[68] BuAer Telegram, November 17, 1954, No Subject", **B1094VA2RG72.3.2NACP**
[69] WAGT Memo, November 9, 1954, "Qualification Test of the Override Selector Assembly 63E750 for the J46-WE-18", **B1100VE8RG72.3.2NACP**

[70] BuAer Memo, December 6, 1954, "Qualification Test of the Override Selector Assembly 63E750 for the J46-We-18 Engine.", **B1100VE8RG72.3.2NACP**
[71] WAGT Memo, November 17, 1954, "J46-WE-18 Improvement Program, Contract NOa(s) 10825 – forwarding of information.", **B1100VE8RG72.3.2NACP**
[72] BuAer Telegram, December 9, 1954, No Subject, **B1096VE4RG72.3.2NACP**
[73] BuAer Memo, December 9, 1954, "Contract NOa(s) - 10825 Qualification test of J46-WE-18.", **B1096VE4RG72.3.2NACP**
[74] BuAer Naval Speed Letter, December 14, 1954, "Qualification Test of J46-WE-18", **B3123V9RG72.3.2NACP**
[75] WAGT Memo, December 15, 1954, "Qualification of J46-WE-18 Engine on Contract NOa(s)-53-556", **B3123V9RG72.3.2NACP**
[76] BuAer Memo, December 17, 1954, "Qualification of J46-WE-18 Engine on Contract NOa(s) 53-556", **B3123V9RG72.3.2NACP**
[77] BuAer Memo, December 16, 1954, "Contract NOa(s) 53-556, XJ46-WE-18 Turbojet engines; status of", **B1099VE15RG72.3.2NACP**
[78] BuAer Memo, December 20, 1954, "Contract NOa(s) 10825 – J46-WE-8 and XJ46-WE-18 Progress Report (November 1954); Forwarding of", **B1101VE9RG72.3.2NACP**
[79] BuAer Memo, December 21, 1954, "Contract NOa(s) 10825 – Westinghouse J46-WE-18 engines", **B1097VE12RG72.3.2NACP**
[80] BuAer Memo, December 21, 1954, "Contract NOa(s) 10825. YJ46-WE-18 Turbojet Engines, disposition of", **B1100VE8RG72.3.2NACP**
[81] BuAer Memo, January 5, 1955, "Contract NOa(s) 10825 – Westinghouse J46-WE-18 engines", **B1098VE13RG72.3.2NACP**
[82] WAGT Report A-1902, Qualification Test of the XJ46-WE-18 Turbo-jet Engine on Contract NOa(s) 53-556, **B3123V9RG72.3.2NACP**
[83] BuAer Memo, January 26, 1955, "Contract NOa(s) 53-556 – 150 hour qualification test of the XJ46-WE-18 engine model, Transmittal of report on", **B3123V9RG72.3.2NACP**
[84] BuAer Memo, March 1, 1955, "Contract NOa(s) 10825, qualification test of J46-WE-18 turbojet engine.", **B1101VE9RG72.3.2NACP**
[85] WAGT Memo, March 21, 1955, "Contract NOa(s) 10825 – Westinghouse Turbo-jet Engine Model YJ46 We-18. Comparison of Engine Configuration with that of J46 WE-8B and -8 Models.", **B1101VE9RG72.3.2NACP**
[86] BuAer Shipping Authorization, November 5, 1956, CHI-001184/57
[87] BuAer Shipping Authorization, November 1, 1956, CHI-001173/57, **B1101VE10RG72.3.2NACP**

Chapter 8
J46-WE-1, -3, -5, -7 Development

The Air Force approached the Navy and requested they facilitate an interagency contract to have Westinghouse produce a limited quantity of the new J46-WE-2 that met the specific needs of the Air Force. WAGT submitted a proposal (AGT-46, date unknown) based on the -2 and BuAer rejected it, asking that it be revised to include some new requirements. Specifically, the engines had to generally conform to the new MIL-E-5007 requirements for specifications, that they had no need for mock-ups, and WAGT had to investigate how to increase the thrust of the engine at 35,000 feet by 500 pounds over the XJ46-WE-2 and at speeds ranging from 0.9 to 1.9 Mach number.

As BuAer requested, WAGT resubmitted their proposal on October 31, 1949, using (confusingly) the same proposal number AGT-46, but now broken into two parts, A and B, the first covering development and the second the production of ten engines.[1]

Part A specifically stated that the Air Force version of the engine was to be tailored to support the Douglas MX-656 project, the research airplane model DS-499D, known today more commonly as the X-3 *Stiletto*. This aircraft was being designed to explore high speed aerodynamic phenomena to speeds of Mach 2 for sustained periods of not less than 30 minutes initially, later reduced to 10 minutes. That flight duration time would allow thermal soaking of the structure to occur, giving designers the data they needed on materials and construction techniques to deal with such heating.

The engine would be developed under specification WAGT-X24C10-2D dated October 3, 1949 along with its Appendix D. Delivery of the ten production engines (the schedule predicated on receipt of authority to proceed was received not later than March 15, 1950) would begin with two engines a month in November 1951. These engines were styled the XJ46-WE-1 and were to be 50 hour engines. Defects found within the 50 hours of operation would be repaired or corrected at no charge to the government. Final acceptance of the engines would be made by BuAer at WAGT's plant in accordance with the specification. The 50 hour acceptance test would be in accordance with Specification MIL-E-5156 (USAF), as modified by Appendix D.

WAGT had to produce a report covering their investigation of how to improve engine thrust by March 31, 1951, provide two sets of experimental laboratory maintenance tools by June 30, 1951, provide engine operating recommendations for the engine by June 30, 1951, provide supplements to the XJ46-WE-2 Service and Overhaul Manuals by August 31, 1951, and provide a set of final design drawings and specifications on or before June 30, 1951.

The specification was initially to be delivered by February 15, 1950 but was later changed to July 31. The Preliminary Flight Rating Test Report was to be delivered on or before May 31, 1951. The tested engine was to be overhauled and delivered to the Air Force Air Material Command on or before June 30, 1951.

Additional requirements that had to be included in Appendix D were:

1. Standard engine operation with AN-F-32 (JP-1) fuel.
2. Be capable of flight at Mach 2 above 35,000 feet under NACA Standard Conditions.
3. Provision for relocation of certain control equipment to increase the installation clearance behind the engine driven accessories and forward of the fuel manifold section.
4. Relocation of the mounting points.
5. Removal of anti-icing provisions.
6. The A/B extension to incorporate an angle between the A/B centerline and the engine centerline and to incorporate a quick-disconnect joint.

Interestingly, the purpose of the thrust improvement study was now stated as the "evaluation of the ways and means of providing maximum additional thrust over the standard XJ46-WE-2 engine at 35,000 feet and at speeds ranging from 0.9 to 1.3 Mach number". The specific goal of achieving a 500 pound thrust increase was no longer a goal, just a general increase in thrust (if possible?).

The development contract (covering proposal AGT-42A) was to be cost plus fixed fee, with the total price to be $942,000. It was negotiated through the interagency contract MIPR (33-038) R-MCREXP-50034-N dated November 29, 1949. BuAer issued a Letter of Intent for Contract NOa(s)-11028 on February 24, 1950 to cover the development contract (which did not include production costs). Within a few months, the LOI contract was replaced by a Cost Plus Fixed Fee Contract dated February 24, 1950 and put in place for $901,654. The price included a fee of 7%. The lower price reflected costs already incurred while the contract went through processing and already paid to Westinghouse under the LOI.[2]

Interestingly, WAGT had to issue a clarification letter to BuAer in April of 1950. It stated that since the LOI contract had not actually been received by WAGT until February 24, 1950, the delivery dates were delayed two and a half months. They further stated that it was their understanding that any investigation into thrust improvement possibilities should not delay the delivery dates of the engine. Assuming that was true, any changes to the airflow path of the basic engine could not be made in the time available and the other basic engine components could only marginally affect engine performance. WAGT argued that higher A/B thrust through higher operating temperature was not certain of success due to the difficulty of maintaining reasonable engine life with a high temperature operation, basically ruling that out before they began.

WAGT then stated they were not saying there was no possibility of performance growth over the -2, but were saying an appreciable increase was not a certainty and recommended that the basic -2 performance figures in the specification (which were from the -2) be used to calculate the aircraft performance.

(In other words, the basic -2 engine performance was what they were going to get given the time constraints imposed due to late execution of the contract but some increase might occur. One wonders if these people were ever on the same page. Why ask for and accept the task to study ways to get more power and then turn around and say you'll never get it within the time constraints of the contract? Surely that was a conversation that should have been held back in 1949 before the contract was executed? I believe the letter was a backhanded way of getting BuAer to negotiate a schedule slip with the Air Force without coming out and stating that was what they needed. – Auth.)

As for reliability, WAGT pointed out that the parts common to the J46-WE-2 already had a demonstrated reliability having passed the reliability test contained in the contract NOa(s)-9670, which was based on the AN-E-32 specification qualification test. They acknowledged that test was not as stringent as the replacement MIL-E-5009 test but that the components would meet the MIL-E-5256 standard in the proposal.[3]

The first Monthly Progress Report (MPR) for April 1950 stated that since the development of the -2 had a direct bearing on the -1 and since evaluation of the -2 was just beginning, considerable testing had to be undertaken (of the -2) before it could be correlated with the XJ46-WE-1.[4]

Jay Miller's chapter on the X-3 in his excellent book on the X-Planes states that as early as October 1949 it was apparent that the J46 engine would never be ready in time for the airframe and the engine was switched to a version of the J34-WE-32.[5] Given the contract dates and the communications already noted, it is likely the time frame of that decision was really October 1950, a year later.

The May 1950 MPR was positive, reporting that many of the engine elements and components had been put on order and the most urgent development projects had been started.[6]

The MPR for June/July continued in the same vein, reporting progress was satisfactory, stating again most of the accessories were the same as the -2 and attaching a copy of the Contract NOa(s) 9670 MPR for the same period for the Air Force's information.[7]

In September, The Chief of BuAer sent a memo over to the Commanding General of Air Material Command asking if a three month slip in delivery schedules for all the items under the contract for the -1 was acceptable.[8]

The August MPR contained more assurances that everything was on track and that the program had been accelerated. It blandly stated that with the background thus far obtained with the XJ46-WE-2 program, the -1 development would be that much easier. The first running -1 was not contemplated by the BAR until March or April 1951.[9]

The rumbles of delay in actual progress appeared the very next month. The MPR for September reports that very little progress had been made on the unique elements of the -1. The compressor housing and compressor vanes and other minor elements were still in layout even though similar parts for the -2 had been released to manufacturing.[10]

BuAer notified WAGT in November that the schedule changes to the development deliverables had been accepted but reminded them that the change would not be considered justification for delays in production XJ46-WE-1 engines, if such materialized.[11]

The next MPR for October had a cover memo that reminded the BuAer Chief that there was no actual commitment to deliver an engine of higher performance than the -2, but that an effort would be made if it did not impact the ability to achieve the qualification of the -1. The reason for the clarification was that WAGT had finished their thrust improvement study (report A-1047) and found that it would not be possible to increase the thrust in time to deliver the engines on schedule, given the amount of development time it would take for the possible improvement paths they had considered in the study. This time, the BAR stated that the progress of the -2 was satisfactory and reminded everyone that many elements of the -2 are "like that used on the XJ34-WE-32" and that engine was "almost fully qualified".

After giving such a confidence booster, the BAR stated that the first -2 engine had suffered a second stage compressor failure and parts from the 2nd and 3rd house engines were being juggled or borrowed to get the first two house engines back into test. Given that, the XJ46-WE-1 program progress was now only rated as fair.

WAGT's detailed report stated the compressor housing would be made of steel (instead of magnesium) and made in one piece (the -2 was in two pieces bolted together longitudinally). The detail drawings were about 50 percent complete. The combustion chamber drawings were complete and were being quality checked. The bend A/B extension design from the layout was complete. The engine would use twenty segment liners in the turbine housing instead of the ten used in the -2. The compressor would be constructed from two (2) aluminum, nine (9) titanium and one (1) steel discs. The first stage turbine nozzles would be hollow. A variable displacement pump for the nozzle actuator would not be used, but the actuator itself would be a heavier (more robust?) unit. An insulating blanket was on order with details to be taken from the engine mock-up. The estimated weight of the engine was 2,001 pounds (vs. 2,074 for the –2 with a 2% contingency factor).

The mock-up was ready for construction with completion expected in December. (Apparently the differences in the engine installation envelope from the -2 were enough to make the Air Force change their mind about the need for a mock-up.)[12]

So, just what did WAGT study as possible paths to increased thrust? Their report lists:

1. Increasing the A/B operating temperature. The NACA was studying that but their results were not yet available. Assuming a 420°F net increase in operating temperature was achievable, it would increase net thrust by 6% if the turbine inlet temperature was held at 1,525°F.
2. Increasing the turbine inlet temperature. Going from the current 1,525°F to 1,800°F would give about a 3% thrust increase at the airspeeds and altitude desired.
3. Increasing the air flow. Approximately 5% more air could be moved through the engine, but it would require all new blading. Degradation of slow speed compressor performance could be expected, although given the MX-656 program focus that might not be a concern. The approach being given to the XJ40-WE-10 (the high power version of the J40 which failed) if applied to the J46 could theoretically increase the sea level thrust of the J46 from 6,100 to 7,850 pounds.
4. Use water-alcohol injection. Such use in the combustor raised the compressor ratio of a given cycle to near the stall line and, although it allowed higher short term A/B thrust increase, would require redesign of the compressor to provide enough engine stall margin. In theory it could increase engine thrust about 6% but would increase the total fluid consumption of the engine by approximately 65% to obtain the short term increase. (Not stated in the report, but they could have added that tankage for the injection fluid and various pumps and controls that would have been needed in the tight MX-656 airframe.)
5. Expect a gradual thrust improvement as the present development programs progressed (no details to support this statement were given).

Two attached charts were included in the report showing thrust increase vs. aircraft speed for both the increased turbine inlet temperature approach and the improved higher temperature A/B. Neither assist the reader in understanding how much faster the aircraft would have been able to go given either approach.[13]

The very next report moves the mock-up completion to the second week of January 1951. Components different from the -2 were still not fully released to manufacture. The control system was changed from the electronic system to the "Simple Control System" which was undergoing evaluation and testing. The bellows booster pump had to be redesigned due to it failing the first -2 attempted qualification test. The BAR still considered the progress to be only fair. However, the basic position was that the commonality with the -2 would somehow make up for the risks to the schedule.[14]

As with the other engines under contract, in November WAGT sent a memo to BuAer stating that the aircraft designer's tasks would be simplified if the specifications were changed to show the maximum fuel consumption of the engine by stating that using the primary control, consumption would not exceed "X" pounds per hour.[15]

It will come as no surprise that the December 1950 MPR reported that the -2 program was suffering from serious engineering problems. The compressor and turbine had not been matched yet with compressor stalling being experienced. The first stage nozzles were changed to see if that would eliminate the stall. The new vaporizing burner had suffered "candy cane" burnout after four hours of testing.

All the parts unique to the -1 still had not been released to manufacture, with the compressor housing and plug type exhaust nozzle still in layout.

In spite of the engineering problems being encountered with the -2, the BAR thought the -1 was progressing satisfactorily.[16]

On February 11, WAGT asked BuAer for information on any starters they planned to use on production engines, including installation drawings and part numbers, characteristic curves for torque and horsepower and installation information for exhaust gas temperature for cartridge starters, power sources for electric starters and air sources for air starters.[17]

The next day, WAGT sent out a survey regarding changing the design of the engine front mount. A full copy of the survey was not found so we do not know exactly what was involved.[18]

The January 1951 MPR states the -2 was still experiencing compressor stall, but release of most -1 engine components to manufacture had occurred. The most important part not yet released was the "fabricated" compressor housing, but this was due out to the shops by the end of February. The matching problem was also holding up the compressor and turbine disc release. The insulation blanket had been positioned on the mock-up and mock-up completion was expected in mid-March.[19]

It was at this time that WAGT asked that high temperature magnet wire supplies be released from the Bureau of Ships to BuAer and WAGT for the manufacture of the relay package in the Auxiliary Electrical Control.[20]

During March surveys on spring-loaded quick disconnect clamps and oil filter removal clearance for the -1, -2 and -4 were sent out.[21, 22]

The February MPR states that vibration being experienced in the titanium compressor discs showed that a heavier web would be needed in the discs but the redesign was not expected to hold up the -1 progress. The problems in the -2 were still unresolved and were the primary reason the -1 could not make further progress. In spite of those issues, the progress on the -1 was considered good.[23]

WAGT simplified the A/B fuel system on the -2 and received BuAer approval for the design change in April, but BuAer told them not to include the changes on the -1 until the Air Force had reviewed and approved them.[24] The Air Force referred the change to Douglas because the change added a new 1.25 inch pipe on the exterior of the engine.[25]

Compressor design in the -2 was still stalled in March, with six different designs having been tried to date. The plan was to modify one of the earlier designs that had demonstrated unacceptable stall characteristics and use it to be the basis to move forward with endurance testing.[26]

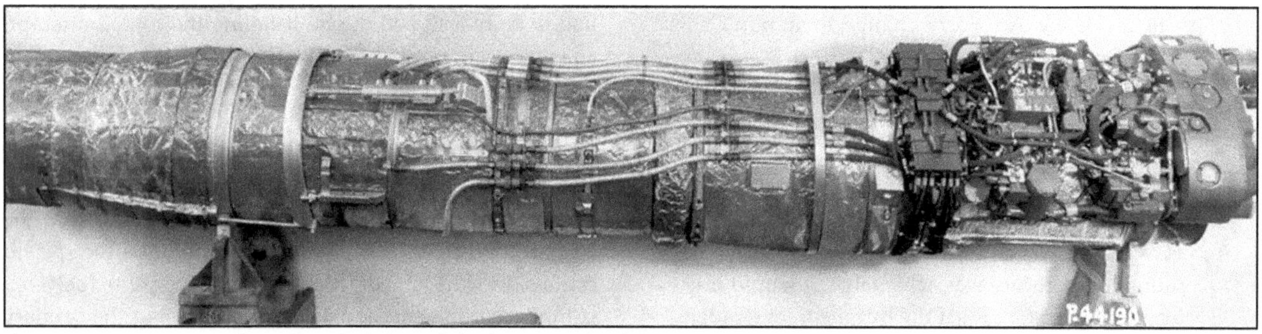

Figure 1: XJ46-WE-1 Mockup. A typical view of the bottom side of the engine. The Air Force could see any piping changes and the planned insulation blanket installation. WAGT P44190, 3/15/1951. ***Courtesy Hagley Museum and Library***

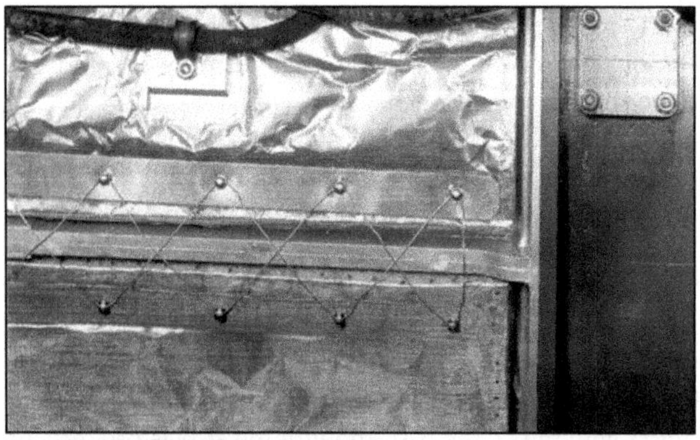

Figure 2: Mockup detail of insulation blanket attachment method around the compressor case, just behind the inlet. WAGT P44200, 3/15/1951. ***Courtesy Hagley Museum and Library***

Unable to produce a -1 with the original performance characteristics, a secret conference was held at Westinghouse in April. There it was proposed that WAGT build an interim de-rated XJ46-WE-1 engine (X24C10-1D) operating on AN-F-58 fuel (JP-3) to Specification MIL-E-5156 as modified by specification WAGT-24C10-1. The engine would be able to operate at Mach 2.0 above 35,000 feet. WAGT proposed such an engine for the additional estimated development cost of $174,665 and resetting the original delivery dates for the -1. With the caveat that the authorization to proceed was received by June 1, 1951, these now became:[27]

1.	Fifty (50) hour test	8/31/52
2.	Overhaul and delivery	9/30/52
3.	Investigation to Improve Engine Thrust	6/30/52
4.	Maintenance Tools	1/30/52
5.	Operating Recommendations	12/30/51
6.	Service & Maintenance Manuals	2/30/52
7.	Final Design Drawings &	9/30/52
8.	Monthly Progress Reports	Monthly
9.	Appendix to Model Specification	7/31/51

The new interim engine dates were to be:

10.	Model Specification & Preliminary Drawings	7/1/51
11.	Fifty (50) Hour Preliminary Flight Rating Test Report	10/15/51
12.	Final Design Drawings & Specifications	11/15/51

The WAGT proposal asking for additional funds to deliver an interim engine caused potential accounting problems for both BuAer and the Air Force. BuAer recommended that both they and the Air Force treat the additional costs entailed to develop the interim engines as "cost overruns" and charge them off to other allocated funds through normal procedures.[28]

The April MPR states that the -1 design and release was 90% complete, being based on the designs for the -2 in large part. The engineering problems with the -2 continued, but progress was being made on the stall problem and there was hope it would be resolved during May. (Details are in the -2 development chapters.) The BAR states that if the WE-2 problems were cleared up by early June, then the "WE-1 50-hour test scheduled for 31 August looks favorable."[29]

The latter part of May found WAGT surveying the power requirement of the emergency control systems for the J46 and J40.[30]

In mid-June, the Air Force rejected the Westinghouse proposal for additional funds to produce an interim de-rated 5,700 pound thrust engine. Instead, they would accept BuAer's recommendation and make the 50-hour test of the interim engine an allowable charge under the NOa(s) 11028 contract. They stated that delivery of six (6) engines was needed to meet the flight schedule of the number two X-3 airplane and asked BuAer to negotiate with WAGT the date and terms of the 50-hour de-rated test and keep the Air Development Force of the Air Force apprised of the progress of negotiations.[31]

Just after this memo was received, Amendment 1 to the contract was signed changing the allowable labor rates for the contract but not changing the estimated cost or fixed fee of the contract. Such renegotiation of reimbursable labor rates was done annually.[32]

The May MPR arrived at that time and it reported that the compressor stall problem was closer to resolution but not completely gone yet. Considerable development needed to be completed before the -1 would be ready for the 50-hour test scheduled for completion in August. However, all detail parts of the engine had been released for manufacture. Based on the timing of the report, it must be the de-rated version of the -1 being discussed.[33]

Another survey was done in June, this on the revised auxiliary electric control to add a magnetic amplifier type circuit for emergency control of the variable displacement oil pump for the operation of the exhaust nozzle. The circuit was going to be moved into the box holding the auxiliary electric control which had to be made larger as a result. Shortly after it was sent out, the survey was made redundant when the control system was changed to the simple hydraulic control system.[34]

The weight of the engine crept up, going to 2,074 pounds. This was due to elimination of the airframe cooling shrouds around the engine, requiring an additional insulation blanket (at a weight of 9 extra pounds) from the A/B quick disconnect clamp to the aft end of the ejector.[35]

BuAer finally got around to notifying WAGT that they could charge the interim 50-hour development test of the interim engine to the contract, which was not going to be modified. If the costs to develop the engine were going to exceed the negotiated contract price, they were to notify BuAer. A specification defining the interim engine was to be prepared and attached to the -1 specification as Appendix A. WAGT was reminded of the importance of meeting the Air Force's schedule and if the schedule could not be met, the bureau was to be advised immediately.[36]

A revised power regulator was surveyed in late June, the revision making the box a bit wider by a little more than a quarter inch. The adjustment points and cable connector were relocated and a new deceleration adjustment pot added. They were needed due to extra circuitry required for combustion control with engines employing the short combustion chamber and the addition of an internal voltage regulator for controlled tube heater voltage.[37]

WAGT sent out a memo in early July notifying both BuAer and the Air Force that they were proceeding with the simplified afterburner fuel system for the J46 engines. This system was usually just called the Simple Hydraulic Control System in future communications.[38]

The July MPR (the June report is not in the archives) was issued August 10. It states 98 percent of the detail parts for the -1 were on order. Work was continuing on the A/B, extension and engine assembly drawings. New compressor discs to avoid critical frequencies in the rear stages were on order. Stronger first and second stage discs were also on order. The finished design for the parts was expected to be complete in October. The change to the simple hydraulic A/B control would slightly reduce operating thrust at some operating conditions and this was under investigation to insure thrust would not be affected at the MX-656 planned operation altitudes. Steel discs for the -1 had been ordered as a backup to the planned titanium discs and these would affect the engine weight if they were used instead of the titanium discs.[39]

The August MPR reports that the interim -1 engine had now been designated the XJ46-WE-3 and the A/B parts assembly drawings were expected to be complete by the middle of September. The -3 was essentially the same as the BuAer -8 engine. The insulation covers were on order and the effort now was to coordinate detail drawings with the insulation manufacturer.

The new Simple Hydraulic Control (SHC) system was under consideration for use on the -1 and the -3, showing it had not definitely been accepted in spite of prior statements in surveys and status reports.

The complete assembly drawing of the -3 would be made during September. The dry weight of the engine was holding at 2,069 pounds.[40]

The decision was made in September to switch the -1 and -3 to the SHC and this created a schedule delay while a version of this system was designed for the AF engines. Any delays in the -8 program were affecting the -3 engine as well. WAGT had reported the electronic control was not ready for the Air Force application and due to their limited laboratory facilities it could not be brought to operational status without delaying the program.

Delivery of the first -3 engine slipped to April 1952 and the first -1 to July 1953.[41]

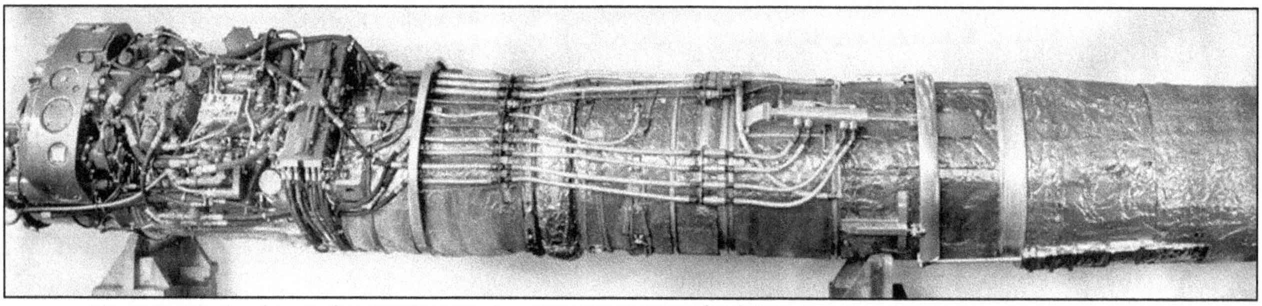

Figure 3: Interim de-rated XJ46-WE-3 Mockup. WAGT P47166, 6/19/1952. *Courtesy Hagley Museum and Library*

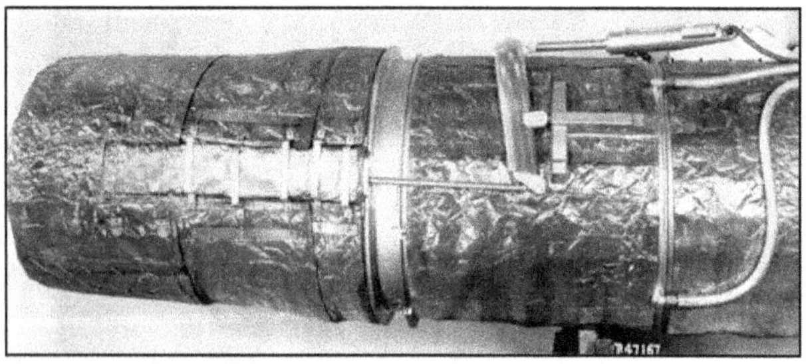

Figure 4: XJ46-WE-3 mockup insulation details on the A/B section. WAGT P47167, 6/19/1952. *Courtesy Hagley Museum and Library*

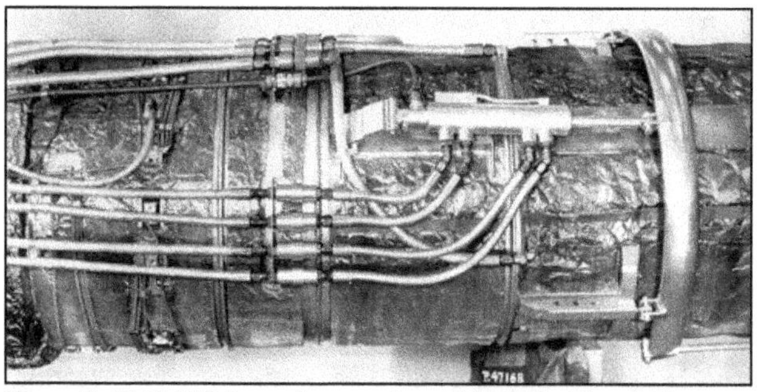

Figure 5: XJ46-WE-3 mockup engine mid-section insulation blanket details. WAGT P47168, 6/19/1952. *Courtesy Hagley Museum and Library*

With the electronic control not totally dead, in November, WAGT surveyed adding a manual override lever installation on the exterior of the variable displacement pump that would add an emergency way of preventing the pump from rapidly closing the nozzle in the event of control failure, creating an over-temp of the engine. This was EC No. 1 for the -1 and 103 for the -8.[42]

The November MPR is missing from the archives.

In December Amendment 3 to the contract was issued, again adjusting the general and administrative expense rate attributable to HQ activities at WAGT.[43]

The Air Force was not giving up on getting more thrust from the A/B of the -1 if it could be done. WAGT in November 26 submitted a proposal for subcontracting the work to Marquardt Aircraft Company. An A/B would be constructed and tested for eight weeks. A technical report would be submitted eleven months from the date of receipt of authority to proceed and the final report within 15 months. The cost for the work would be $422,425. The work was predicated on using the Air Force owned altitude test facility located at the Marquardt Aircraft Company's location. It was further predicated on the use of parts from the J46-WE-6 afterburner evaluation testing to be performed by Marquardt. (The -6 afterburner was to have been an attempt to get A/B power up to 60,000 ft. A contract was initiated but never signed.)

The proposal from Marquardt to WAGT dated June 17 included some details of the J46-WE-6 afterburner (*found nowhere else in the Archives*) that would form the basis of the work on increasing thrust at 35,000 feet.

1. Size would be to WAGT drawings to insure installation compatibility with the MX-656.
2. The assembly would be flight type with the exception of the nozzle exit portion, which would be water cooled to minimize cross section changes due to exit temperature variations.
3. The variable area portion of the exit nozzle would not be fitted.
4. The fuel tubes and inner shroud would be WAGT parts.
5. The flame holders would be "close copies" of welded construction.
6. The testing gas generator would be a TG-180 engine with a 20 foot duct to the A/B test section with suitable provisions in the duct to provide desired variations in the radial distribution of the air flow.
7. The fuel would be MIL-F-5161A weathered (aged in tanks) and MIL-F-5616 (JP-5).

The rest of the document covers the test parameters, the test equipment to be used and the information needed in the resulting reports.[44]

BuAer forwarded the proposal to the Air Force on December 20, 1951, giving a general description of the contents in the cover letter, particularly the proposal's dependence on the high altitude test cell.[45]

Attached to the cover memo to the January status report (the December and January MPR's are not in the archives) was a note saying the expenses to date were covered by the contract funds but that the balance of the funds was not deemed sufficient to complete the contract and negotiations were underway.[46]

WAGT sent out an information request in late December asking for information on the performance of the Auxiliary Electric Control in order to gauge the performance of the Micro-positioner.[47]

On January 8, WAGT sent over an updated cost estimate for the A/B thrust increase study, made necessary due to a change in Marquardt's costing rates. The price rose from $422,425 to $433,999.[48]

After considering the request to use the Air Force altitude test cell at Marquardt, the Air Force negotiated for the testing to be accomplished at the NACA. The Marquardt Jet Laboratory would be busy with checkout tests until March 10 and was not available for the A/B work. The Air Force complained that the request for the use of the Marquardt laboratory should have come from the Navy, not WAGT directly. If such a request was made by the Navy, it would be considered within the overall facility schedule. Except for a small amount of time left on another Navy contract for the facility, the rest of the schedule for 1952 was filled.[49]

In March, BuAer reported that the Air Force had rejected WAGT's A/B proposal and that it was their understanding that WAGT had made arrangements with NACA to do the work on increasing A/B thrust.[50] (The author reviewed the voluminous number of A/B related reports issued by NACA from 1952 through 1956 and found none specifically related to the project in question. NACA tested everything possible about A/B design and operation during that period, including cooling, combustion efficiency, higher temperature operation, ejector designs, flame holder location and designs, fuel or cooling injection, etc. Some of the work clearly influenced the J46, as the A/B diffuser was lengthened after the -8A was designed.)

Work on the -1 and -3 continued, with WAGT responding to Douglas's comments relative to an earlier survey covering the suggested removal of the cross-over valves of the A/B fuel system and replacing them with a cross-feed line from the engine booster pump outlet to the A/B fuel pump inlet. The reason for the proposed change was simplification of the A/B fuel system. The cross-feed line was added to satisfy Douglas's requirement for a fail-safe need on the A/B during take-off of the MX-656 aircraft. Douglas stated there was no room in the MX-656 for the cross-feed line and rejected the survey. WAGT commented that they found that the emergency A/B fuel system, as originally designed for the J34-WE-32 and also used on the Air Force -17 version of the -32 turned out to be very hard to assemble in manufacture and added torque to the power lever operation. Since very few -1/-3 engines were to be built, they agreed to keep the system on the -1 and -3.[51]

In mid-March, in response to an Air Force query on the impact to engine performance if JP-4 was used for the -1/-3 engines instead of JP-1, Douglas compared the two fuels and cryptically reported:

1. Fuel system critical altitude – JP-1 better because if the tank mounted booster pumps failed, the subsonic absolute ceiling on a hot day with non-A/B operation was reduced 2,000 feet. Standard day ceiling in the circumstances would be unaffected.
2. Altitude starting – JP-4 was better because it had a higher altitude limitation than JP-1.
3. Boil-off – JP-1 was better with less boil off at altitude, but since the fuel tanks were pressurized to 5 psi, JP-4 would not be a problem either.
4. Ratio Fuel to Airplane Gross Weight – JP-1 was better, due to slightly higher energy density.
5. Heat of Combustion – JP-4 slightly better.
6. Combustion Efficiency – Either, no difference in combustion.
7. Convert engine to JP-1 – JP-4 better since the engine was already designed (the -3) to use it.
8. Procurement of fuel – JP4 better, fuel was already readily available but JP-1 was not.

Douglas recommended using JP-4 on the grounds that items 1, 3 and 4 did not make a significant difference in engine performance and also recommended JP-4 be tested on the airframe fuel system to verify any operation limits that might be affected.[52]

Looking ahead for a month, BuAer authorized WAGT to use JP-4 in the XJ46-WE-1/-3/-5. This is the first time the XJ46-WE-5 is mentioned.[53]

The contract was amended in April 1952 to add the costs to pay for development, qualification, drawings and specifications for the XJ46-WE-3. The justification was the MX-656 project.[54] On the same day, the Air Force changed their accounting internally to charge two different accounts to cover the contract costs.[55]

When WAGT sent out their survey for the -2 which covered the change to a larger turbine (the T-13), the -1 was affected as well. Douglas commented that the drawings included did not show the routing of lines near the flexible joint and they assumed they would run under the joint near the centerline of the engine to stay within the envelope shown on the Douglas drawings. They had no problem with the larger turbine.[56]

The MPR for June 1952 states the planned qualification test of the -3, scheduled for August, would not be met. October now appeared the more likely time frame for the test to occur. The report includes the information that the three jaw starter clutch shaft was replaced by a twelve jaw type at the Air Force's request. Although new milling fixtures were in design for making titanium compressor discs, the existing machines would be used to make the set to be used in the 50-hour test engine. Total engine weight was now 2,056.9 pounds. The weight reduction was caused by the shift from the electronic control to the simply hydraulic control.[57]

In July, the Air Force asked BuAer if there were additional tests or costs needed to test the -3 engine using JP-4 fuel. The memo covered a broad range of WAGT engines, some being tested on JP-3 and some on JP-4.[58]

On July 28, WAGT surveyed a new eight point fuel injection system for all the J46 models. The purpose of the change was to achieve a more positive and even ignition. The location of the spark plugs and the A/B igniter nozzle was changed slightly to prevent interference between parts.[59]

At the end of July, WAGT sent out the descriptions of the differences between the Simple Hydraulic Control and Hydro-Mechanical Control systems to the Air Force.

It was planned that production -1, -3, and -5 engines that would be delivered beginning in mid-1953 would use the Hydro-Mechanical Control, but the test XJ46-WE-3 engine would use the Simple Hydraulic Control. The first -5 engine would also use the Simple Hydraulic Control.[60]

In the August MPR, WAGT reported that assembly of the -3 engine for green test had been suspended pending a meeting with the Air Force to determine which performance improvement components should be included in the engine. WAGT was taking advantage of the delay to finish the machining of the titanium compressor discs so they could be included in the green run instead of the planned XJ46-WE-8 discs. The latest stiffened splitters in the diffuser were installed as well as the twisted diaphragm vanes complying with the AEL testing findings. The engine weight was now up to 2,085.2 due to the compressor housing turning out to weight 27.5 more pounds than estimated.[61]

On October 1, the Air Force approved the use of the T-13 turbine in the XJ46-WE-1 and XJ46-WE-5 engines. This approval shows that the XJ46-WE-5 would have been essentially a version of the Navy's -18 if it had been built.[62]

The Air Force sent over a memo stating that if Westinghouse was going to recommend using ammonia injection to increase the thrust of the -3, it was their understanding a new A/B fuel control would have to be developed. It also included the statement that the -5 had to stick with the 25.5 diameter A/B for the engine to fit in the X-3 aircraft. The -2 was to move to a 28 inch diameter A/B. The contract could not be redefined until BuAer decided whether to incorporate the 28 inch diameter A/B in lieu of the 25.5 inch on the -2. Since WAGT had stated the 6,100 pound thrust performance could not be obtained with the 25.5 inch A/B, a new engine definition would be required and also new engine model designations.[63]

The Chief of BuAer received a memo from the Air Force stating that Westinghouse had just slipped the schedule for the qualification test of the -3 engine from mid-November 1952 to May 1953. After listing the known reasons given for the prior delays that had slipped the schedule from the original October 1951 date to mid-November 1952, they asked BuAer to explain the new slip of 6.5 months.[64]

WAGT informed BuAer in late October that their failure analysis of the Temperature Control Amplifier and its associated report, due on October 3, was not yet complete.

The actual analysis of the system was finished, but due to changes being made to improve the control circuitry in regards to reliability, the full description of the operation of the circuitry was not yet complete. WAGT now estimated the report would be delivered by December 12.[65]

At the end of October, WAGT sent out a change notice regarding the overspeed relay. Due to reports of various interferences within the airframes in its current location and rotation, WAGT was going to rotate the relay 25 degrees clockwise, looking aft. Initially, mounting brackets would have holes for both rotations as the existing relays used the old bolt pattern. Going forward, all new production engines would begin to have the rotated bolt pattern relays.[66]

BuAer's response to the Air Force request for an explanation for the schedule change on the -3 was sent over on November 18. This document is possibly the best example available explaining the intricate dependencies of the Air Force on BuAer's versions of the engine.

"The scheduled completion of the qualification test in November 1952 was based on the Air Force acceptance of de-rated -3 engines (5,500 lb); however, it is now understood that the Air Force cannot use engines with this thrust. The May 1953 date for fully rated -3 engines was based on availability of turbine improvements which were to be incorporated in the XJ46-WE-8 engine during April 1953. This bureau has informally been advised that the qualification date for the -8 engine has slipped, therefore it is estimated that the -3 will slip the same amount. This is due not to the Navy requiring prior qualification of the -8 but to non-availability of parts.

"Due to basic differences between the XJ46-WE-2 and the XJ46-WE-5 engine, the Bureau of Aeronautics no longer considers it mandatory to qualify the -2 engine prior to starting the qualification of the -5 engine. It is considered that the programs can be carried along in parallel where the two engines differ. The use of the 25.5 inch afterburner for the -5 will probably result in earlier qualification of the -5 engine inasmuch as an afterburner of similar size will be used on the -8 engine."[67]

On November 26, 1952, WAGT surveyed a hinged quick disconnect clamp at the end of the combustion chamber flange. The clamp bolts were repositioned to 45 degrees on either side of the bottom vertical centerline and the hinge joint increased in size to handle the additional stress requirement.[68]

Two days later, WAGT sent out a memo following up on the survey for the new eight point fuel injection system. WAGT stated that the new injection system combined with the new longer engine diffuser would eliminate the difficulty of the fuel feed lines to the ignitor nozzles passing over the engine-mount ring flange and out of the engine envelope. On the -3 engine, the two-point injection system would be retained unless laboratory tests indicated a need for the new system.[69]

WAGT suggested a forward engine mount change for the J46 in a survey on November 28. The change tightened the dimensions between the inner faces from 1.140-1.125 to 1.135 to 1.120 inches.[70]

They also surveyed the removal of the bottom mounted oil level indicator from the oil reservoir at this time and suggested replacing it with sight glasses. The net weight change was +0.3 pounds.[71]

The MPR for November showed up on December 12. As did the prior reports, it only officially reported on the -1 and -3 progress, reminding us that the -5 often referred to was still not officially under a contract. As usual, it relied on the data in the MPRs covering the -2 and -8. The engine weight was now 2,081.4 pounds. The 50-hour test engine was scheduled to start testing in December. It lacked the accessory drive gearbox, which was in disassembly for modernization and redesigned A/B components.[72]

A few days later, WAGT surveyed the moving of the flexible joint to eliminate the interference between the afterburner flexible joint and the external pad weldments covering the exhaust collector supporting struts. The flexible joint would be moved 9/16 of an inch aft of the present location.[73]

The confused state of affairs was captured in an Air Force memo dated January 8, 1953. First, the Air Force stated that the continuing lack of parts, even after the improvements in the Westinghouse organization in the last year, was a major cause for qualification delays.

The Air Force had requested that the 50-hour qualification of an XJ46-WE-3 be moved up to follow the XJ46-WE-8 (5,720 or 5,800 pound thrust) engine and asked for a response to this request. It was of the utmost importance that every effort be made to provide an XJ46-WE-3 5,720 pound thrust engine at the earliest date.

They requested they be notified of the effect additional development of the 25.5 inch diameter A/B might have on the 50-hour qualification date for an XJ46-WE -18 engine incorporating 75°F over temperature operation and NH_3 (ammonia) injection at maximum thrust during

1-1/4 hours of this 50-hour test. The 50-hour test schedule date of the J46-WE-18 engine was stated to be October 1955 during an October 1952 meeting. It was requested that NACA's changes that were made on testing ammonia injection on a J34 be examined to determine if they could be applied to the Block IIA development program. Using that approach, WAGT would not have to await future NACA testing to move forward. The WAGT configuration could then be substantiated by NACA testing and the 50-hour test accelerated. Additional investigations on thrust improvement for the X-3 should continue.

The memo went on to say that with WAGT's inability to attain sufficient estimated altitude thrust on the contract, there might be little basis for continuing the Mach 2 flight requirement of the contract. A firm decision could not be made on that until Douglas's studies were received.

It was requested that BuAer notify the Air Force of the development cost of an alternate XJ46-WE-8 Air Force engine to the XJ46-WE-3 external configuration and the 50-hour schedule dates. Internally, it could use either steel or titanium in the last seven compressor discs. (The Block I engine was the basic -8.)

Lastly, the budget requirement for an additional $1,750,000 for development of the Air Force XJ46-WE-18 version was a serious concern, as funding requests were not submitted in either fiscal year 1953 or 1954 for such work. WAGT had currently ordered a work stoppage on the contract work due to lack of funding but an official funding request from WAGT had not been received, so the Air Force felt the stoppage could not be justified.[74]

BuAer did not respond until January 27 and stated the information requested by the Air Force was not available at BuAer. They requested the BAR at Essington give a copy of the Air Force's memo to WAGT to obtain the information needed to respond to the Air Force.

As for the work stoppage, BuAer had been given several verbal explanations as to the reasons, but nothing official had been received. They pointed out that the contract required WAGT to give at least two months' notice of a pending cost overrun and that no such notice had been received at the Bureau.[75]

On January 21, 1953, WAGT submitted their failure analysis report on the Automatic Temperature Control. The report gives a full description of the electronic circuits of the control and then a detailed analysis of the effect of failure of any component within the circuits. The most interesting thing about this report is that one of the engines listed to use the circuit is the "J46-WE-7", the first reference to this engine.[76]

On January 30, 1953, a stop work order was sent to WAGT stating that all work on the contract should stop pending a further decision of the Air Force to proceed.[77] A week later, the Air Force sent over a memo explaining why they stopped work on the contract. It basically stated that the experimental work on the J46 was to produce the ability to operate with at least 500 extra pounds of thrust at Mach 1 at 35,000 feet in order to be able to push the X-3 to the supersonic speeds needed to obtain the data that was wanted. In spite of many delays, the schedule slipped again to February 1955 and was still for a down-rated thrust engine of 5,800 pounds instead of the desired 6,600 pounds.

A review of the various alternate versions of the engines over time was then covered, with the alternate engine delayed to May 1953, but to achieve at least 6,100 pounds of thrust, the engine would no longer fit in the X-3.

In view of the lack of a suitable engine within the time frame necessary for the airplane to complete its flight research mission, the Air Force had cancelled the entire project.[78]

BuAer issued its memo of contract termination in its entirety in a memo on March 13.[79]

On March 16, WAGT responded to the earlier Air Force memo which the BAR had given them, asking for information on the status of the engine program. The long response, of only historical interest at that point, is summarized here:

1. For the October meeting referenced in the letter, WAGT left the meeting with the impression that a firm program had been mutually agreed upon and that official confirmation from the Air Force would be sent shortly thereafter. Subsequent to the meeting, the WAGT representative in Dayton, Ohio was told the Air Force was considering alternate power plants for the MX-656 project and no decision on the J46 project had been reached. Because of that, and with available funds under NOa(s) 11028 about to run out, they put an internal work stop order on the program until the Air Force made a decision.
2. A complex paragraph relating to the scheduling delays on the 50-hour test of the -3 explained the various project interdependencies and scheduling conflicts between the -3 test and BuAer's -8, with the BuAer test taking precedence, but it appeared the Air Force "seemed to have presumed" otherwise. Given all aspects of the work as WAGT knew it in October, the August 1953 date appeared to be a realistic for the -3 50-hour test.
3. It was implied that the Air Force had requested and WAGT approved parallel development of the Air Force engines along with the Navy development of J46 engines. If so, WAGT was never notified of this decision. "It has been our understanding since the

initiation of the MX-656 project that any development required for Air Force J46 engines was to be done as an additional part, or as an extension to the Navy work." WAGT began the program with a policy that favored the Navy with the basic configuration taking precedence. The Navy -2 was the basic model and the -3 was an alternate, being based on the -8 which was itself an alternate.

4. With the unapproved designation of J46-WE-7, WAGT had departed from the policy as they felt they could handle the development in parallel with the Navy work. That decision was based on the fact the basic engine design and general afterburner work was largely completed, so the timing of the -7 could be independent of Navy work. Depending on NACA work on additional thrust improvement, the schedule date could be anywhere from September 1954 to February 1955, but WAGT had no control over that portion of the program.
5. Ammonia injection was discussed in report A-1482 (not in the Archives) and was the basis for the estimate of the work required for the 50-hour test. WAGT stated A/B development on all previous engines had required changes to the fuel injection systems and flame anchoring system for combustion stability and durability reasons. The NACA tests on the J34 had shown instability problems both in the A/B and main burner.
6. A discussion of steel vs. titanium discs in the -3 for the 50-hour test followed and WAGT recommended staying with the titanium discs, since development on them was complete.
7. The next paragraph covered the detail cost estimates for the -3 and -7, with the $974,000 for the -3 and $1,845,000 for the -7. They recognized the previous estimates had been too low and appreciated the Air Force's budget problem.
8. No action would be taken on budget planning, since all work had been stopped.

This letter at least gives us the information that the J46-WE-7 was to be a version of the -8 with a new A/B featuring ammonia injection, a longer diffuser, modified controls, different bend point and ten stages of titanium discs in the compressor.[80]

WAGT's memo was apparently never forwarded to the Air Force by BuAer, as the Air Force asked in August 1953 for a copy of WAGT's response.[81]

Prior to closing out the contract, BuAer had to request that the Air Force inform them of their estimate of the percentage completion of the work at the time of termination in relation to the total work specified in the contract, including a description of how the Air Force determined the percentage completed.[82] No answer to that request was found. Given that prior to the contract ending, the Air Force had asked BuAer for the status of the development work and BuAer had replied that they did not know, one wonders if anyone at BuAer had paid any real attention to the Air Force contract for a long time.

At final contract settlement, the total obligations were $1,070,178 and the total payments made were $1,017,157.59. The settlement payment owed was determined to be $53,020.41.[83] WAGT elected to adjust the amount owed based on fixed fee payments already made ($41,291) and accept a final payment of $17,696 as full and complete settlement of the contract.[84] The settlement had to be adjusted again to account for some materials that WAGT had retained that were later determined to be government property, the finally adjustment taking place in 1957.

Actual testing of -1, -3 or -5 engines was very limited. The X-3 program had shifted to the J34-WE-17 of 4,850 pounds of A/B thrust and the lower thrust resulted in the program never attaining the supersonic test data it was intended to collect.

The McDonnell XF-88 Voodoo No. 1 (46-525) flew with J34-WE-13's of 3,000 pounds thrust (Military) and the second airframe (46-526) flew with J34-WE-15's of 3,600 pounds thrust (A/B rating), the A/B's being a short stack design McDonnell developed in-house.

The Lockheed XF-90 project considered using the J46, but the delays in engine development left it flying with J34's without A/B's in the first airframe and with Solar A/B's in the second.

Figure 6: XJ46-WE-3 #8 with extra A/B line insulation. WAGT P48464, 1/13/1953. *Courtesy Hagley Museum and Library*

Chapter 8 – J46-WE-1, -3, -5, -7 Development

[1] WAGT Proposal AGT-46A & AGT-46B, October 31, 1949, The Development and Manufacture of the XJ46-WE-1 Turbo-Jet Engine, **B1133V1RG72.3.2NACP**
[2] MIPR (33-038) R50034-N for Letter of Intent for Contract NOa(s)-11028, February 24, 1950, **B1133V1RG72.3.2NACP**
[3] WAGT Memo, April 6, 1950, "Letter of Intent for Contract NOa(s)-11028, Additional Information Concerning the Development and Manufacture of the XJ46-WE-1 Turbo-Jet Engine.", **B1134VERG72.3.2NACP**
[4] BuAer Memo, May 12, 1950, "Contract NOa(s) 11028, monthly progress report (month of April); forwarding of;", **B1134V1RG72.3.2NACP**
[5] Miller, Jay, *The X-Planes*, Aerofax, Inc. 1988, Arlington, TX. Page 47.
[6] BuAer Memo, June 16, 1950, "Contract NOa(s) 11028, monthly progress report (month of May); forwarding of.", **B1134VERG72.3.2NACP**
[7] BuAer Memo, August 16, 1950, "Contract NOa(s) 11028, item 8, monthly progress report (June and July 1950); forwarding of.", **B1134VERG72.3.2NACP**
[8] BuAer Memo, September 28, 1950, "Contract NOa(s) 11028 for development of XJ46-WE-1 turbojet engines; request for change in deliveries", **B1134V1RG72.3.2NACP**
[9] BuAer Memo, September 28, 1950, "Contract NOa(s) 11028, item 8, monthly progress report (August1950); forwarding of.", **B1134V1RG72.3.2NACP**
[10] BuAer Memo, October 27, 1950, "Contract NOa(s) 11028, item 8, monthly progress report (September 50), XJ46-WE-1 engine; forwarding of.", **B1134V1RG72.3.2NACP**
[11] BuAer Memo, November 6, 1950, (No subject), Memo Number AER-CT-43, **B1134V2RG72.3.2NACP**
[12] BuAer Memo, November 24, 1950, "Contract NOa(s) 11028, Item 8 – Monthly Progress Report (October 1950) – Forwarding of.", **B1134V1RG72.3.2NACP**
[13] WAGT Report, A-1047, December 6, 1950, "Investigation for Increased Thrust on the XJ46-WE-1 Engine. Contract NOa(s) 11028, Item 3 (Fulfillment), **B1134VERG72.3.2NACP**
[14] BuAer Memo, December 14, 1950, "Contract NOa(s) 11028, Item 8 – Monthly Progress Report (November 1950) – Forwarding of.", **B1134V1RG72.3.2NACP**
[15] WAGT Memo, November 17, 1950, "Revision of Engine Model Specification to show maximum fuel consumption with Primary Control Contracts NOa(s) 10385, 10067, 10943, 11028 and 9670.", **B1134VERG72.3.2NACP**
[16] BuAer Memo, January 19, 1951, "Contract NOa(s) 11028, item 8, Monthly Progress Report (December 1950) – forwarding of.", **B1134V1RG72.3.2NACP**
[17] WAGT Memo, February 11, 1951, "J46-WE-1, -2 and -4 Contracts NOa(s) 9670 and 11028; Starting of", **B1134V1RG72.3.2NACP**
[18] BuAer Memo, February 12, 1951, "Contracts NOa(s) 9670 and 11028, WEC turbo jet engine models J46WE-1, -2 and -4, change in design of engine front mount; survey of.", **B1134V1RG72.3.2NACP**
[19] BuAer Memo, February 21, 1951, "Contract NOa(s) 11028, item 8, monthly progress report (January 1951); forwarding of.", **B1134V1RG72.3.2NACP**
[20] WAGT Memo, February 21, 1951, "Auxiliary Electrical Control for Westinghouse Turbo Jet Engines on Contract NOa(s) 10067, 10114, 10825, and 11028. Material required for:", **B1134V1RG72.3.2NACP**
[21] BuAer Memo, March 7, 1951, "Contracts NOa(s) 9670 and 11028, WEC turbo jet engine models XJ46WE-1, -2 and -4, spring loaded quick disconnect clamps; survey of.", **B1134V1RG72.3.2NACP**
[22] BuAer Memo, March 8, 1951, "Contracts NOa(s) 9670 and 11028, WEC turbo jet engine models XJ46WE-1, -2 and -4, oil filter removal clearance; survey of.", **B1134V1RG72.3.2NACP**
[23] BuAer Memo, March 20, 1951, "Contract NOa(s) 11028, item 8, Monthly Progress Report (February 1951) – Forwarding of.", **B1134V1RG72.3.2NACP**
[24] BuAer Memo, April 11, 1951, "Contract NOa(s) 11028; proposed simplification of XJ46-WE-1 afterburner fuel system", **B1134V1RG72.3.2NACP**
[25] Air Force Memo, May 3, 1951, "Proposed Simplification of XJ46-WE-1 Afterburner Fuel System (Contract NOa(s)-11028)", **B1134V1RG72.3.2NACP**
[26] BuAer Memo, April 20, 1951, "Contract NOa(s) 11028; Monthly Progress Report (March 1951); forwarding of.", **B1134V1RG72.3.2NACP**
[27] WAGT Memo, April 19, 1951, "Contract NOa(s)-11028 Proposal for Interim XJ46-WE-1 Engine Our Reference WG-65000", **B1134VERG72.3.2NACP**
[28] BuAer Memo, May 14, 1951, "XJ46-WE-1 engines", **B1134VERG72.3.2NACP**
[29] BuAer Memo, May 18, 1951, "Contract NOa(s) 11028, item 8; Monthly Progress Report (April 1951); forwarding of.", **B1134V1RG72.3.2NACP**
[30] BuAer Memo, May 24, 1951, "Contracts NOa(s) 9670, 11028, and 10385, WEC turbo jet engine models XJ46WE-1, -2, -4 and XJ40WE-10; airframe A.C. power requirement for use with emergency control systems; survey of.", **B1134VERG72.3.2NACP**
[31] Air Force Memo, June 12, 1951, "XJ46-WE-1 Engines", **B1134VERG72.3.2NACP**
[32] BuAer Contract Amendment 1 to Contract NOa(s) 11028, June 14, 1951. **B1134V2RG72.3.2NACP**

[33] BuAer Memo, June 15, 1951, "Contract NOa(s) 11028, item 8; Monthly Progress Report (May 1951); forwarding of.", **B1134V1RG72.3.2NACP**
[34] WAGT Memo, June 14, 1951, "Contract NOa(s) 11028, 9670 and 10067 WEC Turbo Jet Engine Models XJ46WE-1, -2, -4 and XJ40WE-10. Revised Auxiliary Electric Control, Survey of. Engine Change Nos. (100) for -2 engines, (14) for -4 engines and (70) for -10 engines.", **B1134VERG72.3.2NACP**
[35] WAGT Memo, June 29, 1951, "Contract NOa(s) 11028, total Dry Weight of J46-WE-1 Engine, Increase of.", **B1134V1RG72.3.2NACP**
[36] BuAer Memo, July 5, 1951, "Contract NOa(s) 11028; proposal for development of interim XJ46-WE-1 engine", **B1134VERG72.3.2NACP**
[37] WAGT Memo, June 27, 1951, "Contract NOa(s) 11028, 9670, 10114 and 10067. WEC Turbo-Jet Engine Models XJ46WE-1, -2, -4, XJ40WE-8, -10 & -12. Increased Power Regulator Envelope Survey of. Engine Change No. 99 for -2 engines, 13 for -4 engines, 83 for -8 engines, 69 for -10 engines and 68 for -12 engines.", **B1134VERG72.3.2NACP**
[38] WAGT Memo, July 5, 1951, "Contract NOa(s) 9670 and 11028 WEC Turbo Jet Engine Models XJ46WE-1 and -2. Current Status of the Design of the Simplified Afterburner Fuel System, Information on.", **B1134V1RG72.3.2NACP**
[39] WAGT Report, A-1199, August 10, 1951, "Monthly Progress Report for July 1951. Contract NOa(s) 11028", **B1134V1RG72.3.2NACP**
[40] WAGT Report A-1219, September 10, 1951, Monthly Progress Report for August, 1951. Contract NOa(s) 11028., **B1134V1RG72.3.2NACP**
[41] BuAer Naval Speed Letter, October 18, 1951, "Delay in XJ46-WE-3 delivery schedule; reasons for", **B1134V1RG72.3.2NACP**
[42] WAGT Memo, November 2, 1951, "Contract NOa(s) 11028, 9670, and 10825 WEC Turbo Jet Engine Models XJ46WE-1, XJ46WE-2, XJ46WE-4, XJ46WE-8, and XJ46-WE-10. Manual Over-ride lever Installation, Survey of. Engine Change Nos. 1 for -1, engines, 1 for-2 engines, 1 for -4 engines, 103 for -8 engines, and 16 for -10 engines.", **B1134V1RG72.3.2NACP**
[43] BuAer Contract Amendment #3 to NOa(s) 11028, December 1, 1951. **B1134V2RG72.3.2NACP**
[44] WAGT Memo, November 26, 1951, "Contract NOa(s)-11028, Proposal for Further Development of J46-WE-1 Turbo-Jet Engine to Obtain Additional Afterburning Thrust, Our Reference WG-65000", **B1134VERG72.3.2NACP**
[45] BuAer Memo, December 20, 1951, "Westinghouse Proposals for development of additional afterburning thrust in J46 engines, forwarding of", **B1134VERG72.3.2NACP**
[46] WAGT Memo, February 9, 1952, "Contract NOa(s) 11028 – Monthly Progress Report (Month of January) – Forwarding of.", **B1134V1RG72.3.2NACP**
[47] WAGT Memo, December 28, 1951, "Contract NOa(s) 11028, 9670, and 10067, WEC Turbo Jet Engine Models XJ45WE-1, -2, -4 and XJ40WE-10; Revised Auxiliary Electric Control, Information on.", **B1134VERG72.3.2NACP**
[48] WAGT Memo, January 8, 1952, "Contract NOa(s) 11028, Proposal for Further Development of J46-WE-1 turbo-Jet Engine to Obtain Additional Afterburning Thrust", **B1134VERG72.3.2NACP**
[49] Air Force Memo, January 31, 1952, No subject, **B1134VERG72.3.2NACP**
[50] BuAer Memo, March 15, 1952, "Proposal for increased afterburner thrust in XJ46-WE-1 engine", **B1134VERG72.3.2NACP**
[51] WAGT Memo, February 29, 1952, "Contract NOa(s) 11028 WEC Turbo-Jet Engine Models J46WE-1 and -3 Simplification of the Afterburner Fuel System, Comments on.", **B1134V1RG72.3.2NACP**
[52] Douglas Aircraft Company Memo, March 21, 1952, "Contract W33-038 ac-10413, Project MX-656, Fuel Requirements for J46 Powered X-3 Airplane", **B1134V1RG72.3.2NACP**
[53] BuAer Memo, May 12, 1952, No Subject, **B1134V2RG72.3.2NACP**
[54] BuAer Contract NOa(s) 11028, Amendment 4, April 2, 1952, **B1134V2RG72.3.2NACP**
[55] Air Force Memo, Air Material Command, April 2, 1952, "(Unclassified) Follow-up, NOA(s) 11028, Westinghouse Electric Corp.", **B1134V2RG72.3.2NACP**
[56] Douglas Memo, May 28, 1952, "Contract W33-038ac-10413 – Project MX-656 Westinghouse Survey Concerning Models J46WE-1 and -2. Incorporation of Larger Diameter Turbine, Survey of.", **B1134V2RG72.3.2NACP**
[57] WAGT Report A-1415, July 10, 1952, Monthly Progress Report for June, 1952. Contract NOa(s) 11028 XJ46-WE-1 and XJ46-WE-3 Engines, **B1134V2RG72.3.2NACP**
[58] Air Force Memo, July 23, 1952, "(Restr) Contracts NOa(s) 11028 and NOa(s)-10385, Use of JP-4 Fuel in Westinghouse XJ40-WE-1 and XJ46-WE-1, -3 and -5 Engines", **B1134V2RG72.3.2NACP**
[59] WAGT Memo, July 28, 1952, "Contract NOa(s) 11028, 10825, 9670; AF-33-(038)-17843; WEC Turbo Jet Engine Models J46-WE-1, -2, -3, -5, -8, -12. Eight Point Fuel Injection System, Survey of. Engine Change Nos. 6,8,7,1,112,3 for the -1, -2, -3, -5, -8 -12 Engines Respectively.", **B1134V2RG72.3.2NACP**
[60] WAGT Memo, July 31, 1952, "Contract NOa(s) 11028; AF-33-(038)-17843. WEC Turbojet Engine Models J46WE-1,-3, -5. Interim Exhaust Nozzle Control System

and Hydro-Mechanical Exhaust Nozzle Control System., Description of.", **B1134VERG72.3.2NACP**

[61] WAGT Report A-1449, Monthly Progress Report For August, 1952 Contract Noa(s) 11028 (XJ46-WE-1 and XJ46-WE-3), **B1134V2RG72.3.2NACP**

[62] Air Force Memo, October 1, 1952, "(Uncl) Contract Noa(s) 11028, Incorporation of Improved T-13 Type turbine on XJ46-WE-1 and -5 Engines, Engine Change No's. 3 and 5.", **B1134V2RG72.3.2NACP**

[63] Air Force Memo, October 22, 1953, "(Restr) Conference on XJ46-WE-1 and -5 Engines, Contract NOa(s)-11028", **B1134VERG72.3.2NACP**

[64] Air Force Memo, October 17, 1952, "(Restr) XJ46-WE-3 50-Hour Flight Qualification Test on Contract NOa(s)-11028", **B1134V2RG72.3.2NACP**

[65] WAGT Memo, October 24, 1952, "Failure Analysis of the Temperature Control Amplifier.", **B1134VERG72.3.2NACP**

[66] WAGT Memo, October 29, 1952, "Contract NOa(s) 11028, 10825, 9670; AF33-(038) 17843. WEC Turbojet Engine Models J46-WE-1, -2, -3, -5, -8, -12. Rotation of Overspeed Relay, Survey of. Engine Change Numbers 10, 13, 9, 4, 116, 5 for the -1, -2, -3, -5, -8, -12 Respectively.", **B1134V2RG72.3.2NACP**

[67] BuAer Memo, November 18, 1952, "Contract NOa(s) 11028 – XJ46-WE-3/-5 engine status", **B1134VERG72.3.2NACP**

[68] WAGT Memo, November 26, 1952, "Contract NOa(s) 11028:AF-33-(038) 17843. WEC Turbojet Engine Models XJ46-WE-1 and J46-WE-1. Hinged Quick-Disconnect Clamp, Survey of. Engine Change No. 10.", **B1134V2RG72.3.2NACP**

[69] WAGT November 28, 1952, "Contract NOa(s) 11028; AF-33(038)17843. WEC Turbojet Engine Models XJ46-WE-1 and J46-WE-1,3. Eight Point Fuel Injection Ignition System, Information on.", **B1134V2RG72.3.2NACP**

[70] WAGT Memo, November 28, 1952, "Contract NOa(s) 11028, 9670; 10825, AF-33(038)-17843. WEC Turbojet Engine Numbers XJ46-WE-1, -2 and J46-WE-1, -2, -3, -8. Forward Engine Mount tolerance Change In. Engine Change Nos. 11,15,19,117 for the J46-WE-1,2,3,8 respectively.", **B1134V2RG72.3.2NACP**

[71] WAGT Memo, November 26, 1952, "Contract NOa(s) 11028, 9670; 10825, AF-33(038)-17843. WEC Turbojet Engine Numbers XJ46-WE-1, -2, -5 and J46-WE-1, -2, -5, -8. Removal of Bottom Mounted Oil Level Indicator from Oil Reservoir, survey of.", **B1134V2RG72.3.2NACP**

[72] WAGT Report, A-1501, Monthly Progress Report for November, 1952 Contract NOa(s) 11028 (XJ46-WE-1 and XJ46-WE-3), **B1134V2RG72.3.2NACP**

[73] WAGT Memo, December 15, 1952, Contract NOa(s) 11028, 9670: 10825; AF-33(038)-17843. WEC Turbojet Engine Models XJ46-WE-1, -2 and J46-WE-1, -2. Relocation of Flexible Joint, Survey of. Engine Change Nos. 12 and 16 for J46-WE-1 and -2 Respectively.", **B1134V2RG72.3.2NACP**

[74] Air Force Memo, January 8, 1953, "(Restr) Contract NOa(s)-11028, XJ46-WE-1, -3, -5 Engine Status", **B1134VERG72.3.2NACP**

[75] BuAer Memo, January 27, 1953, "Contract NOa(s) 11028, XJ46-WE-1/-3 /-5 engine status, Request for", **B1134V2RG72.3.2NACP**

[76] WAGT Report A-1510, Failure Analysis of Automatic Temperature Control-Westinghouse Dwg. No. 61E635. Engines J40-WE-1, -8, -22 and J46-WE-3, -7, -8, -12., **B1134VERG72.3.2NACP**

[77] BuAer Naval Wire, January 30, 1953, No Subject, **B1134V2RG72.3.2NACP**

[78] Air Force Memo, February 1953, "(Restr) Stopping of Work on Contract NOa(s)-11028", **B1134VERG72.3.2NACP**

[79] WAGT Memo, March 13, 1953, "TN-1", **B1134V2RG72.3.2NACP**

[80] WAGT Memo, March 16, 1953, "Contract NOa(s) 11028, Contract Status, Our Reference WG-65000", **B1134V2RG72.3.2NACP**

[81] Air Force Memo, August 4, 1953, "

[82] BuAer Memo, March 25, 1954, "Contract NOa(s) 11028, Westinghouse XJ46-WE-1/-3 turbojet engines", **B1134V3RG72.3.2NACP**

[83] BuAer Memo, June 29, 1955, "Contract NOa(s) 11028 with Westinghouse Electric Corporation", **B1134VCRG72.3.2NACP**

[84] Contract NOa(s) 11028, Amendment 7, December 5, 1955., **B1134V2RG72.3.2NACP**

Chapter 9
WAGT J46 Flight Testing

Flight testing of Westinghouse engines was accomplished under contracts NOa(s) 52-020, 52-621 and 55-198. The surviving flight test reports are filed in the Contract Correspondence records retained at the National Archives in College Park.

The lack of sufficient dedicated test beds for altitude engine testing became more and more of a problem as the scope of WAGT's engine programs grew. While NACA and the AEL helped with some altitude testing, the lack of similar capabilities at WAGT itself was a distinct hindrance to basic research, engine performance verification and problem resolution.

The flight testing of the J34 and J40 series of engines is outside the scope of this book, but a general description of the overall program will touch on them as well, as their tests went on in parallel to J46 work and sometimes overlapped.

At first Navy flight testing of WAGT engines was occurring at Vought's location in Dallas, Texas and McDonnell's location in St. Louis, MO. This was all related to the J34-WE-22 and earlier series. WAGT supported such airframe contractor testing with technical personnel on site.

To get dedicated airframes that would allow WAGT the control to plan and schedule the testing, BuAer became convinced they needed to bail (lease) airframes to WAGT and support them as necessary. Each of the engines needed a different airframe and each posed their own problems.

Setting up flight test programs for the J40 required a flight vehicle capable of taking an engine the size of the afterburner models of that very large engine. No small airframe existed that could handle it, so after a search for a candidate, an obsolete B-45A-5 (Serial 47-049) was bailed from the Air Force and overhauled by North American Aviation (contract NOa(s) 52-621) back to airworthy condition. Three major modifications to accommodate its new testing role were made during the overhaul: 1) addition of test control instruments and controls, 2) installation of an engine pod suspended on a pylon under the bomb bay that could be lowered under the belly of the airplane in flight for engine testing and raised for take-off and landing, and 3) installation of an aft fuel tank to hold a separate fuel supply from the aircraft's. This allowed the engine to be run on a variety of fuel types as necessary.

Westinghouse had no flight test experienced personnel within their organization and no strong relationships with airframe manufacturers to support such testing. Flight qualified personnel had to be located, flight gear purchased and the pilots and mechanics checked out on the airframe and engines. During the years of 1951-1953 obtaining technical support and needed parts was a problem due to the Korean War absorption of skilled personnel. Parts and inspections for this obsolete aircraft would continue to be a problem with much downtime and ferrying time required.

The J40 flight testing was planned to occur at Edwards AFB but was moved to Chance Vought's Dallas facilities and airfield prior to its start. In 1955, all testing was consolidated to Olathe, Kansas, where a servicing pit for the B-45's pod had to be constructed and hanger facilities allocated. With the end of the J40 program in 1955, the engine pod was modified to enable testing of the J46.

The later consolidation of test airframes to Olathe in early 1954 brought its own challenges, as hanger space on the Navy airfield had to be allocated. In addition, WAGT had to build a complete metal working shop, install underground fuel tanks and was beginning to install a complete compressed air storage facility for the J54-WE-2, which used impingement starting, when testing wound down. WAGT maintained a work force of about 70 people at Olathe.

BuAer supported the operation with hanger space, spare parts for the standard engines and airframes including the B-45, and with fuel purchase authorizations granted to WAGT for both aviation and ground motors. (A special permission was given to allow running a tab at the base instead of demanding individual payments per purchase. The base supply officer demanded a good faith deposit from WAGT to ensure timely monthly payments would be made!). BuAer also brought in the necessary ground handling equipment and special tools. The allocated radio test frequencies were transferred from New Castle, Delaware to Olathe with the Navy supplying the crystals for the radios in the airframes.

Once the testing was started in earnest at Olathe, the whole test program became far better organized with detailed monthly reporting for each airframe including the flights, the tests run and any problems encountered during the testing. Fully detailed requests for a given test had to be formally submitted and approved by WAGT engineers, management and the BAR before being sent to Olathe to be put into an engine's test schedule. Unfortunately, the reduction of the masses of collected test data from the recording equipment, both electronic and pictorial, was all done manually, a protracted process. Actual lessons learned from testing disappeared into the engineering teams back in Essington and only glimpses of the results appear in the testing summary reports. BuAer complained several times over the delays in learning test findings, but only near the end of the flight test program in Olathe was automated data reduction equipment beginning to be applied to speeding some parts of the data reduction for later analysis. This problem was not unique to WAGT but was industry wide and soon methods would evolve to transmit the data directly to the ground as the testing was done and computers then used to crunch numbers and generate printouts for analysis. Alas, all of those advances would largely come after WAGT's involvement in the industry.

For the needed J34 flight testing, a Douglas F3D-2 SkyKnight, BuNo. 124607, was bailed to WAGT and ferried from the manufacturer's plant to New Castle, Delaware on February 26-27, 1952 where a test facility was established by WAGT. As received, this engine had two standard production J34-WE-36 engines installed. Other than the paint needing to be touched up, the aircraft was found to be in excellent condition.

Battery carts were used early on to start the engines in the F3D and only three or four starts could be made before the cart's batteries had to be recharged, taking up to 36 hours. Another limiting factor was a battery amperage drop after 3-4 starts that risked hot starts on the engines. This limitation was dealt with by ordering a Hobart motor-generator starter card with a constant current control as a "Brickbat" priority order.

The original engines were replaced with modified J34-WE-36's with compressor outlet mixer vanes to allow high altitude testing and WAGT hired John Griffith and R. F. Mushlit as test pilots in April. The engines were then grounded before test flights could start and the aircraft remained on the ground awaiting a fix for the third stage compressor disc cracking problem. This aircraft would remain a J34-WE-36 test platform until returned to BuAer.

Another F3D-2, BuNo. 124600, was bailed to WAGT in April for J46 testing and preparations for a ferry flight to New Castle were put in hand in that month at El Segundo. Before the flight, the airframe was to be modified to have a dual fuel system to enable the port test engine to operate on JP-4 fuel while the J34 operated on its own fuel type. Douglas reworked the engine mounts and cowling to enable the J46 mockup to be installed. A fire suppression system was designed and installed.

The need for a base of operations radio station operation permit at New Castle resulted in delays and ultimately had to be addressed by WAGT "borrowing" several U.S. Navy tactical frequencies for communications with their aircraft. The frequencies were 133.92 and 150.12.

Finally in the second half of May, four J34-WE-36 engines were received to be used in testing support. All outstanding EC's on the engines were installed except the third stage compressor fix, which was to be accomplished back at the Essington Works. A Zyglo inspection revealed cracked spindles on two of the engines which required the spindles be replaced.

Airframe 607 had to have its cowling grommets replaced to remove a flight speed 400 knot restriction. The original engines were removed with the special instrumentation wiring on the port side being left in place.

Finally in the latter part of June, repaired engines were installed in 607. It was then discovered that the brushes on the installed engine starters were badly crazed with cracks. Inspections were put in place every five hours and the starters were now planned to be replaced when the new motor-generator on order was received and put in service.

Airframe 600 was now fully modified at El Segundo to the latest production configuration but changes for the J46 testing requirements were still being addressed. Lack of an actual J46 test engine on site as well as a lack of skilled labor slowed progress. The work was to drag on for many more months.

On July 10, 1952, 607 finally made it into the air with several flights to accomplish pilot checkouts, with Lt. Blackman from Patuxent River, MD giving instruction. The base radio station using the call sign "Octaroon" was put into service. Other than some high altitude overspeeds on the engines being experienced, the flights were all successful.

On July 24, 607 experienced a compressor failure on the test engine even though the third stage fix had been applied. Confusion then ensued as it was not clear that the engine had actually had the compressor disc fix installed because it had a spindle replacement at the same time. Because of the confusion, the aircraft was grounded. A check of the clearances on the compressor showed it was to specification. A temperature inversion test was planned and a special exhaust collector with 54 thermocouples installed on another test engine. This airframe continued to test -36 engines until August when it was returned to Douglas in El Segundo after F3D-2 BuNo. 124600 arrived in New Castle.

This second F3D-2 would have a long flight test history with WAGT, first at New Castle and then at Olathe, Kansas. It primarily tested various J46 models and at one time was being outfitted to test a J46-WE-20 engine when that program support was ended by BuAer.

One immediate problem after the F3D-2 600 arrived in New Castle was provisioning JP-4 fuel, as no one had figured out how WAGT could purchase JP-4 fuel for use in testing. The problem would dog the New Castle testing efforts for most of the time New Castle operated as a test base. While it was being sorted out, the aircraft operated both the engines on aviation fuel. The J46's operated well on this fuel with some clogging of fuel nozzles and manifolds from the lead content and the engines had to be regularly inspected for buildup.

The F3D-2 was a reliable aircraft and most of the problems were with engine test instrumentation failing in one way or another and requiring a lot of troubleshooting to resolve. It sat on the ground over the next few years for months at a time awaiting test engines for the port nacelle. The intake duct for the test engine on the left side caused a temperature inversion in the combustor of the J46 which the intake of the F7U-3 did not. The root cause for this on the F3D-2 was never isolated as, due to the schedule delays in the J46 program, the J46 was no longer being considered as an engine for the F3D-2 fleet and the F3D-3 program was ended before flight.

After the J46 program was ended, the F3D-2 600 was used to test a prototype XJ81-WE-3 missile engine mounted in a pod on a wing rack. After this testing was complete, the airframe was to be returned to BuAer but WAGT requested the bailment continue so they could test J34-WE-36 engines. As the airframe was already instrumented for engine testing, the Navy agreed. The exact date the airframe bailment ended was not determined, but it was after 1957.

F3D-2's had a very long career in the Navy, far longer than anticipated. Totally reconfigured, they flew electronic jamming missions during the Vietnam War, still using J34-WE-36 engines, these having improved combustors and turbines.

Returning to the B-45, this aircraft did limited testing of J40 engines at Dallas and Olathe, being often grounded for airframe problems or awaiting test engines. The details are covered in the author's companion volume on J40 development. Later it did a few tests with J46 engines then moved on to XJ81-WE-3 testing. At the end of the bailment period, the engine test pod was being modified for PD33/J54-WE-2 testing when that program ended. The B-45 remained on the ground for many months and WAGT argued it should be retained for engine test work, since engines of almost any size could be hung in the test pod and the instrumentation was already installed. BuAer did an engine development survey across themselves, the Air Force and industry and found no engine likely to need the aircraft. At that point, the airframe was ferried to Litchfield Park, Arizona and placed in long term storage.

The first F7U-3 airframe used in engine testing was in fact bailed to Chance Vought Aircraft but operated in Dallas by WAGT using Chance Vought pilots. The airframe was BuNo. 128453, the third production F7U-3 built and always referred to in the reports as "#3". This airframe had a J35-A-29 installed in the starboard side and the test engine in the port side. It tested a whole series of J46 engines in different programs, mostly on A/B behavior improvement and maintenance changes. This aircraft had an engine fire in the second half of August 1954 (exact date not given) when a third stage compressor disc on engine WE040004 burst. Shrapnel penetrated a fuel cell and a fire started. The damage was so severe the airframe was determined to be beyond economical salvage. (In the flight test report covering the period of the engine failure, the standard engine in the airframe was listed as a J34. The author believes this was an error, as no evidence was found of any change to the standard starboard engine configuration after the prior reporting period, which listed the J35 as still installed.)

The damaged F7U-3 was replaced with #22 (BuNo. 128452) in September, which was assigned to testing the prototype house engines supporting the -18 program. It continued testing various engineering changes to the installed engine to verify planned improvements to solve leaks and maintenance issues. This aircraft had a standard J46-WE-8B installed on the right side and the left side had the test engine of the day installed there. Various test recording equipment was installed, including a Brown Recorder and a video filming bay. Because of the center of gravity problems caused by the test engine's weight, ballast was installed in the ammunition cans to retain the cg in the proper location for each flight.

This aircraft continued to fly tests with J46-WE-18 prototype house engines until it was removed from the bailment program and flown to Litchfield Park in early May 1957 and put into long term storage. It cannot be proved with certainty, but the author believes no true YJ46-WE-18 engine ever flew in the aircraft.

Chance Vought flew many test flights of their own house airframes to determine the cause of J46 turbine damage when the guns were fired. This was eventually traced to a standing wave being set up in the intake which fooled the fuel and temperature controls into injecting too much fuel into the engine. The gun circuits were modified to force the guns to fire in such a way they never synchronized and set up the standing pressure wave. The author is indebted to Steve Ginter's book on the Chance Vought F7U for this information. Interestingly, Westinghouse never mentioned the problem in any of their correspondence in any of the J46 related contracts.

F7U-3 #22 Assigned J46-WE-18 Program Flight Test Engines		
Serial	Type	Status
WE40001	J46-WE-18	In Build-up
WE40002	J46-WE-8B	In Build-up
WE40003	J46-WE-18	In Build-up, for test on April 29th
WE40004	J46-WE-8, with C-94 compressor	Shipped to CVA
WE40005	J46-WE-18	In Build-up
WE40006	J46-WE-8	In Test
K.C. 101	J46-WE-8	In Build-up

Table 1: House engines assigned to the -18 program on F7U-3 #22. Note the confusing house engine serial numbers to actual production J46-WE-18 serials listed in Chapter 8.

An F7U-3M (BuNo. 129705) was added to the bailment fleet in May 1955 for short term testing of methods to increase thrust for short periods to assist in go-arounds. The weight of the missiles and their pylons under the wings increased the bring-back weight of the aircraft to the carrier to the point where the normal military thrust of both engines was marginal during a go-around to the point of it being dangerous. The A/B's had to be used and this was considered a risky solution as they burned fuel so fast. An F7U-3 already as low on fuel as necessary to reduce the bring back weight could mean a go-around might be initiated only to lose the aircraft due to fuel starvation before it could make a circuit back to the deck. Simply jettisoning the missiles was not a practical answer due to their cost.

The tests were called the "Hot Shot" program and were to determine the ability of the standard -8B and -8 engine turbines to take operating for short periods at 1,600°F. Numerous flights were made simulating go-arounds to build turbine time at the higher temperature using specially adjusted fuel and temperature governor controls. The 20 minutes total of high temperature testing proved the turbine section could take the additional thermal loads with no short term damage. The pilots reported the extra thrust at military power was noticeable.

The F7U-3M left the program in March 1957 after being returned to an operational configuration with all the test equipment and wiring removed. It too went to Litchfield Park and was put into long term storage.

An F2H-2 Banshee, BuNo. 123284, was added to the fleet in April 1955 and was used for turbine development work on the J34-WE-34 and later for development work on the combustors and Woodward fuel controls for the J34-WE-44/46. When exactly it departed the fleet was not determined, but it was likely in 1957 or later.[1,2,3]

High quality photos of the airframes used in the WAGT test fleet have not been located. One of the flight test report containers at Archives contained a few lesser quality photo reproductions on plain paper and are included here. Note the shots were taken to obscure the airframe serial numbers. All service unit identifications had been removed.

Figure 1: Douglas F3D-2 flight test airframe. Test engines always installed in left (port) instrumented engine nacelle. Actual airframe BuNo. not known. WAGT undated photo. NACP

Figure 2: J46 (possibly a -10) installed in F3D-2. Note extensive test electrical connections down side of engine. WAGT undated photo. NACP

Figure 3: XF7U-3 flight test airframe on first flight 12/20/1951. J-35-A-29 engines. BuNo 128451. WAGT undated photo. NACP

Figure 4: Instrumentation bay on the port side of F7U-3 BuNo. 128472. WAGT undated photo attached to a memo dated February 8, 1956. NACP

Figure 5: Flight Test B-45A-5 serial no. 47-049 with engine test pod in lowered position. It could be raised enough to allow adequate ground clearance for take-offs and landings. The test engine was never operated except in flight or in the test pit. Pod has a J40-WE-8 installed in the photo (above, right). WAGT undated photo. NACP

Figure 6: B-45A-5 positioned over new test pit in Olathe, KS. No engine is installed in the test nacelle. WAGT undated photo appears to have been taken when the pit was just finished and J40 testing was still occurring. The nacelle could not interchangeably test different type engines without being reconfigured. NACP

Figure 7: B-45 test nacelle modified for J46 testing shown on initial installation on the B-45. Pod is in lowered position. WAGT undated photo attached to July 5, 1956 flight test MPR. NACP

Figure 8: B-45 test nacelle modified for J46 from front. Engine mounting rail on left matched by one on right. WAGT undated photo attached to July 5, 1956 flight test MPR. NACP

Figure 9: WAGT test flight hanger 37 at Olathe, Kansas. The F3D-2 has the test pod for a WAGT XJ81-2 installed under its starboard wing. WAGT undated photo but it was likely taken in late 1956 or early 1957. NACP

Figure 10: WAGT Flight Test HQ, Building 56, Olathe, KS Naval Air Station. WAGT undated photo. NACP

Chapter 9 – WAGT J46 Flight Testing

[1] BuAer Contract NOa(s) 52-020b, Record Group 72, 7.3.2 Contract Correspondence, Containers 2453-54, National Archives College Park, Maryland

[2] BuAer Contract NOa(s) 55-198b, Record Group 72, 7.3.2 Contract Correspondence , Containers 3684-86, National Archives College Park, Maryland

[3] BuAer Contract NOa(s) 52-621, Record Group 72, 7.3.2 Contract Correspondence, Container 2584, National Archives College Park, Maryland

Chapter 10
J46 Development and Growth Program

Westinghouse entered into a technical assistance contract with Rolls-Royce in August 1953. It was to be for ten years and included eight years of full technical disclosure, the right to manufacture Rolls-Royce engines in the United States and sub-contract (license) if required by the United States Government. Rolls-Royce was granted reciprocal rights in England.

In exchange, Westinghouse had to pay Rolls-Royce £1,000,000 pounds (1953 values) by January 1954. In addition, Westinghouse would pay royalties on all J34 engines manufactured and sold after March 1st, 1955. Other engines manufactured and sold after March 1st, 1954 would incur royalties of 2.4% until $5,000,000 was paid and then 1.5% thereafter, regardless of whether the engines incorporated Rolls-Royce patents and/or information. The minimum royalties would be $560,000 per year. All licenses to Rolls-Royce were royalty free.

The contract offered no better terms to Westinghouse than had been offered to United Aircraft Corporation (makers of Pratt and Whitney engines).

Rolls-Royce's initial review of the current situation in the USA was that Westinghouse was not in sufficiently close contact with the airframe constructors. A further review with the major airframe companies found extensive positive impressions of Rolls-Royce products but revealed that confidence in Westinghouse was very low and it would take a lot to restore it. Another review two years later showed little if any progress had been made to improve that opinion and it was not until Westinghouse left the aviation gas turbine business in the latter part of 1959 that Rolls-Royce was released from the agreement and could then begin to market their engines directly to contractors in the USA.

About the time the assistance contract was signed, WAGT's failure to produce a high power version of the J40 all the while seeing Pratt and Whitney make rapid advances in such engines caused them to formulate a marketing strategy of building mid-range, lower power engines, a niche in which they perceived no one else to be competing.

Aside from a limited number of J34 engines to be produced and continuing J34 overhaul contracts, the J46 program was the production program about to come on line and with three main airframes as users (F7U-3, A2U-1 and F2Y-1) it was vital that an improvement program be put in place that would let the airframe contractors know where the engine program would be in the years ahead.

WAGT used the opportunity to study the Rolls RA.14 (Avon) engine with the intent of raising the efficiencies of the J46. This study led directly into the evolution of an engine development and growth improvement program under the leadership of Reinout P. Kroon in his new organizational role as Development Manager at the Kansas City site.

Initially, the growth program was presented based solely on the J46-WE-18 and RA-24. The earliest chart found that shows the outlook at that time is below.

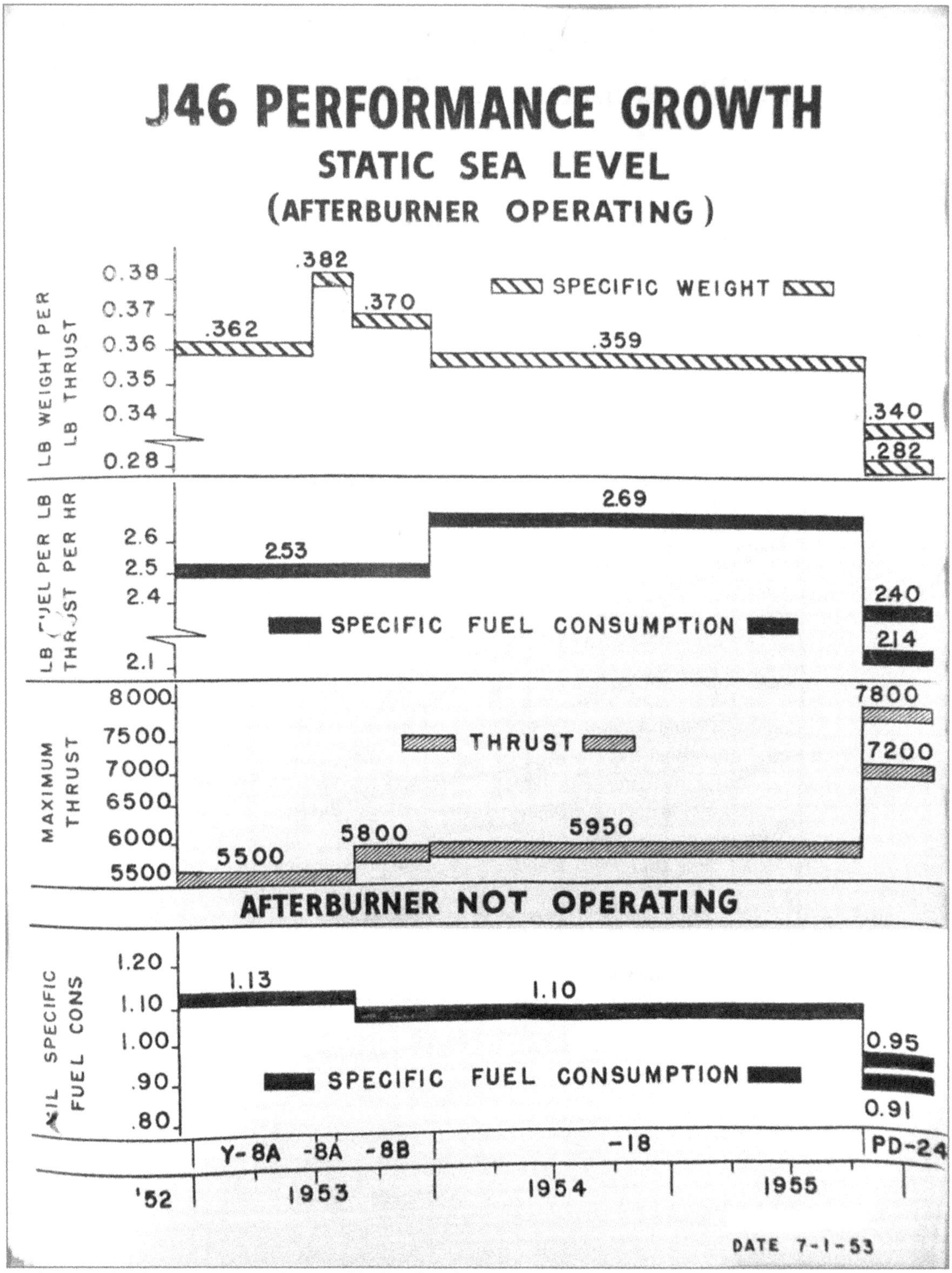

Figure 1: First known presentation graphic showing the emergence of an engine beyond the J46-WE-18. WAGT chart 7/1/1953.[1] ***Courtesy Hagley Museum and Library***

The basic program was later expanded and updated, now being stated as a continuation of the existing engine models using new "block" designators as follows:

J46 Development and Growth Program (abt. 12/1953)[2]					
Block	Designation	Status/Outlook	A/B Thrust (lb)	Military Thrust (lb)	Military SFC
Early Block I	J46-WE-8B	Qualified at Essington and verified at Kansas City, in production.	5,725	3,940	1.14
Block I	J46-WE-8	Identical to -8B except for the hydro-mechanical nozzle control system. One 150 hour unofficial endurance test completed. The official test is scheduled for February 1954, production to start April 1954.	5,725	3,940	1.14
Block II	J46-WE-18	Three major changes from -8B and -8: Compressor has an extra stage to eliminate high-referred speed stall, a larger more efficient turbine and larger diameter afterburner. Qualification scheduled for March 15, 1954 with production to start in May 1954.	5,950	3,980	1.10
Block III	PD-24	J46-WE-18 compressor operating at 5.6 pressure ratio (compared to 5.3 of the -18), scaled Rolls-Royce RA.14 turbine, longer diffuser. One hundred pounds lighter than -18.	6,400	4,180	1.028
Block IV	PD-24A	Compressor scaled from the RA.14 (Avon), 7.56 CR, improved A/B with flame holder and diffuser design derived from the terminated J40-WE-24.	7,900	5,250	.938
Block V	PD-24A-PT	Adds a zero stage to the compressor and air-cooled turbine discs and blades. 8.35 CR. Higher turbine inlet temperature raises SFC slightly.	9,450	6,550	.970
Block VI	PD-19 (cycle #23)	Same physical installation size. Two spool compressor, turbine and mechanical construction scaled from the Rolls-Royce Conway, deleting the Conway's by-pass stage. 12.48 CR.	11,250	7,500	.828

The "PD" prefix was from a WAGT "Preliminary Design" designation tracking control list which started with a PD-1 design that was for a Ducted Fan Study compiled in WAGT report AGT-632. The WAGT reports related to the various PD studies have not been located and may have only been used internally. Only the PD-24 and PD-19 will be covered here.

The additional stage to be added to the PD-24A-PT was another stage added in front of the original first stage of the compressor to raise the compression ratio and the airflow. Such an additional stage is normally called a "zero stage".

The scaled compressor for the RA-24A and RA-24A-PT was based on the Avon's RA.14 compressor, which was itself based on the original Metrovick F2 Sapphire design.

The hand drawn charts from which the data is derived lists the following availability of the engines above:

Engine Production Availability	
Block I	1953-54
Block II	1954-55
Block III	1955-56
Block IV	1956-57
Block V	1957-58
Block VI	1959

The accompanying charts were hand drawn and may not have been the ones actually used in a presentation, but are interesting nonetheless in showing us a bit of the internal WAGT sales approach as they developed it.[3]

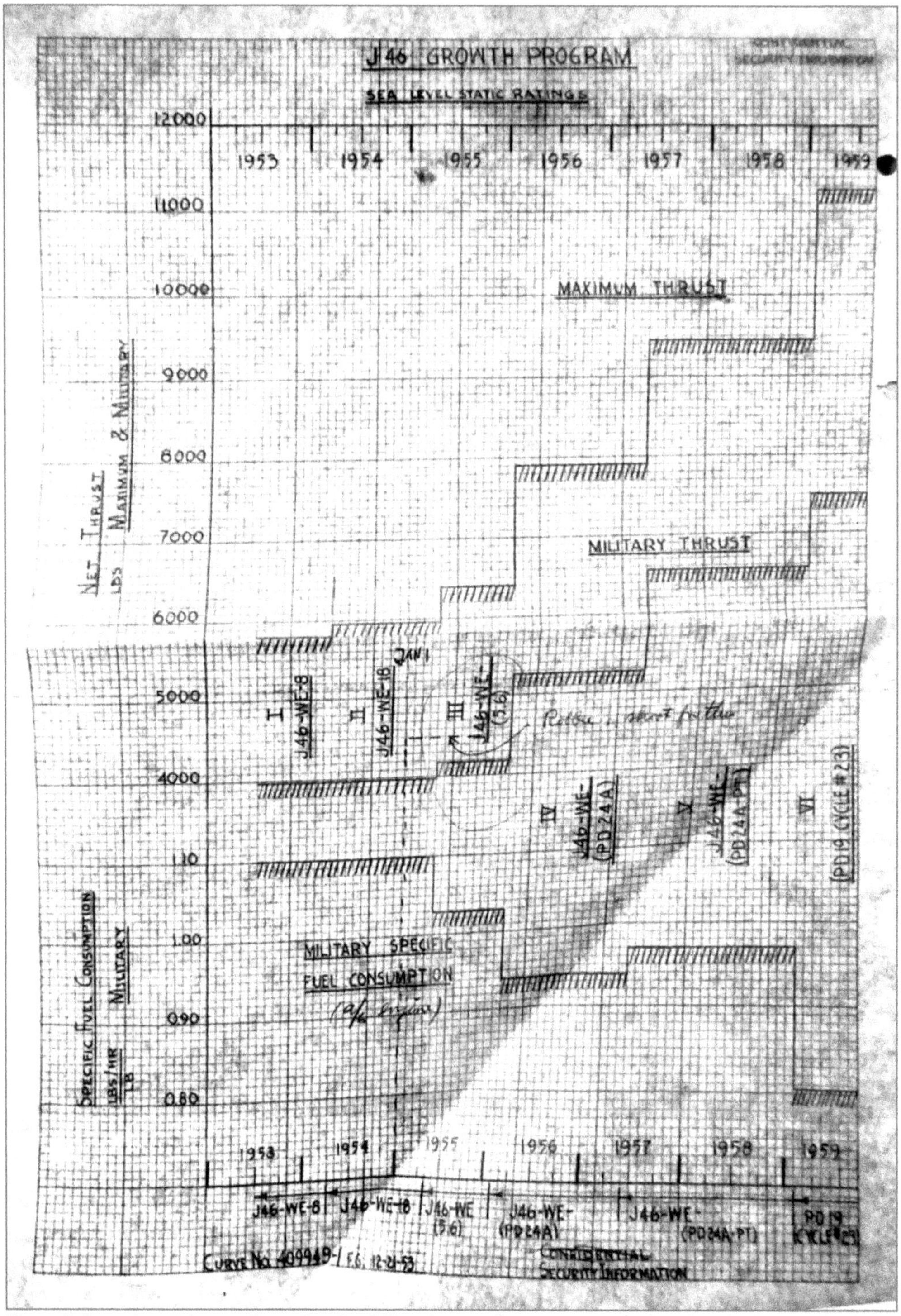

Figure 2: Later hand drawn chart showing fuller picture of program. Note that the Block III engine is called the "J46-WE-(5.6)". It is not believed any were built, the program going directly to the RA-24A. WAGT, 12/21/1953.[4] ***Courtesy Hagley Museum and Library***

The sales pitch was to be accompanied by other charts (next page) that were intended to show BuAer the practical benefits to mission performance of the improved engines, focusing on mission range. (The handwritten marks are on the original documents).

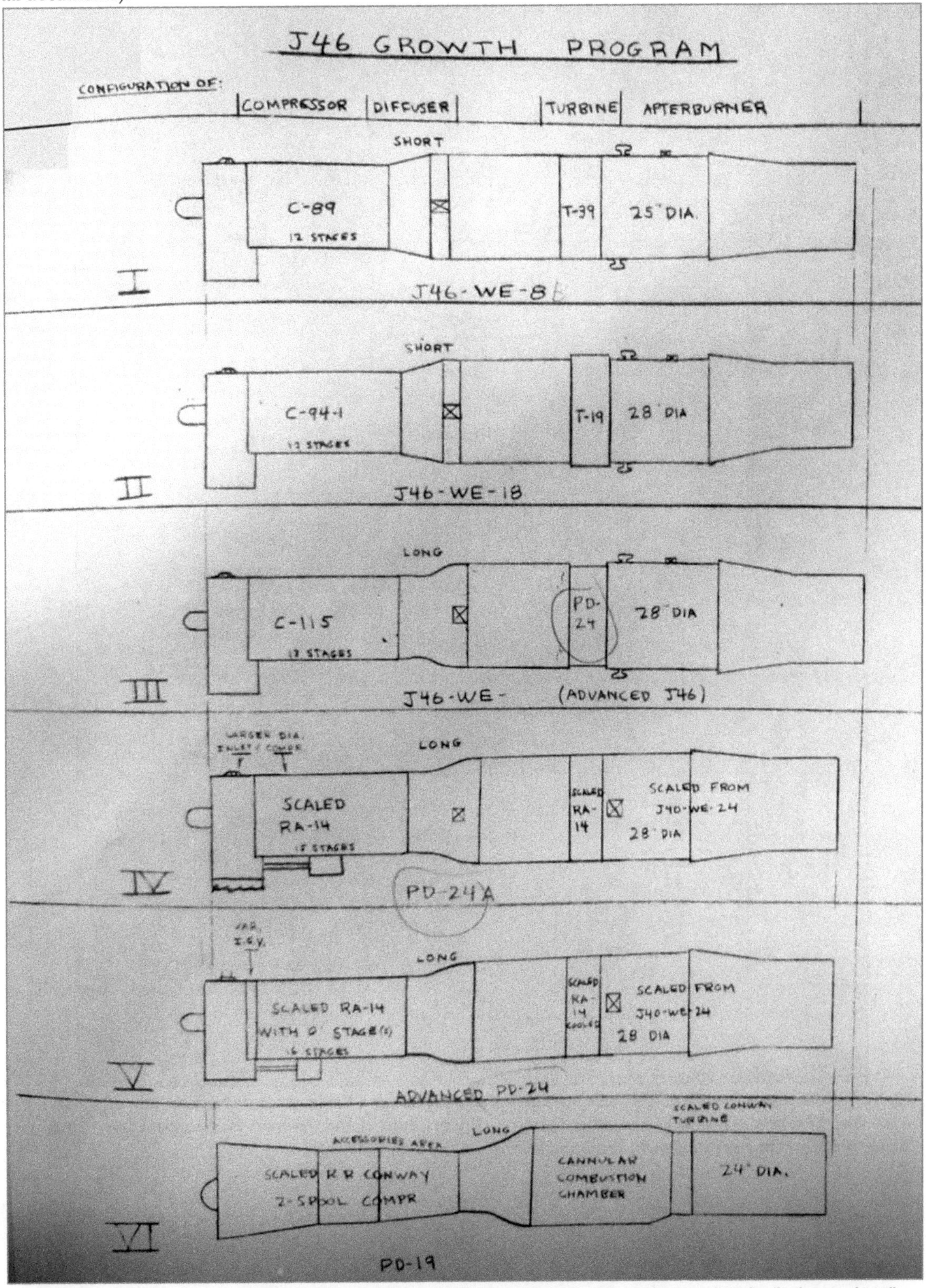

Figure 3: This chart now calls the Block III engine the "RA-24" and other documents call it the "Advanced J46". WAGT hand drawn chart dated 12/24/1953.[5] *Courtesy Hagley Museum and Library*

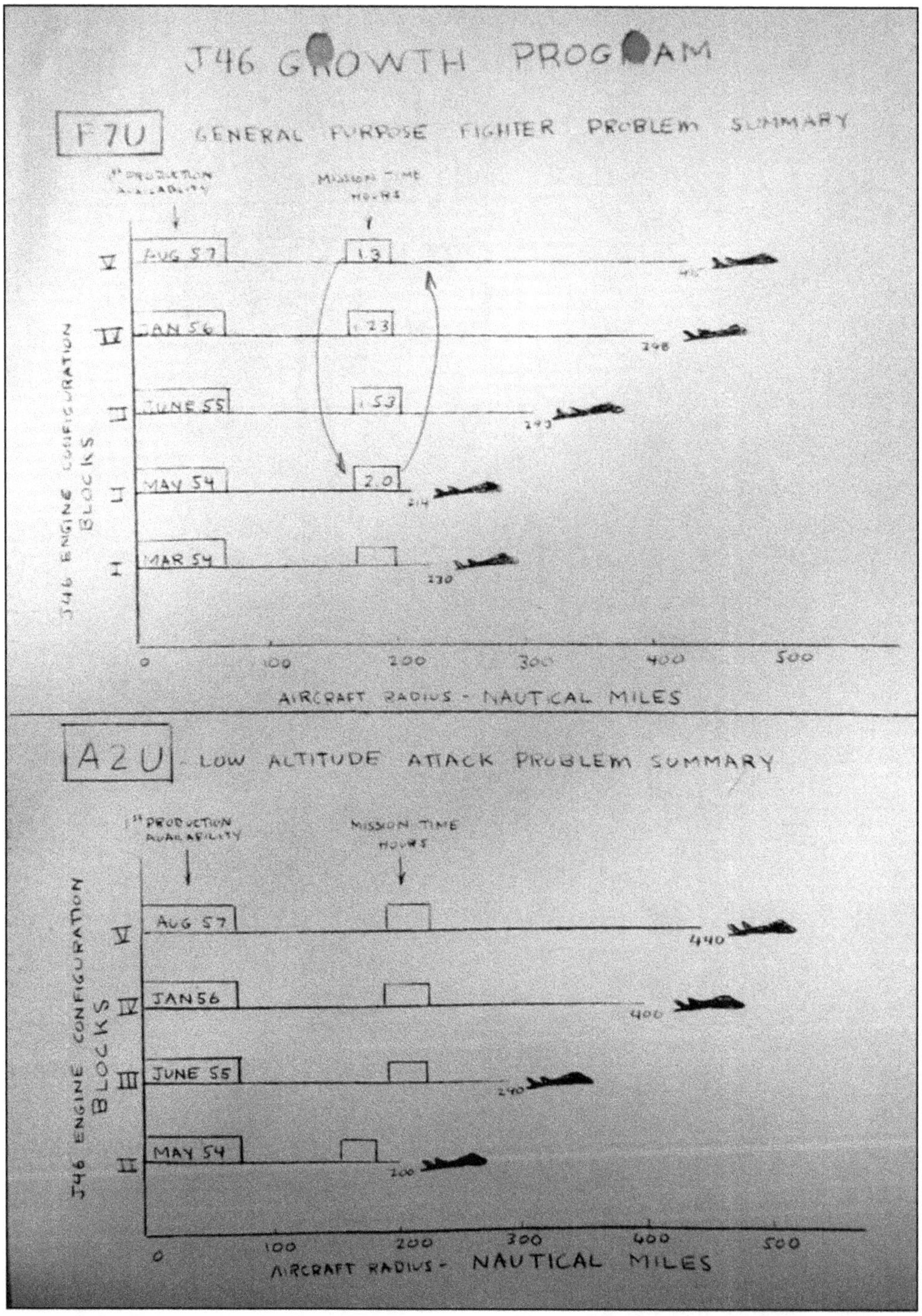

Figure 4: Note the correction on the figures at the top of the chart and the incomplete mission data on the A2U. The F2Y program was still active when these charts were drafted and the chart for it follows. (abt. 1954)
Above and next Courtesy Hagley Museum and Library

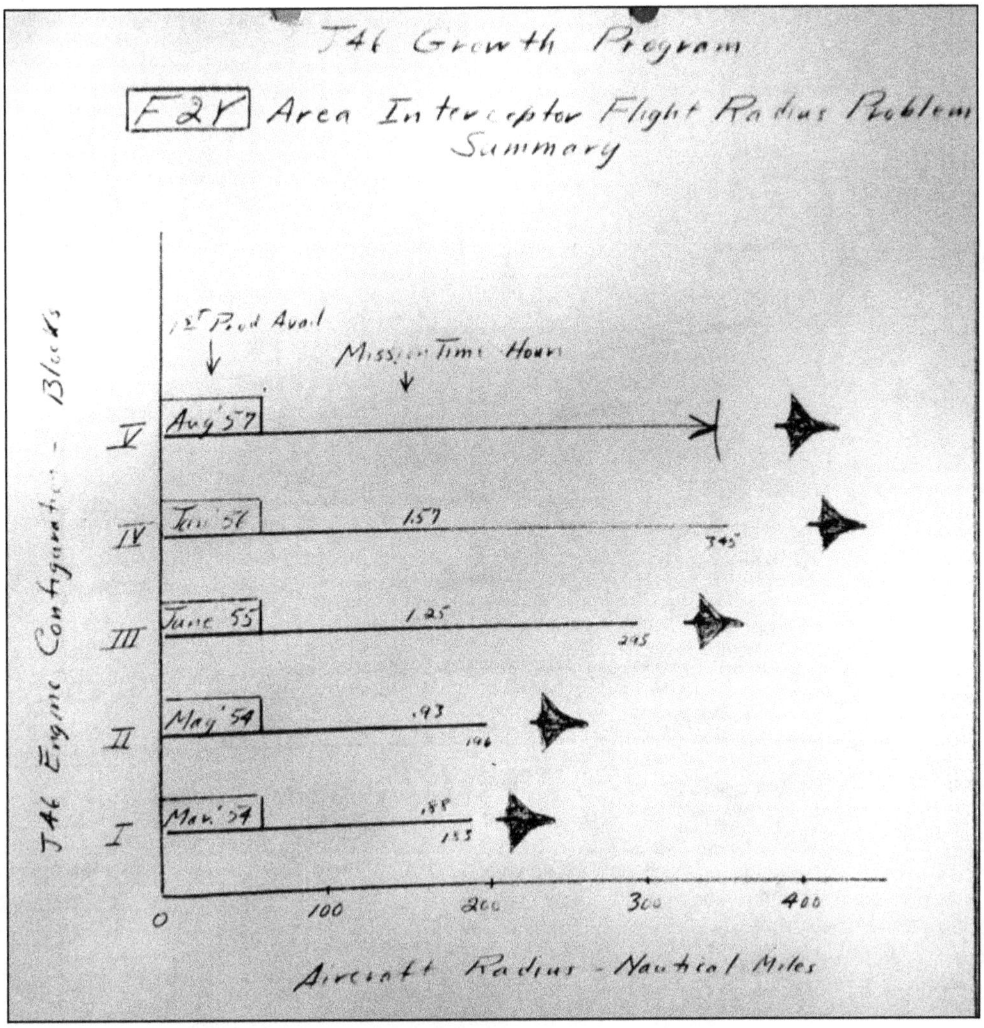

The Block I-II engines were covered in detail in the earlier chapters. The Block III engine compression ratio improvement from 5.3 to 5.6 was based on test work at the AEL *(no report was issued)* on the -18 compressor that showed the compressor could operate successfully at up to 5.6 CR without stall being encountered. The improved turbine was to be based on the Rolls-Royce Avon RA-14 design and scaled down to the needed size for the smaller engine. This engine was apparently never built, with WAGT actually moving on to the higher performance RA-24A, building at least one and testing it. The images of this engine are being included in large format as it is not believed they have ever been published before. The test results were a bit below the estimates. Getting reliable data below the operating speed of 10,100 RPM was problematic and the report indicates difficulty was experienced in getting duplicate bleed valve and inlet guide vane operation. It was suggested in future tests the guide vane ram be operated on fuel as designed *(instead of oil? – auth.)* or be set up to operate manually.[6]

Initial PD-24A #1 Military Test Results

	Estimated	Actual
Thrust (SL)	5,540 lb	5,240
SFC	.867 lb/hr/lb	.887
Air Mass	93.5 lb/sec	89.8
CR	7.5:1	7.23:1
Compressor Efficiency	82.0%	80.7
Turbine Efficiency	88.5%	89.0

Figure 5: The first PD-24A test engine. Note the blow-off valves. The WAGT photo card identifies this as the PD-24A #1. The compressor casing is significantly different from any known picture of a J46-WE-18 compressor, leading to the conclusion this could not be a Block III PD-24. The author does not believe any Block III RA-24 engines were built. WAGT P51317, 3/20/1954. *Courtesy Hagley Museum and Library*

Figure 6: No. 1 PD-24A bottom view. Still a plumbing rat's nest. WAGT P51318, March 15, 1954. *Courtesy Hagley Museum and Library*

Figure 7: Port side of inverted PD-24A test engine. Note ram on the right side to drive the interlinked variable stator vanes (arrow) and on the left the similarity of the oil/fuel cooler to the J46 design. This shows some WAGT features made it onto the engine. WAGT P51320, 3/20/1954. *Courtesy Hagley Museum and Library*

Figure 8: Mock-up of PD-24 "IB" appears to be the PD-24A. WAGT P50433, 10/23/1953. *Courtesy Hagley Museum and Library*

The profusion of confusing nomenclature carried over into the development of new engines, with the experimental shop apparently using different model ID's than the marketing team for the engines, as these three images show.

Figure 9: Mock-up of PD-24 "IB" right side view. WAGT P50429, 10/23/1953. *Courtesy Hagley Museum and Library*

Figure 10: Mock-up of PD-24 "IB" left side view. WAGT P50434, 10/23/1953. *Courtesy Hagley Museum and Library*

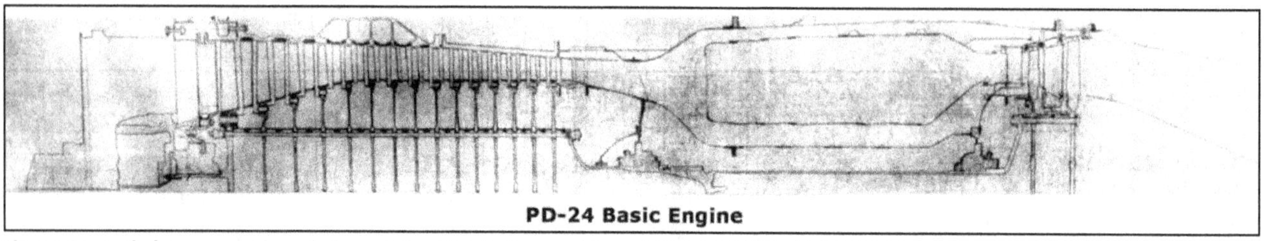

Figure 11: Only known sectioned view of the RA-24. Cropped from WAGT P50501, 10/30/1953. Internal layout appears almost identical to the PD-33/J54 (below) except for compressor construction, 15 stages vs 16 below. The low-contrast drawing was pinned to the wall above tables where all the parts for a PD-24 were laid out for review. In spite of the drawing being labelled an "RA-24", it is almost certianly an RA-24A. Scale of the two drawings is unknown. *Courtesy Hagley Museum and Library*

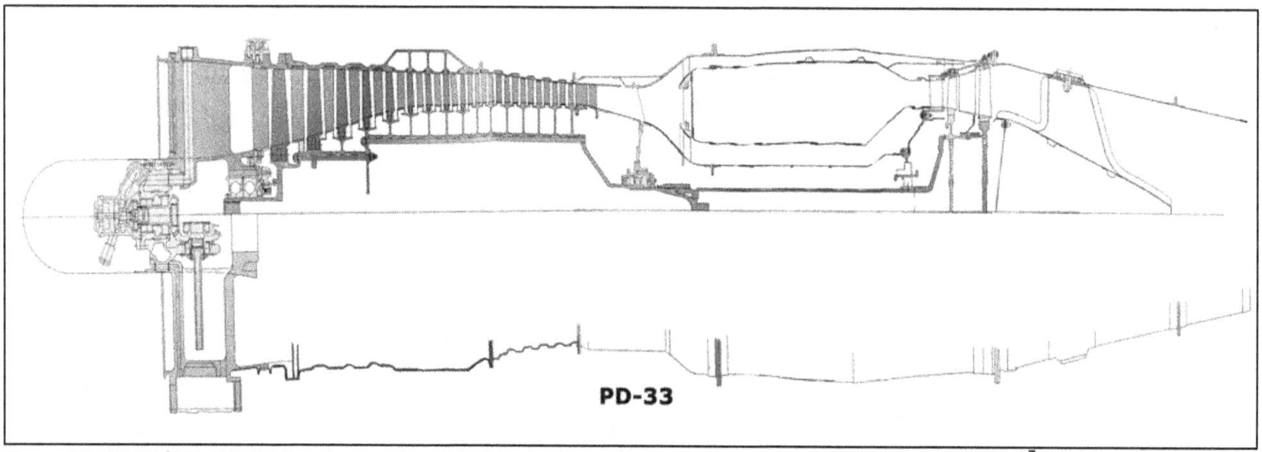

Figure 12: PD-33/J54. WAGT drawing, undated. Original color-coded to indicate the construction materials.[7] Not dated. *Courtesy Hagley Museum and Library*

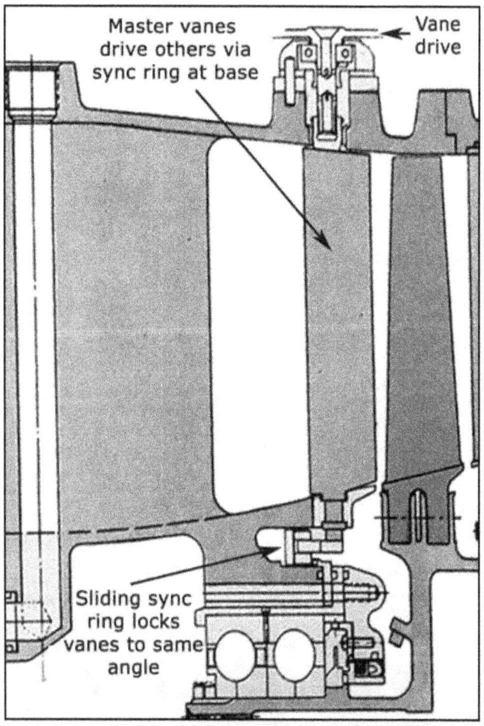

Figure 13: Variable stator vane synchronizing mechanism. Cropped from PD-33 image above.

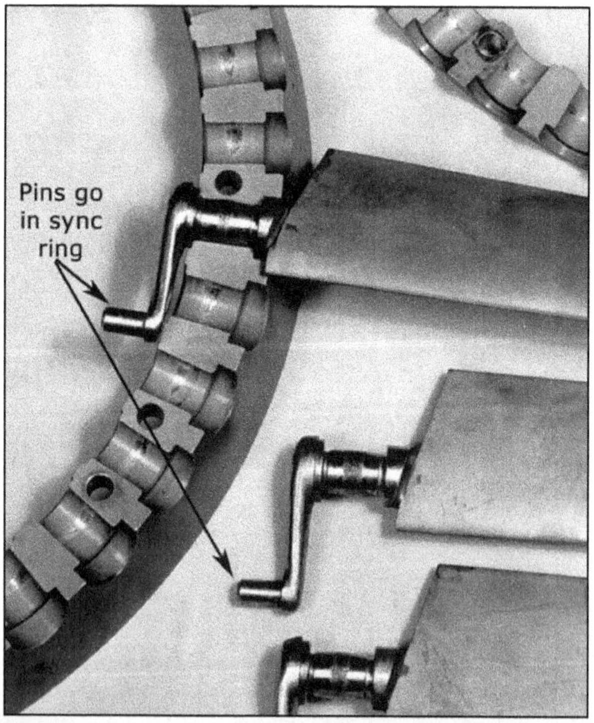

Figure 14: PD-24A variable stator vane sync pivots after testing. WAGT P53133, 11/12/1954. *Courtesy Hagley Museum and Library*

In spite of Westinghouse's investment and their efforts to keep BuAer informed as to their progress, BuAer never expressed any interest in extending the J46 in any form. As the F2Y program shifted to using the J57 in single engine form for production, Westinghouse argued that by the time the airframe was redesigned for the J57, the PD-24A or PD-24PT would be ready for production. The F2Y was cancelled for other reasons than engine suitability or availability and that potential airframe opportunity ended. The A2U program ended as a result of the problems with the F7U-3 from which it was derived. And then the F7U program completed its production run and the J46 new production window closed as well.

It appears the development focus then moved to other PD engines, with the PD-24A effort resulting in a PD-24PT version called internally the PD-33 and the J54 by BuAer. Space precludes the telling of the complete story of the various PD engines here, except for the J46's growth program's PD-19.

The PD-19 was to be a scaled version of the Rolls-Royce Conway, modified to eliminate the bypass feature and fitting inside the J46 installation envelope. The performance specifications for this engine were listed in the table on the third page of this chapter. A cleaned up copy of the only known drawing of the PD-19 is on the right. All air coming into the intake would go through the core of the engine. The Conway was a dual-spool compressor engine, with a low pressure spool supercharging the high pressure spool. It was a two shaft engine with each spool driven by its own turbine stage. It has a single stage of variable stator vanes. No blow-off valves are shown as these are typically not needed on a dual spool engine. The engine would not be expected to suffer from stalling during startup, a nice feature of multiple spool engines. Notice the elegant bearing mounting systems for both shafts, typical of the design skill of the Rolls-Royce engineers. No one would ever think this was a Westinghouse engine if they saw this drawing without knowledge of its source. Combustion was to be "can-annular", where the pressurized diffuser air from the high pressure spool was led into multiple combustion cans arranged around the shaft system with excess air bled around the cans being used to cool them. Both the combusted and cooling air would then flow through the turbine stages. The A/B drawn seems a notional one, not a detail design.

It is not believed any PD-19 engines were ever built.

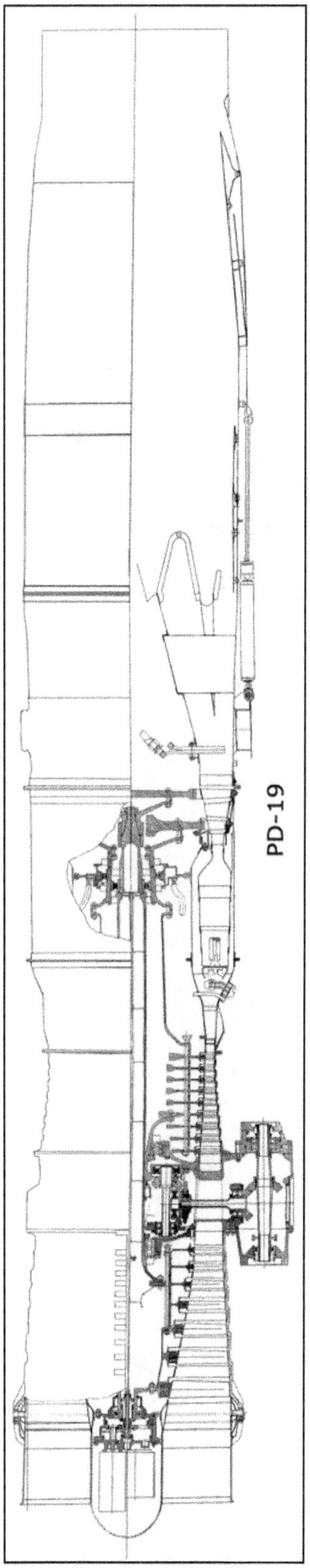

Figure 15: PD-19 Drawing. Not dated.
Courtesy Hagley Museum and Library

Chapter 10 – J46 Development and Growth Program

[1] Projected growth graphic, July 1, 1953. Accession 1459, Box 14, Hagley Museum and Library. **A1495, B14, Hagley**
[2] WAGT spreadsheet dated December 21, 1953 and attached engine descriptions. **A1495, B14, Hagley**
[3] Hand drawn chart, December 21, 1953, **A1495, B14, Hagley**
[4] Growth chart, December 24, 1953. **A1495, B14, Hagley**
[5] Hand drawn program grown engine profile chart, December 24, 1953, **A1495, B14, Hagley**
[6] WAGT Report A-1887 (summary page), June 10, 1954. **A1495, B14, Hagley**
[7] Figures 11,12,13,14, and 15 Drawings, undated, **A1495, B14, Hagley**

Chapter 11
Analysis and Conclusions

It would seem that this chapter would be straightforward, the many problems at Westinghouse combining to an obvious answer of incompetence leading to failure. Yet the challenges of the J46 program certainly did not appear to be so large in the beginning years to immediately conclude that BuAer's sponsorship of the engine was a mistake.

The early J34 series certainly seemed amenable to the constant development being applied to it. It was becoming increasingly reliable and the active development of afterburner technology by WAGT and their exposure to the early A/B of Solar Corporation running on the back of their J34-WE-22 in the Vought F6U-1 seemed to have placed them in an ideal position to move ahead aggressively.

Was there an obvious early red flag that BuAer should have seen as an excessive risk warning? One might have been the fact that the historical suppliers of aircraft engines such as Pratt and Whitney, General Electric and Wright were quoting much longer development schedules for new turbojets not based on imported technology. This was seen during the bidding for the engine contract that would become the J40. Westinghouse bid a much shorter schedule, lower cost and lower weight to boot. It was a temptation too great and BuAer succumbed.

At the same time, WAGT was bidding on what would be their first production version of the J34 with an integrated A/B and a "low technical risk" growth A/B equipped engine of roughly the same size as the J34, basing both on A/B technology and a new yet to be designed electronic engine control that on paper promised significant improvements in SFC. Additionally, the engine aerodynamics were assumed to be largely knowns because the engines were planned as simple aerodynamically scaled up versions of the J30/J34 series. Given multiple airframe programs now resting on these engines and a critical path for all of them leading through Westinghouse, prudence should have given BuAer pause.

Things that BuAer should have expected to see happening at WAGT to mitigate risk were rapid engineering staff expansion at WAGT along with extension plant investment well in advance of detail design. WAGT's refusal to invest their own money in a manufacturing plant and the negotiations to have BuAer supply such a plant should have raised a very strong reconsideration of how seriously WAGT was committed to the aviation business. With large numbers of engines under contract and a large sub-contracting network in the process of being assembled, this was big business for sure, even in peacetime.

In the first two years of the contracts, the difficulty of designing an integrated engine control on an A/B equipped engine intended to achieve what is known today as carefree handling eventually caused a complete rethink of the control. The electronic advantage was a reach too far at that time and a simpler purely mechanical control had to be designed even as production and pre-production engines were about to go into qualification testing. The lower SFC's BuAer was counting on were only marginally achieved and the airframe mission performance suffered.

The simple aerodynamic scale-up also proved an illusion, as the higher compression ratios matched to the higher turbine efficiencies needed to achieve the guaranteed thrusts went beyond WAGT's design abilities. The many combinations of compressors and turbines having to be tested to achieve the performance desired is a clear indication of WAGT's weakness in design and their falling back on prototyping by example, their common approach. WAGT's practice of designing to the requested performance point and not including contingency and/or a potential growth factor meant any shortfall in efficiencies left them with limited options available to solve the problem. Adjustment of one area meant trade-offs in the others. Proof of the result was the need to solve the compressor and turbine efficiency short-falls on the J46. They had to add another stage and a larger turbine which increased the engine weight and changed the installation envelope.

Chance Vought was to argue later in contract settlement correspondence that they expected some sort of "normal" engine performance increases to occur over the life of the J46 engine. As their airframe weight increases ate away at their own ability to achieve the F7U-3 performance guarantees, the lack of engine growth blocked them from recovering their airframe performance shortfalls.

While WAGT did in fact recognize the need for more power (albeit too late), their growth program was based on pushing the J46 design a bit more to operate it at a slightly higher compression ratio as a first step. The next step was to use a new compressor based on a scaled down Rolls-Royce Avon RA.14 design matched to the original WAGT designed turbine. The main thing that would have been accomplished by these changes was to finally provide adequate stall margin in the engine, finally resolving the engine stalling problem. Bigger increases in performance would have come when they replaced the turbine itself with a turbine scaled down from the RA.14 design. The end result would essentially have been an aerodynamically scaled down RA.14 with a WAGT designed accessories package from the J46 that could fit in the F7U-3.

Beyond this, these improved engines were blocked in as the F7U-3 was not a true all weather fighter, even in the F7U-3M version. It was not supersonic nor would it have been even with more powerful engines. Its complexity, high maintenance costs and lack of robust systems were exactly the opposite of the type of aircraft needed on an aircraft carrier. Operating it from a carrier brought a high accident rate. WAGT's proposed PD-24PT engine with a zero stage on the compressor and air-cooled turbine blades would have offered even more thrust and, more importantly, significantly improved F7U-3 mission radius. It was not enough by itself in the end to save the F7U-3 program and no evidence exists that the Navy considered the potential of these improved engines in their decision to withdraw the F7U-3 from service.

If everything was working, the F7U-3 has been reported to have been a pleasant aircraft to fly and many pilots liked it. Somewhere along the line, it picked up the nickname "the gutless Cutlass". Was the nickname justified? A table of airframe weights and engine thrusts show the following thrust to weight ratios for various F7U versions and engine combinations as well as contemporary aircraft in the same category:

Airframe Combat Weights vs Engine Thrust Analysis

	Combat	Engines						Total Thrust		Military	A/B
Aircraft	Weight	Type	Military	A/B	Type	Military	A/B	Military	A/B	Th/Wt	Th/Wt
XF7U-1	15,860	J34-WE-34	3,240	0	J34-WE-34	3,240	0	6,480		0.41	
F7U-1	17,707	J34-WE-34	3,240	4,160	J34-WE-34	3,240	4,160	6,480	8,320	0.37	0.47
F7U-1	17,707	J34-WE-32	3,370	4,900	J34-WE-32	3,370	4,900	6,740	9,800	0.38	0.55
F7U-1	17,707	YJ34-WE-32	3,170	4,700	YJ34-WE-32	3,170	4,700	6,340	9,400	0.36	0.53
F7U-1	17,707	YJ34-WE-32	3,170	4,700	J34-WE-34	3,240	4,160	6,410	8,860	0.36	0.50
XF7U-3	23,672	J35-A-29	5,600	0	J35-A-29	5,600	0	11,200	0	0.47	
YF7U-3	23,672	J35-A-29	5,600	0	YJ46-WE-8A	3,900	5,300	9,500	10,900	0.40	0.46
YF7U-3	23,672	J46-WE-8A	3,905	5,500	J46-WE-8A	3,905	5,500	7,810	11,000	0.33	0.46
F7U-3	23,672	J46-WE-8B	3,980	5,800	J46-WE-8B	3,980	5,800	7,960	11,600	0.34	0.49
F7U-3	23,672	J46-WE-8	3,980	5,800	J46-WE-8	3,980	5,800	7,960	11,600	0.34	0.49
F7U-3P	26,840	J46-WE-8	3,980	5,800	J46-WE-8	3,980	5,800	7,960	11,600	0.30	0.43
A2U-1	29,375	J46-WE-18	4,100	6,100	J46-WE-18	4,100	6,100	8,200	12,200	0.28	0.42

This chart indicates that the Cutlass F7U-3 variants after the first twelve powered by J35's were seeing a steady decrease in their power to weight ratio. Only when using their afterburners was the early performance promised by the early prototypes restored. And there we seem to have the issue in a nutshell. Most of the flying would have been at less than military power settings with only minimal time using the afterburner. The issue that all afterburner equipped aircraft have faced since the afterburner was developed: you can go fast, or you can go far, but you can't do both. Any pilot experiencing the performance of the airframe when the afterburners were used was going to compare that to what he had in cruise and feel he had a lead sled.

The Cutlass's contemporaries, particularly the non-afterburner equipped ones, fared a bit better, as this next table indicates:

Contemporary Aircraft to the F7U-3 and a Bit Beyond						
Aircraft	Combat Weight	Engine Type	Military	A/B	Military Th/Wt	A/B Th/Wt
F-86-F-30	14,857	J47-GE-27	6,000		0.40	
FJ-2	15,813	J47-GE-27	6,000		0.38	
FJ-3	15,669	J65-W-4B	7,650		0.49	
FJ-4	17,845	J65-W-16A	7,700		0.43	
H. Hunter F.6	22,000	Avon 203	10,000		0.45	
F-100A	24,996	J57-P-7/-39	9,700	14,800	0.39	0.59
F3H-2	31,000	J71-A-2E	9,700	14,000	0.31	0.45
F8U-1	27,000	J57-P-4A	10,500	16,200	0.39	0.60

Our historical perspective allows us to see that in the F7U the Navy had a transonic day fighter that was costly to maintain, had low serviceability and was slightly underpowered (vs. contract). Due to airframe and engine delays, it emerged into a world where a supersonic all-weather airframe was now the platform that was needed. Even pushing as hard as they could to make it work, the problems of the Cutlass and the risks it posed to the pilots were too high and it was removed from service. History records that only the emergence later of the McDonnell F-4 Phantom would give the Navy the weapon they truly needed.

So, did the Navy get any benefit at all from the J46 (and its precursor, the J34-WE-32) program? It can be argued that they learned a lot of hard lessons regarding jet engine development in general and that the effort to produce a reliable control system matched to the engine can be as great an effort as the development of the basic mechanical components. They should also have learned that continually changing the requirements rarely gets the benefit hoped for, but the author wonders about that.

The endless battle over engine weight was almost a fixation with BuAer. The number of changes to the engines to save a few pounds here and there all the while struggling to achieve high volume production indicates a lack of focus on everyone's part. A simple order to "save the weight ideas" for a later improved model and a refusal to process the weight related surveys would have stopped it. Ironically, the weight turned out to be whatever WAGT was able to achieve, regardless of how much arguing occurred.

The Air Force effort to obtain greater high altitude thrust from a variant of the J46 for their test aircraft degenerated into a muddle at best. The passage of time made the effort redundant as the march of engine and airframe technology made the X-3 program obsolete. There is no evidence this program made any contribution, direct or indirect, to the Navy's J46 models, except possibly to drain badly needed resources from the BuAer engines. The confused state at the end of the Air Force contract speaks for itself.

These things were symptoms of bigger problems - either the parties were unable to recognize them or the press of events left them unable to actually address them in a meaningful way. The price in the end for the Navy and WAGT was the funding and production of an engine of marginal thrust and reliability that had no future in military or civilian aviation. As its program ended, WAGT's own future was sealed, even if it took them a few years to face up to it and withdraw from the aviation gas turbine business.

Appendix A

Ratings and Specifications

The compiled ratings and specifications for the various models of the J46 are presented for those models where sufficient data are available. The last known specification version for a model has been used, supplemented with various data in correspondence and reports.

J34-WE-32 Ratings and Specifications (WAGT-24C8-2E)

Appendix B (YJ34-WE-32)**

Rating	Thrust (lb)	RPM*	SFC Specification
Afterburner (10 Mins.)	4,900	12,500	2.60
Military (30 Mins.)	3,370	12,500	1.08
Normal	6,500	12,500	1.03
90% Normal	5,850	12,500	1.00
75% Normal	4,880	11,950	0.97
Idle	450	5,000	5.7

* Engine operated as a constant speed engine above 75% thrust levels.

** Thrust not less than 90% of guaranteed values, SFC not more than 107.5% of guaranteed values

Physical Details

Compressor	Axial, Single Spool, 11 Stages
Pressure Ratio	Design 4.0:1
Airflow	58 lb/sec
Turbine	Reaction, 2 Stages
Exhaust Nozzle	Variable Eyelid Type (161-300 in$^{2)}$
Fuel	Avgas 100/130 Octane, Alt - JP-3 (MIL-F-5624A)
Power Control	Single Lever, Electronic Power Control
Ignition	Low Tension, 2 Spark Plugs (50 Hr), A/B Hot Streak
Oil Consumption	Less than 0.7 lb/hr
Oil Type	AN-O-9 Grade 1010, 5 Gal Max, Normal 4.5 Gal
Length	184 in Room Temp / 186 in Max Operating Temp
Width	30.0 in Room Temp / 30.25 in Max Operating Temp
Height	38.0 in Room Temp / 38.25 in Max Operating Temp
Diameter	27.0 in Room Temp / 27.25 in Max Operating Temp
Weight	Guaranteed 1,698 pounds

Operating Limits

Turbine Inlet Temperature, Idle / Max	1,350°F / 1500°F
Accelerate from Idle to Military	5 seconds above 1,400°F, Never Exceed 1,700°F
Maximum Altitude	40,000 ft non-A/B
Maximum Relight Altitude	40,000 ft/ Ram Pressure Ratio 1.2 or Higher
Maximum Altitude in A/B	40,000 ft +
Maximum A/B Relight Altitude	35,000 ft / Ram Pressure 1.2 or Higher
Continuous A/B Operation Limit	10 Minutes
Inverted Negative "G" Flight	10 Seconds (Fuel Flow May be Interrupted)
Overhauls	Initial engines 50 hours, raised to 150 hr later
Operating Environment Limits	SL Ram 1.38 at -65°F; Ram 1.89 at +100°F / 165°F Maximum

Production and Survivors

Total Production: Serials WE023001 to WE023125

Survivors: None known.

J34-WE-38 Ratings and Specifications (WAGT-24C8-2E)

Appendix A

Appendix C (YJ34-WE-38)**

Rating	Thrust (lb)	RPM*	SFC
Military	3,500	12,500	1.04
Normal	3,150	12,500	0.99
90% Normal	2,840	12,500	0.95
75% Normal	2,360	11,950	0.905
Idle	210	5,000	5.70

* Engine operated as a constant speed engine above 75% thrust levels.
** Thrust not less than 90% of guaranteed values, SFC not more than 107.5% of guaranteed values

Physical Details

Compressor	Axial, Single Spool, 11 Stages
Pressure Ratio	Design 4.0:1
Airflow	58 lb/sec
Turbine	Reaction, 2 Stages
Exhaust Nozzle	Variable Eyelid Type (164.5–259.0 in^2)
Fuel	JP-4 (MIL-O-5624A), Alt. JP-3, Avgas 94 Octane
Power Control	Single Lever modulated by electronic control to prevent over-speeds and over-temps
Ignition	Low Tension, 2 Spark Plugs
Oil Consumption	Less than 0.7 lb/hr
Oil Type	AN-O-9 type, 1010, 5 Gal Max, Normal 4.5 Gal
Length	112.0 in Room Temp / 113.0 in Max Operating Temp
Width	28.0 in Room Temp / 28.25 in Max Operating Temp
Height	38.0 in Room Temp / 38.25 in Max Operating Temp
Diameter	27.0 in Room Temp / 27.25 in
Weight	Guaranteed 1,448 lb

Operating Limits

Turbine Inlet Temperature, Idle / Max	1,350°F / 1500°F
Accelerate from Idle to Military	5 seconds above 1,400°F, Never Exceed 1,700°F
Maximum Altitude	40,000 ft
Maximum Relight Altitude	35,000 ft/ Ram Pressure Ratio 1.2 or Higher
Inverted Negative "G" Flight	10 Seconds (Fuel Flow May be Interrupted)
Overhauls	Initial engines 50 hours, raised to 150 hr later
Operating Environment Limits	Sea Level Ram 1.6 Standard Day

Production and Survivors

Total Production: Serials WE024001 to WE024531

Survivors: None known.

J46-WE-8 Ratings and Specifications (WAGT-24C-6C)

Rating	Thrust (lb)	At Max. Operating Limits	RPM*	SFC Specification	SFC Actual**
Afterburner	5,800	5,593	10.100	2.53	2.40
Military	3,980	4,240	10,100	1.100	1.088
Normal	3,560	3,830	10,100	1.097	1.072
90% Normal	3.205	3,447	10,100	1.106	1.078
75% Normal	2,665	2,873	10,100	1.186	1.145
Idle	245	146	4,000	980 lb/hr	781 lb/hr

* Engine operated as a constant speed engine at and above 75% thrust levels. **J46-WE-8 102 Acceptance

Physical Details

Compressor	Axial, Single Spool, 12 Stages
Pressure Ratio	Design 5.32
Airflow	73 lb/sec
Turbine	Reaction, 2 Stages
Exhaust Nozzle	Iris Type, 400 in^2 Full Open
Fuel	JP-4 (MIL-F-5624A, Alt. 100/130 (MIL-F-5572)
Power Control	Single Lever, Hydro-mechanical Control, Auto Temperature Trim
Ignition	High Tension, 2 Spark Plugs, A/B Hot Streak
Oil Consumption	Less than 0.7 lb/hr
Oil Type	MIL-L-7808 type, Normal 3.25 Gal
Length	197.3 in Room Temp / 198.42 in Max Operating Temp
Width	31.6 in Room Temp / 37.6 in Max Operating Temp
Height	37.3 in Room Temp / 37.6 in Max Operating Temp
Diameter	28.2 in Room Temp / 28.5 in Max Operating Temp
Weight	Guaranteed 2,116 lb

Operating Limits

Turbine Inlet Temperature, Operating	1,325°F
Accelerate from Idle to Maximum	20 secs Below 20,000 ft / 10 secs Above
Maximum Altitude	60,000 ft
Maximum Relight Altitude	40,000 ft/ Ram Pressure Ratio 1.2 or Higher
Maximum Altitude in A/B	48,000 ft / Ram Pressure 1.5 or Higher
Maximum A/B Relight Altitude	43,000 ft / Ram Pressure 1.23 or Higher
Continuous A/B Operation Limit	Unlimited
Inverted or Vertical Negative "G" Flight	30 Seconds
Overhauls	Initial engines 150 hours, (Actual ~100)
Operating Environment Limits	SL Ram 1.40 at -65°F; Ram 1.89 at +100°F

Production and Survivors

Total Production: Serials WE404685-WE405120

Survivors: Some -8 engines survive in museum collections. One cutaway exists

J46-WE-8B Ratings and Specifications (WAGT-24C10-6B)					
Rating	Spec. Thrust (lb)	At Max. Operating Limits	RPM*	SFC Specification	SFC Actual**
Afterburner	5,800	5,920	10,100	2.53	2.40
Military	3,980	4,090	10,100	1.100	1.097
Normal	3,560	3,780	10,100	1.097	1.078
90% Normal	3,205	3,780	10,100	1.106	1.091
75% Normal	2,665	---	10,100	1.186	1.176
Idle	245	---	4,000	980 lb/hr	675 lb/hr

* Engine operated as a constant speed engine at and above 75% thrust levels. **X24C10-13 Acceptance

Physical Details

Compressor	Axial, Single Spool, 12 Stages
Pressure Ratio	Design 5.32
Airflow	73 lb/sec
Turbine	Reaction, 2 Stages
Exhaust Nozzle	Iris Type, 400 in^2 Full Open
Fuel	JP-4 (MIL-F-5161B), Alt. 100/130 (MIL-F-5572)
Power Control	Single Lever, Hydro-mechanical Control, Auto Temperature Trim
Ignition	High Tension, 2 Spark Plugs, A/B Hot Streak
Oil Consumption	Less than 0.7 lb/hr
Oil Type	MIL-L-7808 type, Normal 3.25 Gal
Length	197.3 in Room Temp / 198.42 in Max Operating Temp
Width	31.6 in Room Temp / 37.6 in Max Operating Temp
Height	37.3 in Room Temp / 37.6 in Max Operating Temp
Diameter	28.2 in Room Temp / 28.5 in Max Operating Temp
Weight	Guaranteed 2,063 lb

Operating Limits

Turbine Inlet Temperature, Operating	1,325°F
Accelerate from Idle to Maximum	9.0 secs Below 20,000 ft / 10 secs Above
Maximum Altitude	60,000 ft
Maximum Relight Altitude	40,000 ft/ Ram Pressure Ratio 1.2 or Higher
Maximum Altitude in A/B	48,000 ft / Ram Pressure 1.5 or Higher
Maximum A/B Relight Altitude	43,000 ft / Ram Pressure 1.23 or Higher
Continuous A/B Operation Limit	Unlimited
Inverted or Vertical Negative "G" Flight	30 Seconds
Overhauls	Initial engines 150 hours, (Actual ~100)
Operating Environment Limits	SL Ram 1.40 at -65°F; Ram 1.89 at +100°F

Production and Survivors

Total Production: Serials WE040013 – WE040080 WE404516-WE404884
J46-WE-12B serials allocated: WE041114-WE041121

Survivors: Some -8B engines survive in museum collections. One cutaway exists.

J46-WE-8A Ratings and Specifications (WAGT-X24C10-6B)

Rating	Spec. Thrust (lb)	At Max. Operating Limits	RPM*	SFC Specification	SFC Actual**
Afterburner	5,500	5,870	10,100	2.53	2.492
Military	3,905	4,140	10,100	1.13	1.091
Normal	3,490	3,815	10,100	1.12	1.077
90% Normal	3,140	3,815	10,100	1.13	1.090
75% Normal	2,620	---	10,100	1.20	1.193
Idle	241	---	4,000	983 lb/hr	679 lb/hr

* Engine operated as a constant speed engine at and above 75% thrust levels. **X24C10-12 Acceptance

Physical Details

Compressor	Axial, Single Spool, 12 Stages
Pressure Ratio	Design 5.32
Airflow	73 lb/sec
Turbine	Reaction, 2 Stages
Exhaust Nozzle	Iris Type, 400 in² Full Open
Fuel	JP-4 (MIL-F-5161B), Alt. 100/130 (MIL-F-5572)
Power Control	Single Lever, Simple Hydraulic Control, Auto Temperature Trim
Ignition	Low Tension, 2 Spark Plugs, A/B Hot Streak
Oil Consumption	Less than 0.7 lb/hr (0.9 on test)
Oil Type	MIL-L-7808 type, Normal 3.25 Gal
Length	197.3 in Room Temp / 198.42 in Max Operating Temp
Width	31.6 in Room Temp / 37.6 in Max Operating Temp
Height	37.3 in Room Temp / 37.6 in Max Operating Temp
Diameter	28.2 in Room Temp / 28.5 in Max Operating Temp
Weight	Guaranteed 2,069 lb

Operating Limits

Turbine Inlet Temperature, Operating	1,325°F
Accelerate from Idle to Maximum	9.0 secs Below 20,000 ft / 10 secs Above
Maximum Altitude	60,000 ft
Maximum Relight Altitude	40,000 ft/ Ram Pressure Ratio 1.2 or Higher
Maximum Altitude in A/B	48,000 ft / Ram Pressure 1.5 or Higher
Maximum A/B Relight Altitude	43,000 ft / Ram Pressure 1.23 or Higher
Continuous A/B Operation Limit	Unlimited
Inverted or Vertical Negative "G" Flight	30 Seconds
Overhauls	Initial engines 150 hours, (Actual ~100)
Operating Environment Limits	SL Ram 1.40 at -65°F; Ram 1.89 at +100°F

Production and Survivors

Total Production: YJ46-WE-8A Serials WE040001-WE040006
J46-WE-8A Serials WE040007-WE040012; WE404501-WE404515

Survivors: No -8A engines are known to exist. Most likely all were converted to -8B's at overhaul.

J46-WE-18 Ratings and Specifications (WAGT-124D-A)

Rating	Spec Thrust (lb)	At Max. Operating Limits	RPM*	SFC Specification	SFC Actual**
Afterburner**	6,100	6,445	10,000	2.60	2.46
Military**	4,100	4,435	10,000	1.080	0.985
Normal	3,690	4,020	10,000	1.065	0.976
90% Normal	3,320	---	10,000	1.060	0.985
75% Normal	2,765	---	10,000	1.090	1.035
Idle	245	---	4,000	980 lb/hr	753 lb/hr

* Engine operated as a constant speed engine at and above 75% thrust levels.**XJ46-WE-18 #27 Acceptance

Physical Details

Compressor	Axial, Single Spool, 13 Stages
Pressure Ratio	6.0:1
Airflow	75 lb/sec
Turbine	Reaction, 2 Stages
Exhaust Nozzle	Iris Type, 388 in^2 Full Open
Fuel	JP-4 (MIL-F-5624B), Alt. 100/130 (MIL-F-5572)
Power Control	Single Lever, Hydro-Mechanical
Ignition	Low Voltage High Energy, 2 Spark Plugs, A/B Hot Streak
Oil Consumption	Less than 0.7 lb/hr
Oil Type	MIL-L-7808B type, 1010, 3.25 Gal Max
Length	197.3 in Room Temp / 198.4 in Max Operating Temp
Width	31.6 in Room Temp / 37.6 in Max Operating Temp
Height	37.3 in Room Temp / 37.6 in Max Operating Temp
Diameter	28.2 in Room Temp / 28.5 in Max Operating Temp
Weight	Guaranteed -18- 2,163 lb / -20 - 2,168 lb

Operating Limits

Turbine Inlet Temperature, Idle	1,519°F
Accelerate from Idle to Military	5 seconds above 1,400°F, Never Exceed 1,700°F
Maximum Altitude	60,000 ft
Maximum Relight Altitude	45,000 ft/ Ram Pressure Ratio 1.2 or Higher
Maximum Altitude in A/B	NLT 45,000 ft / Ram Pressure 1.25 or Higher
Maximum A/B Relight Altitude	40,000 ft / Ram Pressure 1.2 or Higher
Continuous A/B Operation Limit	Unlimited
Inverted or Vertical Negative "G" Flight	30 secs
Overhauls	150 Hours
Operating Environment Limits	Ram 1.38 at -65°F; Ram 1.89 at +100°F / 165°F Maximum, Sea Level – No Limit

Production and Survivors

Total Production: YJ46-WE-18 Serials: WE040001–WE040008 (No -20's built)

Survivors: None. Two partial engines sent to schools not located.

J46-WE-1 Ratings and Specifications (WAGT-X24C10-2D)

(Same as J46-WE-2. See text for differences.)

Rating	Thrust (lb)	RPM*	SFC
Afterburner**	6,100	10,100	2.50
Military**	4,080	10,100	1.01
Normal	3,670	10,100	0.96
90% Normal	3,300	10,100	0.927
75% Normal	2,755	9,800	0.903
Idle	245	4,000	----

* Engine operated as a constant speed engine above 75% thrust levels.

Physical Details

Compressor	Axial, Single Spool, 12 Stages
Pressure Ratio	5.32:1
Airflow	73.0 lb/sec
Turbine	Reaction, 2 Stages
Exhaust Nozzle	3° Cant, Clamshell type, 400 in² Max
Fuel	AN-F-58 (MIL-F-5624) (JP-3)
Power Control	Single Lever, Electronic
Ignition	Low Tension, 3 Spark Plugs
Oil Consumption	Less than 0.7 lb/hr
Oil Type	AN-0-9, 1010 Grade, 3.25 Gal
Length	191.7 in Room Temp / 194.4 in Max Operating Temp
Width	30.5 in Room Temp / 30.75 in Max Operating Temp
Height	37.25 in Room Temp / 37.5 in Max Operating Temp
Diameter	29.0 in Room Temp / 29.25 in Max Operating Temp
Weight	Guaranteed 1,863 lb

Operating Limits

Turbine Inlet Temperature, Idle	1,300°F
Accelerate from Idle to Military	5 seconds above 1,400°F, Never Exceed Temp 1,500°F
Maximum Altitude	60,000 ft
Maximum Relight Altitude	45,000 ft/ Ram Pressure Ratio 1.2 or Higher
Maximum Altitude in A/B	NLT 45,000 ft / Ram Pressure 1.25 or Higher
Maximum A/B Relight Altitude	40,000 ft / Ram Pressure 1.2 or Higher
Continuous A/B Operation Limit	1 hr
Inverted Negative "G" Flight	10 secs
Overhauls	150 Hours
Operating Environment Limits	Ram 1.78 at -65°F; Ram 1.89 at +100°F / 165°F Maximum, Sea Level – No Limit

Production and Survivors

Total Production: None House Engine Serials WE004027-31

Survivors: None are known to have been assembled.

J46-WE-3 Ratings and Specifications (WAGT-X24C10-3)

(Same as J46-WE-8A but tested to MIL-E-5009. See text for differences.)

Rating	Thrust (lb)	RPM*	SFC
Afterburner**	5,500	10,100	2.53
Military**	3,905	10,100	1.13
Normal	3,490	10,100	1.12
90% Normal	3,140	10,100	1.13
75% Normal	2,620	10,100	1.20
Idle	241	4,000	4.08

* Engine operated as a constant speed engine above 75% thrust levels.

Physical Details

Compressor	Axial, Single Spool, 12 Stages
Pressure Ratio	5.32:1
Airflow	73.0 lb/sec
Turbine	Reaction, 2 Stages
Exhaust Nozzle	3° Cant, Iris type, 400 in^2 Max
Fuel	AN-F-58 (MIL-F-5624) (JP-3)
Power Control	Single Lever, Simple Hydraulic
Ignition	Low Tension, 3 Spark Plugs
Oil Consumption	Less than 0.7 lb/hr
Oil Type	AN-0-9, 1010 Grade, 3.25 Gal
Length	203.0 in Room Temp / 205.0 in Max Operating Temp
Width	29.4 in Room Temp / 29.7 in Max Operating Temp
Height	36.9 in Room Temp / 37.2 in Max Operating Temp
Diameter	29.0 in Room Temp / 29.3 in Max Operating Temp
Weight	Guaranteed 2,074 lb

Operating Limits

Turbine Inlet Temperature, Idle	1,300°F
Accelerate from Idle to Military	Never Exceed Temp 1,300°F
Maximum Altitude	60,000 ft
Maximum Relight Altitude	35,000 ft/ Ram Pressure Ratio 1.15 to 1.5 40,000 ft Ram Pressure Ratio 1.2
Maximum Altitude in A/B	NLT 40,000 ft / Ram Pressure 1.4 or Higher
Maximum A/B Relight Altitude	40,000 ft / Ram Pressure 1.2 or Higher
Continuous A/B Operation Limit	1 hr
Inverted Negative "G" Flight	60 secs
Overhauls	150 Hours
Operating Environment Limits	Ram 1.78 at -65°F; Ram 1.89 at +100°F / 165°F Maximum, Sea Level – No Limit

Production and Survivors

Total Production: None built. House Engine Serials (for -5 as well) WE004021-26

Survivors: None known, only house engines were partially tested.

Appendix B

Engine Operating Limits Charts

The charts presented are the best available copies after being cleaned up. Charts were not found for some models and it is likely they were never produced. Drafting such a chart in the late 1940's and 1950's era was a significant effort and the originals were in a 24 inch by 24 inch layout. The remaining copies in the BuAer contract files had been reduced to fit on 8.5x11 inch paper with one inch margins, the readability suffering considerably.

It is important to remember that all of the data points on such charts were calculated using formulas and do not represent actual data taken during altitude test chamber running or actual flight testing. An enormous amount of manual calculation using slide rules and mechanical calculator machines was required to determine the various data points. When actual flight and tunnel tests returned different values from those calculated, it brought the formulas being used into question. The use of such charts, however, represented the state of the art available at the time to aircraft designers in predicting airframe performance capabilities.

XJ34-WE-32 and XJ34-WE-38

March 8, 1949

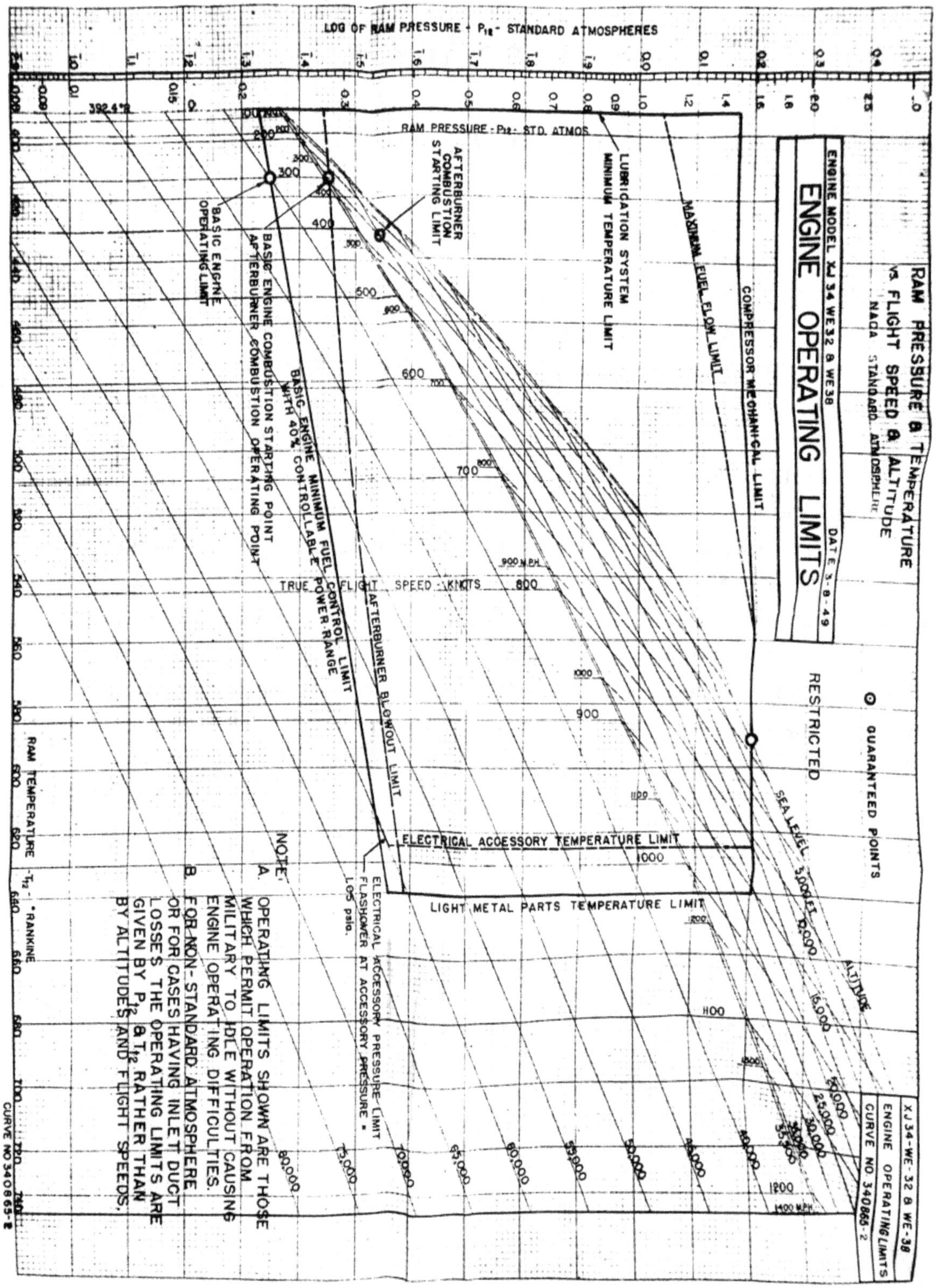

XJ46-WE-8, -10

February 3, 1953

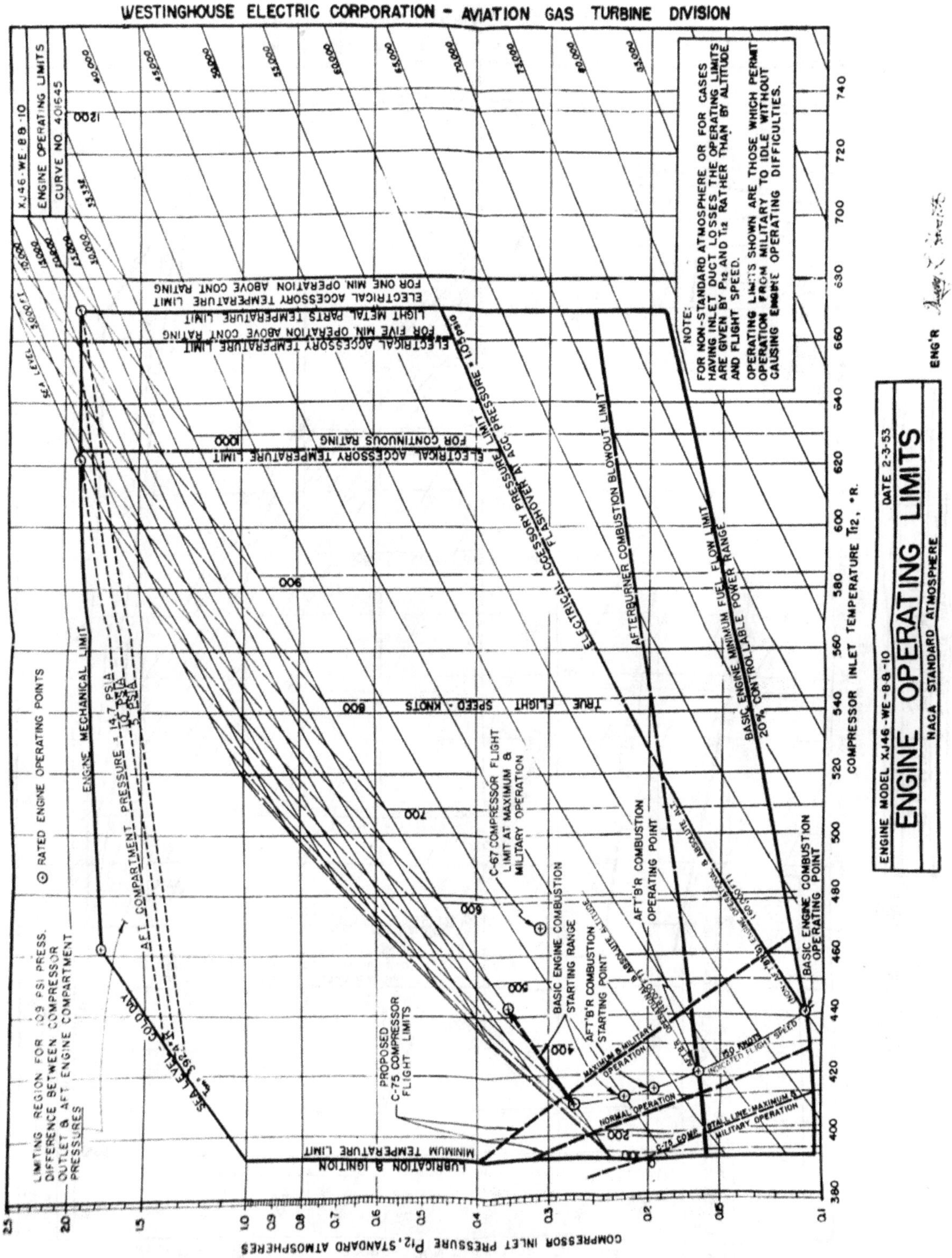

XJ46-WE-8, -8B,-10, -12B, -12B

Revision 2 – December 12, 1953

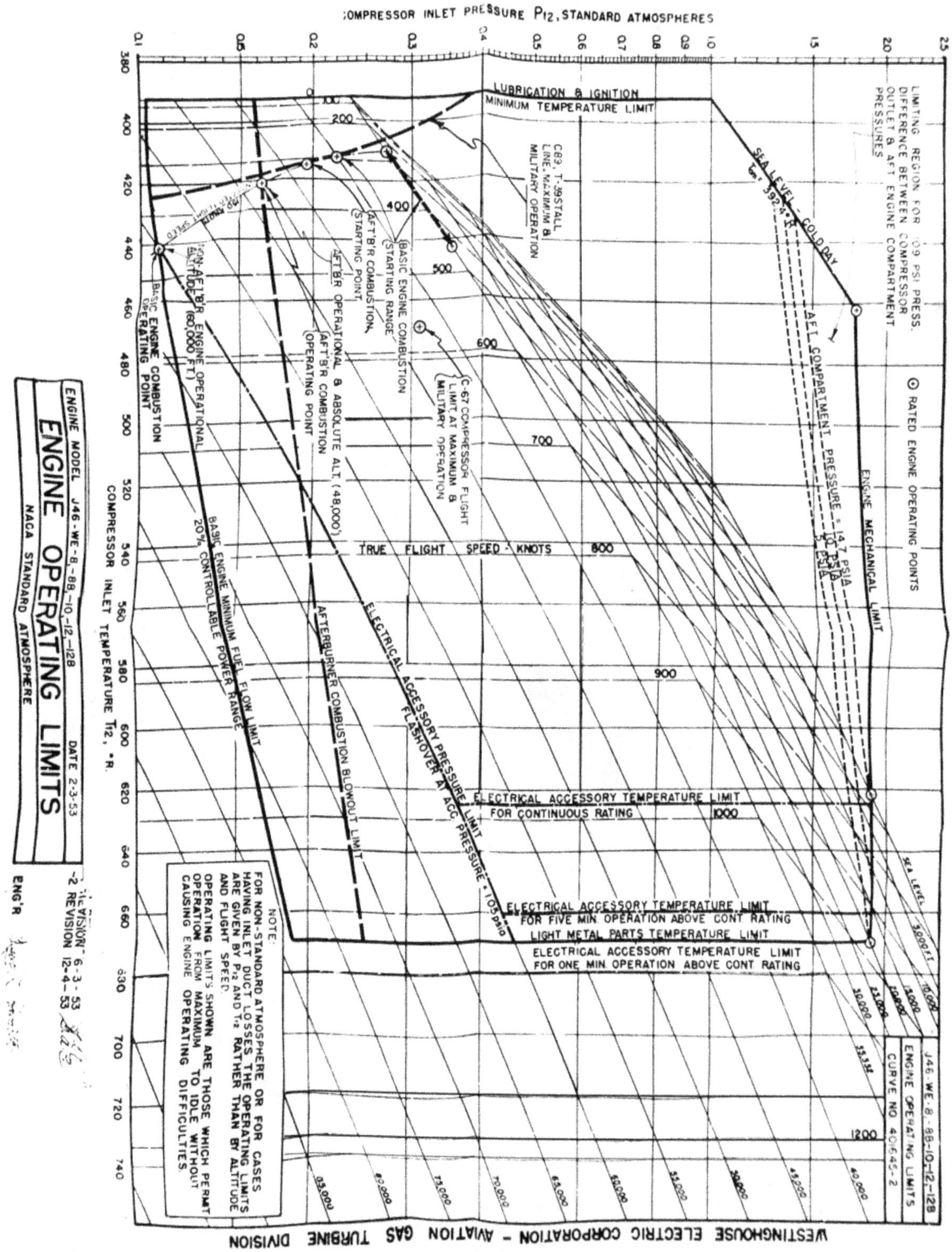

J46-WE-8A, -12A; YJ46-WE-8A

May 1, 1953

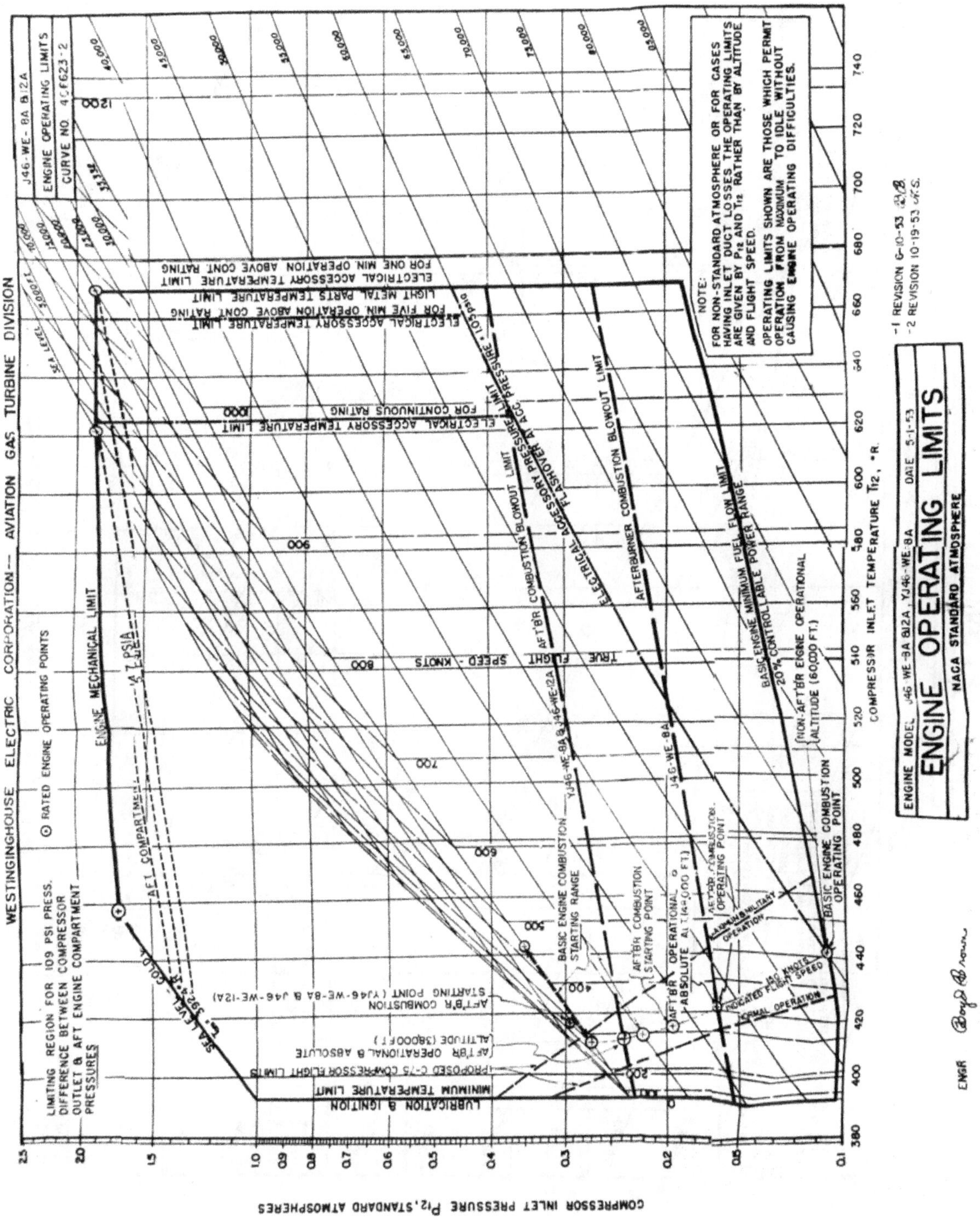

J46-WE-18

March 1, 1954

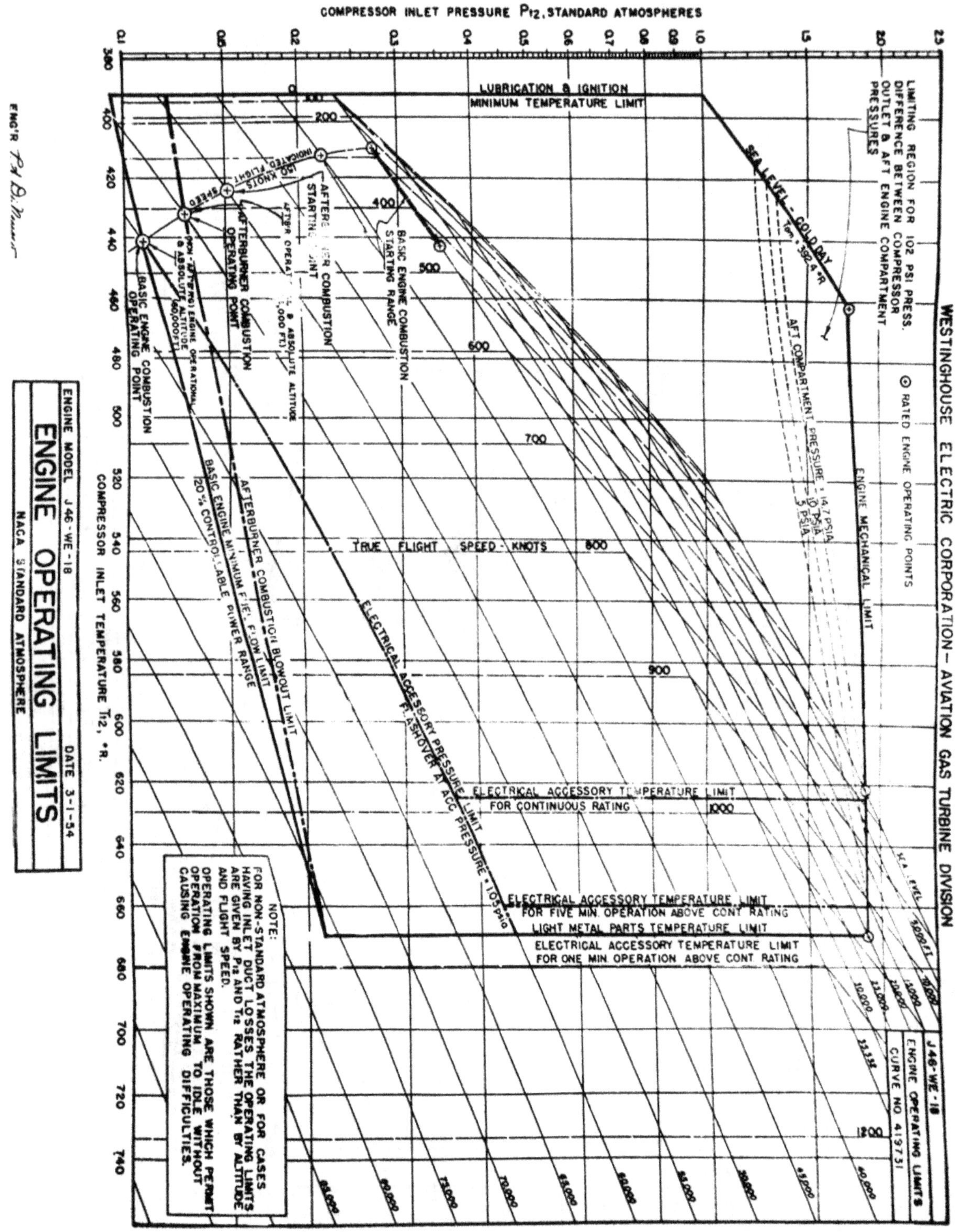

XJ46-WE-1

October 28, 1949

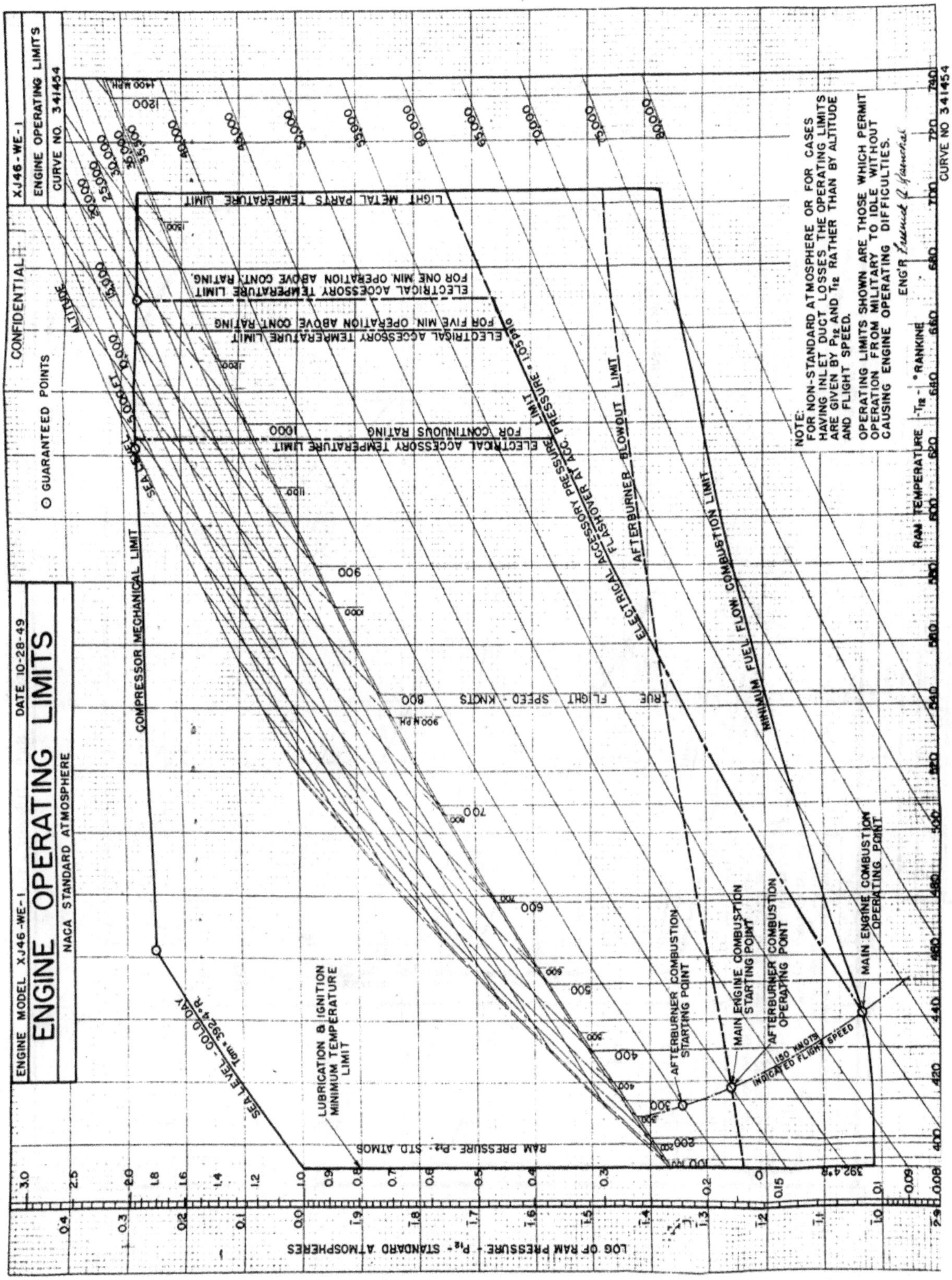

Proposed PD-24A

October 1, 1953

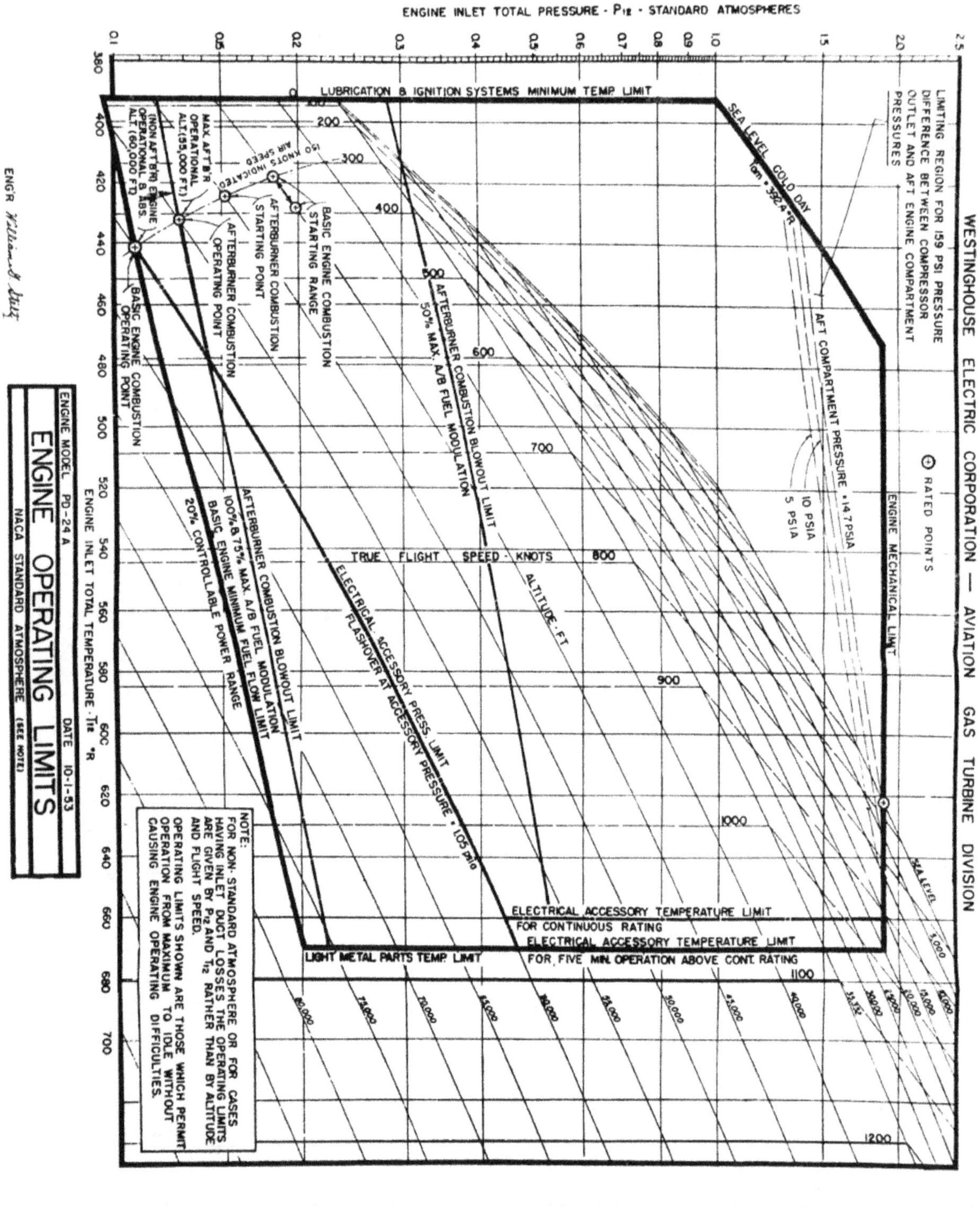

Appendix C

YJ46-WE-8A Field Problems Reports

When the early YJ46-WE-8A engines reached Chance Vought, they were subjected to extensive ground running and installation testing before they were allowed to power an aircraft. WAGT kept track of the problems reported and sent BuAer reports of the incidents experienced, the local corrective actions taken and the planned permanent changes to be made to eliminate the issue. Chance Vought expressed the opinion that the engines were not safe for use in flight and WAGT argued the opposite. Only the first five of the six YJ46-WE-8A engines were included in the reports and not all the reports were saved in the archives, but those found are quite revealing.

YJ46-WE-8A Field Problems Reports As of 05/20/1953					
Engine Serial: WE040001					
Date	**Engine TTD**	**Trouble**	**Cause**	**Correction**	**Comment**
01/19/53	17.88	Aerotec P3 Line Failed	Mishandling during transit	Replaced Line	More rugged line to be used.
01/24/53	17.89	Adapter gear box (starter) Modification	Unsuitable for flight	Installed latest gearbox version	Improved bushings and tab washers.
01/24/53	17.89	Compressor bleed boss covers warped during acceptance test at W.	Aluminum covers not suitable for flight	Installed stainless steel covers	None
01/27/53	18.00	Compressor buffet in the 55-72% rpm range	Airflow discontinuity in this range with the C-67 compressor	Re-twisted compressor blading to C-75 configuration	None
01/29/53	21.81	Oil accumulation in the accessories drive gearbox at low rpm	Pump scavenge capacity below 60% rpm is not sufficient to obtain proper scavenging	None. Accumulation of 3-4 quarts of oil in the gearbox but is not detrimental to engine operation.	Higher scavenge capacity pump to be designed and made available on 21st Essington produced engine.
02/02/53	21.7	A/B over-temperatures with a complete electrical failure.	Bleed solenoid opens during D.C. power failure allowing A/B discharge pressure to bleed indirectly into the A/B fuel regulator control pressure return line. This causes the A/B throttle to move to the full open position and therefore a high A/B fuel flow was experienced.	CID 16823B installed adding a separate line from the solenoid directly to the fuel booster pump discharge.	None.
02/03/53	21.81	Accumulation of oil in fuel dump tank.	A/B pump seal leak allowed engine oil to enter the pump and then leak out through the surge check valve into the dump tank at a rate of 75 cc/min.	Installed new pump assembly and "O" rings. Rejected original assembly and sent it back to W.	Replacement has improved seal assembly to eliminate future troubles.
02/04/53	24.26	Exhaust nozzle failed to open fully.	Shuttle valve failed at the junction of the valve seat and stem at the end with the screw-driver slot. One ear of the screw driver slot was missing	Installed new shuttle valve.	Examination revealed the lack of the designed radius at the point of failure. This concentrated stress at the point of failure. Sheared ear was the result of over-torqueing during valve assembly

YJ46-WE-8A Field Problems Reports As of 05/20/1953					
Engine Serial: WE040001					
Date	**Engine TTD**	**Trouble**	**Cause**	**Correction**	**Comment**
02/06/53	24.26	Aerotec malfunctioned during A/B lite-off.	Aerotec P1 line (metal-flex) failed approx. 5″ from a support bracket.	A new line was manufactured locally from steel tubing because replacement line was unavailable.	Minimum bend radius of the hose had been exceeded. Tighter braid on hose will restrict bends.
02/09/53	26.28	Aerotec P1 pressure reading dropped off.	Aerotec P1 rake broken off at the first pressure hole at outer periphery.	Replaced the rake.	Rake of increased cross section needed to withstand temperatures and pressures.
02/10/53	28.52	Oil leaking from exhaust nozzle actuator shuttle valve assembly.	"O" ring on valve cap was nicked at outer edge.	Replaced the "O" ring.	"O" ring had been possibly damaged during replacement of the shuttle valve on 2/4/53.
02/12/53	30.25	When the manual trim is set for military temperature during A/B and the A/B blows out, the engine may over-temp.	K3 relay (trim relay) in the Auxiliary Electric Control will not drop out on blowout due to the nature of the circuit.	Installed a new control with modified circuits to prevent the condition.	Problem was known prior to operation at CVA but paperwork for the fix was not completed prior to this time.
02/16/53	30.25	Routine "C" inspection of 2nd stage turbine revealed a crack on a blade from trailing edge and extending forward approx. 11/16″.	Analysis at W revealed a cold shut at this point which initiated a fatigue failure.	Affected blade and one directly opposite replaced with ones having weight within 0.01 of each other.	Engine run showed acceptable vibration levels. New electro-polish inspection method expected to reveal this type of defect.
02/21/53	31.28	Oil leakage at the fittings connecting the two high pressure relief valves on exhaust nozzle actuating system.	"O" rings of improper size (too small) may have been installed during initial assembly.	Installed new "O" rings of proper size. No further leakage observed.	Later engines to have a one piece valve assembly eliminating the connecting fittings.
02/21/53	31.28	Routine inspection revealed that snap rings on several of the iris nozzle leaf actuating roller bearings had dropped out of their retaining grooves.	Snap rings relaxed due to heat.	Removed snap rings. Brackets on either side of the bearing retain the rollers and races in proper alignment.	A CID calling for a heavier snap ring is being qualified by test. Their primary purpose is to hold the leaf in position during removal or installation of the bearing.
02/23/53	31.28	Fuel leakage around the distributor valve.	Two bottom center mounting studs and stbd tapped hole which attaches to oil cooler inlet and outlet boss leaking under pressure test.	New distributor installed.	Possible porous casting.

YJ46-WE-8A Field Problems Reports As of 05/20/1953					
Engine Serial: WE040001					
Date	Engine TTD	Trouble	Cause	Correction	Comment
02/23/53	31.28	Leakage at oil cooler adapter plate flange.	Some nicks on face of adapter plate flange. Leakage under pressure test. Stbd oil cooler inspection revealed a thin shield pressed into the "O" ring grooves which would prevent proper seating of the "O" rings. Shield part of production process and should have been removed.	Refaced the adapter plate flange by grinding but leak persisted. No replacement cooler was available. Leakage was minor and reassembly allowed ground runs while replacement was procured.	New cooler design calls for welding the adapter plate to the cooler. Supervisors advised of the deficiency in manufacture and will ensure the shield will be removed.
02/26/53	31.28	Fuel distributor valve mounting leg broken at the intersection of the horizontal and vertical components of the leg.	Four legs being slightly out of plane by approx. 1/16" and the engine mounting box being 0.0020" out of plane. Undercut on the face of one leg creates a high stress concentration at point of failure.	No replacement assembly available. Unit from WE040002 installed (leg on this unit also broke during installation). Engine mounting pad resurfaced to remove out of plane condition. New distributor installed without a problem	Mounting legs found to be be damaged during shipment causing them to be bent and out of parallel. Better packaging being designed for shipments.
03/05/53	32.38	Accumulation of magnetic chips found in the bottom of the oil reservoir. Only one large piece found and it was approx. 1/8" long and irregular triangular cross section with sides of approximately 1/16".	Undetermined. Believed to have entered the reservoir during manufacturing processes.	All cover plates removed but source of chips not located. Oil filter and bearing edge filter clean. Engine was ground run and inspections at 2 and 5 hours found no chips. Engine released for flight.	None.
03/07/53	32.38	Oil seepage at oil reservoir dipstick assembly.	A sliver of metal was found under "O" ring. Caused by misalignment of the retainer during installation causing shearing of the metal when tightened, also bending the ears on the retainer. Bent ears prevented the cover from seating properly.	Retainer and "O" rings reinstalled. Bent ears realigned.	Was caused upon reassembly of unit after inspection of reservoir for metal chips.
03/14/53	33.78	Engine driven fuel boost pump bearing inspected for grease.	To ascertain that the bearing was in satisfactory condition and contained correct quantity of grease in order to evaluate the effectiveness of the A/C vent system.	Fuel boost pump inspected.	Unit was satisfactory.

YJ46-WE-8A Field Problems Reports As of 05/20/1953					
Engine Serial: WE040001					
Date	**Engine TTD**	**Trouble**	**Cause**	**Correction**	**Comment**
03/15/53	33.78	Fuel boost pump mounting stud failed at the necked down section separating the coarse and fine threads during reinstallation of the pump.	Examination of the sheared section revealed that a flaw existed in the stud at this point. Excessive torque had not been applied.	No spare studs available. "AN" stud altered to fit.	Further investigation into cause of chronic abrasion which appears to be chronic suggested to prevent similar failures in the future.
03/17/53	34.45	Oil leaking from exhaust nozzle actuator shuttle valve housing.	"O" ring on valve cap appeared to be damaged by abrasion.	Leather back-up washer appeared to be OK. New "O" ring installed.	Further W investigation into cause of chronic abrasion in future.
03/18/53	34.85	A/B ignitor relay valve leakage was 530 cc/min during A/B operation. Limit is 125 cc/min.	Excessive clearance between valve and body.	New valve installed. Leakage was 40-50 cc/min.	None.
03/26/53	38.08	Fuel distributor valve control diaphragm pressure was zero. Turbine outlet temperature was approximately 10°C higher than normal.	Control diaphragm pressure damping orifice was plugged with a gritty sand-like substance. With orifice plugged the distributor valve would move open to maximum stop causing some change in fuel and temperature distribution.	Orifice cleaned out. Subsequent operation normal	Source of dirt not determined. Some consideration being given to increasing orifice design or size to prevent plugging.

YJ46-WE-8A Field Problems Report As of 05/20/1953					
Engine Serial: WE040002					
Date	**Engine TTD**	**Trouble**	**Cause**	**Correction**	**Comment**
02/05/53	19.42	Plug in the generator-oil pump shaft in accessories drive gearbox dislodged.	Pressure check of gearbox at W possibly exceeded retaining limits of plug.	Installed new "O" ring on plug and reinstalled the plug.	Pressure check calls for 9 psi. Operating pressure is approx. 3 psi. W CID processed to have the plug pin-locked.
02/19/53	19.42	Minimum 2nd stage turbine tip clearance was 0.062" on one housing shoe at 6:30 o'clock position. Minimum allowed is 0.063".	Warpage of turbine housing at this point.	Removed shoe, ground approx. 0.015" maximum and I.D. of shoe. Minimum was 0.065" upon reassembly.	Engine vibration check showed within limits. W investigating possibility of eliminating housing warping by increasing radial clearance between the housing and combustion chamber shell
02/20/53	19.42	Shuttle valve inspected and found to have improper radius at junction of valves and stem.	Inspection was result of failure of shuttle valve on WE040001.	New shuttle valve with minimum 1.32" radius at junction of valves and stem installed.	None
02/20/20	19.42	Snap rings on several of the iris nozzle leaf actuating roller bearings had dropped out of their retaining grooves.	Snap rings relaxed due to heat.	Removed defective snap rings. Brackets on either side of the bearing serve to retain the roller and races in proper alignment.	W CID calls for a heavier snap ring to be incorporated pending completion of qualification test.
02/23/53	19.42	Compressor buffet in 55% to 72% range.	Airflow discontinuity in this range with the C-67 compressor configuration.	Re-twist compressor blading to C-75 configuration.	None.
02/23/53	20.42	Accumulation of oil in the fuel dump tank.	A/B fuel pump seal leak allowed oil to enter pump from gearbox an drain into fuel dump tank via the surge check valve at rate of 75 cc/min.	Installed modified pump assembly and "O" rings.	Modified assembly has improved seal assembly to eliminate further leakage.
02/23/53	20.42	Oil accumulates in the accessories drive gearbox at low rpm.	Lube pump scavenge capacity below 60% rpm is not sufficient to obtain proper scavenging.	None taken. Approx 3-4 quarts of oil will accumulate in the gearbox. This condition is not detrimental to engine operation but does present a problem of uncertain oil reservoir level measurement.	A pump with a higher capacity scavenge element is to be installed when it becomes available. The 21st So. Phila. Engine is scheduled to have the new pump.

YJ46-WE-8A Field Problems Report As of 05/20/1953					
Engine Serial: WE040002					
Date	**Engine TTD**	**Trouble**	**Cause**	**Correction**	**Comment**
03/03/53	23.62	A/B ignitor relay valve leakage was 540 – 600 cc/min during A/B. Limit is 125 cc/min.	Excessive clearance between valve and body.	New valve installed. Leakage rate within limits.	None.
03/07/53	23.68	A/B fuel flow was approximately 5,000 lb/hr which is 3,000 lb/hr below normal.	A/B fuel control shims used to position the fuel control metering pin with respect to the bellows had become displaced and were lying in the bellows cavity. When electrical failure was simulated, this caused the recharge solenoid valve to open and force a large amount of fuel into the engine return system causing the bellows chamber to become pressurized. The bellows then contracted to allow the metering pin to become disengaged allowing the shims to fall out of position.	New fuel control installed.	Defective control was returned to W for recalibration and installation of one piece shim in replacement of several thin shims to increase margin before shim will be displaced. CID processed to modify the recharge piping to prevent pressurizing the return system when a power failure occurs.
03/12/53	23.65	A/B over-temps when a complete electrical failure occurs.	When an electrical failure occurs, the A/B fuel regulator bleed solenoid opens allowing the A/B discharge pressure to bleed indirectly into the A/B fuel regulator control pressure return line. This causes the A/B throttle to move to the full open position and therefore a high A/B fuel flow was experienced.	Incorporated W CID which added a separate line from the solenoid directly to the fuel booster pump discharge.	None.
No reports found covering the period 3/13/53 to 5/8/53					
05/09/53	27.12	Interference between the channel stiffeners on the A/B iris nozzle inner leaf and the inner surface of outer leaf when at or towards the closed nozzle position thus preventing proper sealing between them.	Slope on the stiffeners too great.	A/B serial D-002. Edge of the stiffeners ground down enough so that a 0.002" feeler gage would pass between it and the inner face of the adjacent leaf.	Westinghouse initiated an EA to inspect all A/B's for the condition and grind as required to relieve the interference.

YJ46-WE-8A Field Problems Report As of 05/20/1953					
Engine Serial: WE040002					
Date	**Engine TTD**	**Trouble**	**Cause**	**Correction**	**Comment**
05/12/53	27.12	Cooler – Fuel distributor assembly borrowed for installation on WE040001, 2-26-53.	No spares for GFE WE010001.	Installed assembly.	None.
05/12/53	27.12	Installation of engine in production aircraft required removal of the 3 degree bend (cant) in the A/B.	Aircraft design change	Removed 3 degree bend, installed short activator side rods.	None
05/12/53	27.12	Iris nozzle side push rod clevis fitting not hardened to required Rockwell C-18 to C-25 range.	Improper heat treatment by supplier.	Installed clevis fittings treated to Rockwell c-18 to c-25 range.	None
05/12/53	27.12	When manual trim is set for military temp during A/B and the A/B blows out the engine may over temp.	K3 relay (trim relay) in the Auxiliary Electric Control will not drop out on blowout due to the nature of the circuit.	Installed new AEC that had modifications to circuit.	Unit returned to Westinghouse for modification.
05/16/53	27.12	Fuel boost pump not modified to conform with J46 Notice S-1.	Fuel Boost Pump modification being made to eliminate possibility of the impeller shaft lock nut coming off and to inspect the bearing for grease.	Installed fuel boost pump from WE040005 to expedite delivery engine for aircraft installation.	Replace pump was modified locally for installation in WE040005.
05/18/53	27.57	Oil leak at union in inlet to the high pressure relief valve on the "open" side of the exhaust nozzle actuating system.	Oil relief valve "O" ring frayed slightly for unknown reason.	Installed new "O" ring and washer.	None.
05/18/53	29.47	Threads on Aerotec P2 tap seized and stripped during attempt to remove it from A/B D-002.	P2 tap subjected to high temperature during A/B operation which eventually leads to subject condition.	Installed a P2 tap from A/B D-001. No spares in GFE.	Defective unit returned to <u>W</u> for inspection.
05/20/53	29.57	Engine stalled during A/B lite-off.	Aerotec switch travel approx.. ¼"; normally 1/8". Excessive travel delayed signal to open exhaust nozzle.	Installed new switch.	Defective unit returned to <u>W</u> for inspection
05/20/53	29.57	With engine operating, the throttle could not be advanced beyond 78 degrees travel.	Interference existed at 78 degrees on power lever gearbox between the first tooth on the sector gear (drive) and the slanted 2nd tooth on the A/B fuel control distributor valve gear.	Removed 0.010" material from the slanted tooth. Subsequent operation O.K.	None.

YJ46-WE-8A Field Problems Report As of 06/03/1953					
Engine Serial: WE040003					
Date	**Engine TTD**	**Trouble**	**Cause**	**Correction**	**Comment**
05/07/53	46.37	When manual trim is set for military temperature during A/B and the A/B blows out, the engine may over-temp.	K3 relay (trim relay) in the Auxiliary Electric Control will not drop out on blowout due to the nature of the circuit.	Installed AEC from WE040001.	Unit returned to Westinghouse for modification.
05/07/53	46.37	Igniter relay valve exceeded maximum allowed leakage during afterburning was 125 cc/min. The fuel was discharged into the A/B via the relay torch ignitor.	Design.	Installed surge check valve to bypass this fuel to boost pump discharge.	None.
05/11/53	48.91	Several cracks 3/16″ long in iris nozzle leaf retaining hinges at 6, 9 and 12 o'clock positions.	Not determined.	1/16″ stop holes drilled at base of cracks and cracks were welded up.	None.
05/23/53	53.85	Spot welds pulled at 12 places where radial supports are attached to squeal baffle plus two longitudinal cracks 4″ long about 2″ apart at the 5 o'clock position from the wall of the squeal baffle.	Attempted A/B lite-off at 30,000 ft at 0.5 MN resulted in a deep compressor stall of approx. 1.5 minutes with T5 ranging from 750 to 1,000°C.	Installed combustion chamber assembly from A/B D-001.	Tests conducted to overcome compressor problem on A/B lite-offs.
05/23/53	53.85	Iris nozzle side push rod clevis fittings not hardened to required Rockwell c-18 to C-25 range.	Improper heat treatment by supplier.	Installed clevis fittings heat treated to Rockwell c-18 to c-25 range.	None.

YJ46-WE-8A Field Problems Report As of 06/03/1953					
Engine Serial: WE040004					
Date	**Engine TTD**	**Trouble**	**Cause**	**Correction**	**Comment**
05/09/53	31.84	Interference between the channel stiffeners on the A/B iris nozzle inner leaf and the inner surface of outer leaf when at or towards the closed nozzle position thus preventing proper sealing between them.	Slope on the stiffeners too great.	A/B serial D-004. Edge of the stiffeners ground down enough so that a 0.002″ feeler gage would pass between it and the inner face of the adjacent leaf.	Westinghouse initiated an EA to inspect all A/B's for the condition and grind as required to relieve the interference.
05/15/53	31.84	Installation of engine in production aircraft required removal of the 3 degree bend (cant) in the A/B.	Aircraft design change	Removed 3 degree bend, installed short activator side rods.	None

05/25/53	31.84	Iris nozzle side push rod clevis fittings not hardened to required Rockwell c-18 to C-25 range.	Improper heat treatment by supplier.	Installed clevis fittings heat treated to Rockwell c-18 to c-25 range.	None.
06/01/53	31.84	Oil accumulates in the accessories drive gearbox at low rpm.	Lube pump scavenge capacity below 60% rpm is not sufficient to obtain proper scavenging.	None taken. Approx. 3-4 quarts of oil will accumulate in the gearbox. This condition is not detrimental to engine operation but does present a problem of uncertain oil reservoir level measurement.	A pump with a higher capacity scavenge element is to be installed when it becomes available. The 21st So. Phila. Engine is scheduled to have the new pump.

YJ46-WE-8A Field Problems Report As of 06/03/1953

Engine Serial: WE040005

Date	Engine TTD	Trouble	Cause	Correction	Comment
05/19/53	14.89	Interference between the channel stiffeners on the A/B iris nozzle inner leaf and the inner surface of outer leaf when at or towards the closed nozzle position thus preventing proper sealing between them.	Slope on the stiffeners too great.	A/B serial D-004. Edge of the stiffeners ground down enough so that a 0.002" feeler gage would pass between it and the inner face of the adjacent leaf.	Westinghouse initiated an EA to inspect all A/B's for the condition and grind as required to relieve the interference.
05/19/53	14.89	Iris nozzle side push rod clevis fittings not hardened to required Rockwell c-18 to C-25 range.	Improper heat treatment by supplier.	Installed clevis fittings heat treated to Rockwell c-18 to c-25 range.	None.
05/281/53	15.14	Oil accumulates in the accessories drive gearbox at low rpm.	Lube pump scavenge capacity below 60% rpm is not sufficient to obtain proper scavenging.	None taken. Approx. 3-4 quarts of oil will accumulate in the gearbox. This condition is not detrimental to engine operation but does present a problem of uncertain oil reservoir level measurement.	A pump with a higher capacity scavenge element is to be installed when it becomes available. The 21st So. Phila. Engine is scheduled to have the new pump.

Appendix D
Surviving J46 Engines

List includes all known engines and engine data as of the time of publication.

Type	Serial	Accepted	A/B	Location	Status
J46-WE-8B	WE404950	03/10/1955	YES	Phoenix, AZ	Owned by Al Casby, Mid-Century Aviation. Engine is operable with about 35 hours since its last overhaul.
J46-WE-8B	WE404966	03/04/1955	YES	Phoenix, AZ	Owned by Al Casby, Mid-Century Aviation. Engine is operable with about 35 hours since its last overhaul.
J46-WE-8A	WE040048	03/15/1954	NO	Phoenix, AZ	Owned by Al Casby, Mid-Century Aviation. Only Essington built engine known. Non-runnable. Parts source.
J46-WE-8?	Unknown	Unknown	NO	Phoenix, AZ	Owned by Al Casby, Mid-Century Aviation. No manufacturer's plate. Modified for dragster use. Non-runnable. Parts source.
J46-WE-8	WE405599[t]	12/29/1954	NO	Phoenix, AZ	Owned by Al Casby, Mid-Century Aviation. From Dunn, NC. A zero-time engine since overhaul.
J46-WE-8?	Unknown	Unknown	NO	Phoenix, AZ	Owned by Al Casby, Mid-Century Aviation. Just Purchased.
J46-WE-8	WE405754[t]	09/26/1955	NO	Denver, CO	In possession of the Wings Over the Rockies Air & Space Museum. In storage.
J46-WE-??	Unknown	Unknown	NO	Denver, CO	In possession of the Wings Over the Rockies Air & Space Museum. Motorized cutaway display engine on the exhibit floor. No manufacturer's plate.
J46-WE-8B	WE404588	02/24/1954	YES *	Charlotte, NC	In possession of the Carolinas Aviation Museum. In storage.
J46-WE-8B	WE404979	11/30/1955	YES *	Charlotte, NC	In possession of the Carolinas Aviation Museum. In storage.
J46-WE-8	WE405778[t]	01/07/1955	NO	Charlotte, NC	In possession of the Carolinas Aviation Museum. In storage.
J46-WE-8	WE404963	08/31/1954	NO	Charlotte, NC	In possession of the Carolinas Aviation Museum. In storage.

Type	Serial	Accepted	A/B	Location	Status
J46-WE-8	WE405773[t]	10/17/1955	YES	Chantilly, VA	In possession of the Smithsonian Air & Space Museum. Catalog Number: 1971-0911. In storage. Flew on 3 acceptance flights. Engine log exists. Total Engine Running Time: 12.72 hrs. Engine can has never been opened after receipt into NASM collection.
J46-WE-8?	Unknown	Unknown	NO	San Carlos, CA	In possession of the Hiller Aviation Museum. Motorized cutaway display engine on the exhibit floor. No manufacturer's plate.
J46-WE-8?	Unknown	Unknown	NO	Tampa, FL	Installed in "Green Mamba" dragster owned by Doug Rose.
J46-WE-8B	Unknown	Unknown	NO	Phoenix, AZ	Owned by Al Casby, Mid-Century Aviation. Came from jet-powered dragster from York, PA.
J46-WE-8B	WE404782	07/26/54	NO	Phoenix, AZ	Owned by Al Casby, Mid-Century Aviation. From Australia still in its engine can.
J46-WE-8B	WE404957	03/14/1955	NO	Phoenix, AZ	Owned by Al Casby, Mid-Century Aviation. From Australia still in its engine can.
J46-WE-8B	Unknown	Unknown	UN	Australia	Reported to be in a museum.
J46-WE-8B	WE404515	06/16/1956	NO	Akron, OH	Belonged to Embry-Riddle Aeronautical University.
J46-WE-8B	WE404546	01/05/1954	NO	Akron, OH	Belonged to Embry-Riddle Aeronautical University.
J46-WE-8B	WE405568[t]	12/22/1954	NO	Akron, OH	Belonged to Embry-Riddle Aeronautical University.

* A/B's not attached to any of the four engines and arbitrarily listed here as they can be mounted on any J46-WE-8 or -8B engine.

[t] The -8 serial numbers going beyond WE405120 appear to be remanufactured 8B's brought up to the full modernization standard and then given new serial numbers by Westinghouse. In support of this, the manufacturer's plate on WE405568 has no entries stamped in the "Engine Bulletins Incorporated" section.

Glossary

50-Hour Flight Certification Test – a jet engine had to run through this test and demonstrate no significant mechanical failures or weaknesses either during the test or upon post-test teardown. The performance and operational control of the engine had to be adequate for flight with fuel and oil consumption at levels acceptable for normal operational test flying. Engines built to this standard were limited to 50 hours of operation in aircraft between overhauls.

150-Hour Qualification Test – a longer version of the 50-Hour Flight Certification Test. Passing this test meant the engine achieved the performance guarantees for thrust, specific fuel consumption and oil consumption with no significant mechanical failures or weaknesses occurring during the test or upon post-test teardown. Engines passing the test were cleared for volume production using the test engine build configuration, initially with a 150 hour overhaul limit. They usually also had a shorter maximum thrust operating time limit before rejection to overhaul.

AEL – Aeronautical Engine Laboratory, Naval Aircraft Factory, Philadelphia, Pennsylvania. Maintained by the Bureau of Aeronautics. It had many of the altitude and temperature testing capabilities of the NACA high altitude laboratory in Cleveland which WAGT lacked.

Aerotec switch – a pressure sensing switch that was used to detect excess turbine inlet pressure and then closed a switch in the fuel control to limit the fuel delivered until the pressure dropped to safe limits. This was done to protect from over-temperature damage.

Afterburner – a section added to a jet propulsion engine downstream from the turbine section. Additional fuel is added here and burned in the hot gas stream coming back from the engine, gaining more thrust. Sometimes referred to in memos of this period as "tail pipe burning". In the UK, it is more generally known as "reheat".

Atomized combustion – WAGT's term for evaporation fuel delivery to the combustion chamber. Now commonly known as "walking stick" evaporators because of the shape of the evaporator tubes. This was a large step forward in even fuel delivery, eliminating the fuel spray nozzles and high pressure pumps with all of their uniform delivery problems at different engine speeds and altitudes. Some WAGT reports call these tubes "candy canes".

BAR – U.S. Navy Bureau of Aeronautics Area Representative. He was delegated oversight and approval authority over specific contractor operations within a given geography. The BAR was charged with being the Bureau Chief's "eyes-on" representative with a direct line to the Chief as necessary. They generally had engine acceptance responsibilities for both project and production engines coming out of acceptance testing for service usage.

BLC – Boundary Layer Control – sometimes called boundary layer blowing. Air is bled from the compressor of the engine and blown over a wing, flap or control surface at high, sometimes sonic, speeds to improve lift and/or control effectiveness at low speeds. Its use allowed smaller wings and flaps to be used even as airframe weights increased by producing lower stall speeds. The J46 was not designed to accommodate the additional bleed capacity needed by BLC.

Boss – a circular ring inserted into the side of a larger pipe or case. It is threaded on the inside edge to take a plug, valve, lug, or other instrument. Bosses surrounding the after end of the combustion chamber allow ganged thermocouples to be lowered to various parts of the gas flow to measure temperatures and build temperature distribution patterns.

Bureau of Aeronautics – (BuAer) the U.S. Navy organization (1921–1959) that had responsibility for the procurement, development and operational service of both aircraft engines and airframes.

Cold Shut – 1: The freezing of the surface of liquid metal during the pouring of an ingot or casting due to interrupted or improper pouring; also : an imperfection thus caused. 2 : the imperfect weld caused in a forging by the inadequate heat of one surface under working or by an oxide film. *(Merriam-Webster dictionary)*

C.I.D. – Change in Design. The official change process within the Power Plant Division of BuAer that was utilized after a contractor engine change was surveyed and approved by BuAer.

Curvic Clutch – WAGT's term for a ring on each side of a compressor or turbine disc that was machined with interlocking teeth so that when the adjoining disk is mated to it and bolted tight, the disks are locked together.

Drain – any opening in the engine to allow excess or trapped fluids to escape to avoid danger of ignition on engine restarts or in operation. Confusingly called "scuppers" (the naval shipboard name for a drain) on the J40 contracts, but rarely so in the J34 or J46 material.

Fireseal – a firewall in ring form that surrounds the combustion chamber and helps to prevent a fire in the aft end of the engine from migrating forward in the airframe. In the documentation, this was sometimes written as fire-seal or fire seal.

FOD – Foreign Object Damage. A major hazard for all jet engines. "FODing" an engine can be caused by any object from any source passing through the engine while it is operating. Hard and larger objects can totally wreck the engine. Smaller objects may pass through but cause enough damage to lower performance and require unscheduled repairs.

GFE – Government Furnished Equipment. As a rule, things that are needed by a contractor to satisfy a contract that are to be supplied by the government procuring agency. The WAGT engines were GFE to the airframe companies using them in their airplanes.

Guide Vanes – Curved vanes between the compressor disc stages that slow the air striking them and then guide the airflow into the spinning compressor blades of the next compressor stage at the correct angle. The slowing of the air adds to the compression of the air. The modern term for these vanes is "stators".

Hastelloy – Brand name of a series of Westinghouse developed high temperature alloys, produced at their plant in Pittsburg, PA.

Hot Streak – An ignition method for afterburners. A fuel injector squirts a small amount of liquid fuel into the hot gas exiting the aft end of the combustion chamber and then ignites it. The burning fuel flows through the turbine still burning where it ignites the fuel vapor in the afterburner. It is highly reliable and works at all altitudes.

Iris nozzle – as applied to afterburners, it is a design using overlapping petals (modern term "feathers") that are opened and closed by some form of ring on runners that pull/push them "open" or "closed" by forcing a change in the overlap amount at the ends. It is a more accurate gas flow control than a hinged eyelet design as the nozzle shape is always circular. The eyelet design is only a circle when fully open or in one position.

Isochronous governor – a governor that maintains the same speed in the mechanism controlled regardless of the load. The J46 was a constant speed engine from 60% power setting to Military.

Micro-jet – a form of ignition used in the J46-WE-2 and later versions.

MUB – Mock-up Board. A board that issues an official report recording the mock-up inspection discussion items and assigning action items to bring the engine mock-up to a satisfactory state. After the mock-up is modified to address the board's approved recommended actions, it becomes the installation envelope baseline. All proposed changes affecting the envelopment afterwards have to be BuAer approved after a survey review by potential engine users and BuAer itself.

NACA – National Advisory Committee for Aeronautics They ran research facilities around the USA and possessed many laboratories capable of conducting engine tests that WAGT did not possess. The NACA later became NASA.

Nicrobrazing – a brazing technique where the atmosphere is solely nitrogen. This eliminates any oxidation defects in the weld. It is usually done in a special chamber.

Nozzles – WAGT's term for the guide vanes in front of the turbine blades, guiding the hot combusted area into the turbine at the correct angle for maximum turbine efficiency with the lowest pressure loss possible. WAGT tested variable nozzles on the first stage turbines of the XJ46-WE-2 and XJ40-WE-12 without measureable performance improvements.

Quick Disconnect – any method of mechanically connecting two components that reduces or eliminates multiple bolts or screws. On the WAGT engines that used them, the major assembles and accessories used bands that held the two components together by gripping the edges of the flanges, the bands being tightened at their ends with one or two bolts. Rapid replacement of a component is possible by simply loosening the band.

Refractalloy – Brand name of series of Westinghouse developed high temperature alloys, produced at their plant in Pittsburg, PA.

SHC – Simple Hydraulic Control – This replaced the electronic controls on all J46 engines once developed to an acceptable standard. It was to be modified to provide A/B control if the airframe electrical system failed (Interim Control System) and then modified further into the Hydro-Mechanical Control to eliminate all dependence on the airframe electrical system for control functions either primary or emergency.

Snap Acceleration/Deceleration – Movement of the power lever (throttle) from one position to the maximum power or idle position in one second or less. Today such a movement to full power (non-A/B or A/B) is termed a "throttle slam".

Standard Day – A measure of pressure and temperature to which performance measurements can be adjusted to allow actual performance comparisons and determine compliance against the specification. For BuAer, a standard day was 14.7 lb/in^2 at 59°F.

Survey – A formal detailed description of a change to the approved installation envelope of the engine. It is sent to all prospective engine users for comments and approval. Once consensus is reached, BuAer would issue a formal letter of approval. A C.I.D. was then issued to the Power Plant Division.

TBO – Time Between Overhauls – A durability measure of the time in cumulative engine running hours before an engine has to be completely dismantled and some parts routinely replaced and the others inspected for wear and/or damage. After overhaul, the engine is now a "zero-time" engine, meaning the running hours to the next overhaul count start at zero again.

Turbojet – a pure jet engine where all the air entering the intake (less air bleed) flows through the combustor and turbine. In the period of the book, usually written turbo-jet or sometimes just turbo jet. The WAGT photography department always referred to these engines as "turbo compressors".

WAGT – the author's acronym for the Westinghouse Gas Turbine Division. Sometimes referred to as the Westinghouse Aviation Gas Turbine Division, BuAer called them "WEC", "WECO" or just "the contractor". On rare occasions when the BuAer Chief wrote directly to the CEO of Westinghouse, they were called "Westinghouse Electric".

Water Injection – Water is injected into the engine combustion chamber to add both mass and a cooling effect on the combustion. This allows the engine to run at slightly higher compressor ratings for a short period (typically take off) and the injection of more fuel for more power. If done on an engine not designed for it, the thermal shocks and/or rotation stresses can shorten the turbine blade life.

Throttle Slam – The rapid movement of the throttle from any point below full non-afterburner power settings to full afterburner. The control system must reach the full power setting as quickly as possible without exceeding the maximum speed or temperature limits of the engine and without stalling or hesitating.

Trunnion – Very strong attachment points on external points on the engine used to both lift the engine and suspend the installed engine within the airframe. The trunnion(s) on the front of the engine had to carry the thrust load into the airframe as well as support the engine.

Thryratron - a type of gas-filled tube used as a high-power electrical switch and controlled rectifier. They can handle much greater currents than similar hard-vacuum tubes. Electron multiplication (increased current) occurs when the gas becomes ionized, producing a phenomenon known as Townsend discharge.

Tubes – most commonly called vacuum tubes in the USA, they were used in parts of the electronic circuitry of the primary control of the planned control system. In the UK, they are more commonly called valves.

Bibliography

Hagley Museum and Library, Wilmington, Delaware
Westinghouse Documents, Accession 1459
Paper documents and reports in 15 Boxes

National Advisory Committee for Aeronautics (NACA) Reports, Internet database:

Altitude Investigation of Gas Temperature Distribution at Turbine of Three Similar Axial Flow Turbojet Engines, W.R. Prince and F.W. Schulze, RM E52H06, August 18, 1952

An Evaluation of Turbojet Engine Thrust Control by Exhaust-Nozzle-Area Modulation and Compressor-Inlet Throttling, James L. Harp, Jr., Wallace W. Velie, and William E. Mallett, RM E54F21, August 10, 1954

Analysis of Stage Matching and Off-Design Performance of Multistage Axial-Flow Compressors, Harold B. Finger and James F. Dugan, Jr., RM E52D07, June 27, 1952

Effect of Blade-Root Fit and Lubrication on Vibration Characteristics of Ball-Root-Type Axial-Flow-Compressor Blades, Morgan P. Hanson, RM E50C17, June 15, 1950

Effect of Mach Number on Over-All Performance of Single-Stage Axial-Flow Compressor Designed for High Pressure Ratio, Charles H. Voit, Donald C. Guentert and James F. Dugan, RM E50D26, July 14, 1950

Effect of Rotor- and Stator-Blade Modifications on Surge Performance of an 11-State Axial-Flow Compressor II – Redesigned Compressor for XJ40-WE-6 Engine, E. William Conrad, Harold B. Finger, and Robert H. Essig, RM E52I10, May 25, 1953

Effects of Obstructions in Compressor Inlet on Blade Vibration in 10-Stage Axial-Flow Compressor, Andre J. Meyer, Jr., Howard F. Calvert and C. Robert Morse, RM E9L05, February 13, 1950

Investigation of Performance of Typical Inlet Stage of Multistage Axial-Flow Compressor, Jack R. Burtt, RM E9E13, July 18, 1949

Comparison of Performance of AN-F-58 Fuel and Gasoline in J34-WE-22 Turbojet Engine, Harry W. Dowman and George G. Younger, RM E8L10a, April 7, 1949

Preliminary Performance Data on Westinghouse Electronic Power Regulator Operating on J34-WE-32 Turbojet Engine in Altitude Wind Tunnel, James R. Ketchum, Darnold Blivas, and George J. Pack, RM SE50J11, October 27, 1950

Preliminary Transient Performance Data for Afterburner Operation of Westinghouse Electronic Power Regulator on XJ34-W-32 Turbojet Engine in Altitude Wind Tunnel, George Vasu, Glennon V. Schwent, and James R. Ketchum, RM SE50L29, February 14, 1951

High-Temperature Lubricants and Bearings for Aircraft Turbine Engines, NACA Subcommittee on Lubrication and Wear, RM E54D27, July 19, 1954

A Comparison of Carrier Approach Speeds as Determined from Flight Tests and from Pilot – Operated Simulator Studies, Maurice D. White and Fred J. Drinkwater III, RM A57D30, June 19, 1957

Experimental Investigation of Tail-Pipe-Burner Design Variables, W.A. Fleming, E. William Conrad, and A. W. Young, RM E50K22, March 5, 1951

Full-Scale Investigation of Cooling Shroud and Ejector Nozzle for a Turbojet Engine – Afterburner Installation, Lewis E. Wallner and Emmert T. Jansen, RM E51J04, December 17, 1951

National Aeronautics and Space Administration (NASA)

Presentation: Fundamentals of Aircraft Turbine Engine Control, Dr. Sanjay Garg, Glenn Research Center at Lewis Field, (Not Dated)

Paul D. Lagasse, Master of Arts Thesis: The Westinghouse Aviation Gas Turbine Division 1950-1960: A Case Study in the Role of Failure in Technology and Business (unpublished). Aircraft Engine Historical Society, www.enginehistory.org

U.S. Air Force, Air Force Research Laboratory, Propulsion Sciences and Advanced Concepts Division, "History of Aviation Fuel Development in the United States", (date unknown)

United States Navy, Bureau of Aeronautics Contract Correspondence, Record Group 72.3.2, National Archives at College Park, Contracts:

NOa(s) 3962, Containers 1295-1308
NOa(s) 5382, Containers 1546-1555
NOa(s) 9051, Containers 256-260
NOa(s) 9212, Containers 303-310
NOa(s) 9237, Containers 327-328
NOa(s) 9433, Containers 444-446
NOa(s) 9670, Containers 552-558
NOa(s) 9791, Containers 685-695
NOa(s) 9937, Containers 740-760
NOa(s) 10592, Container 1061
NOa(s) 10653, Containers 1076-1078
NOa(s) 10825, Containers 1093-1101
NOa(s) 10943, Containers 1116-1117
NOa(s) 11028, Containers 1133-1135
NOa(s) 52-020b, Containers 2453-2454
NOa(s) 52-621, Container 2584
NOa(s) 53-556, Containers 145-147, 3121-3124
NOa(s) 55-198b, Containers, 64, 3684-3686

United States Navy, Bureau of Aeronautics Contract Correspondence, Record Group 72.3.2, National Archives at College Park, Technical Reports

F7U-1, A2U Airframe
F7U-1, F7U-3, A2U-1 Flight Related

United States Navy, Bureau of Naval Weapons, Naval Air Material/Engineering Center, Aeronautical Engine Laboratory, Technical Reports 1922-1965

AEL-1092, Cranking and Starting Tests of Various Westinghouse Model J34 Turbo-Jet Engines
AEL-1106, Evaluation of Holley Model R-46 Turbo-Jet Engine Control
AEL-1110, Lubricating Oil Requirements of J-34 Turbo-Jet Engines Under Low Temperature Conditions
AEL-1126, Tests of Eclipse Solid Propellant Starter
AEL-1135, Qualification Test of Westinghouse 23F856-2 Jet Engine Ignition Coils
AEL-1140, Evaluation of Woodward Model X-770 turbo-Jet Fuel Control
AEL-1144, Evaluation of the Model R-46 Independent Emergency Fuel Systems for Turbo-Jet Engine Control
AEL-1146, Qualification Test of Pesco C0-Axial Fuel Pump for J34-WE-30 Engine
AEL-1150, Altitude Calibration of Westinghouse X24C-4B Turbo-Jet Engine
AEL-1151, Condensation Tests of a J34-WE-2 Turbo-Jet Engine
AEL-1152, Step Response Characteristics of the J-34 Turbojet Engine
AEL-1162, Evaluation of the Holley Part Number A5462 Fuel Controls, Westinghouse Part Numbers 22E473-4 and -5 on the XF2H-1 Airplane
AEL-1179, Development and Test of an Automatic Starting System for Turbo-Jet Engines
AEL-1188, Determination of Control Requirements of the WECO J34-WE-22 Turbo-Jet Engine
AEL-1189, Performance Characteristics of a J34-WE-34 Engine for Simulator Application (Phase I – Sea-Level Static Investigation
AEL-1192, Turbo-Jet Engines – Investigation of Starting by Direct Impingement
AEL-1196, Performance Characteristics of a J34-WE-34 Engine for Simulator Application (Phase II-Altitude Investigation)
AEL-1213, Investigation of Windmilling Cranking and Starting Characteristics of a J34-WE-34 Engine (Phase A)
AEL-1221, Performance of J34-WE-34 Engine with JP-4, AVGAS, and AVGAS-Kerosene Fuels
AEL-1224, Comparison of Performance Characteristics of Westinghouse P/N A28A8544, Eclipse Type 36E12-2-A, and Jack and Heintz Model D-27 Starters on J34 Turbo-Jet Engines
AEL-1227, Evaluation of WECO J34-WE-34 Engine and Control System Under Engine "B" Condition of MIL-E-5009 Specification
AEL-1229, Altitude Calibration of WECO J34-WE-36 Engine
AEL-1246, Qualification Tests of Eclipse Model 856843 and 856844 Solid Propellant Starters – Phase A
AEL-1247, Altitude Performance Test of WECO J34-WE-36 Engine Using Specification MIL-F-5624A (JP-3) Fuel
AEL-1260, Cold Starting Tests of Westinghouse J34-WE-36
AEL-1270, The Effect on J-34 Engine Operation of Using Avgas with Tricesyl Phosphate and Ethylene Dibromide Additives
AEL-1274, Evaluation of Electronic Control for J34 Engine with Variable Area Exhaust Nozzle
AEL-1282, Ignition Characteristics of Special Jet Engine Fuels
AEL-1284, Preturbine Fuel Injection for J34 Afterburner
AEL-1303, Determination of a Method for Measuring Flight Gross Thrust of Turbojet Engines

AEL-1308, Effect of a Modified Cowling for a J34 Type Engine Starter on Engine Performance and Compressor Inlet Velocity Distribution at Static Sea Level Conditions

AEL-1321, Reduced voltage Starting of J34 Engines Using Eclipse Type 36E12-2-B Starters

AEL-1338, Cold Starting Tests of a Westinghouse J34-WE-36 Engine (Phase B)

AEL-1343, Altitude Performance of WECO J46 Turbojet Engine with Various Compression Configurations

AEL-1349, Sea Level Tests of Westinghouse XJ46-T-30 Air Cooled Turbine Engine

AEL-1358, Cranking Tests of a Westinghouse J46-WE-8B (X24C10 #2) Turbo-jet Engine (Phase A)

AEL-1364, Performance of a WECO J34 Engine with an Afterburner Using a Fixed Area Exhaust Nozzle

AEL-1378, Gas Turbine Icing Tests Under Project Summit

AEL-1406, Verification Test of Westinghouse J46-WE-8 Engine Using Grade JP-5 Fuel

AEL-1411, Investigation of the Quasi-Static Test Method for

AEL-1417, Altitude Investigation of WECO J34-WE-36 Engine Compressor Blade Vibratory Stresses

AEL-1421, Performance and Altitude Endurance Tests of Westinghouse Model 58J105-5 Dual Fuel Pump

AEL-1451, Evaluation of Idle Characteristics of a J46-WE-8B Engine with WECO P/N 64J317-8 Fuel Regulator

AEL-1464, Cranking and Starting Tests of a Westinghouse J46-WE-8B (XC10 #2) Turbo-Jet Engine Using a Starter and Various Starter-Generators

AEL-1475, Sea Level Starting Tests of a Westinghouse J46-WE-8B (XC10 #2) Using Specification MIL-F-5624C Fuel, Grade JP-5, and Specification MIL-F-5161B Fuel, Grade JP-4

AEL-1520, Altitude Calibration of WECO J46-WE-8B Engine

AEL-1541, Development of Combustor for WECO J34-WE-46 Engine

AEL-1585, Development of Combustor for WECO J34-WE-46 Engine

AEL-1586, Special Sea Level Tests of a Westinghouse J46-WE-8B (XC10 #2) Engine in the Starting Speed Range

Westinghouse Electric Corporation , Aviation Gas Turbine Division Reports (does not include progress or flight test reports)

A-459, February 26, 1947, Design and Test of a Shipping Crate for 24C Jet Engine

A-490, April 10, 1947, Development Testing Report for February 1947

A-551, July 14, 1947, 24C4B Electrical Ignition System Testing to Specification AN-E-32

A-583, October 15, 1947, J34-WE-22 Verification Testing

A-587, October 10, 1947, Thrust Measuring Kit for Flight Test Work Item 3, Task III

A-589, October 22, 1947, Airflow Nozzle Calibration Project N. 758

A-593, October 27, 1947, Mechanical Test of Thermocouple Project 535

A-594, October 29, 1947, Statistical Engine Tests

A-604, November 11, 1947, Removable Fuel Nozzle Block

A-611, December 12, 1947, Blade Frequency Increase

A-648, April 26, 1948, Power Control Unit Tests

A-669, April 26, 1948, Interim Report on Engine Materials for Item 3 Task III Subtask

A-673, April 13, 1948, Forged Bearing Housing

A-675, April 13, 1948, Test of New Coupling Material

A-685, April 28, 1948, Turbine Nozzle Shroud Redesign

A-697, May 14, 1948, Qualification Testing of the J34 Turbo-jet Engine with Electrical Ignition

A-723, August 20, 1948, Investigation of J34 (24C4B) Roller Bearing Failures

A-724, June 21, 1948, Design Study of J34-WE-32 Engine Accessary Gearbox Arrangements

A-726, June 28, 1948, Flight Test Measuring Kit

A-727, July 23, 1948, Final Report on Development of an Exhaust-Gas Temperature Measuring Device for J-34 Turbojet Engine

A-728, June 25, 1948, Compressor Outlet Seal Support

A-729, July 8, 1948, Cleaning Compressor Blading

A-734, July 8, 1948, Manufacture and Test of Rolled Section Stationary Vanes

A-735, July 21, 1948, WAGT Model XJ_WE-32 Turbo-Jet Engine Study of Methods of Improving Cold Starting Limit of Lubrication System

A-737, August 5, 1948, Thermal Expansion Tests on the XJ34-WE Engine

A-741, August 6, 1948, J46-WE-2 Engine Arrangement Design Study

A-741, August 29, 1948, Design Study of Accessories Layout for the J46-WE-2
A-746, September 3, 1948, Governor Time Constant and Transient Response of Engine Controls
A-752, January 17, 1949, Compressor Failure on Engine BuAer Serial No. WE021001
A-758, October 15, 1948, Power Take off Gear Box Test
A-770, November 11, 1948, Preliminary Development of J34 Afterburner Control System
A-777, December 22, 1948, Tailburner AB Model Size Tests
A-781, December 9, 1948, Design Manufacture and Test of May 1948 Afterburner
A-785, December 20, 1948, Test of WAGT Afterburner Design No. 1
A-789, December 13, 1948, Fatigue Test on 24C Trunnions
A-790, January 1, 1949, Substituting Test of Control Timer
A-796, January 20, 1949, Afterburner Progress Report
A-801, January 21, 1949, Cast Diffusion Chamber Outer Cone
A-811, March 15, 1949, Final Report on Undercurrent Relay and Starter Contactor Assembly
A-813, March 8, 1949, Engine Installation Features and Accessory Arrangements in the J46-WE-2
A-814, March 21, 1949, Final Report on J34 Ignition System Development Tests
A-818, March 17, 1949, XJ34-WE-32 Turbo-Jet Engine Design Study of Turbine Outlet Temperature Override for AB Fuel Control
A-832, May 18, 1949, Additional Anti-icing Tests on a J34 Turbojet Engine on Mt. Washington
A-848, May 31, 1949, Final Report of Water Injection Tests on the J34 Turbojet Engine using the Chance Vought Injection Ring
A-851, June 7, 1949, Combustion Development, Final Development
A-852, June 15, 1949, Final Report on Compressor Development
A-855, June 17, 1949, Interim Technical Report of J34 Compressor Development
A-861, June 17, 1949, Summary Report of Project 1004 (J34-WE-30 Fuel Pumps)
A-863, June 28, 1949, Modification to J34-WE-22 Engines to Accommodate an Afterburner
A-864, June 29, 19 49, Afterburner Fuel Control Development
A-867, June 27, 1949, Final Technical Report on the Ceramic Coating of a Carbon Steel Combustion Chamber Liner
A-877, August 5, 1949, Compressor Bleed Off Air Contamination in J34-WE Engines
A-949, April 4, 1950, Qualification Test J34-WE-38 Contract Noa(s) 9670 Engine No X24C7-1
A-1005, October 5, 1950, Verification Test of Kansas City Manufactured J34-WE-34 Engine WE200000, Contract Noa(s) 10592
A-1019, November 20, 1950, Verification Test of Kansas City Manufactured J34_WE_34 Engine, Serial WE2000000-3, Contract Noa(s) 10592
A-1033, January 4, 1951, Verification Test of Kansas City Manufactured J34-WE-34 Engine, Serial WE200000-2, Contract Noa(s) 10592
A-1034, January 5, 1951, Evaluated Tests of J34-WE-34 Engine Compressor Blade Platform Fits.
A-1047, December 16, 1950, Investigation for Increased Thrust on the XJ46-WE-1 Engine
A-1048, November 17, 1950, Final Report on Starting Requirements for the XJ46-WE-2 Engine
A-1063, April 6, 1951, Verification Test of Kansas City Forged J34-WE-34 Compressor Blades and Modified J34-WE-34 Engine Parts
A-1111, April 18, 1951, Summary Report of Verification Test of Kansas City Manufactured J34-WE-34 Engines
A-1274, November 11, 1951, Interim Report on the Simple Hydraulic Controls
A-1302, February 2, 1952, Exhaust Nozzle Control System For The J46 Turbojet Engine
A-1317, April 2, 1952, Calculated Starting Torque and Speed Requirements for -1, -3, -8 and -10 Turbojet Engines
A-1318, February 18, 1952,"Summary Report of Contract 9670-25 Items 11 (and12) XJ34-WE-32 and -38 Engines
A-1379, May 13, 1952, Study of Non-Electrical Afterburner Fuel Control System, J46-WE-8 Engines
A-1424, March 9, 1954, Failure Analysis of AB Pressure Switches (Table Update Only)
A-1510, January 30, 1953, Failure Analysis of the Temperature Control Amplifier
A-1519, January 12, 1953, Contract 9670 MPR A-1519 December 1952
A-1520, January 21, 1953, Failure Analysis of Automatic Temperature Control, Engines J40-WE-1, -8, -22 and J46-WE-3, -7, -8, -12
A-1543, April 19, 1953, Qualification Test of YJ46-WE-8A Turbo-jet Engine

A-1565, August 12, 1953, Final Report of Flight Testing J32-WE-32 in the F7U-1
A-1571, April 19, 1953, Preliminary Windmilling Data for the J46-WE-8
A-1582, May 22, 1953, 150 Hour Laboratory Qualification Test of J46 Accessory Gearbox
A-1618, January 13, 1954, Qualification Test of the J46-WE-8A Turbojet Engine
A-1625, May 1, 1953, 150 Hr Full Load Test of a Starter Generator Adapter Gearbox
A-1630, January 15, 1953, Cold Starting of the J46-WE-8 Lubrication System
A-1634-1, October 29, 1953, Qualification Test of Fuel Booster Pump for Westinghouse J46 (All series) Engines
A-1650, September 11, 1953, J46 Fuel Regulator Component Qualification Test
A-1657, October 16, 1953, Component Qualification Testing of the J46-WE-8A, 8B, 12A, 12B Afterburner Fuel System
A-1677, November 11, 1953, Investigation of Stresses in Rotating Blades and Stationary Vanes of XJ46-WE-8A Engine with C-75 Compressor Configuration
A-1696, June 10, 1954, Verification Test of J46-WE-12A and -12B Afterburner on Contract 10825
A-1697, October 28, 1954, Qualification Test of J46-WE-8B Turbojet Engine on Contract Noa(s) 10825
A-1699, November 27, 1953, Deposit Tests on Gummy Fuels in the J46 Combustor
A-1700, January 22, 1954, A Study of the Removal of Power Limits from the Variable Delivery Oil Pump
A-1701, January 20, 1954, Verification Test of Kansas City Manufactured -8B Inlet Housing Assembly
A-1708, March 22, 1954, Qualification Test of J46-WE-8 Turbo-Jet Engine
A-1727, November 16, 1954, Endurance Test of a J46-WE-8A Turbo Jet Engine in Accordance with MIL-E-5009A Specification
A-1740, March 2, 1954, Pages 7-8 of Report A-1740 Curvic Clutch Sealing
A-1741-Z, March 15, 1954, 500 Hour Laboratory Qualification Test of the Microjet Pressure Switch
A-1748, January 12, 1954, Failure Analysis of Electrical Controls on J46-WE-8B Engines
A-1763, February 19, 1954, J46 Fuel Regulator Low Temperature Test with Parker PS10-13L -O- Ring Seals
A-1764Z, April 1, 1954, Qualification Testing of Components for J46-WE-8, 8A, 8B, 12A, 12B, 18, J40-WE-22
A-1767, March 8, 1954, Electric Control Assembly and Power Scheduler Test Report
A-1778, July 26, 1954, Control Transients, Compressor air Bleed, and Alternate Fuel Qualification Tests for the J46-WE-8 Turbo-Jet Engine
A-1798, December 2, 1954, J46-WE-8B 50-hour Alternate Fuel Test
A-1811, July 22, 1954, Accelerated Low Temperature Test of Parker 443-12 and Goshen 1120 O-Rings in a Jet Fuel Regulator
A-1863, April 29, 1955, Altitude Temperature Inversion Testing on J46-WE-8B Engine
A-1878, February 1, 1955, Throttle Shaft Seal Leakage Investigation of the 58J846 Fuel Control
A-1889, March 7, 1955, Ignition Components Qualification Test For High Energy Ignition System Westinghouse PN 64E719 Contract Noas 10825
A-1890, November 15, 1954, Qualification Test of the XJ46-WE-18 Turbo-Jet Engine
A-1891, October 26, 1953, Air Starting Tests of Woodward X838 All Speed Governor
A-1893, November 29, 1954, 150 Hr Verification Test of Indiana Gearworks Manufactured 61E370-8 Starter Generator Adapter Gearbox
A-1902, January 21, 1955, Qualification Test of the XJ46-WE-18 Turbo Jet Engine
A-1904, April 14, 1955, J46-WE-8 50-Hour Alternate Fuel Test Using JFC Fuel, Contract NOa(s) 10825
A-1907Z, October 1, 1955, Laboratory Qualification Test of Afterburning Fuel System Components for J46-WE-8, -8A, -8B, -12A and -12B
A-2140-1 February 29, 1956, J46-WE-8 Phase II Second Stage Turbine Disc Failure
A-2166, March 26, 1956, Removal of Fuel Boost Pump
A-2186, April 16, 1956, Flight Test of the Effects of J46 Relay Hot Streak Removal
A-2186, May 8, 1956, Flight Test of the Effects of J46 Relay Hot Steak Removal (Updated)

Vought Aircraft Heritage Foundation, www.vought.org
- Various documents related to Westinghouse

Illustrations

Westinghouse Electric Corporation Steam Division, Accession 1969.170, Hagley Museum and Library, Wilmington, DE 19807

Photographs by Westinghouse "P" reference number/ Date/Box Number

P40602A, 12/8/1949, 260
P42491, 3/28/1950, 260
P42536, 4/7/1950, 260
P43747, 1/3/1951. 260
P43750, 1/3/1951, 260
P44190, 3/15/1951, 260
P44200, 3/15/1951, 260
P45044, 8/9/1951, 261
P45181, 4/7/1950, 261
P45343, 9/25/1951, 261
P45882, 12/21/1951, 261
P45979, 1/8/1952, 261
P46005, 1/9/1952, 240
P46035, 1/11/1952, 261
P46048, 1/15/1952, 261
P46122, 1/18/1952, 261
P46126, 1/28/1952, 241
P46151, 1/31/1952, 261
P46181, 8/30/1951, 261
P46193, 2/1/1952, 261
P46198, 2/4/1952, 261
P46376, 2/29/1952, 261
P46679, 4/14/1952, 261
P46819, 4/29/1952, 261
P46977, 5/20/1952, 261
P46998, 5/27/1952, 261
P47021, 5/29/1952, 264
P47098, 6/11/1952, 261
P47101, 6/11/1952, 261
P47166, 6/19/1952, 262
P47167, 6/19/1952, 262
P47168, 6/19/1952, 262
P47627, 9/4/1952, 262
P47629, 9/4/1952, 262
P48097, 11/14/1952, 262
P48105, 11/17/1952, 242
P48464, 1/13/1953, 263
P48795A, 12/4/1953, 263
P49258, 4/24/1953, 263
P49317, 5/4/1953, 263
P49954, 8/15/1953, 265
P50333-14, 10/12/1953, 265
P50433, 10/23/1953, 270
P50434, 10/23/1953, 270
P50429, 10/23/1953, 270
P50676-1A, 12/18/1953, 266
P50856, 1/11/1954, 266
P50895, 1/13/1954, 266
P50896, 1/13/1954, 266
P50897, 1/13/1954, 266
P51317, 3/20/1954, Box 271
P51318, 3/20/1954, Box 271
P51320, 3/20/1954, Box 271
P51444, 3/31/1954, 267
P51449, 3/31/1954, 267
P51712, 5/3/1954, 268
P53133, 11/12/1954, 271

Accession 1459, Hagley Museum and Library, Wilmington, DE 19807

P32653, 5/29/1945, Box 14
P49513, undated, Box 14
P49288, undated, Box 14
P50317, 3/20/1954, Box 14
P51318, 3/15/1954, Box 14
P51320, 3/20/1954, Box 14
P50501, 10/30/1953, Box 14

Westinghouse Electric Corporation, Aviation Turbine Division Photos by WAGT "P" or "KC" reference number sourced from other than Hagley Museum and Library

P39259 (WAGT Report A-724)
P39259 (WAGT Report A-724)
P39824 (MUB Report)
P39827 (MUB Report)
P40732 (WAGT Memo, 9/30/49)
P40733 (WAGT Memo, 9/30/49)
P40851 (WAGT Memo, 9/30/49)
KC14034, 1/20/1954, (WAGT Report A-1701)
KC 14035, 1/20/1954 (WAGT Report A-1701)
KC15800, 3/21/1955, (WAGT Memo 3/21/55)

Publications

Chance Vought Aircraft, *Service Training Manual, Powerplant, F7U-3, J46-WE-8, -8A, -8B*, October 6, 1954

Ginter, Steve. *Naval Fighters Number Six, Chance Vought F7U Cutlass.* Simi Valley, CA: S. Ginter, 1982. Print

Gunston, Bill. *Fighters of the Fifties.* Osceola, WI: Speciality Press Publishers and Wholesalers, Inc., 1981. Print

Miller, Jay. The X-Planes. Arlington, TX: Aerofax, Inc., 1988. Print

Thomason, Tommy H. *Naval Fighters Number Ninety-Four, Chance Vought F7U-1 Cutlass.* Simi Valley, CA: Steve Ginter, 2012. Print

United States Navy, Bureau of Aeronautics, *Flight Handbook, Navy Model F7U-3, -3M and -3P Aircraft (BuNo. 128467 and Subsequent),* AN 01-45HFD-1, July 1, 1956

United States Navy, Bureau of Aeronautics, *Handbook Maintenance Instructions, Navy Model F7U-3 Aircraft (BuNo. 128451-128466),* November 1, 1954

www.ingramcontent.com/pod-product-compliance
Lightning Source LLC
Chambersburg PA
CBHW080520220326
41599CB00032B/6148
* 9 7 8 0 6 9 2 7 6 4 8 8 6 *